Encyclopaedia of Mathematical Sciences
Volume 110

Probability Theory

Subseries Editors:
A.-S. Sznitman   S.R.S. Varadhan

Springer
*Berlin
Heidelberg
New York
Hong Kong
London
Milan
Paris
Tokyo*

Harry Kesten (Editor)

Probability on Discrete Structures

Springer

Harry Kesten
Cornell University Department of Mathematics
Malott Hall 310
14853-4201 Ithaca, NY
USA
e-mail: kesten@math.cornell.edu

Founding editor of the Encyclopaedia of Mathematical Sciences:
R. V. Gamkrelidze

Mathematics Subject Classification (2000): 60B99, 60C05, 60F17, 60G50,
60J10, 60J27, 60K35

ISSN 0938-0396
ISBN 3-540-00845-4 Springer-Verlag Berlin Heidelberg New York

Springer-Verlag Berlin Heidelberg New York
a member of BertelsmannSpringer Science+Business Media GmbH
http://www.springer.de
© Springer-Verlag Berlin Heidelberg 2004
Printed in Germany

Typeset by LE-TEX Jelonek, Schmidt & Vöckler GbR, Leipzig
Cover Design: E. Kirchner, Heidelberg, Germany
Printed on acid-free paper   41/3142 db   5 4 3 2 1 0

Preface

Probability on discrete structures covers a wide area. Most probability problems involve random variables indexed by space and/or time. Almost always these problems have a version in which space and/or time are taken discrete. Roughly speaking this volume deals with some areas in which the discrete version is more natural than the continuous one, or perhaps even the only one which can be formulated without complicated constructions and machinery. Clear examples of this situation can be found in the articles in this volume on the random cluster model (by Grimmett) and on first-passage percolation (by Howard) and in most of the problems in the forthcoming book "Probability on Trees and Networks" by R. Lyons and Y. Peres. The article by Howard actually also discusses a continuous variant – called Euclidean first-passage percolation – but this came later and even though this continuous version has some clear advantages, its analysis brings in extra difficulties.

Problems on discrete structures can often be stated with minimal prerequisites, and sometimes only "elementary" (but by no means easy) probability theory is needed for their solution. Often the arguments have more of a combinatorial flavor than an analytic one, but the articles here certainly do not shun the use of the tools of analysis.

Since the subject matter of this volume is so broad and varied, it is not surprising that it does not lend itself to a simple linear ordering. It did not seem possible to me to produce a volume which introduced a reader to much of the field in textbook fashion, in which one goes through the chapters in order. Instead, the present volume introduces a reader to the problems and progress so far, in various representative directions and subjects in which there is considerable activity, and which have seen recent successes. The various articles are not dependent on each other.

There is one obvious omission from the list of possible topics in this volume, namely percolation. This subject was omitted here, because its classical aspects have been reviewed only two years ago by G. Grimmett in an encyclopedia article (see Development of Mathematics, 1950–2000, Jean-Paul Pier ed.), while its very recent successes by Lawler, Schramm, Smirnov and Werner are still evolving while this volume is being prepared.

I hope this volume will give a reader a solid introduction to the flavor and excitement of probability on discrete structures and encourage her to work in the subject herself.

Ithaca, NY   *Harry Kesten*
July 25, 2003

Contents

List of Contributors

David Aldous
Department of Statistics,
University of California Berkeley,
Berkeley, CA, 94720-3860.
aldous@stat.berkeley.edu

Geoffrey Grimmett
Statistical Laboratory,
University of Cambridge,
Wilberforce Road,
Cambridge CB3 0WB,
United Kingdom.
g.r.grimmett@
statslab.cam.ac.uk

C. Douglas Howard
City University of New York,
Baruch College.
dhoward@baruch.cuny.edu

Fabio Martinelli
Dipartimento di Matematica,
Universita' di Roma Tre,
L.go S. Murialdo 1,
00146 Roma, Italy.
martin@mat.uniroma3.it

Laurent Saloff-Coste
Cornell University,
Department of Mathematics,
Ithaca, NY 14853-4201.
lsc@math.cornell.edu

J. Michael Steele
Department of Statistics,
Wharton School,
University of Pennsylvania,
Philadelphia, PA 19104-6302.
steele@wharton.upenn.edu

The Objective Method:
Probabilistic Combinatorial Optimization and Local Weak Convergence

David Aldous and J. Michael Steele

1 Introduction

This survey describes a general approach to a class of problems that arise in combinatorial probability and combinatorial optimization. Formally, the method is part of weak convergence theory, but in concrete problems the method has a flavor of its own. A characteristic element of the method is that it often calls for one to introduce a new, infinite, probabilistic *object* whose *local properties* inform us about the *limiting properties* of a sequence of finite problems.

The name *objective method* hopes to underscore the value of shifting ones attention to the new, large random object with fixed distributional properties and way from the sequence of objects with changing distributions. The new object always provides us with some new information on the asymptotic behavior of the original sequence, and, in the happiest cases, the constants associated with the infinite object even permit us to find the elusive limit constants for that sequence.

1.1 A Motivating Example: the Assignment Problem

The assignment problem for the $n \times n$ cost matrix (c_{ij}) is the task of determining the permutation $\pi : [n] \to [n]$ that minimizes the total cost $c_{1,\pi(1)} + c_{2,\pi(2)} + \cdots + c_{n,\pi(n)}$ of assigning one column to each row. This problem arises in many applied contexts, and it has spawned an extensive algorithmic literature. Here we are more specifically concerned with the value of the objective function

$$A_n = \min_\pi \sum_{i=1}^n c_{i,\pi(i)} \, ,$$

where the costs c_{ij}, $1 \le i, j \le n$, are assumed to be independent and identically distributed random variables.

Investigation of A_n seems to have begun in 1962, when Kurtzberg [42] used heuristic methods like greedy matching to obtain upper bounds on $E[A_n]$ for uniformly distributed costs. Kurtzberg's bounds were of order $\log n$, and these were not improved until 1979 when Walkup [66] showed that $E[A_n]$ is bounded independently of n, which was quite a surprising discovery at the time. Eight years later, Karp [38] introduced a new approach to the estimation of $E[A_n]$ that was based on linear programming. Karp exploited the explicit bases that were known to be optimal for the assignment problem, and he obtained the elegant bound $E[A_n] \le 2$. Inspired in part by Karp's result, Dyer, Frieze and McDiarmid [19] developed a general bound for the objective function of linear programming problems with random costs, and they were able to recapture Karp's bound without recourse to special bases. A probabilist's interpretation of the Dyer-Frieze-McDiarmid inequality forms the basis of Chapter 4 of Steele [62] where one can find further information on the early history of the assignment problem with random costs.

A new period in the development of the random assignment problem began in 1987 with the fascinating article of Mézard and Parisi [51] which offered a non-rigorous statistical mechanical argument for the assertion that

$$\lim_{n \to \infty} E[A_n] = \frac{\pi^2}{6} = \zeta(2) \, . \tag{1.1}$$

The desire for a rigorous proof of this limit has influenced much of the subsequent work on the assignment problem, and the critical first step is simply to show that actually $E[A_n]$ converges as $n \to \infty$. Convergence would follow immediately if one could show that $E[A_n]$ is nondecreasing, but it is still not known if $E[A_n]$ is monotone. Nevertheless, in 1992 Aldous [3] used the objective method to show that $E[A_n]$ does indeed converge.

In 1998 Parisi [53] added further interest to the random assignment problem when he advanced the remarkable conjecture that for independent exponentially distributed random cost c_{ij} with mean 1, one has an *exact formula*:

$$E[A_n] = \sum_{k=1}^{n} k^{-2} \ . \qquad\qquad (1.2)$$

Alm and Sorkin [9] proved the conjecture for all values of n up to $n = 5$. Then in 2003, as this paper was about to go to press, two groups (Linusson and Wästlund [46] and Nair, Prabhakar and Sharma [52]) announced different proofs of the conjecture. These papers use arguments which are closely tied to the specifics of the random assignment problem, and whose range of applicability is therefore as yet uncertain. Earlier, Aldous [4] had proved – by means of the objective method – the $\zeta(2)$ limit formula that Mézard and Parisi [51] first brought to light in 1987. This method does turn out to be applicable to a fairly wide range of asymptotic problems, as this survey will show.

1.2 A Stalking Horse: the Partial Matching Problem

One of the aims of this survey is to show how the objective method helps to solve problems such as the determination of the limiting value of $E[A_n]$, but the assignment problem itself is too burdened with individual nuance for it to serve as our basic guide. For this reason, we introduce a new problem, the *maximal partial matching problem.*

This problem does not have the long history of the random assignment problem, but it is especially well suited for illustrating the objective method. In particular, it builds on the theory of random trees, and the limit theory of such trees provides the prototype for the objective method. Also, the maximal partial matching problem leads inexorably to the notion of a distributional identity. Such identities have always had a role in probability theory, but, along with allied notions like self-similarity and subadditivity, distributional identities are now a topic of emerging importance.

The relevant identities for the maximal partial matching problem are also much more tractable than the corresponding identities for the random assignment problem, yet many of the same tools come into play. In particular, we will find that one often does well to begin by guessing the solution of a distributional identity. After one has a good guess, then classic tools like Kolmogorov's consistency theorem and contemporary tricks like "coupling from the past" can be used to confirm the existence, uniqueness, and stability of the solutions of the distributional identity.

1.3 Organization of the Survey

The maximal partial matching problem provides our introductory case study, but before it can be addressed we need to deal with some foundational issues. In particular, Section 2 develops the notion of *local weak convergence*. Formally, this is nothing more than the weak convergence of probability measures on a certain metric space of "rooted geometric graphs," but the attending intuition differs substantially from the classic weak convergence theory. Section 2 also introduces the "standard construction," which is a general recipe for building rooted geometric graphs. This construction turns out to have a subtle, yet pervasive, influence on the applications of the objective method.

After we have dealt with the essential metric space formalities, we see what local weak convergence tells us about the *simplest* model for random trees. In particular, we review Grimmett's lemma, and we develop an understanding of the convergence of the large and the small parts of a random tree that has been cut at a randomly chosen edge. This analysis provides us with the essential distribution theory for random trees that is needed later.

The theory of the maximum partial matching problem developed in Section 3 is concrete and self-contained. Nevertheless, it faces most of the issues that one meets in more complex applications of the objective method, and it offers the best introduction we know to the essential ideas.

In Section 4 we introduce the mean-field model of distance which is a physically motivated probability model designed to gain insight into problems for point processes in $\mathbb{R}^d$. In this model the distribution of some of inter-point distances of $\mathbb{R}^d$ are captured precisely while other (hopefully less essential) inter-point distance distributions are distorted in comparison to $\mathbb{R}^d$. The mean-field model leads us to the PWIT, or Poisson weighted infinite tree, which is arguably the most important infinite tree model. To illustrate the close connection of the theory of the PWIT and the objective method, we give a reasonably detailed proof of the $\zeta(3)$ theorem of Frieze.

The relationship of the PWIT to problems of combinatorial optimization is continued in Section 5 by developing the limit theory for the minimum cost C_n of a perfect matching of the complete graph K_n with independent edge weights ξ_e having a common distribution F. We provide a reasonably detailed (and hopefully friendly) sketch of the fact that $E[C_n]$ converges to $\zeta(2)/2$ as $n \to \infty$ when F is the exponential distribution with mean one; this result is the direct analog of the $\zeta(2)$ limit theorem for $E[A_n]$. Understandably, some non-trivial details must be omitted from the sketch, but the section should still provide a useful introduction to Aldous's proof of the $\zeta(2)$ limit theorem.

All of the problems in Sections 3 – 5 call on distance models with substantial intrinsic independence; for example, in Section 3 we have independent weights on every edge, and we face the infinite Poisson-Galton-Watson tree in the limit, while in Sections 4 and 5, we have the tempered independence that one inherits from the mean-field model, and we face the PWIT in the limit. The problems of Section 6 are different; they deal directly with inter-point distances in $\mathbb{R}^d$ rather than with edge weights that are blessed *ex cathedra* with independence properties, and in the limit we often need to deal with the Poisson process.

The probability theory of Euclidean combinatorial optimization has grown quite large, and in Section 6 we provide very few proofs. Nevertheless, we hope to provide a useful up-date of the survey of Steele [62]. In particular, we address progress on the minimal spanning tree problem, including a recent weak

law of large numbers developed in Penrose and Yukich [55] which has close ties
to the objective method. The section also contrasts the objective method and
the subadditive method that has served for years as a principal workhorse in
the probability theory of Euclidean combinatorial optimization. In the closing
subsection, we break into entirely new territory and describe a remarkable
new result of Benjamini and Schramm on the recurrence properties of the
local weak limits of planar graphs.

In Section 7, the last section, we first summarize some of the circumstances
that seem to be needed for the objective method to be successful. We then develop the background of an attractive conjecture on the independence number
of a random regular graph. This problem is used in turn to illustrate several
of the basic challenges that appear to be critical to the deeper development
of the objective method.

Finally, we should note that even though our main intention is to provide
a survey and a tutorial, this exposition also contains new results. In particular, the material in Section 3 on the maximum partial matching problem is
new, including the basic limit theorem (Theorem 3.3) and the theorem that
determines the limit constants (Theorem 3.4). Another new result is the Convergence Theorem for Minimal Spanning Trees (Theorem 5.4). This theorem
both generalizes and simplifies much earlier work; it is also applied several
times during the course of the survey.

2 Geometric Graphs and Local Weak Convergence

Before the theory of local weak convergence can be brought face-to-face with
the concrete problems of combinatorial optimization, one is forced to introduce an appropriate complete metric space. This introduction has been made
several times before on a purely *ad hoc* basis, but now there is enough cumu-
lative experience to suggest a general framework that should suffice for most
applications. After describing this framework, we will give it a quick test run
by discussing a prototypical result from the limit theory of random trees –
Grimmett's lemma on the convergence of rooted Cayley trees.

2.1 Geometric Graphs

If $G = (V, E)$ is a graph with a finite or countable vertex set V and a corresponding edge set E, then any function $\ell : E \to (0, \infty]$ can be used to define
a distance between vertices of G. Specifically, for any pair of vertices u and v,
one just takes the distance between u and v to be the infimum over all paths
between u and v of the sum of the lengths of the edges in the path.

Definition 2.1 (Geometric Graphs and the Two Classes $\mathcal{G}$ and $\mathcal{G}_\star$).
*If $G = (V, E)$ is a connected, undirected graph with a countable or infinite
vertex set V and if ℓ is an edge length function that makes G **locally finite**
in the sense that for each vertex v and each real $\varrho < \infty$ the number of vertices
within distance ϱ from v is finite, then G is called a **geometric graph**. When
there is also a distinguished vertex v, we say that G is a **rooted geometric
graph** with root v. The set of geometric graphs will be denoted by $\mathcal{G}$, and the
set of rooted geometric graphs will be denoted by $\mathcal{G}_\star$.*

2.2 $\mathcal{G}_\star$ as a Metric Space

The set $\mathcal{G}_\star$ of rooted geometric graphs provides our basic workspace, but
before honest work can begin we need to say what we mean for a sequence
G_n of elements of $\mathcal{G}_\star$ to converge to a G in $\mathcal{G}_\star$. The idea one wants to capture
is that for large n, the rooted geometric graph G_n looks very much like G in
an arbitrarily large neighborhood of the root of G.

To formalize this idea, we first recall that an isomorphism between graphs
$G = (V, E)$ and $G' = (V', E')$ is a bijection $\phi : V \to V'$ such that
$(\phi(u), \phi(v)) \in E'$ if and only if $(u, v) \in E$. Also, given any such isomorphism, one can extend the domain of ϕ to E simply by defining $\phi(e)$ to be
$(\phi(u), \phi(v))$ for each $e = (u, v) \in E$.

Finally, we say that two geometric graphs $G = (V, E)$ and $G' = (V', E')$
are isomorphic provided that (1) they are isomorphic as ordinary graphs and
(2) there is a graph isomorphism ϕ between G and G' that also preserves edge
lengths (so $\ell'(\phi(e)) = \ell(e)$ for all $e \in E$). In the case of two *rooted* geometric
graphs $G = (V, E)$ and $G' = (V', E')$, we will say they are isomorphic provided
that there is a graph isomorphism ϕ that preserves edges lengths and that also
maps the root of G to the root of G'.

Next we consider a special rooted geometric graph that one may view
intuitively as the "neighborhood of radius ϱ about the root" of the rooted
geometric graph G. Specifically, for any $\varrho > 0$ we let $N_\varrho(G)$ denote the graph
whose vertex set $V_\varrho(G)$ is the set of vertices of G that are at a distance of at
most ϱ from the root of G and whose edge set consists of just those edges of G
that have both vertices in $V_\varrho(G)$, where, as before, the distance between any
two vertices u and v in G is taken to be the infimum over all paths between
u and v of the sum of the lengths of the edges in the path. We again view
$N_\varrho(G)$ as an element of $\mathcal{G}_\star$ whose the edge length function and root are just
those of G. Finally, we say that $\varrho > 0$ is a *continuity point* of G if no vertex
of G is exactly at a distance ϱ from the root of G.

Definition 2.2 (Convergence in $\mathcal{G}_\star$). *We say that G_n converges to G_∞ in
$\mathcal{G}_\star$ provided that for each continuity point ϱ of G_∞ there is an $n_0 = n_0(\varrho, G_\infty)$
such that for all $n \geq n_0$ there exists a isomorphism $\gamma_{n,\varrho}$ from the rooted
geometric graph $N_\varrho(G_\infty)$ to the rooted geometric graph $N_\varrho(G_n)$ such that for
each edge e of $N_\varrho(G_\infty)$ the length of $\gamma_{n,\varrho}(e)$ converges to the length of e as
$n \to \infty$.*

With a little work, one can show that this definition determines a topology that makes $\mathcal{G}_\star$ into a complete separable metric space. As a consequence, all of the usual tools of weak convergence theory apply to sequences of probability measures on $\mathcal{G}_\star$, and we may safely use the conventional notation and write

$$\mu_n \xrightarrow{\;d\;} \mu \quad \text{to mean that} \quad \int_{\mathcal{G}_\star} f\, d\mu_n \to \int_{\mathcal{G}_\star} f\, d\mu$$

for each bounded continuous function $f : \mathcal{G}_\star \to \mathbb{R}$.

2.3 Local Weak Convergence

The topology on the metric space $\mathcal{G}_\star$ turns out to give weak convergence of probability measures on $\mathcal{G}_\star$ a local character that sharply differs from the traditional weak convergence such as one finds in the weak convergence of scaled random walk to Brownian motion. Weak convergence of measures on $\mathcal{G}_\star$ never involves any rescaling, and the special role of the neighborhoods $N_\varrho(G)$ means that convergence in $\mathcal{G}_\star$ only informs us about behavior in the neighborhood of the root.

In practical terms, this means that weak convergence in $\mathcal{G}_\star$ can tell us about local features such as the degree of the root, the length of the longest edge incident to the root, and so on; yet it cannot convey detailed information on a global quantity such as the length of the longest path. To underscore this difference, one sometimes speaks of weak convergence in $\mathcal{G}_\star$ as *local weak convergence*.

2.4 The Standard Construction

Most of the random processes considered here are associated with a standard construction that has some lasting consequences, even though it may seem rather bland at first. The construction begins with a probability measure that is concentrated on the subset of $\mathcal{G}$ consisting of geometric graphs with exactly n vertices. We then consider a random element $\mathbf{G}_n$ of $\mathcal{G}$ that is chosen according to this measure, and we choose a vertex X_n at random according to the uniform distribution on the n vertices of $\mathbf{G}_n$. We then make $\mathbf{G}_n$ into a random rooted geometric graph $\mathbf{G}_n[X_n]$ by distinguishing X_n as the root vertex. The distribution of $\mathbf{G}_n[X_n]$ is then a probability measure on the set of n-vertex elements in the set of rooted geometric graphs. Finally, if a sequence $\{\mathbf{G}_n[X_n] : n = 1, 2, ...\}$ of such $\mathcal{G}_\star$-valued random variables converges weakly in $\mathcal{G}_\star$ to a $\mathcal{G}_\star$-valued random variable $\mathbf{G}_\infty$, then we say that the distribution of $\mathbf{G}_\infty$ is obtained by the *standard construction*.

One might think that virtually any measure on $\mathcal{G}_\star$ might be obtained by the standard construction, but the measures given by the standard construction are not unconstrained. Later we will find that they must satisfy a modest symmetry property that we call *involution invariance*. This property may sometimes be used to rule out what would otherwise be a tempting candidate for a limiting object.

2.5 A Prototype: The Limit of Uniform Random Trees

A classical formula of Cayley tells us that if V is a set of n distinguishable elements, then the set $\mathcal{S}_n$ of rooted trees with vertex set V has cardinality n^{n-1}. If T_n denotes a tree that is chosen from $\mathcal{S}_n$ according the uniform distribution, then one can show that T_n converges in distribution to a random variable T_∞ that takes values in $\mathcal{G}_\star$. Remarkably, one can characterize the distribution of T_∞ by a direct construction that relies on the classical theory of branching processes.

The Infinite "Skeleton Tree" T_∞ – or, PGW$^\infty$(1)

To begin, we consider a Galton-Watson branching process with one progenitor and an offspring distribution that has the Poisson distribution with mean one. With probability one, any such branching process has a finite number of vertices, and the finite trees generated in this way are said to have the *Poisson Galton-Watson distribution* with mean one. This distribution on the finite elements of $\mathcal{G}_\star$ is denoted by PGW(1).

Now consider an infinite sequence of independent PGW(1) distributed trees $T_0, T_1, T_2, ...$, and let $v_0, v_1, v_2, \ldots$ denote their roots. Finally, to make this collection of trees into one infinite rooted tree, we add all of the edges $\{(v_i, v_{i+1}), \ 0 \leq i < \infty\}$, and we declare v_0 the root of the new infinite tree. The tree T_∞ that one builds this way is said to have the PGW$^\infty$(1) distribution, and it is illustrated in Figure 1.

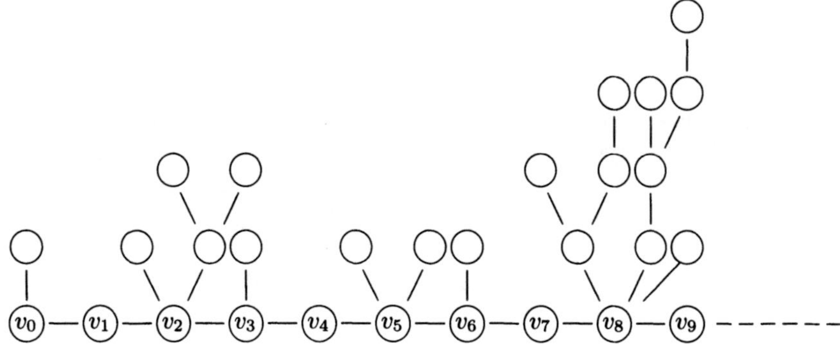

Fig. 1. The "Skeleton Tree" T_∞ – also known as PGW$^\infty$(1). Here, one should note that only the root v_0 of T_∞ is labelled; the chimerical notations $v_1, v_2, \ldots$ are only given to focus the eye on the unique path from the root to infinity. The existence of such a path is one of the most fundamental features of PGW$^\infty$(1).

One might imagine that T_∞ could be an undistinguished member of the multitude of infinite trees that anyone might paste together from finite trees, but this intuition is faulty. As the next lemma explains, the skeleton tree T_∞ has an inevitable – and highly distinguished – role in the theory of random trees. We should note that in the next lemma, and throughout the survey, a graph without its own edge length function is given one by taking the length of each edge to be 1.

Lemma 2.3 (Grimmett's Lemma).   *The randomly rooted uniformly distributed labelled tree T_n on n vertices converges weakly in $\mathcal{G}_\star$ to the "skeleton tree" T_∞; that is, one has*

$$T_n \xrightarrow{\ d\ } T_\infty \quad as\ n \to \infty\,.$$

This result was first formalized and proved in Grimmett [26], and further proofs and generalizations are given by Devroye [18] and Aldous [2]. In some respects, Lemma 2.3 is the most fundamental fact about random trees under the model of uniform distribution. Grimmett's Lemma also has an interesting *self-generalizing* quality, and shortly we will see that the convergence in distribution of T_n actually implies that certain special parts of T_n must converge to corresponding special parts of T_∞.

This *convergence of parts* seems to be a recurring theme in the theory of local weak convergence. In particular, it provides one key to the appearance of the distributional identities that have had a crucial role in some of the most interesting applications of the objective method.

Special Parts of the Skeleton Tree PGW$^\infty$(1)

Our first observation about the skeleton tree of Figure 1 is that it has a pleasantly recursive structure. From the root v_0 there is just one edge (v_0, v_1) that is part of the unique infinite path $v_0 \to v_1 \to \cdots$ of T_∞, and if we delete that edge we find two subtrees – one finite and one infinite. If T_∞^{small} denotes the finite tree rooted at v_0 and T_∞^{big} denotes the infinite tree rooted at v_1, then we see from the definition of T^∞ that we can identify their distributions

$$T_\infty^{small} \overset{d}{=} \text{PGW}(1) \quad \text{and} \quad T_\infty^{big} \overset{d}{=} \text{PGW}^\infty(1)\,.$$

This decomposition and the definition of weak convergence of probability measures on $\mathcal{G}_\star$ promptly suggest that one may give a more detailed interpretation of Grimmett's Lemma.

Convergence of the Corresponding Parts

Consider again the randomly rooted uniform tree T_n on n vertices, and let r denote the root of T_n. Any child c of r determines an edge $e = (r, c)$, and, when e is deleted from T, we let $T_{n,c}$ denote the remaining component of T_n that contains c. We again view $T_{n,c}$ as a rooted geometric weighted graph by taking its root to be c.

Now, we let c^* denote a child of r that is chosen at random from the set of all children of the root for which the cardinality of $T_{n,c}$ is maximal; in fact, for large n, one finds that with high probability there is a unique maximal subtree. If we remove the edge (r, c^*) from T_n we get two subtrees that we now label T_n^{small} and T_n^{big} according to their cardinality (and with any tie being broken by a coin flip). Finally, we view T_n^{small} and T_n^{big} as rooted trees by letting the end point of the deleted edge determine the new roots.

Next, by our observations on the structure of the skeleton tree, we can give a more detailed view of the convergence that is guaranteed by Grimmett's Lemma. Specifically, Grimmett's Lemma automatically implies both the convergence of the small component

$$T_n^{small} \xrightarrow{d} T_\infty^{small} \overset{d}{=} \mathrm{PGW}(1)$$

and convergence of the large component

$$T_n^{big} \xrightarrow{d} T_\infty^{big} \overset{d}{=} \mathrm{PGW}^\infty(1) \,.$$

Moreover, Grimmett's Lemma even implies the joint convergence

$$\left(T_n^{small}, T_n^{big} \right) \xrightarrow{d} \left(T_\infty^{small}, T_\infty^{big} \right) ,$$

where the two processes $T_\infty^{small} \overset{d}{=} \mathrm{PGW}(1)$ and $T_\infty^{big} \overset{d}{=} \mathrm{PGW}^\infty(1)$ are *independent*. This independence is often useful, and in the Section 3 its contribution is essential.

Beyond the Skeleton Trees

Many models for random trees have been introduced in combinatorics and the theory of algorithms ([48], [65]), and for the majority of these models one finds that there is a natural weak limit. Moreover, the analysis of Aldous [2] shows that for a large class of these models the limit object is a random rooted tree which shares many of the qualitative features of the Skeleton Tree given in Grimmett's Lemma.

Specifically, one finds that each of these limit objects has a unique infinite path from the root, and, as we saw in our discussion of Grimmett's Lemma, when one removes this skeleton one finds a sequence of finite trees that have a useful description. In general, these finite trees are no longer independent and identically distributed; instead, this sequence trees is more closely described as a hidden Markov process. In fortunate circumstances, the Markov structure that one finds can serve as a useful substitute for the independence that we use here.

3 Maximal Weight Partial Matching on Random Trees

Like most methods, the objective method is best understood through the examination of a concrete example, and, for all the reasons mentioned in the introduction, we believe that the maximal weight partial matching problem is the right place to begin. It is tractable enough to be developed completely, yet rich enough to demonstrate the structural features of the objective method that are important in more complex problems such as the random assignment problem.

3.1 Weighted Matchings of Graphs in General

For any graph $G = (V, E)$ a *partial matching* S is simply a subset of the set of edges of G such that no pair of edges of S share a common vertex. Any function $w : E(G) \to \mathbb{R}$ may be viewed as a weight function on the edge set of G, and the *weight of the partial matching* S is defined simply as

$$w(S) \stackrel{\text{def}}{=} \sum_{e \in S} w(e) \, .$$

If S^* is a partial matching of G such that $w(S^*)$ is equal to the supremum of $w(S)$ over all partial matchings of G, then S^* is called a *maximal weight partial matching* of G. Such matchings are important in many problems of combinatorial optimization, and methods for computing maximum weight partial matchings have an important role in the theory of algorithms (cf. Lovász and Plummer [47]).

3.2 Our Case: Random Trees with Random Edge Weights

To each fixed finite rooted tree T and to each edge e of T, we now associate a random variable ξ_e. Moreover, we assume that the ensemble $\{\xi_e : e \in T\}$ is independent and that ξ_e has distribution F for each $e \in T$. We view $e \mapsto \xi_e$ as a weight function on T, and a T with this weight function will be called an *F-weighted rooted tree*. Finally, for each F-weighted rooted tree T we let $M(T)$ denote the maximum weight over all partial matchings of T.

In particular, if T_n denotes an F-weighted rooted tree that is chosen at random according to the uniform distribution on the set of n^{n-1} rooted trees with n-vertices, then we write M_n as shorthand for $M(T_n)$. The random variable M_n is therefore the maximum weight of a partial matching of a random n-vertex F-weighted rooted tree, and it will be at the center of our attention for the rest of this section.

3.3 Two Obvious Guesses: One Right, One Wrong

When the edge weight distribution F has a finite mean, one immediately suspects that there is an asymptotic relationship of the form

$$E[M_n] \sim \gamma n \quad \text{as } n \to \infty$$

where $\gamma = \gamma(F)$ is a constant that depends only on F. What one further suspects is that even for the nicest choices of F the calculation of γ might be an almost impossible task. By comparison, one can list perhaps a dozen problems where subadditive methods yield a similar asymptotic relationship, yet the limit constants have gone unknown for decades.

In the partial matching problem we face a happier circumstance. For continuous F the objective method not only yields a proof of the asymptotic relation, it also provides a concrete characterization of the limit constant. This characterization is not necessarily simple, but at least in the leading case of exponentially distributed edge weights it does lead to an explicit integral representation for γ that can be calculated numerically.

Our Assumptions on the Distribution F

The intuition that leads one to guess that $E(M_n)/n$ converges does not impose any constraint on the edge-weight distribution F, except the trivial constraint that the expectation of $\xi_e \sim F$ should be well defined. Nevertheless, here we will always assume that F is continuous and that $P(\xi_e \geq 0) = 1$; these assumption guarantee that the maximal weight partial matching exists and is unique with probability one.

We do not believe that the continuity of F is needed for the convergence of $E(M_n)/n$, but, without this assumption, the possibility of multiple optima would forced us to face many irritating complications. Since our main intention here is to demonstrate the fundamental features of the objective method in the simplest realistic light, the issue of discontinuous F is best left for another time.

3.4 Not Your Grandfather's Recursion

The first thought of anyone interested in the asymptotics of $E[M_n]$ is to look for a relation between $E[M_n]$ and the earlier expectations $E[M_i]$, $1 \leq i < n$. Here we will also hunt for a recurrence relation, but what we find differs radically from the recursions one commonly meets in discrete mathematics.

If we remove an edge e from the edge set of T_n, then T_n is broken into two connected components. We can then view these components as rooted trees where we take the old vertices of e to be our new roots. Next, we label these trees $T_n^{small}(e)$ and $T_n^{big}(e)$ according to their cardinality (with any tie being broken by taking the labels at random). This process is similar to the discussion of the refinement of Grimmett's Lemma, except that here the cut edge e can be any edge of T_n.

Now, we consider whether or not the edge e is in the maximal partial matching of T_n. If one does not use the edge e, then the maximum weight of a partial matching of T_n is also equal to

$$M\big(T_n^{small}(e)\big) + M\big(T_n^{big}(e)\big) \, . \tag{3.1}$$

To go deeper we need some additional notation. If T is any weighted rooted tree, we define $B(T)$ by letting $M(T) - B(T)$ denote the maximum weight of a partial matching of T where the root is *not* in an edge of the partial matching. By optimality of $M(T)$ we see that $B(T) \geq 0$, and we think of $B(T)$ as the *bonus* one gets from the *option* of being allowed to use edges that meet the root. With this notation, we see that the maximal weight of a partial matching that is *required* to use e is given by the sum

$$\xi_e + \big\{ M(T_n^{small}(e)) - B(T_n^{small})(e)\big\} + \big\{M(T_n^{big}(e)) - B(T_n^{big})(e)\big\} \, . \tag{3.2}$$

When we compare the weights (3.1) and (3.2), we see that with probability one the edge e is in the maximum weight partial matching *if and only if*

$$\xi_e > B\big(T_n^{small}(e)\big) + B\big(T_n^{big}(e)\big) \, . \tag{3.3}$$

This inclusion criterion naturally gives us a nice way to write M_n as a sum over the edges of T_n. If we use $\mathbf{1}(A)$ to denote the indicator function for an event A, then the inclusion criterion (3.3), tells us that

$$M_n = \sum_{e \in T_n} \xi_e \mathbf{1}\Big(\xi_e > B(T_n^{small}(e)) + B(T_n^{big}(e)) \Big) \, . \tag{3.4}$$

Now, if $\mathbf{e}$ denotes an edge chosen uniformly from the edge set of T_n, we see from the sum (3.4) that the expectation $E[M_n]$ can be written as

$$E[M_n] = (n-1)E\big[\xi_\mathbf{e} \mathbf{1}\big(\xi_\mathbf{e} > B(T_n^{small}(\mathbf{e})) + B(T_n^{big}(\mathbf{e}))\big)\big] \, .$$

Finally, since the distribution of $\xi_\mathbf{e}$ does not depend on $\mathbf{e}$ and since $\xi_\mathbf{e}$ is independent of $B(T_n^{small}(\mathbf{e}))$ and $B(T_n^{big}(\mathbf{e}))$, the last equation may be written a bit more crisply as

$$E[M_n] = (n-1)E\big[\xi \mathbf{1}\big(\xi > B(T_n^{small}(\mathbf{e})) + B(T_n^{big}(\mathbf{e}))\big)\big] \, , \tag{3.5}$$

where we understand that $\xi \overset{d}{=} F$ and that ξ is independent of the pair $\big(B(T_n^{small}(\mathbf{e})), B(T_n^{big}(\mathbf{e}))\big)$.

3.5 A Direct and Intuitive Plan

The representation for $E[M_n]$ given by equation (3.5) is hardly a conventional recursion, yet it is still encouraging. The first factor $n-1$ hints loudly at the gross behavior one expects from $E[M_n]$. Moreover, from our discussion of the refinement of Grimmett's Lemma, one can guess that two tree processes $T_n^{small}(\mathbf{e})$ and $T_n^{big}(\mathbf{e})$ will converge in distribution, and this strongly suggests that the second factor of formula (3.5) will converge to a constant as $n \to \infty$.

Certainly, one should worry about the fact that $B(\cdot)$ is not a bounded continuous function on $\mathcal{G}_\star$; after all, $B(\cdot)$ is not even well defined on all of $\mathcal{G}_\star$. Nevertheless, $B(\cdot)$ is well defined on the finite graphs of $\mathcal{G}_\star$, and any optimist surely suspects it to nice enough for the random variables $B(T_n^{small}(\mathbf{e}))$ and $B(T_n^{big}(\mathbf{e}))$ to inherit convergence in distribution from the weak convergence of $T_n^{small}(\mathbf{e})$ and $T_n^{big}(\mathbf{e})$.

To refine these suspicions, we first note that $T_n^{small}(\mathbf{e})$ and $T_n^{big}(\mathbf{e})$ differ from the trees T_n^{small} and T_n^{big} in our discussion of Grimmett's Lemma only in that the roots of $T_n^{small}(\mathbf{e})$ and $T_n^{big}(\mathbf{e})$ are determined by the random edge $\mathbf{e}$ while the roots of T_n^{small} and T_n^{big} were determined by a random uniform choice from the respective vertex sets.

As Figure 2 suggests, the root r of $T_n^{small}(\mathbf{e})$ is not quite uniformly distributed on the vertex set of T_n when $\mathbf{e}$ is chosen at uniformly at random from the edge set of T_n. In fact, if n is odd, then there is always one vertex of T_n that has probability 0 of being the root of $T_n^{small}(\mathbf{e})$, and, when n is even, then there are always two vertices that have probability $1/(2(n-1))$ of being the root of $T_n^{small}(\mathbf{e})$. In both cases, all of the other vertices have probability $1/(n-1)$ of being the root of $T_n^{small}(\mathbf{e})$, so, even though r is not uniformly distributed on the vertex set of T_n, it is *almost* uniformly distributed. In fact, the total variation distance between the distribution of r and the uniform distribution is always bounded by $1/2n$, and for odd values of n this bound is exact.

From this observation and a traditional coupling argument, one finds that there is only a small change in the distribution of the bonus when the omitted edge is used to determine the root; specifically, for all $n \geq 1$ and all $x \in \mathbb{R}$ we have the inequality

$$\left| P\big(B(T_n^{small}(\mathbf{e})) \leq x \big) - P\big(B(T_n^{small}) \leq x \big) \right| \leq \frac{1}{2n} \, . \tag{3.6}$$

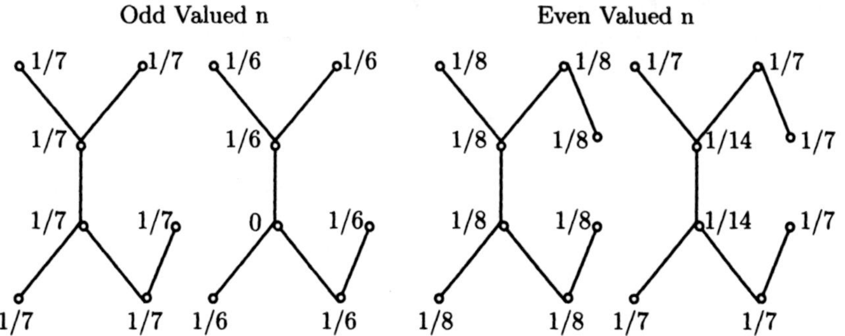

Fig. 2. In this figure, the value next to a vertex is the probability of that vertex being the root of T_n^{small}. One should note that the uniform choice of an edge typically leads to a non-uniform distribution on the vertices.

A parallel analysis of the root of $T_n^{big}(\mathbf{e})$ yields an identical bound for the distributions of the bonuses of the big tree components, and together these bounds tell us that if the distribution F of ξ is not too long-tailed, then the asymptotic behavior of the right hand side of our exact formula (3.5) for $E[M_n]$ will not be changed if we replace the edge biased trees $T_n^{small}(\mathbf{e})$ and $T_n^{big}(\mathbf{e})$ by their randomly rooted cousins T_n^{big} and T_n^{small} whose asymptotic behavior is well understood.

We now see that along the path to proving that $E[M_n]/n$ converges, we have the natural intermediate task of showing that the bonuses $B(T_n^{small})$ and $B(T_n^{big})$ converge in distribution. Further, if we hope to have a serious chance of calculating the value of the limit of $E[M_n]/n$, we also need a concrete characterization of the limiting distributions of $B(T_n^{small})$ and $B(T_n^{big})$.

3.6 Characterization of the Limit of $B(T_n^{small})$

From our discussion of Grimmett's Lemma, we know that T_n^{small} converges in distribution to a tree T with the Poisson Galton-Watson distribution PGW(1). When we view T as an F-weighted tree, then the bonus $B(T)$ is well defined since T is almost surely finite. Moreover, one can check directly from the definition of weak convergence in $\mathcal{G}_\star$ that we have

$$B(T_n^{small}) \xrightarrow{\ d\ } B(T)\,, \tag{3.7}$$

so the real question is whether there is an effective way to characterize the distribution of $B(T)$. This is where the notion of a distributional identity enters the picture.

The basic idea is that the recursive definition of the tree T should translate into a useful self-referencing identity for the distribution of $B(T)$; we just need to put this idea to work. If C denotes the set of children of the root of T, then for each $i \in C$ there is a subtree T^i of T that is determined by the descendants of i, and we may view T^i as a rooted tree with root i. If ξ_i denotes the weight of the edge from the root of T to the child i, then the maximum weight of a matching on T that *does not* meet the root is given by the sum

$$\sum_{i \in C} M(T^i)\,;$$

so, to get an identity for $B(T)$ we just need to calculate the maximum weight of a partial matching that *does* meet the root.

If $j \in C$, then the maximum weight for a partial matching that uses the edge from the root of T to j is equal to

$$\xi_j + \left\{ M(T^j) - B(T^j) \right\} + \sum_{i \in C,\, i \neq j} M(T^i) = \xi_j - B(T^j) + \sum_{i \in C} M(T^i)\,,$$

so the maximal weight for a partial matching that *does* include the root is equal to

$$\max_{j \in C} \left\{ \xi_j - B(T^j) \right\} + \sum_{i \in C} M(T^i) \, .$$

We now see that the difference $B(T)$ between the overall maximum and the constrained maximum $\sum_{i \in C} M(T^i)$ is given by

$$B(T) = \max \left\{ 0, \, \xi_i - B(T^i) : \, i \in C \right\} \, . \tag{3.8}$$

Finally, the cardinality of C is Poisson with mean one, $T \stackrel{d}{=} T^i$ for all $i \in C$, and all of the random variables on the right hand side of the identity (3.8) are independent, so we have before us a very powerful constraint on the distribution of $B(T)$. In fact, we can check that any random variable that satisfies a distributional recursion like that defined by equation (3.8) must be equal in distribution to $B(T)$.

Formalization

To formalize this assertion, we let F and G denote any two distributions, and we define $D_F(G)$ to be the distribution of the random variable

$$\max\{0, \, \xi_i - Y_i : 1 \leq i \leq N\} \tag{3.9}$$

where random variables in the collection $\{N, Y_i, \xi_i : i = 1, 2, ...\}$ are independent, N has the Poisson(1) distribution, and we have $\xi_i \sim F$ and $Y_i \sim G$ for all $1 \leq i < \infty$. The next proposition tells us that one may characterize the distribution of the bonus $B(T)$ as a fixed-point of the mapping $D_F(\cdot)$.

Proposition 3.1. *If T is an F-weighted PGW(1) tree where F is continuous and $F([0, \infty)) = 1$, then the distribution*

$$G(x) = P\big(B(T) \leq x\big)$$

is the unique solution of the fixed-point equation

$$D_F(G) = G \, . \tag{3.10}$$

Since the discussion leading up to this proposition shows that the distribution of $B(T)$ is a solution of the fixed-point equation, we see that the characterization will be complete if we show that any solution of equation (3.10) must be the distribution of $B(T)$ for some F-weighted PGW(1) tree T. This will be done in a way that is particularly probabilistic.

A Uniqueness Proof by a Probabilistic Construction

Let G denote a solution of the fixed-point equation $D_F(G) = G$, so our task is to show that G is the distribution function of $B(T)$ where T is an F-weighted

18   David Aldous and J. Michael Steele

PGW(1) tree. Our plan is simply to use G to construct such a T, and Figure 3 suggests how we start.

In the first step, we just consider the tree consisting of a root and a set C of N children. We then attach a random variable Y_i to each child $i \in C$ where the Y_i are independent and $Y_i \sim G$. We then *define* Y by the identity

$$Y = \max\{0, \xi_i - Y_i : i \in C\} ,$$

where the ξ_i are independent and $\xi_i \sim F$. We describe this process by saying that the root has been *expanded*.

Next, take any unexpanded vertex v (at the second step this would be any of the elements of C) and expand v in the same way that we expanded the root. Specifically, we give v a set $C(v)$ of children with cardinality $N(v) = |C(v)|$ that is independent Poisson(1), attach a random variable $Y_{v,i}$ to each child $i \in C(v)$ where the $Y_{v,i}$ are independent and $Y_{v,i} \sim G$, and then define Y_v by the identity

$$Y_v = \max\{0, \xi_{v,i} - Y_{v,i} : i \in C\} , \tag{3.11}$$

where as before the $\xi_{v,i}$ are independent and $\xi_{v,i} \sim F$.

When we expand the root, the fixed-point property of G guarantees that Y has the distribution G. Later, when we expand v, the values of Y_v and Y will be changed, but their distribution will not be changed. Thus, to complete the construction, we simply repeat the process of expanding any unexpanded vertex until no such vertices remain. The tree we construct in this way is precisely a Galton-Watson tree with the Poisson(1) off-spring distribution, so, with probability one, the expansion process stops after finitely many steps. Moreover, even though each vertex expansion changes the values of the random variables that are attached to the vertices along the path back to the root, all of the distributions will be unchanged.

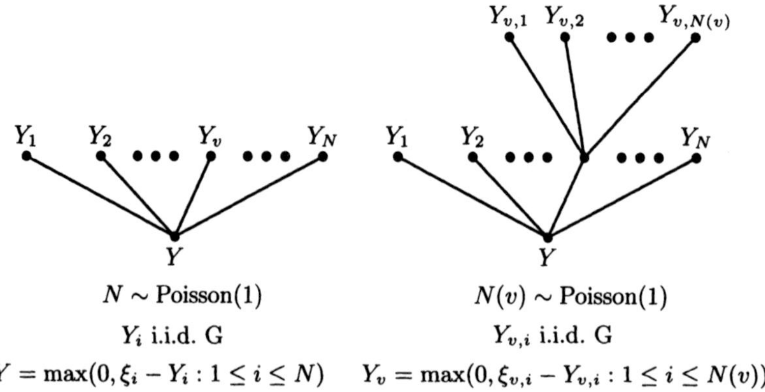

Fig. 3. Expansion of the root and expansion of a descendant vertex.

By this construction, we see that for any leaf ℓ of the resulting tree T the value Y_ℓ associated to ℓ by the value rule (3.11) is zero. The bonus $B(T_\ell)$ is also zero for any leaf, and the recursion (3.8) for the bonus is the same as the recursion that defines the associated values Y_v, $v \in T$. We therefore find that $B(T) = Y$, and consequently we see that any solution G of the fixed-point equation $D_F(G) = G$ is equal to the distribution of $B(T)$ for an F-weighted PGW(1) tree T.

An Analytical Question

This proof of Proposition 3.1 is quite natural given the way that we came to the fixed-point equation $D_F(G) = G$, but the uniqueness question for this equation is in fact a purely analytical problem. It seems natural to ask if the uniqueness of the solution of $D_F(G) = G$ may be proved directly without recourse to the probabilistic interpretation of G, but we do not address that question here. For the moment the more pressing question is whether we might also be able to characterize $B(T_n^{big})$ by a distributional identity like the one we used to characterize $B(T_n^{small})$.

3.7 Characterization of the Limit of $B(T_n^{big})$

To be sure, one can begin just as before; specifically, we can note from our discussion of Grimmett's Lemma that T_n^{big} converges weakly in $\mathcal{G}_\star$ to a random tree, although this time the limit is T_∞, the skeletal tree with distribution $PGW^\infty(1)$. We would dearly love to conclude that

$$B(T_n^{big}) \xrightarrow{d} B(T_\infty) , \tag{3.12}$$

but as soon as one writes this equation trouble appears.

Since T_∞ is almost surely infinite, one sees that the total weight of the maximal partial matching of T_∞ is infinite. For the same reason, the total weight of the maximal partial matching that does not meet the root of T_∞ is also infinite. The bottom line is that the bonus $B(T_\infty)$, which is nominally the difference of these two quantities, is not well defined.

Nevertheless, one should not lose heart. If indeed there is a finite random variable Z such that

$$B(T_n^{big}) \xrightarrow{d} Z , \tag{3.13}$$

then we still may be able to use our intuition about the undefined quantity $B(T_\infty)$ to help us to understand Z. Moreover, if we successfully guess a distributional identity that characterizes Z, we may even be able to use the stability of that characterization to build an honest proof of the conjectured convergence (3.13).

Guessing the Distributional Identity for Z

As physicists sometimes do, we cross our fingers and calculate with $B(T_\infty)$ even though it represents the difference of two infinite quantities. Since we only want to guess an identity for Z that will be subsequently justified by a rigorous argument, there is no cost to proceeding heuristically for the moment.

If we label T_∞ as we did in Figure 1 and if we remove the edge (v_0, v_1), then we obtain two subtrees with roots v_0 and v_1. If T denotes the subtree with root v_0, then T has the PGW(1) distribution, and, if T'_∞ denotes the subtree with root v_1, then T'_∞ has the PGW$^\infty$(1) distribution. By the "definition" of $B(T_\infty)$, the optimal partial matching of T_∞ that *does not match the root* v_0 has weight $M(T_\infty) - B(T_\infty)$, and our first task is to get an alternative representation for this quantity.

To begin, we let C denote the set of all of the children of v_0 *except* for the special child v_1 that is on the path from v_0 to infinity in T_∞. For each $i \in C$ we view the vertex i and its descendants as a subtree T^i of T, and we take i as the root of T^i. With this notation, we can write the weight of the optimal matching on T_∞ that does not match the root v_0 to a second vertex and obtain the bookkeeping identity

$$M(T_\infty) - B(T_\infty) = M(T'_\infty) + \sum_{i \in C} M(T^i) . \tag{3.14}$$

The next step is to find an appropriate expression for the maximal weight of a partial matching of T_∞ that *does* meet the root. We first note the maximal weight of a partial matching of T_∞ that contains the edge (v_0, j) with $j \in C$ is given by

$$\xi_j + \left\{ M(T^j) - B(T^j) \right\} + M(T'_\infty) + \sum_{i \in C, i \neq j} M(T^i)$$

which we may simplify to

$$\left\{ \xi_j - B(T^j) \right\} + M(T'_\infty) + \sum_{i \in C} M(T^i) .$$

Now, if ξ denotes the weight of the edge (v_0, v_1), then the maximal weight of a partial matching of T_∞ that contains the edge (v_0, v_1) can be written as

$$\sum_{i \in C} M(T^i) + \xi + \left\{ M(T'_\infty) - B(T'_\infty) \right\} ,$$

so, choosing the best of all of the possibilities, we see that $M(T_\infty)$ equals

$$\sum_{i \in C} M(T^i) + M(T'_\infty) + \max \left\{ 0, \max_{j \in C} \{ \xi_j - B(T^j) \}, \xi - B(T'_\infty) \right\} .$$

When we subtract the value of $M(T_\infty) - B(T_\infty)$ given by equation (3.14), we find

$$B(T_\infty) = \max\{0, \max_{j \in C}\{\xi_j - B(T^j)\}, \xi - B(T'_\infty)\} \ .$$

We know $\max_{j \in C}\{0, \xi_j - B(T^j)\}$ has the same distribution as $B(T)$ by the basic distributional identity (3.8) for the bonus, and we also know that T'_∞ and T_∞ have the same distribution, so at last we have a nice, simple, and – lamentably heuristic – identity for $B(T_\infty)$:

$$B(T_\infty) \overset{d}{=} \max\{B(T), \xi - B(T_\infty)\} \ .$$

Conversion to an Honest Proposition

For any given continuous distribution F, we know that the distribution G of the bonus $B(T)$ for the F-weighted PGW(1) tree T can be characterized as the unique solution of the fixed-point equation $D_F(G) = G$, and our heuristic derivation suggests that we can characterize the limiting distribution of $B(T_n^{big})$ in a way that is perfectly analogous. Specifically, given any distribution H and independent random variables

$$Z \sim H, \quad Y \sim G, \quad \text{and} \quad \xi \sim F \ ,$$

we let $\widetilde{D}_F(H)$ denote the distribution of $\max\{Y, \xi - Z\}$, and we consider the fixed-point equation

$$\widetilde{D}_F(H) = H \ .$$

This time we do not know *a priori* that there is a solution to this equation, but we do expect the limiting distribution of $B(T_n^{big})$ to exist, and for large n we also expect $B(T_n^{big})$ to be much like $B(T_\infty)$. Therefore, we suspect that the limiting distribution of $B(T_n^{big})$ will satisfy $\widetilde{D}_F(H) = H$. The next proposition confirms these suspicions while also confirming the crucial fact that the solution of the fixed-point equation $\widetilde{D}_F(H) = H$ is unique.

Proposition 3.2. *For any continuous distribution F with support on $[0, \infty)$, the fixed-point equation*

$$\widetilde{D}_F(H) = H \tag{3.15}$$

has a unique solution H; moreover, one has

$$\lim_{n \to \infty} P\big(B(T_n^{big}) \leq x\big) = H(x) \quad \text{for all } x \in \mathbb{R} \ .$$

Before proving this proposition, we will first show that it really does complete the program with which we began.

3.8 The Limit Theorem for Maximum Weight Partial Matchings

Our whole discussion of $E[M_n]$ has been motivated by the formula

$$E[M_n] = (n-1)E\big[\xi \mathbf{1}\big(\xi > B(T_n^{small}(\mathbf{e})) + B(T_n^{big}(\mathbf{e}))\big)\big] \ , \tag{3.16}$$

and now with the help of Proposition 3.1 and Proposition 3.2 we are ready to take the limit. If we assume that ξ has a finite expectation, then the integrands in formula (3.16) are uniformly integrable and they converge in distribution to $\xi\mathbf{1}(\xi > Y + Z)$, so without further concern we can take the limit under the expectation to obtain the main theorem of this section.

Theorem 3.3. *If F is a continuous distribution with support on $[0, \infty)$ and a finite mean, then the maximum weight M_n of an F-weighted random tree on n vertices satisfies*

$$\lim_{n \to \infty} \frac{1}{n} E\left[M_n\right] = E\left[\xi\mathbf{1}(\xi > Y + Z)\right] \tag{3.17}$$

where the random variables ξ, Y, and Z are independent,

$$\xi \sim F, \quad Y \sim G, \quad and \quad Z \sim H,$$

and the distributions G and H are the unique solutions of the fixed-point equations

$$D_F(G) = G \quad and \quad \tilde{D}_F(H) = H. \tag{3.18}$$

The expression $E[\xi\mathbf{1}(\xi > Y + Z))]$ for the limit (3.17) in Theorem 3.3 always provides some structural insight into the value of the limit, but, since the distributions of Y an Z are only defined through the solutions of two fixed-point equations, one might think that Theorem 3.3 is still rather abstract. The next result tells us that in one of the cases that matters most, the limit $E[\xi\mathbf{1}(\xi > Y + Z))]$ may be calculated explicitly.

Theorem 3.4. *If $F(x) = 1 - e^{-x}$ for $x \geq 0$, then the maximum weight M_n of a partial matching on an F-weighted random tree on n vertices satisfies*

$$\lim_{n \to \infty} \frac{1}{n} E[M_n] = \int_0^\infty \int_0^s c(e^{-y} - be^{-s}) \exp(-ce^{-y} - ce^{-(s-y)})se^{-s} \, dy \, ds,$$

where the constants b and c are defined by equations (3.21) and (3.22) of Lemma 3.5. Moreover, by numerical integration one finds

$$\lim_{n \to \infty} \frac{1}{n} E[M_n] = 0.239583....$$

To prove Theorem 3.4 we just need to solve two fixed-point equations, and the lack-of-memory property of the exponential distribution makes this easier than anyone has any right to expect. Later we will find several other situations like this one where distributional identities lead us to calculations that are mysteriously pleasant.

Lemma 3.5. *If $F(x) = 1 - e^{-x}$ for $x \geq 0$, then the solutions G and H of the fixed-point equations (3.18) are*

$$G(y) = \exp(-ce^{-y}) \quad \text{for } y \geq 0 \tag{3.19}$$

$$\text{and} \quad H(z) = (1 - be^{-z}) \exp(-ce^{-z}) \quad \text{for } z \geq 0 , \tag{3.20}$$

where b and c are related by

$$b = c^2/(c^2 + 2c - 1) \tag{3.21}$$

and c is the unique strictly positive solution of

$$c^2 + e^{-c} = 1 . \tag{3.22}$$

Numerically, one finds $b = 0.543353...$ and $c = 0.714556....$

To prove Lemma 3.5, we first consider $Y \sim G$ along with an independent pair of independent sequences $\{Y_i : i = 1, 2, ...\}$ and $\{\xi_i : i = 1, 2, ...\}$ with $Y_i \sim G$ and $\xi_i \sim F$. In terms of these variables, the equation $D_F(G) = G$ simply means

$$P(Y \leq y) = P\big(\xi_i - Y_i \leq y \text{ for all } 1 \leq i \leq N\big) = \exp\big(-P(\xi - Y > y)\big) ,$$

where the second identity follows by conditioning on the independent Poisson variable N. Now, if we take logarithms and apply the lack-of-memory property of the exponential, then we find

$$\log P(Y \leq y) = -P(\xi > y + Y) = -e^{-y}P(\xi > Y) , \tag{3.23}$$

and this identity gives us the explicit formula (3.19) when we set c equal to $P(\xi > Y)$.

Next, to get the determining equation (3.22) we only need to calculate $c = P(\xi > Y)$ with help from the distribution (3.19) of Y; specifically, we find

$$c = P(\xi > Y) = \int_0^\infty e^{-y}P(Y \leq y)dy = \int_0^\infty \exp(-ce^{-y}) \, e^{-y}dy$$

$$= \int_0^1 \exp(-cz) \, dz = c^{-1}(1 - e^{-c}) .$$

Finally, to get the distribution of Z, we use the defining relation (3.15) and the lack-of-memory property of the exponential to see for $z > 0$ that

$$P(Z \leq z) = P(Y \leq z)P(\xi - Z \leq z)$$

$$= \exp(-ce^{-z}) \, P(\xi \leq z + Z) = \exp(-ce^{-z}) \left\{1 - e^{-z}P(\xi > Z)\right\} ,$$

so if we set $b = P(\xi > Z)$, then $P(Z \leq z) = (1 - be^{-z}) \exp(-ce^{-z})$, just as we hoped. Now, all that remains is to express $b = P(\xi > Z)$ as a function of $c = P(\xi > Y)$, and we begin by noting

$$b = P(\xi > Z) = \int_0^\infty P(Z < z)e^{-z}\,dz$$

$$= \int_0^\infty (1 - be^{-z})\,\exp(-ce^{-z})\,e^{-z}dz = \frac{c-b}{c^2} + \frac{b-c+bc}{c^2}e^{-c}\,.$$

Now, if we use the identity (3.22) to replace e^{-c} by $1 - c^2$, we see that the last equation may be solved for b to obtain the promised relation (3.21). Finally, we note that the uniqueness of c follows from the convexity of the map $x \mapsto x^2 + e^{-x}$.

3.9 Closing the Loop: Another Probabilistic Solution of a Fixed-Point Equation

To complete our investigation of the maximal partial matching problem, we only need to prove Proposition 3.2. Specifically, we need to show that the fixed-point equation

$$\widetilde{D}_F(H) = H$$

has a unique solution, and we need to show that this solution is the limiting distribution for the bonuses $B(T_n^{big})$.

Existence Follows from General Theory

The existence of a solution to the equation $\widetilde{D}_F(H) = H$ is almost obvious once one draws a connection between the fixed-point equation and the theory of Markov chains. Specifically, if we set

$$K(z, A) = P\big(\max\{\xi - z, Y\} \in A\big)$$

where $\xi \sim F$ and $Y \sim G$ are independent, then $K(x, A)$ defines a Markov transition kernel on $\mathbb{R}^+$. One then sees at that a distribution is a fixed point of $\widetilde{D}_F$ if and only if it is a stationary distribution for the Markov chain with kernel K.

Finally, since the kernel K has the Feller property and since it is dominated by the distribution of $\max\{\xi, Y\}$, the general theory of Markov chains on $\mathbb{R}^+$ tells us that there is a stationary distribution for K (cf. Meyn and Tweedie [50], Theorem 2.1.2).

Uniqueness – Observed and Confirmed by Coupling

While we might also be able to show the uniqueness of the solution of the fixed-point equation with help from Markov process theory, there is another approach via coupling that is particularly informative. Moreover, the argument anticipates a useful stability result that gives us just the tool we need to prove that $B(T_n^{big})$ converges in distribution.

To begin, we assume that the variables $\{\xi_m, Y_m : 1 \leq m \leq \infty\}$ are independent with $\xi_m \sim F$ and $Y_m \sim G$ for all m. Next, simultaneously for each $z \geq 0$, we construct a Markov chain $Z_m(z)$, $m = 1, 2, ...$ by taking $Z_0(z) = z$ and by setting

$$Z_m(z) = \max\{\,\xi_m - Z_{m-1}(z),\, Y_m\,\} \quad \text{for } m \geq 1 \,.$$

Now, for all z and z' in $\mathbb{R}^+$, we see that

$$Z_m(z) \geq Z_m(z') \quad \text{implies} \quad Z_{m+1}(z) \leq Z_{m+1}(z') \,,$$

and from this anti-monotone property we see that $Z_m(z) \geq Z_m(0)$ for all even values of m. Therefore, if we let

$$T = \min\{\,m \text{ odd } : Z_m(0) = Y_m\,\} \,,$$

then we see that $Z_T(z) = Y_T$ for all z. As a consequence, we see that T is a *coupling time* in the exceptionally strong sense that

$$Z_m(z) = Z_m(0) \quad \text{for all} \quad z \geq 0 \quad \text{and all} \quad m \geq T \,; \tag{3.24}$$

moreover, the definition of $Z_m(0)$ gives us the simple bound

$$P(Z_m(0) = Y_m \mid Z_{m-1}(0) = z) \geq P(\xi_m < Y_m) \quad \text{for all } z \geq 0 \,.$$

Now, if we set $\varrho = 1 - P(\xi_m < Y_m)$ then we can easily check that $\varrho < 1$. To see why this is so, we first note that in a tree $\mathcal{T}$ that has exactly one edge with weight ξ' the bonus $B(\mathcal{T})$ is equal to ξ'. Now, if $\mathcal{T}$ is a PGW(1) tree, then the probability that $\mathcal{T}$ has exactly one edge is equal to e^{-1}, so for $Y \overset{d}{=} B(\mathcal{T})$ we have

$$P(Y \in \cdot) \geq e^{-1} P(\xi' \in \cdot) \,.$$

Now, ξ is independent of Y and has a continuous distribution, so we have the bound

$$P(Y \leq \xi) \geq e^{-2} P(\xi' \leq \xi) = \frac{1}{2e} \,.$$

Thus, we find $\varrho \leq 1 - 1/2e < 1$, so the bound

$$P(T \geq 2k + 1) \leq \varrho^k \quad \text{for all } k \geq 0$$

gives us more than we need to show that T is finite with probability one.

Now, if H is the distribution function for a stationary measure for the kernel K and if $Z \sim H$, then by stationarity we have $Z_n(Z) \sim H$ for all $n \geq 1$. Therefore, by the bound on the coupling time, we find for all $x \in \mathbb{R}$ that

$$|H(x) - P(Z_m(0) \leq x)| \leq \varrho^k \quad \text{for all } m \geq 2k + 1 \,,$$

so we see that there is at most one stationary distribution.

A Coupling Connection

Here we should note that this simultaneous coupling argument shares several elements in common with the "coupling from the past" technique that has become a popular tool in the theory of Markov chain Monte Carlo (cf. Propp and Wilson [56]). In particular, the argument given above shows that for *any* random initial value X we have

$$|P(Z_m(X) \in A) - P(Z \in A)| \leq \varrho^k \quad \text{for all } n \geq 2k + 1 \ ;$$

and, remarkably enough, one can even choose X so that it depends on the elements of $\{\xi_m, Y_m : m \geq 1\}$.

3.10 From Coupling to Stability – Thence to Convergence

If one repeats this coupling argument while relaxing the condition on the distribution of the variables Y_m so that they are only required to have a distribution that is close to G, one is led to bounds on distributions that can be viewed the solutions of an "approximate" fixed-point equation. Any such stability result must face some notational clutter because many quantities are constantly changing, but the next lemma should be easy to parse if one keeps in mind that n is an index that increases as the distribution of Y_m^n comes closer to G. Also, in the lemma we choose the time index m so that we can focus our attention on X_0^n, a quantity that one may view as the dependent variable at the "time" 0 when the "approximation quality" equals n.

Lemma 3.6. *Fix the integer $k \geq 1$ and suppose for each n we have non-negative random variables $\{X_m^n : -2k + 1 \leq m \leq 0\}$ that are linked to the variables $\{Y_m^n : -2k + 2 \leq m \leq 0\}$ by the relation*

$$X_m^n = \max\{\, \xi_m^n - X_{m-1}^n, \, Y_m^n \,\} \quad \text{for all } m = -2k + 2, -2k + 3, \ldots, 0 \ ,$$

where the random variables $\{\xi_m^n : -2k + 2 \leq m \leq 0\}$ are independent, have distribution F, and are independent from

$$X_{-2k+1}^n \quad \text{and} \quad \{Y_m^n : -2k + 2 \leq m \leq 0\} \ .$$

Further, suppose that one has

$$(Y_m^n : -2k + 2 \leq m \leq 0) \ \overset{d}{\longrightarrow}\ (Y_m : -2k + 2 \leq m \leq 0) \ \text{as } n \to \infty \ ,$$

where the random variables $\{Y_m : m = 0, -1, -2, \ldots\}$ are independent and have distribution G. If H is the unique solution of

$$\tilde{D}_F(H) = H \ ,$$

then we have the bound

$$\limsup_{n\to\infty} |\, P(X_0^n \leq x) - H(x)\,| \leq \varrho^k \quad \textit{for all } x \in \mathbb{R} \textit{ and } k = 1, 2, \ldots$$

where ϱ is given by

$$\varrho = \iint_{y<x} dF(x)\, dG(y) < 1 \ .$$

From Stability to Convergence

First, we let T_∞ denote an F-weighted $\mathrm{PGW}^\infty(1)$ tree, and let $v_0, v_1, v_2, \ldots$ denote the sequence of vertices along the unique path to infinity from the root v_0 according to the definition of $\mathrm{PGW}^\infty(1)$ and the familiar Figure 1. Next, we let $k \geq 1$ be an arbitrary integer, and we note from the local weak convergence of T_n to T_∞ that there exists an integer $N(k,\omega)$ so that for all $n \geq N(k,\omega)$ there is a path π with $2k$ vertices and the following properties:

1. the first vertex of π is the root v_0^n of T_n
2. if π is written as $v_0^n \to v_1^n \to v_2^n \to \cdots \to v_{2k-1}^n$ then the deletion of the edges of π creates a forest of subtrees T_i^n, $i = 1, 2, \ldots, 2k-1$, of T_n where T_i^n is rooted at v_i^n, and
3. we have the weak convergence relation

$$(\, T_i^n : 0 \leq i \leq 2k - 2\,) \ \xrightarrow{\ d\ } \ (\, T_i^\infty : 0 \leq i \leq 2k - 2\,)$$

where all of the righthand components T_i^∞, $1 \leq i \leq 2k-2$, are independent $\mathrm{PGW}(1)$ trees.

Next, for each $n \geq 1$ and each $1 \leq i < 2k$, we let R_i^n denote the subtree of T_n that is rooted at v_i^n that one obtains by removing the single edge (v_{i-1}^n, v_i^n) from the path π. We think of R_i^n as a "remainder subtree" that contains everything "past" v_{i-1}^n and to complete the picture we write $R_0^n = T_n$.

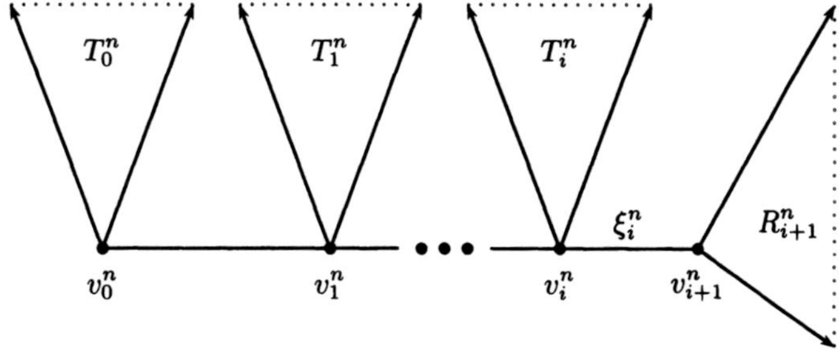

Fig. 4. The Residual R_{i+1}^n.

Finally, if we let ξ_i^n denote the weight on edge (v_i^n, v_{i+1}^n), then we may use *exactly* the same argument that gave us the heuristic derivation of the fixed-point equation $\widetilde{D}_G(H) = H$ to show that

$$B(R_i^n) = \max\bigl\{\, \xi_i^n - B(R_{i+1}^n),\ B(T_i^n)\,\bigr\} \quad \text{for all } 0 \le i \le 2k - 2 . \qquad (3.25)$$

In this case the argument is also rigorous since all of the trees involved are finite, and $B(\cdot)$ is well defined at each stage.

Finally, we note that the recursion (3.25) sets up the direct application of Lemma 3.6. We only need to take $m = -i$, to find that the lemma gives us

$$\limsup_{n \to \infty} |P(B(T_n) \le x) - H(x)| \le \varrho^k .$$

Since $\varrho < 1$ and since k is arbitrary, the last inequality is more than we need to complete the proof of Proposition 3.2.

3.11 Looking Back: Perspective on a Case Study

When one looks back on the maximal partial matching problem, there are several themes that seem to be worth recording for future use. First of all, the inclusion criterion (3.3),

$$\xi_e > B\bigl(T_n^{small}(e)\bigr) + B\bigl(T_n^{big}(e)\bigr) ,$$

helped drive the analysis almost from the beginning. In particular, this necessary and sufficient condition for the edge e to be in the maximal partial matching gave us the succinct representation (3.5) for $E[M_n]$, which in turn encouraged us to study the limit theory of $B\bigl(T_n^{small}(e)\bigr)$ and $B\bigl(T_n^{big}(e)\bigr)$.

The asymptotic analysis of $B\bigl(T_n^{small}(e)\bigr)$ was relatively straightforward, but the analysis of $B\bigl(T_n^{big}(e)\bigr)$ forced us to work harder. The key to its behavior turned out to be the random variable recursion (3.25),

$$B(R_i^n) = \max\{\, \xi_i^n - B(R_{i+1}^n),\ B(T_i^n)\,\} \quad \text{for all } 0 \le i \le 2k - 2 ,$$

but this equation might have left us empty handed if we had not first studied its infinite analog. For the analog, we found there was a unique solution, and this was encouraging. Such equations often have "stabilized" versions, and this principle guided us to the proof of the weak convergence of $B(T_n^{big}(e))$.

One is unlikely to find too many problems that may be solved with a step-by-step application of the method used here, but we will soon see several other examples where most of the same elements are present. Certainly, the heuristic derivation and rigorous solution of a distributional identity emerges as a basic theme. Sometimes this identity is followed by stabilization, and at other times it is exploited to provide an analog of an inclusion rule similar to that provided by equation (3.3). Typically, the new inclusion rule would apply only in a more elaborate model that one constructs with help from the distributional identity.

We will see from the analysis of more complicated problems that the solution of the maximal partial matching problem is a hybrid; it is initiated and organized by the objective method, but it is completed with help from a traditional stochastic recursion. Nevertheless, the maximal partial matching problem offers persuasive evidence of the effectiveness of the objective method, and it provides a quick introduction to several of the method's main themes. In the next three sections, we hope to make this case more complete and more inclusive.

4 The Mean-Field Model of Distance

In physics, and increasingly often in probability theory, one meets geometric problems where one tries to gain insight by building an analytic model that only captures *part* of the structure of the original set of inter-point distances. For example, instead of studying a random variable that depends on all of the inter-point distances of a point process in $\mathbb{R}^d$, one may directly model just the distances from a given point to its successive neighbors and then study an appropriate analog of the original variable.

By preserving the distributions of the most essential inter-point distances while distorting the distribution of less important distances, one hopes to achieve a significant simplification while still gaining some insight into the phenomena originally of interest. Models that aim for this kind of simplification are commonly called *mean field* models following an earlier tradition in physics where complicated fields were replaced by simpler ones that were calibrated to match the average behavior of the more complicated fields.

In this section, we will describe several problems of probabilistic combinatorial optimization where mean-field models have been developed. These models have a natural kinship with the objective method, and local weak convergence turns out to be a most natural tool.

4.1 From Poisson Points in $\mathbb{R}^d$ to a Simple Distance Model

Consider for a moment a Poisson process $\mathcal{P}$ on $\mathbb{R}^d$ that has uniform intensity $1/v_d$ where v_d is the volume of the unit ball in $\mathbb{R}^d$. For such a process the expected number of points in the unit ball is equal to one, and, if we look at the successive distances from the origin to the points of $\mathcal{P}$, we have a new point process

$$0 < \Delta_1 < \Delta_2 < \Delta_3 < \dots. \tag{4.1}$$

This process is again Poisson, although now we have nonconstant intensity, and the expected number of points in $[0, x]$ equal to x^d.

In a world where one is more concerned about the distances to one's nearest neighbors than about other distances, this simple reduction suggests a promising line of inquiry. Can one assign random distances to the edges of a complete graph K_n with n vertices so that for large n the successive distances from a point r in K_n will mimic the distances that one finds for the sequence of neighbors of 0 in the d-dimensional Poisson process $\mathcal{P}$?

A Finite Distance Model: The d-Exponential K_n.

The answer to this question is affirmative, and it is easy to describe. For each edge e of K_n, we simply introduce an independent random variable ξ_e such that

$$\xi_e \overset{d}{=} n^{1/d}\Delta_1, \quad \text{or equivalently,} \quad P(\xi_e \geq n^{-1/d}x) = e^{-x^d} \text{ for } x \geq 0 . \quad (4.2)$$

Now, if we fix a vertex r of K_n and consider the distance from this root r to its nearest neighbor, next nearest neighbor, and so on, then one finds a set of $n - 1$ increasing distances

$$0 < D_1^n < D_2^n < \cdots < D_{n-1}^n.$$

The key feature of the distribution (4.2) is that one may now easily prove that

$$(D_i^n : i \geq 1) \overset{d}{\longrightarrow} (\Delta_i : i \geq 1) \quad \text{as } n \to \infty . \quad (4.3)$$

This useful result yields some geometric insight, but it is possible to go much further. One does not need long to guess that the distributional limit (4.3) may be strengthened to prove the *local weak convergence* of the finite distance models K_n. That is, one hopes to find that K_n satisfies the kind of limit theorem that we found for T_n in Section 2.

Where to Start.

Anytime one hopes to prove a weak limit theorem in $\mathcal{G}_*$, the battle begins by searching out a reasonably explicit candidate for the *limit object*. Here for example, one should minimally note that any reasonable candidate for the local weak limit of K_n must have infinite degree at every vertex. Moreover, from the motivating limit (4.3) we also know that in the candidate object the distances from a vertex to its neighbors must behave like the successive values of the Poisson process (4.1).

After these easy observations, some deeper reflection is needed. In due course, one finds that the description of the candidate is almost complete; only one further observation is needed. Any candidate for the limit of graphs K_n must be a *tree* – despite the fact that K_n is more full of cycles than any other graph!

4.2 The Poisson Weighted Infinite Tree – or, the PWIT

Even with a good limit object in mind, one needs to confirm that the candidate exists as a clearly defined process. Here this is easily done by an inductive procedure. One begins with a single root vertex that we denote by r. This root vertex is then given an *infinite* number of children, and the edges from the root to the children are assigned lengths according to a realization of a Poisson process $(\xi_i^r : 1 \leq i < \infty)$ with mean function x^d. The children of the root are said to form *generation one*.

Now, recursively , each vertex v of generation k is given an infinite number of children, and the edges to these children of v are again assigned lengths according to an independent realization of a Poisson process $(\xi_i^v : 1 \leq i < \infty)$ with mean function x^d. This procedure is then continued *ad infinitum*.

The resulting rooted infinite tree T is then a well defined random variable with values in $\mathcal{G}_\star$. The tree T is said to be a *Poisson weighted infinite tree*, or, a PWIT. Still, the nickname intends no disrespect. PWITs have an inevitable role in the limit theory of trees, and the next theorem tells us that they are precisely the limits of appropriately weighted complete graphs on n vertices.

Theorem 4.1 (Convergence Theorem for K_n). *If K_n is a randomly rooted complete graph on n vertices with independent edge lengths ξ_e that satisfy*

$$P(\xi_e \geq n^{-1/d}x) = e^{-x^d} \quad \text{for all } x \geq 0 \,,$$

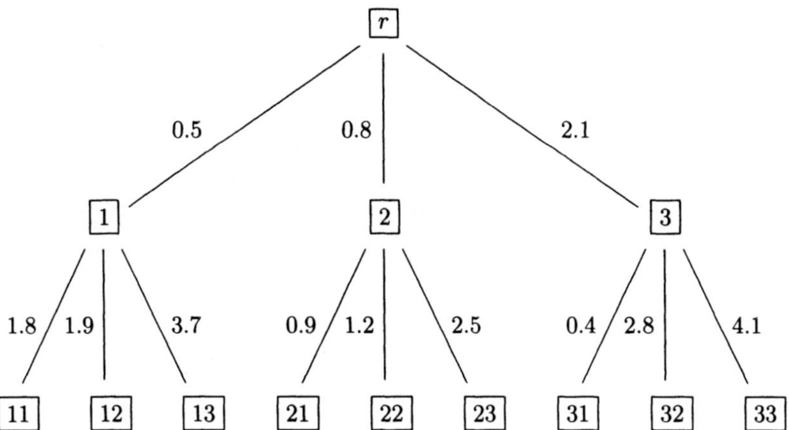

Fig. 5. Part of a realization of a PWIT that shows just the first three children of each vertex; one needs to keep in mind that in the full PWIT each vertex has an infinite number of children. Here the values written next to the edges that descend from a fixed vertex are intended to reflect an independent realization of a Poisson process with mean function x^d. Finally, the labels of the vertices in this figure are only for convenience; in a PWIT the vertices are unlabelled except for the root.

then we have the local weak convergence

$$K_n \xrightarrow{d} T \quad as\ n \to \infty\,.$$

where T is a PWIT with edge weight mean function x^d.

This theorem from Aldous [3] permits one to study optimization problems for K_n in the same way that Grimmett's Lemma permits one to study optimization problems for the uniform random tree T_n. Nevertheless, there is one noteworthy distinction; the exponential weights on K_n are tied to the Poisson processes that generate the PWIT, but the F-weights for the edges of T_n may be chosen with great freedom.

We will not repeat the proof of Theorem 4.1 here, but we should indicate why the cycles of K_n disappear in the limit. If we introduce a random variable that measures the length of the shortest cycle containing the root,

$$C_n = \min_C \left\{ \sum_{e \in C} \xi_e : C \text{ is a cycle of } K_n \text{ that contains the root } r \right\},$$

then the key issue is that one can prove that

$$\lim_{n \to \infty} P(C_n \le \ell) = 0 \quad \text{for all } \ell > 0\,. \tag{4.4}$$

Intuitively this result says that in the limit there is no cycle of finite length that contains the root, and, with some attention to detail, this leads to a proof that the weak limit of K_n in $\mathcal{G}_\star$ must be a random tree.

The Dimension Parameter and a Voluntary Restriction

For the rest of this section, we will restrict our attention to the case when $d = 1$, and the Poisson process (4.1) of successive distances $(\Delta_i : 1 \le i < \infty)$ reduces to a plain vanilla Poisson process with constant rate 1. Nevertheless, we should emphasize that this restriction is voluntary; it is only taken to shorten and simplify our exposition.

For all of the problems studied in this section and the next, there are parallel results even for noninteger $0 < d < \infty$, and, in most cases, one will find complete proofs of these results in the references. Moreover, one should note that for the problems studied here, the logic of the analysis is identical for all values of $0 < d < \infty$, even though the explicit calculations often depend on the specific value. These calculations are almost always simplest for $d = 1$ where one has access to the lack-of-memory property of the exponential.

4.3 The Cut-off Components of a Weighted Graph and a PWIT

For a moment, we consider a perfectly general weighted graph G, and we let G_s denote the subgraph G_s that one obtains from G by deleting all of the edges of length s or greater. Also, if v is a vertex of G we let $c(G, v; s)$ denote the connected component of G_s that contains v.

Now, if T is a dimension-one PWIT with root r we may view the component $c(T, v; s)$ as a rooted tree with root r, and, rather pleasantly, this tree can identified with a Galton-Watson branching process with a single progenitor and an offspring distribution that is Poisson with mean s. In other words, for all $s > 0$ we have

$$c(T, r; s) \stackrel{d}{=} \mathrm{PGW}(s) \,,$$

and this simple observation has several useful consequences.

In particular, if $q(s)$ denotes the probability that the component $c(T, r; s)$ is *infinite*, then classical formulas for the extinction probability of a branching process (such as those of Harris [30], p.7) tell us that we have $q(s) = 0$ when $0 \le s \le 1$, and otherwise we find that $q(s)$ is the unique strictly positive solution of

$$1 - q(s) = \exp(-sq(s)) \quad \text{when } s > 1 \,. \tag{4.5}$$

By inverting this relationship, one also sees that the value of s for which we have extinction probability $0 < q < 1$ is given by

$$s(q) = -q^{-1} \log(1 - q) \,, \tag{4.6}$$

a formula that will soon come in handy.

4.4 The Minimum Spanning Forests of an Infinite Graph

We want to define an analog of the minimal spanning tree for the PWIT and other infinite graphs, but some care must be taken. Naive attempts to mimic the usual greedy algorithms certainly will not work; in most cases, they can never get started.

Instead, we are guided by a well-known criterion for an edge $e = (v_1, v_2)$ with length s to be in the MST of a finite graph G with distinct edge weights. For such a graph, e is in the MST of G if and only if there does not exist a path in G from v_1 to v_2 such that each edge of the path has weight less than s. This criterion is easy to prove, and it is occasionally useful in correctness proofs for MST algorithms, such as the five discussed in Section 4.4 of Matoušek and Neštřil [49].

Definition 4.2 (Minimal Spanning Forest). *The minimal spanning forest of an infinite graph G that has all distinct edge lengths is the subgraph $MSF(G)$ of G with the same vertex set as G and with an edge set that contains each edge $e = (v_1, v_2)$ of G for which*

1. $c(G, v_1; s)$ and $c(G, v_2; s)$ are disjoint, and
2. $c(G, v_1; s)$ and $c(G, v_2; s)$ are not both infinite

when s is taken to be the length of the edge $e = (v_1, v_2) \in G$.

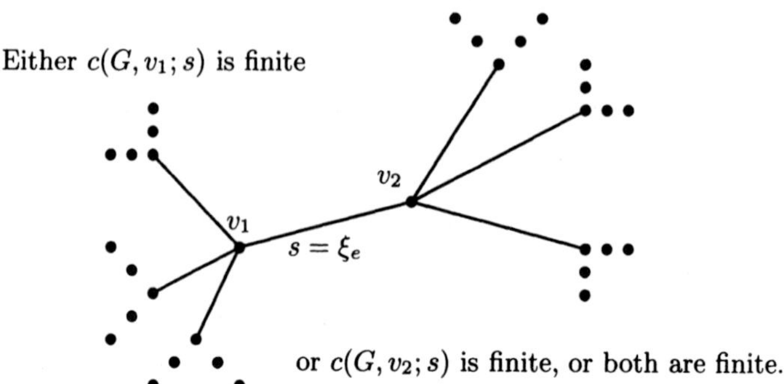

Fig. 6. The edge $e = (v_1, v_2)$ of G is in the minimal spanning forest if and only if at least one of the trees $c(G, v_1; s)$ and $c(G, v_2; s)$ is finite when $s = \xi_e$.

One can easily check that the graph $MSF(G)$ defined this way will have no cycles, and, since it trivially spans the vertex set of G, we see that $MSF(G)$ is certainly a *spanning forest*. Nevertheless, some work is needed to see why $MSF(G)$ is the right analog to the MST of a finite graph and to justify the name *minimal* spanning forest.

A first step, one should note that all of the connected components of $MSF(G)$ are infinite. To see this, suppose to the contrary that G_0 is a finite component of $MSF(G)$ of minimal cardinality. Next, consider the shortest edge $e = (v_1, v_2)$ between G_0 and its complement. Since one cannot get out of G_0 without using an edge length greater than s, we see that $c(G, v_1; s)$ is contained in G_0 and hence that $c(G, v_1; s)$ is finite. From the definition of the $MSF(G)$, we then have $e \in MSF(G)$, and this contradicts the minimality of the component G_0.

4.5 The Average Length Per Vertex of the MSF of a PWIT

If T is a PWIT with root r then the sum of the edges of $MSF(T)$ that are incident to r can be written as

$$S_{\mathrm{MSF}}(T) = \sum_{e:r \in e} \xi_e \mathbf{1}(e \in MSF(T)) \, ,$$

and, if we think of each vertex as owning half of the length of each edge that meets the vertex, then it is reasonable to think of

$$A = \frac{1}{2} E[S_{\mathrm{MSF}}(T)] \tag{4.7}$$

as an *average length per vertex* of the minimal spanning forest of T.

One might guess that A could be almost impossible to compute, but such a view underestimates the natural grace of the PWIT. The next lemma and its proof show that the PWIT supports some almost magical possibilities.

Lemma 4.3. *The average cost A per vertex for the MSF of a PWIT satisfies*

$$A = \zeta(3) = \sum_{k=1}^{\infty} k^{-3} .$$

To prove the lemma, we first recall that if one conditions a Poisson process to have a point at s, then the set of complementary points is again a Poisson process. Thus, if we condition on the existence of an edge e of length s from the root to some child, then the subtrees obtained by cutting that edge are again independent PGW(s) trees.

The probability that at least one of these is finite is equal to $1 - q^2(s)$, so this is also the probability that e is an edge in the minimal spanning forest and we find

$$A = \frac{1}{2} \int_0^{\infty} s(1 - q^2(s)) \, ds . \tag{4.8}$$

Calculation of this integral is quite a pleasing experience. We first integrate by parts and then we use the defining equation (4.5) for $q(s)$ to find

$$A = \frac{1}{2} \int_0^{\infty} s^2 q(s) q'(s) \, ds = \frac{1}{2} \int_1^{\infty} s^2 q(s) q'(s) \, ds = \frac{1}{2} \int_0^1 \frac{\log^2(1 - q)}{q} \, dq ,$$

so the substitution $u = -\log(1 - q)$ lets us finish with

$$A = \frac{1}{2} \int_0^{\infty} u^2 \frac{e^{-u}}{1 - e^{-u}} \, du = \frac{1}{2} \int_0^{\infty} u^2 \sum_{k=1}^{\infty} e^{-ku} \, du = \sum_{k=1}^{\infty} \frac{1}{k^3} = \zeta(3) .$$

4.6 The Connection to Frieze's $\zeta(3)$ Theorem

The emergence of $\zeta(3)$ as the average cost per vertex of the minimal spanning forest of a PWIT is at least a little mysterious. Still, there is a related precedent, and when one sees $\zeta(3)$ in the context of trees, the famous $\zeta(3)$ theorem of Frieze [21] immediately comes to mind.

Theorem 4.4. *Let each edge e of the complete graph on n vertices be assigned an independent cost ξ_e with distribution F where F has a finite variance and where $F'(0)$ exists and is nonzero. If $C_n^{MST}(F)$ denotes the minimal cost of a spanning tree of this graph, then as $n \to \infty$, we have*

$$C_n^{MST}(F) \xrightarrow{p} \zeta(3)/F'(0) \quad and \quad E[C_n^{MST}(F)] \to \zeta(3)/F'(0) . \tag{4.9}$$

By a simple coupling argument, Steele [59] showed that the convergence in probability still holds without any moment assumptions, and later by a martingale argument Frieze and McDiarmid [22] showed that convergence in probability can be strengthened to almost sure convergence.

Nevertheless, what matters here are simply the exponential costs, and to make the connection with our earlier calculations explicit, we note that Frieze's theorem implies that if the edge weight distribution F_n is exponential with mean n, then we also have

$$\lim_{n \to \infty} \frac{1}{n} E[C_n^{MST}(F_n)] = \zeta(3) . \tag{4.10}$$

This limit says the average cost per vertex of the minimal spanning tree of the appropriately weighted complete graph tends to the same value as the expected average cost per vertex of the PWIT with $d = 1$. Conceivably, the reappearance of $\zeta(3)$ is only a coincidence, but it is not. Quite to the contrary, the limit (4.10) is a consequence of the fact that $A = \zeta(3)$.

The Local Weak Convergence Link

As a special case of Theorem 4.1, we know that for the complete graph K_n on n vertices with edge lengths that are independent and exponentially distributed with mean n, one has the local weak convergence

$$K_n \xrightarrow{d} T \quad \text{as } n \to \infty \tag{4.11}$$

where T denotes a PWIT with $d = 1$. We will also see later in Theorem 5.4 of Section 5 that just local weak convergence (4.11) and the fact that the PWIT is an infinite graph with all distinct edge weights is enough to give us

$$\left(K_n, MST(K_n) \right) \xrightarrow{d} \left(T, MSF(T) \right) . \tag{4.12}$$

This limit is typical of the way the theory of local weak convergence informs one about a problem in combinatorial optimization. At a conceptual level, the limit (4.12) answers almost any question one may have about the qualitative behavior of $MST(K_n)$ for large n, provided that the question meets one restriction; it be expressed in terms of a function of some neighborhood of the root of $MST(K_n)$.

A Randomized Root and a Total Sum

The expectation $E[C_n^{MST}(F_n)]$ can be written as a function of the neighborhood of the root, yet this representation seems to be a lucky break. Eventually one calls on this break routinely, but at least once it seems useful to examine the full logical chain; we begin by noting

$$C_n^{MST}(F_n) \overset{d}{=} \sum_e \xi_e \mathbf{1}(e \in MST(K_n))$$

so now if we let R denote a root chosen at random from the vertex set of K_n we also have

$$\frac{1}{n}E\Big[\sum_e \xi_e \mathbf{1}(e \in MST(K_n))\Big] = \frac{1}{2}E\Big[\sum_{e:R\in e} \xi_e \mathbf{1}(e \in MST(K_n))\Big]$$

and from the local weak convergence $MSF(K_n) \to MSF(T)$ we have

$$\sum_{e:R\in e} \xi_e \mathbf{1}(e \in MST(K_n)) \overset{d}{\longrightarrow} \sum_{e:R\in e} \xi_e \mathbf{1}(e \in MST(T)) . \qquad (4.13)$$

Now, once one checks uniform integrability (which is a not particularly easy, but which is a doable technical exercise), one finds from the basic distributional limit (4.13) that

$$E\Big[\sum_{e:R\in e} \xi_e \mathbf{1}(e \in MST(K_n))\Big] \to E\Big[\sum_{e:R\in e} \xi_e \mathbf{1}(e \in MST(T))\Big] = 2\zeta(3) ,$$

where the last equality comes from Lemma 4.3. From the links of the chain we find

$$\frac{1}{n}E\big[C_n^{MST}(F_n)\big] \to \zeta(3) ,$$

just as we hoped to show.

5 Minimal Cost Perfect Matchings

A perfect matching of a finite graph is a collection of disjoint edges with the property that at most one vertex of the graph fails to be contained in an edge of the collection. Perfect matchings are among the most intensively studied structures of graph theory, and they suggest many natural random variables. Here we focus on the prince among these, the minimal cost C_n of a perfect matching of the complete graph K_n with independent edge weights ξ_e with distribution F.

The assignment problem discussed in the introduction is precisely the perfect matching problem for the bi-partite graph $K_{n,n}$, and the analysis of C_n is intimately connected to that of A_n. Thus, the following theorem is a direct analog of the $\zeta(2)$ limit theorem for the expected assignment cost $E[A_n]$, but here we rescale the F-weights to be consistent with our discussion of the local weak limit theory of K_n.

Theorem 5.1. *If C_n denotes the minimal cost of a perfect matching of the complete graph K_n on n vertices with independent edge costs ξ_e having the exponential distribution with mean n, then*

$$\lim_{n\to\infty} \frac{1}{n} E[C_n] = \frac{1}{2} \sum_{k=1}^{\infty} k^{-2} = \frac{1}{2}\zeta(2) . \tag{5.1}$$

Given the earlier experience with the maximum partial matching on the weighted uniform tree T_n, one has every reason to expect that the local weak convergence of K_n should provide the key to the proof of this theorem. This view is well founded, but the earlier analyses only provide one with a proper orientation. There are many devils in the details, and a complete proof of Theorem 5.1 would take more space than we have here.

Nevertheless, we can provide the better part of one-half of the proof; that is, we can provide most of the argument that proves the limit infimum of $E[C_n]/n$ is not less than $\zeta(2)/2$. While doing so, we can also underscore the main structural features of the full proof. In particular, we develop a basic distributional identity and solve the identity to find an unexpected connection to the logistic distribution. We can then use the distributional identity to construct a special stochastic process which can be used in turn to define an inclusion rule for the edges of a PWIT to be in a minimal cost matching. Finally, given just a few properties of the freshly constructed process, we can complete the analysis of the limit infimum.

In the course of this development, we can also complete a promise made in Section 2; specifically, we can formalize the *involution invariance* property that one finds for all limit objects that are given by the standard construction discussed in Section 2. Even though involution invariance seems perfectly innocuous, it has an unavoidable part in the proof of Theorem 5.1. In the earliest stage, it helps weed out defective candidates for the limit objects, and at a later stage it guides us to a proper understanding of the matchings on the PWIT that *can* be attained as limits of matching on K_n.

5.1 A Natural Heuristic – Which Fails for a Good Reason

The local weak convergence of K_n almost compels one to think of using good matchings on the PWIT to help in the hunt for good matchings on K_n. The natural hope is that a perfect matching in the PWIT with minimal average cost per vertex should be close to a perfect matching of K_n for large n; thus, weak convergence theory might set us on the trail of a proof of Theorem 5.1. This hope can be justified, but only after one restricts attention to an important subset of the matchings on the PWIT that satisfy a natural symmetry property. If one ignores this restriction, there are matchings on the PWIT that are simply too good to be tied to matchings on K_n.

A Symbolic PWIT and its Greedy Matching

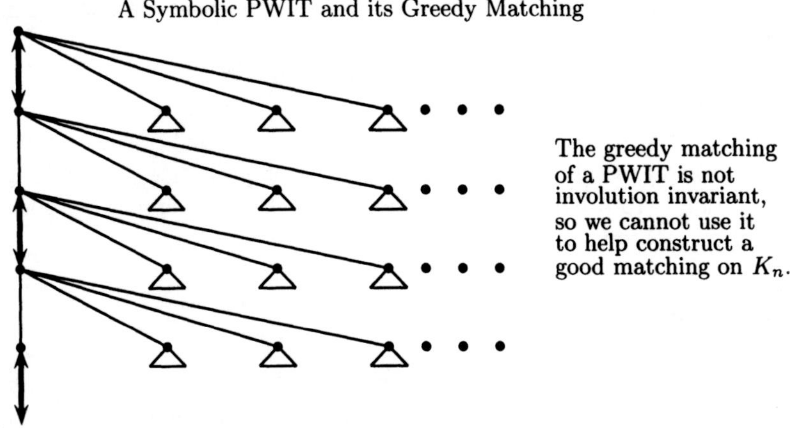

Fig. 7. To construct the greedy matching of a PWIT one takes the shortest edge e out of the root, removes all of the vertices incident to e, and proceeds recursively. The expected cost per vertex of the greedy matching is equal to one-half, which beats the unbeatable limiting average cost of $\zeta(2)/2$ per vertex for K_n.

An Example: Greedy Matching on the PWIT

In a finite weighted graph one reasonable way to hunt for a good matching is the *global greedy algorithm* where one picks the edge e with minimal weight, removes all of the edges incident to either endpoint of e, and proceeds recursively. Unfortunately, in a PWIT this method does not have a chance to start – a PWIT has no edge of minimal cost.

In the PWIT, a reasonable alternative to the global greedy algorithm is the *local greedy algorithm* where one starts by choosing the edge of minimal cost that is incident to the root. In a PWIT such an edge e exists, and we even know that it has expected value equal to 1. One can then delete e, remove the edges incident to e, and proceed recursively along the infinite forest produced by this first step.

The local greedy algorithm produces a matching in the PWIT with an expected cost per vertex equal to one-half, and this is intriguing since it is even smaller than the value $\zeta(2)/2$ that we expected to show to be asymptotically optimal for the K_n.

Unfortunately, the greedy matching for the PWIT simply cannot be the limit of a sequence of matchings for K_n. Moreover, one can even see this fact from first principles. The problem is that the greedy matching lacks a symmetry property that one must find in every limit object like the limit of the optimal matchings of K_n.

5.2 Involution Invariance and the Standard Construction

As the discussion of the greedy matching suggests, the measures on $\mathcal{G}_\star$ that one obtains by the standard construction of Section 2 are not completely general. The random rooting process gives these measures a basic symmetry that we call *involution invariance*.

This property turns out to be fundamental in the deeper applications of the objective method. With some experience, one finds that involution invariance serves as a graph theoretical surrogate for the translation invariance that is central to the theory of stationary point processes. The main goal of this section is to provide a patient development of the simplest properties of involution invariance.

The Defining Elements

If $\widetilde{\mathcal{G}}_\star$ denotes the set of pairs (G, v), where $G \in \mathcal{G}_\star$ and where $v \in V$ is a neighbor of the root of G, then one can view an element $\widetilde{G} = (G, v)$ of $\widetilde{\mathcal{G}}_\star = \mathcal{G}_\star \times V$ as a graph with a *distinguished directed edge*; specifically, the distinguished edge $(\mathrm{root}(G), v)$ may be viewed as a directed edge from the root of G to the neighbor vertex v. One can therefore define an involution

$$\iota : \widetilde{\mathcal{G}}_\star \to \widetilde{\mathcal{G}}_\star$$

by taking $\iota(\widetilde{G})$ to be the element of $\widetilde{\mathcal{G}}_\star$ that one obtains simply by *reversing* the direction of the distinguished edge of $\widetilde{G}$; that is, one makes v the new root and takes the old root to be the new distinguished neighbor.

Now, given any probability measure μ on $\mathcal{G}_\star$, one can define a new positive measure $\widetilde{\mu}$ on $\widetilde{\mathcal{G}}_\star$ by (1) taking the marginal measure of $\widetilde{\mu}$ on $\mathcal{G}$ to be μ and by (2) taking the conditional measure $\widetilde{\mu}_G$ on the neighbors of the root of G to be the counting measure. The measure $\widetilde{\mu}$ one obtains in this way is said to be *induced* on $\widetilde{\mathcal{G}}_\star$ by μ.

One should note that $\widetilde{\mu}$ is *not* a probability measure, except in some special cases. Moreover, since the definition of $\mathcal{G}_\star$ permits infinite vertex sets and permits vertices with infinite degree, $\widetilde{\mu}$ is often an infinite measure. Nevertheless, $\widetilde{\mu}$ is always a nonnegative σ-finite measure on the complete metric space $\widetilde{\mathcal{G}}_\star$. Thus, with proper care, all of the usual results of weak convergence theory may be applied to sequences of induced measures.

Definition 5.2 (Involution Invariance). *A probability measure μ on $\mathcal{G}_\star$ is said to be* **involution invariant** *provided that the σ-finite measure $\widetilde{\mu}$ induced on $\widetilde{\mathcal{G}}_\star$ by μ is invariant under the involution map ι; that is to say,*

$$\widetilde{\mu}(A) = \widetilde{\mu}(\iota^{-1}(A)) \quad \text{for all Borel } A \subset \widetilde{\mathcal{G}}_\star .$$

One of the consequences of involution invariance is that it essentially characterizes the measures on $\mathcal{G}_\star$ that one can obtain via the standard construction. The next lemma formalizes this assertion for finite graphs.

Lemma 5.3. *If μ is the probability distribution of $\mathbf{G}[X]$ where $\mathbf{G}[X]$ is given by the standard construction and $\mathbf{G}[X]$ is finite with probability one, then μ is involution invariant. Conversely, if the probability measure μ is involution invariant and concentrated on the finite connected graphs in $\mathcal{G}_\star$, then μ is the distribution of a $\mathcal{G}_\star$-valued random variable $\mathbf{G}[X]$ that is given by the standard construction.*

To prove the direct implication, we first condition on the event that the underlying graph of $\mathbf{G}[X]$ is the fixed finite graph G. If G has n vertices and m undirected edges, then the marginal measure $\widetilde{\mu}_G$ puts mass $1/n$ on each of the $2m$ directed edges of G, so we trivially get involution invariance. To prove the converse, one notes by the captioned discussion of Figure 8 that $p_G(\cdot)$ is constant on each connected component of G. Finally, if we assume that G is connected, we see $p_G(\cdot)$ is constant on G. This is precisely what it means to say that the root X of G is chosen uniformly from its vertex set, so the proof of the lemma is complete.

Connection to the Objects of the Objective Method

One of the basic features of involution invariance is that it is preserved under local weak convergence in the metric space $\mathcal{G}_\star$. To prove this fact, one first shows that the topology on $\mathcal{G}_\star$ extends in a natural way to a topology on $\widetilde{\mathcal{G}}_\star$ that makes $\widetilde{\mathcal{G}}_\star$ a complete separable metric space. Next one shows that the transformation $\mu \to \widetilde{\mu}$ is a continuous mapping with respect to weak convergence; specifically

$$\mu_n \xrightarrow{d} \mu_\infty \quad \Rightarrow \quad \widetilde{\mu}_n \xrightarrow{d} \widetilde{\mu}_\infty \, .$$

Finally, one shows that the involution map $\iota : \widetilde{\mathcal{G}}_\star \to \widetilde{\mathcal{G}}_\star$ is continuous.

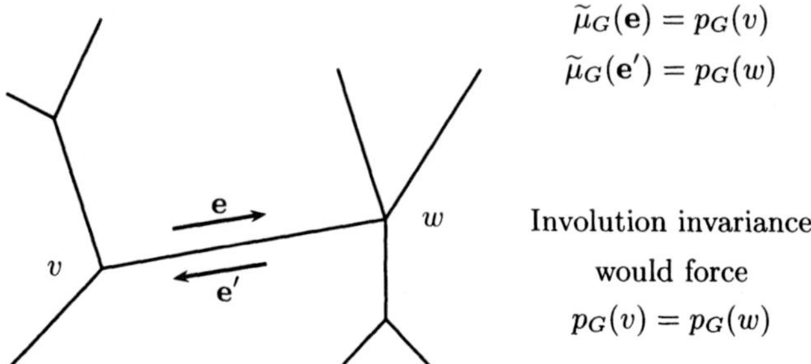

Fig. 8. Given $G = (V, E)$ we let $p_G(v)$ denote the probability that v is the root of G, and if $\widetilde{\mu}$ is the induced measure, we write $\widetilde{\mu}_G$ for the marginal measure given G. For any directed edge (v, w) of G, the measure $\widetilde{\mu}_G$ puts mass $p_G(v)$ on the edge (v, w) and puts mass $p_G(w)$ on the edge (w, v). If one has involution invariance, then $\widetilde{\mu}_G(v, w) = \widetilde{\mu}_G(w, v)$, and we see that $p_G(v) = p_G(w)$. Thus, $p_G(\cdot)$ is constant on each connected component of G whenever one has involution invariance.

5.3 Involution Invariance and the Convergence of MSTs

The formalities of involution invariance give one no hint that the notion has much force, but the next theorem shows that involution invariance can be genuinely powerful. In fact, there is something fundamental about involution invariance, since, in a way, it replaces the spatial invariance that one uses constantly in the theory of homogeneous point processes. The pleasantly general Theorem 5.4 contains as special cases Frieze's $\zeta(3)$ theorem which was discussed earlier and the convergence theorem for MSTs of random samples in the cube which we will discuss in Section 6.

Theorem 5.4 (MST Convergence Theorem). *Let G_∞ denote a $\mathcal{G}_\star$-valued random variable such that with probability one G_∞ has infinitely many vertices and no two of the edges of G have the same length. Further, let $\{G_n : n = 1, 2, ...\}$ denote a sequence of $\mathcal{G}_\star$-valued random variables such that for each n the distribution of G_n is given by the standard construction and such that for each n the vertex set of G_n has cardinality n with probability one. If*

$$G_n \xrightarrow{d} G_\infty \quad \text{as } n \to \infty , \tag{5.2}$$

then one has the joint weak convergence in $\mathcal{G}_\star \times \mathcal{G}_\star$,

$$\left(G_n, MST(G_n) \right) \xrightarrow{d} \left(G_\infty, MSF(G_\infty) \right) . \tag{5.3}$$

Further, if N_n denotes the degree of the root of $MST(G_n)$ and N denotes the degree of the root of $MSF(G_\infty)$

$$N_n \xrightarrow{d} N \quad \text{and} \quad E[N_n] \to E[N] = 2 , \tag{5.4}$$

and, if L_n denotes the sum of lengths of the edges incident to the root of $MST(G_n)$ and L denotes the corresponding quantities for $MSF(G_\infty)$, then

$$L_n \xrightarrow{d} L . \tag{5.5}$$

The applications of the limit (5.3) are easy to understand without working through the convergence proof. Nevertheless, a survey on local weak convergence should give at least one such proof in reasonable detail, and the limit (5.3) is a attractive candidate. The proof demonstrates several basic facets of involution invariance, and the net result is genuinely useful.

To begin the proof, we first note that the hypothesis (5.2) tells us the sequence $\{G_n\}$ is tight, and if we consider the sequence of pairs $(G_n, MST(G_n))$ one can check that they determine a sequence $\{\mu_n\}$ of measures on $\mathcal{G}_\star \times \mathcal{G}_\star$ that is also tight. We then let μ denote a measure on $\mathcal{G}_\star \times \mathcal{G}_\star$ that one obtains as the weak limit of a subsequence $\{\mu_{n_j}\}$.

By Skorohod's embedding theorem, we may suppose the $\mathcal{G}_\star \times \mathcal{G}_\star$-valued random variables $(G_{n_j}, MST(G_{n_j}))$ are all defined on a common probability space and that there is a $\mathcal{G}_\star$-valued random variable S such that

$$\big(G_{n_j}, MST(G_{n_j})\big) \to (G_\infty, S) \quad \text{with probability one .} \tag{5.6}$$

From the definition of μ as the weak limit of the sequence $\{\mu_{n_j}\}$, we see that the proof of the limit (4.12) boils down to showing

$$S(\omega) = MSF(G_\infty)(\omega) \quad \text{with probability one .} \tag{5.7}$$

This identity will be proved in two steps – one for each direction of the set inclusions.

The First Inclusion is Easy

To show that $MSF(G_\infty)(\omega) \subset S(\omega)$ for almost all ω, we begin by considering an edge $e = (u, v) \in MSF(G_\infty)$ that has length s. By the definition of the minimal spanning forest, we may assume without loss of generality that the component $c(G_\infty, u; s)$ is finite. For convenience of notation, we will also assume that the subsequence n_j is in fact the full sequence; when the argument is done, one will see that no generality was lost.

Now take a neighborhood H of u that is large enough to contain $c(G_\infty, u; s)$ and note by the almost sure convergence of G_n to T that there is a finite $N(\omega)$ such that for all $n \geq N(\omega)$ there is an isomorphism between H and a neighborhood of the root of G_n. Also, since $c(G_\infty, u; s)$ is finite, there is an $\varepsilon > 0$ such that each edge of $c(G_\infty, u; s)$ has length less than $s - \varepsilon$.

If $e^n = (u^n, v^n)$ denotes the edge associated with e by the neighborhood isomorphism, then for large n the length of e^n is at least $s - \varepsilon/2$ and all edges of $c(G_n, u^n; s)$ have length strictly less than $s - \varepsilon/2$. Therefore, there can be no path in G_n from u^n to v^n that has all of its edges shorter than the length of e^n. Thus, we find that there a finite $N'(\omega)$ such that e^n is in $MST(G_n)$ for all $n \geq N'(\omega)$. Since $MST(G_n)$ converges to S in $\mathcal{G}_\star$, we get our first inclusion

$$MSF(G_\infty)(\omega) \subset S(\omega) \text{ with probability one .} \tag{5.8}$$

A More Subtle Inclusion

To complete the proof we just need to show that the set difference

$$D(\omega) = S(\omega) \setminus MSF(G_\infty)(\omega) \tag{5.9}$$

is empty with probability one. By the general lemma below, we also see that it suffices to show that the number of edges in D that meet the root of G has expectation zero.

Lemma 5.5. *Let G be an involution invariant element of $\mathcal{G}_\star$ with root r and let W be an involution invariant subset of the edges of G. If*

$$E\Big[\sum_{e: r \in e} \mathbf{1}(e \in W)\Big] = 0 ,$$

then W is empty with probability one.

This lemma is proved in Aldous and Steele [7], and, since the way that it uses involution invariance is more simply illustrated by the proof of Lemma 5.6, we will not repeat the proof Lemma 5.5 here. Instead, we will go directly to its application.

Naturally, we want to apply Lemma 5.5 where W is taken to be the difference set D defined by equation (5.9), so our task is to estimate the corresponding expectation. If we let N_n denote the degree of root of G_n in $MST(G_n)$ and let N denote the degree of the root of G_∞ in $MSF(G_\infty)$, then we have the usual Skorohod embedding and the inclusion relation (5.8) give us the bound

$$N \leq \liminf_{n \to \infty} N_n \, , \tag{5.10}$$

and the definitions of D, N_n and N also give us

$$\sum_{e:r \in e} \mathbf{1}(e \in D) \leq (\liminf_{n \to \infty} N_n) - N \, . \tag{5.11}$$

Since an n-vertex tree has exactly $n-1$ edges, we always have the trivial limit

$$E[N_n] = 2(n-1)/n \nearrow 2 \, ,$$

so, by the bound (5.10) and Fatou's Lemma, we have

$$E[N] \leq E[\liminf_{n \to \infty} N_n] \leq \liminf_{n \to \infty} E[N_n] = 2 \, .$$

This observation almost completes the proof of Theorem 5.4, and the next lemma does the rest. It is all we need to confirm that the expectation of the sum (5.11) is zero.

Lemma 5.6.

$$E[N] = 2 \, . \tag{5.12}$$

In fact, only half of the lemma remains to be proved; we just need to show $E[N] \geq 2$. To begin, we consider an edge $e = (r, v)$ that is incident to the root r of $MSF(G)$. If we write f for "finite" and ∞ for "infinite," then each such edge must be one of four possible types

$$(f, f), \quad (f, \infty), \quad (\infty, f), \quad \text{and} \ (\infty, \infty)$$

according to the cardinality of the respective components of $MSF(G_\infty) \setminus \{e\}$ that contain r and v. We will write N_{ff}, $N_{f\infty}$, $N_{\infty f}$, and $N_{\infty\infty}$ for the numbers of edges of the corresponding types.

From the definition of the minimal spanning forest, we know that every vertex is in an infinite component of the MSF, so N_{ff} is always zero. A more important observation is that unless r has infinite degree, the argument given in the caption of Figure 9 gives us

$$2N_{f\infty} + N_{\infty\infty} \geq 2 \, . \tag{5.13}$$

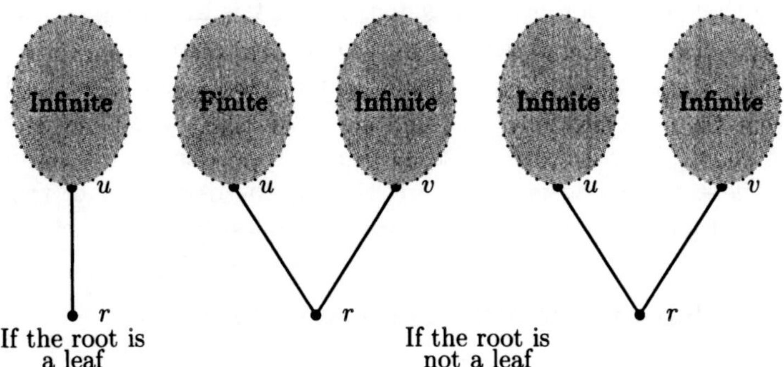

Fig. 9. If the root of G_∞ has finite degree, we are sure to have $2N_{f\infty} + N_{\infty\infty} \geq 2$ because no edge of the MSF of G_∞ can connect two finite components of the MSF and because in the three remaining cases one always has either $N_{f\infty} \geq 1$ or $N_{\infty\infty} \geq 2$.

The key step turns out to be a very simple involution invariance argument that shows that $N_{f\infty}$ and $N_{\infty f}$ have the same expectation. Once this is confirmed, one can complete the proof of the lemma by noting that the bound (5.13) gives us

$$E[N] = E[N_{f\infty}] + E[N_{\infty f}] + E[N_{\infty\infty}]$$
$$= 2E[N_{f\infty} + E[N_{\infty\infty}] \geq 2 .$$

Finally, to establish the equality of $E[N_{f\infty}]$ and $E[N_{\infty f}]$, we first recall the measure $\tilde{\mu}$ that one uses to define involution invariance. For any directed edge $e^{\rightarrow}$ we have

$$E[N_{f\infty}] = \tilde{\mu}\{e^{\rightarrow} \text{ is of type } (f, \infty)\} ,$$

and we also have

$$E[N_{\infty f}] = \tilde{\mu}\{e^{\rightarrow} \text{ is of type } (\infty, f)\}$$
$$= \tilde{\mu}\{e^{\leftarrow} \text{ is of type } (f, \infty)\} ,$$

where $e^{\leftarrow}$ is the reversal of $e^{\rightarrow}$. Now, since G_∞ is involution invariant, we have equality of reversals, so

$$\tilde{\mu}\{e^{\leftarrow} \text{ is of type } (f, \infty)\} = \tilde{\mu}\{e^{\rightarrow} \text{ is if type } (f, \infty)\} ,$$

and we find at last that $E[N_{\infty f}] = E[N_{f\infty}]$. This completes the proof of Lemma 5.6, and hence completes the proof of Theorem 5.4 as well.

A Thought on Involution Invariance

The way that involution invariance does its work is by telling us that "whatever is true from the perspective of the root r" is also true from the perspective of the root's neighbor v. In the case of a connected graph, one can chain this argument from vertex to vertex and conclude that "whatever is true from the perspective of one vertex is true from the perspective of any other vertex." This statement is far from precise, but the derivation of the identity $E[N_{\infty}f] = E[N_{f\infty}]$ suggests that it provides honest guidance.

Results Related to the MST Convergence Theorem

The MST Convergence Theorem given here is new, and it generality serves to underscores the benefit of working with the local weak convergence of sequences of $\mathcal{G}_\star$-valued random variables. The simplicity of Theorem 5.4 should not conceal the fact that it contains as special cases both the basic convergence theory for MSTs for weighted complete graphs developed by Aldous [1] and the corresponding theory for random Euclidean points developed by Aldous and Steele [6]. Avram and Bertsimas [12] independently studied both of these cases, and one further finds that their calculations are also much more transparent when viewed in the light of local weak convergence.

5.4 A Heuristic That Works by Focusing on the Unknown

The greedy algorithm gave us an excellent matching on the PWIT, but it lacked involution invariance and did not help a bit in our hunt for a good matching on K_n. Moreover, the symmetry of the PWIT makes it hard to see how one can do better; almost any matching one might find by good guessing is likely to face the same failings that were met by the greedy matching. Nevertheless, a good involution invariant matching can be found, provided that one takes a hint from the analysis of the maximal partial matching problem.

The overarching plan is to consider the minimal cost per vertex of an involution invariant perfect matching on the PWIT and to treat this cost as an "unknown." We then begin the hunt for an equation that contains this unknown, and, in due course, heuristic considerations will lead us to a distributional identity. This identity can then be solved rigorously, and its solution can be used in turn to build a joint process with one marginal being a PWIT and another marginal being random variables that satisfy the distributional identity. These auxiliary variables can then used to construct an "inclusion rule" that provides a test for an edge's membership in the matching on the PWIT.

One small wrinkle this time is that we will focus on a simple difference, rather than the more subtle bonus that we used before. Specifically, we focus on the difference

$$D(T) = C(T) - C\big(T \setminus \{r\}\big),$$

where $C(T)$ is the minimal cost of a perfect matching of T and $C(T \setminus \{r\})$ is the minimal cost of a perfect matching of the graph $T \setminus \{r\}$. Naturally, both $C(T)$ and $C(T \setminus \{r\})$ are infinite, so $D(T)$ is not really well defined. Nevertheless, in our analysis of the maximal partial matching problem we met a similar difficulty, and we still managed to prevail. We have every reason to hope for a similar outcome this time.

If j is a child of the root, then we let T^j denote the descendants of j and we view T^j as a rooted graph with root j. We then introduce the analogous heuristic difference $D(T^j)$, and note that even though $D(T)$ and $D(T^j)$ are not honestly defined, one can continue in the brave tradition of physical scientists and argue that nevertheless they should satisfy the distributional identity

$$D(T) = \min_{1 \leq j < \infty} \left\{ \xi_j - D(T^j) \right\} . \tag{5.14}$$

To see this, we first note that the minimal cost of a matching that contains the edge (r, j) from the root r to the vertex j has cost

$$\xi_j + C\big(T^j \setminus \{j\}\big) + \sum_{i:i \neq j} C(T^i) \tag{5.15}$$

and that $C(T)$ is just the minimum of all of these values. Also, we have the direct identity

$$C\big(T \setminus \{r\}\big) = \sum_{1 \leq i < \infty} C(T^i) . \tag{5.16}$$

The difference of equations (5.15) and (5.16) is just $\xi_j - C(T^j) + C(T^j \setminus \{j\})$, so in the end we see that

$$C(T) - C\big(T \setminus \{r\}\big) = \min_{1 \leq j < \infty} \left\{ \xi_j - C(T^j) + C\big(T^j \setminus \{j\}\big) \right\} . \tag{5.17}$$

5.5 A Distributional Identity with a Logistic Solution

Equation (5.14), or its expanded equivalent (5.17), motivates us to begin our rigorous analysis by first considering the distributional identity

$$X \stackrel{d}{=} \min_{1 \leq i < \infty} \left\{ \xi_i - X_i \right\} \quad -\infty < X < \infty , \tag{5.18}$$

which we interpret to be an identity for a distribution F that is fully specified by supposing (1) $X \sim F$, (2) the X_i are independent, (3) $X_i \sim F$ for all $i = 1, 2, ...$, and (4) the sequence $\{X_i\}$ is independent of the process $0 < \xi_1 < \xi_2 < \cdots$ which is a Poisson process with constant intensity equal to one.

Naturally, one wants to know about the existence and uniqueness of a distribution F that satisfies this distributional identity. The next lemma provides an answer of striking simplicity.

Lemma 5.7 (Logistic Solution of a Distributional Identity). *The unique solution of the identity (5.18) is the logistic distribution*

$$F(y) = 1/(1 - e^{-y}), \quad -\infty < y < \infty \qquad (5.19)$$

corresponding to the density function

$$f(y) = (e^{y/2} + e^{-y/2})^{-2}, \quad -\infty < y < \infty .$$

Better Than a Proof – A Derivation

Rather than just give a direct verification of this lemma, one does better to see how the probability theory of the distributional identity leads one to discover the connection to the logistic distribution. Thus, for the moment, we assume that the distribution F satisfies the identity (5.18), and we will try to calculate F.

We first note that the distributional identity (5.18) asserts that

$$1 - F(y) = P\left(\min_{1 \leq i < \infty} \{\xi_i - X_i\} \geq y \right)$$

where the $\{X_i : i = 1, 2, ...\}$ on the righthand side are i.i.d. with distribution F and $\{\xi_i : i = 1, 2, ...\}$ is an independent Poisson process. Therefore, if we consider the collection of points $\mathcal{P}$ in $\mathbb{R}^2$ given by $\{ (\xi_i, X_i) : i = 1, 2, ...\}$ we see that $\mathcal{P}$ is a Poisson process on $(0, \infty) \times (-\infty, \infty)$ with intensity $dz\,dF$. As a consequence, we have

$$1 - F(y) = P\big(\text{no point of } \mathcal{P} \text{ is in } \{(z, x) : \ z - x \leq y\}\big)$$

$$= \exp\left(- \iint_{z-x<y} dz\,dF(x) \right)$$

$$= \exp\left(- \int_0^\infty \{1 - F(z - y)\}\,dz \right)$$

$$= \exp\left(- \int_{-y}^\infty \{1 - F(u)\}du \right) .$$

When we differentiate this identity we find

$$F'(y) = \big(1 - F(-y)\big)\big(1 - F(y)\big) , \qquad (5.20)$$

and the symmetry of this equation implies that the density $F'(\cdot)$ is symmetric. As a consequence, we see that equation (5.20) is equivalent to

$$F'(y) = F(y)\big(1 - F(y)\big),$$

which is easily seen to be separable. We can rewrite it as

$$dy = \frac{dF}{F(1-F)} = \frac{dF}{F} + \frac{dF}{1-F}$$

and solve by inspection to find

$$y = \log\frac{F}{1-F} \quad \text{or} \quad F(y) = 1/(1-e^{-y})\,.$$

Thus, F is simply is the logistic distribution. It is famous for popping up in odd places, but its appearance here still a bit surprising. The remainder of our analysis does not call on any theory of the logistic distribution beyond the easily checked fact that it has variance $\pi^2/3$.

5.6 A Stochastic Process that Constructs a Matching

With the explicit solution of the distributional identity (5.18) in our hands, the next step is to construct a joint process where the distributional identity (5.18) suggests an inclusion criterion that can be used to define a good matching on the PWIT. Specifically, we will construct simultaneously (1) a PWIT and (2) a special collection of random variables that are associated with the edges of the PWIT. The variables in this special collection are intended to mimic the differences $D(T^j)$ that led us to the logistic recursion. To put this plan into action, we will need to introduce some notation that makes more explicit the consequences of cutting a PWIT at one of its edges.

Notation for a Split PWIT

When we take a PWIT and delete the edge (u,v) we get two disjoint trees that we can view as being rooted at u and v respectively. To help keep track of these trees (and the difference functionals that we will associate with them), we assign to each edge of the PWIT a pair of *directed* edges.; formally, we just distinguish between (u,v) and (v,u). Given any graph G we let $\mathbf{E}_G$ denote the set of edges obtained by directing the edges of G, and, if T is a PWIT with edge weights ξ_e, then we give the same weight to each of the directed edges associated with e. Finally, if $\mathbf{e} = (u,v)$ is a directed edge and $X(\mathbf{e})$ is a random variable associated with that edge, we will often simply write $X(u,v)$. In the case of the PWIT, the weight of the directed edge $\xi_\mathbf{e}$ will also be written as $\xi(u,v)$. By definition we have $\xi(u,v) = \xi(v,u)$, but, typically, $X(u,v)$ and $X(v,u)$ will be different.

Lemma 5.8 (Triple Tree Process). *There exists a process $(\mathcal{T},\xi_\mathbf{e}, X(\mathbf{e}) : \mathbf{e} \in \mathbf{E}_\mathcal{T})$ where $\mathcal{T}$ is a PWIT with associated edge lengths $\{\, \xi_\mathbf{e} : \mathbf{e} \in \mathbf{E}_\mathcal{T}\}$ and where $\{\, X(\mathbf{e}) : \mathbf{e} \in \mathbf{E}_\mathcal{T}\}$ is stochastic process indexed by the directed edges of $\mathcal{T}$ with the three following properties:*

(i). for each directed edge $\mathbf{e} = (u,v) \in \mathbf{E}_\mathcal{T}$ we have

$$X(u,v) = \min\{\xi(v,w) - X(v,w) : (v,w) \in \mathbf{E}_\mathcal{T},\ w \neq u\} \tag{5.21}$$

(ii). for each $\mathbf{e} \in \mathbf{E}_T$ *that is directed away from the root of* T *the random variable* $X(\mathbf{e})$ *has the logistic distribution, and*

(iii). for each pair of oppositely directed edges (u,v) *and* (v,u) *in* $\mathbf{E}_T$, *the random variables* $X(u,v)$ *and* $X(v,u)$ *are independent.*

The proof of Lemma 5.8 is easy and instructive; it shows how one can exploit a distributional identity to construct useful processes on trees and other graphs. To begin, we fix an integer $g \geq 1$ and then for each directed edge (u,v) with u in generation g and v in generation $g+1$ (relative to the root), we create by fiat an independent logistic random variable $X(u,v)$. We next use the identity (5.21) to define $X(\mathbf{e})$ for each edge within g generations of the root. This construction gives us a collection $\mathcal{C}_g$ of random variables whose joint distribution satisfies properties (i), (ii) and (iii) for all vertices in the first g generations. Moreover, the distribution of the collection $\mathcal{C}_g$ is equal to the marginal distribution of the larger collection $\mathcal{C}_{g+1}$ restricted to the first g generations, so by the Kolmogorov consistency theorem there is a collection $\mathcal{C}_\infty$ such that for each g the marginal distribution of $\mathcal{C}_\infty$ restricted to first g generations is equal to the distribution of $\mathcal{C}_g$. $\quad\square$

A Useful Addendum to the Triple Tree Lemma

Once the triple process is in hand, one can go further; in fact, one can show that the process has some intrinsic independence properties that are not an explicit part of the construction. Moreover, we will find that this additional independence has a key role in our later computations.

First we take a fixed $0 < x < \infty$, and we condition on the event that there exist an edge a the root that has length x. We call this edge (r,v) and note that as in Figure 10, the edge determines two subtrees of the PWIT that one could label $T(r,v)$ and $T(v,r)$. As we have noted before, a Poisson process conditioned to have a point at x is again a Poisson process when this point is deleted, so by the definition of the PWIT we see that $T(u,v)$ and $T(v,u)$ are conditionally independent copies of the original PWIT. We therefore find an addendum to Lemma 5.8:

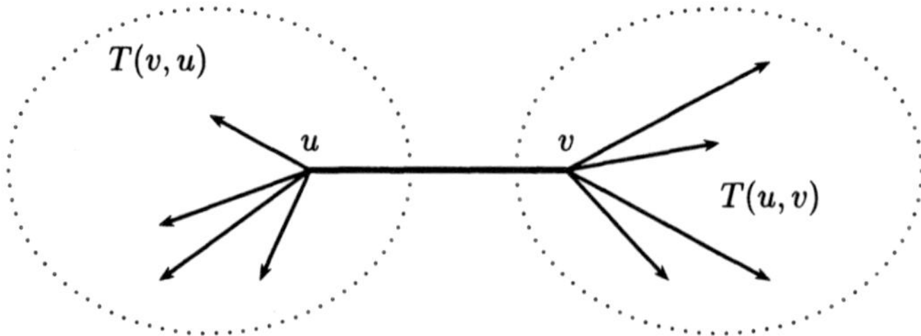

Fig. 10. When one cuts a PWIT at an edge, one obtains two trees that one can label by using the directed edges associated with the cut edge. The tree $T(u,v)$ labelled by the directed edge (u,v) has the root v and contains all of the edges that follow v in the direction of (u,v).

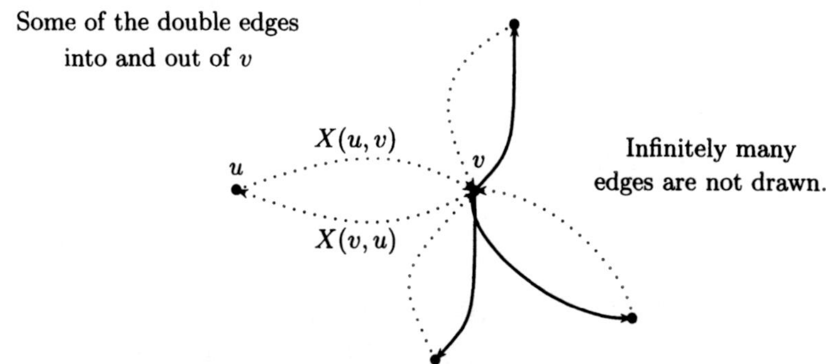

Fig. 11. The edges that feature in the identity for $X(u,v)$ are those directed edges out of v indicated by the solid arrows; note that the edge (v,u) is not one of these. Note also that the weight ξ_e of an edge does not depend on its direction, but the value of X does.

(iv). conditional on the existence of an edge (r,v) from the root having length x, the random variables $X(r,v)$ and $X(v,r)$ are independent random variables with the logistic distribution.

Construction of an Excellent Matching on $\mathcal{T}$

For the triple process $(\mathcal{T},\xi,X)$ given by Lemma 5.8, one can use the additional information carried by the X process to specify a good matching on $\mathcal{T}$. As before, the idea is to use an *inclusion criterion* that is motivated by the heuristic derivation of the distributional identity (5.18).

To define this matching, we first consider the map $\varphi : V \to V$ on the set V of vertices of the PWIT that is given by $\varphi(v) = v^*$ where v^* is the vertex for which $\xi_e - X(\mathbf{e})$ is minimized when $\mathbf{e} = (v,v^*)$. The next lemma confirms that the set of edges (v,v^*) is a matching; we will subsequently show that it is a good one. More precisely, one can show that this matching is optimal in the sense that no other involution invariant matching of a PWIT can have a lower average expected length per edge.

Lemma 5.9 (Matching Lemma). *The set of edges (v,v^*) defined by the rule $\varphi(v) = v^*$ is a matching on the special PWIT $\mathcal{T}$.*

To prove Lemma 5.9, we first note that it is exactly equivalent to showing that φ is idempotent; that is, we must show that $(v^*)^* = v$ for all $v \in V$. Now, by the definition of φ and the recursion (5.21) for X, we have

52   David Aldous and J. Michael Steele

$$\xi(v, v^*) - X(v, v^*) = \min\{\xi(v, y) - X(v, y) : (v, y) \in \mathcal{T}\}$$
$$< \min\{\xi(v, y) - X(v, y) : (v, y) \in \mathcal{T}, y \neq v^*\}$$
$$= X(v^*, v),$$

or, a bit more simply

$$\xi(v, v^*) < X(v, v^*) + X(v^*, v). \qquad (5.22)$$

On the other hand, if z is a neighbor of v other than v^*, then we have

$$\xi(v, z) - X(v, z) > \min\{\xi(v, y) - X(v, y) : (v, y) \in \mathcal{T}\}$$
$$= \min\{\xi(v, y) - X(v, y) : (v, y) \in \mathcal{T}, y \neq z\}$$
$$= X(z, v).$$

This inequality tells us $\xi(v, z) > X(v, z) + X(z, v)$ whenever $z \neq v$, so we see that v^* is the *unique* neighbor of v that satisfies the inclusion criterion (5.22).

We can therefore see that we could have just as well *defined* $\varphi(v)$ as the unique v^* satisfying the criterion (5.22). Finally, if one notes that both sides of the criterion (5.22) are symmetric we see that $\varphi(\varphi(v)) = v$, just as we hoped to show.

Wisdom of a Matching Rule

Given the maneuvering that led us to the matching rule $r^* = \varphi(r)$, one has a right to expect that there is no invariant matching that has a lower expected cost per vertex. The next lemma asserts that this expectation is justified.

Lemma 5.10 (Optimality on the PWIT). *Let M denote any matching of $\mathcal{T}$ that is involution invariant. If (r, s) is the edge of M that contains the root r of $\mathcal{T}$, then one has*

$$E[\xi(r, s)] \geq E[\xi(r, r^*)]. \qquad (5.23)$$

The proof of Lemma 5.10 is unfortunately too lengthy to be included here, and even a convincing sketch would run beyond our boundaries, so for most details we must refer the reader to Proposition 18 of Aldous [4], which additionally proves that inequality (5.23) is strict except when the matching M is almost surely identical to the matching defined by φ.

5.7 Calculation of a Limiting Constant: $\pi^2/6$

If M_{opt} is the matching given by Lemma 5.9 and if $r^* = \varphi(r)$ is the vertex that is matched to the root r by M_{opt}, then the expected value per edge of the weight of the M_{opt} is simply equal to $E[\xi(r, r^*)]$. This expected value turns out to be surprisingly easy to compute.

Lemma 5.11.

$$E[\xi(r, r^*)] = \frac{\pi^2}{6}$$

To prove the lemma we first fix $0 < x < \infty$ and then we condition on the event A_x that there exists an edge (r, v) incident to the root of $\mathcal{T}$ that has length x. The inclusion criterion (5.22) then tells us r is matched to v if and only if we have

$$x < X(r, v) + X(v, r) \ .$$

By the addendum to Lemma 5.8 we know that conditional on the event A_x the pair $\big(X(r, v), X(v, r)\big)$ same distribution as a pair (X, Y) of independent logistics, so by the uniform intensity of the Poisson process on $[0, \infty)$ we have

$$E[\xi(r, r^*)] = \int_0^\infty x\, P(x < X + Y)\, dx = \frac{1}{2} E[(X + Y)^2 \mathbf{1}(X + Y > 0)] \ .$$

Finally, the symmetry of the logistic distribution lets us rewrite the last expectation as

$$\frac{1}{4} E[(X + Y)^2] = \frac{1}{2} E[X^2] = \pi^2/6 \ ,$$

where in the last step we used the fact noted earlier that the logistic distribution has variance $\pi^2/3$.

5.8 Passage from a PWIT Matching to a K_n Matching

We are now in position to sketch the connections that complete the proof of Theorem 5.1, and we begin by noting that the observations in hand already give us one half of the theorem. For each n we let M_n denote the optimal matching on K_n and we view M_n as a $\mathcal{G}_\star$-valued random variable by setting the root of M_n to be the root of K_n. By the tightness of the joint process (M_n, K_n) we know that for every subsequence of $n = 1, 2, \ldots$ there is a further subsequence n_j such that (M_{n_j}, K_{n_j}) converges weakly in $\mathcal{G}_\star \times \mathcal{G}_\star$. By Skorohod's theorem and our earlier analysis of the weak limit of K_n, we may even assume that

$$(M_{n_j}, K_{n_j}) \to (\mathcal{M}, \mathcal{T}) \quad \text{with probability one} \ , \tag{5.24}$$

where $\mathcal{T}$ is a PWIT and $\mathcal{M}$ is a matching on $\mathcal{T}$. Moreover, from the fact that M_n is randomly rooted, we see that $\mathcal{M}$ is involution invariant.

Now, if $c(M_n)$ denotes the cost of matching the root in M_n and $c(\mathcal{M})$ denotes the cost of the edges that meets the root of $\mathcal{T}$, then from the limit (5.24) and the definition of the topology on $\mathcal{G}_\star$, we have

$$c(M_{n_j}) \to c(\mathcal{M}) \quad \text{with probability one} \ .$$

From this limit, Fatou's Lemma, and Lemma 5.11 we then find

$$\liminf_{n \to \infty} E[c(M_{n_j})] \geq E[c(\mathcal{M})] \geq \pi^2/6 \ .$$

Finally, by the generality of the subsequence $\{n_j\}$, we therefore find in the notation of Theorem 5.1 that

$$\liminf_{n \to \infty} \frac{2}{n} E[C_n] \geq \pi^2/6 \ . \tag{5.25}$$

This proves half of the equality (5.1) of Theorem 5.1; the bad news is that this is the easy half.

The Harder Half – and Its Two Parts

The harder half of Theorem 5.1 breaks into two natural steps. First one needs to show that there is a low-cost ε-feasible matching for K_n for large values of n, and then one needs to show that such an ε-feasible matching can be modified to provide an honest matching without substantially increasing the total cost.

To make this plan precise, we first recall that an ε-feasible matching of a graph G with n vertices is a subset M^* of edges of G with the property that each vertex of G is in some edge of M^* and at most εn elements of M^* fail to be isolated from the other elements of M^*. The fact that one can cheaply convert a low-cost ε-feasible matching of K_n to an honest low-cost complete matching of K_n is (almost) purely combinatorial. We do not deal here with that part of the argument, but one can refer to Proposition 2 of Aldous [3] for a proof of an analogous result in the context of bipartite matching.

How to Use a Matching on a PWIT to Find One on K_n

We continue to work with the special PWIT $\mathcal{T}$ that was constructed as part of Triple Tree Process of Lemma 5.8, and we let $\mathcal{M}$ denote the matching defined on $\mathcal{T}$ that was constructed in Lemma 5.9. Next, we fix a $\varrho > 0$, and we let $N_\varrho(r, \mathcal{T})$ denote the neighborhood of radius ϱ about the root r of $\mathcal{T}$. We view $N_\varrho(r, \mathcal{T})$ as an element of $\mathcal{G}_\star$ with root r, and we let $N_\varrho(r, K_n)$ denote the analogously defined ϱ-neighborhood of the root of K_n. Here one should note that $N_\varrho(r, \mathcal{T})$ is always a tree since it is a subtree of $\mathcal{T}$, but the corresponding neighborhood of K_n given by $N_\varrho(r, K_n)$ will not necessarily be a tree. Nevertheless, since ϱ is fixed, the probability that $N_\varrho(r, K_n)$ is a tree will go to one as n becomes large.

Now, for any $G \in \mathcal{G}_\star$ that is a feasible realization of $N_\varrho(r, \mathcal{T})$ and for any $w \in \mathcal{T}$, we let $q(w|G)$ denote the conditional probability that r is matched to w in $\mathcal{M}$ given the event that the neighborhood $N_\varrho(r, \mathcal{T})$ is isomorphic to G as an element of $\mathcal{G}_\star$. We will use $q(w|G)$ to guide a sampling process on K_n that we can use to help build a matching.

Specifically, for any $1 \leq n < \infty$ we define a subset $\mathcal{M}(\varrho, n)$ of the *directed edges* of K_n by the following process: for each $u \in K_n$ we choose (u, v) with probability $q(v \mid N_\varrho(u, K_n))$. Also, in just the same way, we define a subset $\mathcal{M}(\varrho, \infty)$ of the *directed edges* of $\mathcal{T}$ – that is, for each $u \in \mathcal{T}$ we choose the directed edge (u, v) with probability $q(v \mid N_\varrho(u, \mathcal{T}))$. The key observation is that the event $A(\varrho, \infty)$ defined by

$$A(\varrho, \infty) = \{\, r \text{ is in exactly two edges of } \mathcal{M}(\varrho, \infty) \,\}$$

satisfies the identity

$$P(A(\varrho, \infty)) = 1 - \eta(\varrho) \quad \text{where } \eta(\varrho) \to 0 \text{ as } \varrho \to \infty . \tag{5.26}$$

Thus, if we collapse each directed loop $\{(u, v), (v, u)\}$ of $\mathcal{M}(\varrho, \infty)$ to a single undirected edge, then for large ϱ we have a high probability of obtaining an honest matching on $\mathcal{T}$.

Now, we consider the analogous events for K_n. Specifically, we let $A(\varrho, n)$ denote the event that (1) the root r of K_n is in exactly two directed edges of $\mathcal{M}(\varrho, n)$ and (2) these edges have the form (r, v) and (v, r) for some v. The local weak convergence of K_n to $\mathcal{T}$ and the definition of $\eta(\varrho)$ then tells us that

$$\lim_{n \to \infty} P(A(\varrho, n)) = 1 - \eta(\varrho) . \tag{5.27}$$

Next, we let $\mathcal{M}^u(\varrho, n)$ denote the set of *undirected* edges one obtains by collapsing each directed loop $\{(v, w), (w, v)\}$ of $\mathcal{M}(\varrho, n)$ into a single undirected edge. One can then show that the limit (5.27) implies that there is an $\varepsilon(\varrho)$ such that $\varepsilon(\varrho) \to 0$ as $\varrho \to \infty$ for which we have

$$\liminf_{n \to \infty} P(\mathcal{M}^u(\varrho, n) \text{ is } \varepsilon(\varrho)\text{-feasible}) \geq 1 - \varepsilon(\varrho) . \tag{5.28}$$

Finally, to complete our sketch of the proof of Theorem 5.1, we just note that one can now use local weak convergence to relate the costs of the edges of $\mathcal{M}^u(\varrho, n)$ to the costs of those in the matching $\mathcal{M}$ on $\mathcal{T}$, and, in due course, one can use this correspondence to show that the asymptotic mean cost of $\mathcal{M}^u(\varrho, n)$ is at most $n/2$ times $\pi^2/6$. This is precisely what one was brought to expect from the calculation in Lemma 5.11, and thus it completes the sketch.

When one looks back on this construction, one can confirm that the basic idea of sampling from a locally defined conditional distribution is pleasantly general. In fact, there are many problems where a similar sampling method can be applied, and this method of conditional sampling seems to express one of the more fundamental ways that one can exploit local weak convergence. Further elaboration of the sampling construction is expected to be part of Aldous and Steele [7].

5.9 Finally – Living Beyond One's Means

One of the original motivations for the study of the assignment problem was the discovery of Karp [37] that under the model of independent uniform costs c_{ij} the value of A_n provides an impressively tight lower bound on the minimal cost of a *tour* of the vertex set of the complete directed graph where the cost of using the directed edge (i, j) is c_{ij}. This discovery led in turn to sustained exploration of algorithms that use the assignment problem and branch and bound techniques for the travelling salesman problem (cf. Balas and Toth[13]), and it also motivated more detailed investigations of the precision of A_n as an approximation for the TSP cost (cf. Karp and Steele [39]).

While the objective method is particularly well focused on issues such as the convergence of the means $E(A_n)$ and the determination of this limit, it does not provide any information on the variance of A_n or other measures of concentration of A_n. For such information, other methods must be applied.

For example, Karp and Steele [39] used combinatorial and coupling methods to show that with high probability the greatest cost of an edge in an optimal assignment of size n is $O(\log^2 n/n)$, and Talagrand [64] used this observation to illustrate one of his isoperimetric inequalities to show that there is a constant $0 < c < \infty$ such that

$$\mathrm{Var}(A_n) \leq \frac{c}{n}(\log n)^4 \quad \text{for all } n \geq 1 .$$

More recently, Frieze and Sorkin [23] improved the bounds of Karp and Steele [39] to show that with high probability the longest edge in $O(\log n/n)$, and the same technique was then further refined by Lee and Su [45] to show that there are positive constants c_1 and c_2 such that

$$c_1 n^{-5/2} \leq \mathrm{Var}(A_n) \leq \frac{c_2}{n}(\log n)^2 . \tag{5.29}$$

Despite the wide gap between the bounds (5.29) they are the best that are currently known. Nevertheless, one certainly expects that in fact there is $\sigma^2 > 0$ such that

$$\mathrm{Var}(A_n) \sim \frac{\sigma^2}{n} \quad \text{as } n \to \infty . \tag{5.30}$$

Unfortunately, the objective method is powerless to help, and for the moment at least, the conjecture (5.30) seems to be far over the horizon.

6 Problems in Euclidean Space

The objective method is especially well suited for many problems of Euclidean combinatorial optimization. Typically, one begins with an optimization problem that is driven by a sample $\{X_i : 1 \leq i \leq n\}$ of independent random variables with the uniform distribution on the unit d-cube $[0, 1]^d$, so after choosing a random "root" I that is uniformly distributed on $\{1, 2, ..., n\}$ and independent of the $\{X_i\}$, one can take immediate advantage of the classical fact that

$$n^{1/d}(X_i - X_I : 1 \leq i \leq n) \xrightarrow{d} \mathcal{P}_0 \quad \text{as } n \to \infty , \tag{6.1}$$

where $\mathcal{P}_0 = \mathcal{P} \cup \{0\}$ and $\mathcal{P}$ is a homogenous Poisson process on $\mathbb{R}^d$ with uniform intensity equal to 1.

Now, throughout this survey, we have interpreted the indicated convergence as local weak convergence as defined in Section 2, so it may seem unfair to call this result classical. Nevertheless, once one agrees to interpret the two sides of the limit (6.1) as rooted geometric graphs where the root is the vertex at the origin of $\mathbb{R}^d$ and where the edge lengths are given by Euclidean distances, then (up to an irrelevant rotation) local weak convergence (6.1) is in fact equivalent to the traditional weak convergence of point process theory.

The bottom line is that for problems of combinatorial optimization for points in $\mathbb{R}^d$, the Poisson limit (6.1) assumes the role that Grimmett's Lemma 2.3 played for the uniform trees and that Aldous's Theorem 4.1 played for the exponentially weighted complete graphs. Thus, for Euclidean problems the first step of the objective method comes *gratis*; the game only begins with the introduction of an optimization problem.

6.1 A Motivating Problem

One of the early motivations for the objective method was a problem sparked by an empirical observation of R. Bland, who noted from simulations of an algorithm for the MST that for n independent uniformly distributed points in the unit square that one consistently finds that

$$\sum_{e \in MST} |e|^2 \to c > 0 \quad \text{as } n \to \infty ,$$

where $|e|$ denotes the usual Euclidean distance from u to v when $e = (u, v)$. Stimulated by Bland's observation, Steele [60] proved that if the random variables $\{X_i\}$ are independent and have a distribution μ with compact support, then for all $0 < \alpha < d$ one actually has

$$n^{(\alpha-d)/d} \sum_{e \in MST} |e|^\alpha \to c(\alpha, d) \int_{\mathbb{R}^d} f(x)^{(d-\alpha)/d} \, dx \quad \text{a.s. ,} \qquad (6.2)$$

where f is the density of the absolutely continuous part of μ and $c(\alpha, d)$ is a constant that depends only on α and d. This result goes beyond Bland's conjecture in several respects, but, ironically, the subadditive method used in Steele [60] falls short of covering the motivating case $\alpha = d$. This is where the objective method entered the picture.

Benefits of Hindsight – and A Proper Point of View

Fifteen years ago it was not clear how one could make further progress on Bland's conjecture, but with the benefit of hindsight a good plan is obvious. One simply needs to extend the Poisson convergence (6.1) to the joint convergence the point sets and the corresponding spanning trees.

When this was done in Aldous and Steele [6], there was some mystery to emergence of a *forest* as the limit of a sequence of minimal spanning trees. Now the reasons for the appearance of the minimal spanning forest are understood more deeply, and, if one simply applies the MST Convergence Theorem of Section 5.3 to the Poisson limit (6.1), one obtains a theorem that explains most of the qualitative behavior of the MST of a random sample from the d-cube.

Theorem 6.1. *For the randomly rooted normalized sample*

$$S_n = n^{1/d}(X_i - X_I : 1 \leq i \leq n)$$

of independent random variables with the uniform distribution on $[0,1]^d$, one has the joint local weak convergence

$$\left(S_n, MST(S_n) \right) \xrightarrow{d} \left(\mathcal{P}_0, MSF(\mathcal{P}_0) \right). \tag{6.3}$$

A Question of Means

On a conceptual level, the limit (6.3) tells one almost everything there is to say about $MST(S_n)$ for large n. Nevertheless, there are inevitably technical hurdles to be cleared when one applies Theorem 6.1 to a concrete problem. For example, to obtain the convergence of the mean of $MST(S_n)$, one needs some information on the uniform integrability of the set of variables $\{MST(S_n) : n = 1, 2, ..\}$.

Uniform integrability of this sequence was proved in Aldous and Steele [6] by direct estimates, and later the "lens geometry" of the MST and spacefilling curves were used in Chapter 5 of Steele [62] to give more inclusive moment bounds. In the end, one finds for all $p \geq 1$ and all d there is a constant $\beta_d > 0$ such that for random uniform samples of size n from the d-cube one has

$$\sum_{e \in MST} |e|^d \to \beta_d \quad \text{in } L^p \text{ as } n \to \infty.$$

Moreover, one can identify the limit constant β_d in terms of the limit object by the formula

$$\beta_d = \frac{1}{2} E\left[\sum_{e \in R} |e|^d \right] < \infty \tag{6.4}$$

where R is the set of edges of the MSF of $\mathcal{P}_0$ that are incident to the root at the origin of $\mathbb{R}^d$.

The objective method almost always gives one a characterization of the limit constant in terms of the limit object, and we have seen examples in Sections 3 through 5 where this characterization could be pressed to provide an exact determination of the limit constant. In this respect, Euclidean problems are different. Despite tantalizing characterizations such as that provided by formula (6.4), none of the limit constants of finite dimensional Euclidean optimization have been determined analytically. The reason for this failing seems to be that in the Euclidean case, one does not have the self-similarity that is crucial to the recursions that lead us to the limit constants in results like Theorem 3.4 and Theorem 5.1.

6.2 Far Away Places and Their Influence

Throughout the theory of Euclidean combinatorial optimization, one finds hints that tractable problems must be *local* in some sense; or, looked at the other way around, one finds that in tractable problems the points that are far away cannot have much influence on the problem's solution in the neighborhood of a near-by point. This intuition can be formalized in several ways, but the recently introduced notion of *stability* is perhaps most in tune with the philosophy of the objective method.

Sums of Local Contributions

To begin, we consider nonnegative functions of the form $\xi(x; \mathcal{X})$ where $x \in \mathbb{R}^d$ and $\mathcal{X}$ is a subset of $\mathbb{R}^d$ that is locally finite in the sense that each bounded subset of $\mathbb{R}^d$ contains only finitely many elements of $\mathcal{X}$. Also, if a is a positive scalar and $y \in \mathbb{R}^d$ we denote the usual dilations and translations of $\mathcal{X}$ by

$$a\mathcal{X} := \{ax : x \in \mathcal{X}\} \quad \text{and} \quad y + \mathcal{X} := \{y + x : x \in \mathcal{X}\} \, .$$

If we assume ξ is translation invariant in the sense that

$$\xi(y + x; y + \mathcal{X}) = \xi(x; \mathcal{X}) \quad \text{for all } y \text{ and } \mathcal{X} \, ,$$

then the sum $H_\xi(\mathcal{X})$ defined by setting

$$H_\xi(\mathcal{X}) := \sum_{x \in \mathcal{X}} \xi(x; \mathcal{X}) \tag{6.5}$$

is also translation invariant in the sense that $H_\xi(y + \mathcal{X}) = H_\xi(\mathcal{X})$ for all $y \in \mathbb{R}^d$. Moreover, sums of this form (6.5) may be used to represent almost all of the functionals of interest in Euclidean combinatorial optimization.

For example, the length of the MST of $\mathcal{X}$ may be written as $H_\xi(\mathcal{X})$ if we take $\xi(x; \mathcal{X})$ to be half of the sum of the edges of the MST that are incident to x, so

$$\xi(x; \mathcal{X}) = \frac{1}{2} \sum_{y \in \mathcal{X}} |x - y| \mathbf{1}((x, y) \in MST(\mathcal{X})) \, .$$

For a less obvious example, consider the representation $H_\xi(\mathcal{X})$ of the number of connected components of the nearest neighbor graph of $\mathcal{X}$. In this case we can take

$$\xi(x; \mathcal{X}) = 1/\mathrm{card}(C_x)$$

where C_x is the component of the nearest neighbor graph of $\mathcal{X}$ that contains the vertex x. Essentially all of the Euclidean functionals discussed in Steele [62] or Yukich [67] may be represented as $H_\xi(\mathcal{X})$ for an appropriate choice of $\xi(x; \mathcal{X})$.

Stable Functionals and the Influence Function

One benefit of the representation $H_\xi(\mathcal{X})$ is that the summands $\xi(x; \mathcal{X})$ suggest a natural way to formalize the idea of local dependence. To explain this idea, we first take $r > 0$ and let $B(x; r)$ denote the Euclidean ball of radius r about x, then, for any locally finite point set $\mathcal{S} \subset \mathbb{R}^d$ and any integer $m \geq 1$ we set

$$\overline{\xi}(\mathcal{S}; m) := \sup_{n \in \mathbb{N}} \left(\mathrm{ess\ sup}_{m,n}\{\xi(0; (\mathcal{S} \cap B(0; m)) \cup \mathcal{A})\} \right)$$

and where $\mathrm{ess\ sup}_{m,n}$ denotes the essential supremum with respect to Lebesgue measure on $\mathbb{R}^{dn}$ and where $\mathcal{A} \subset \mathbb{R}^d \setminus B(0; m)$ is a set of cardinality n. Analogously, we define

$$\underline{\xi}(\mathcal{S}; m) := \inf_{n \in \mathbb{N}} \left(\mathrm{ess\ inf}_{m,n}\{\xi(0; (\mathcal{S} \cap B(0; m)) \cup \mathcal{A})\} \right),$$

and we think of and $\underline{\xi}(\mathcal{S}; m)$ and $\overline{\xi}(\mathcal{S}; m)$ as measures of the changes that one may make by a finite additions to $\mathcal{S}$ outside a ball of radius m. The interesting case occurs when these two possibilities are comparable for large m.

Definition 6.2 (Stable Functionals and the Influence Function). *The functional ξ is said to be* stable *on the locally finite set $\mathcal{S}$ provided that one has*

$$\lim_{m \to \infty} \overline{\xi}(\mathcal{S}; m) = \lim_{m \to \infty} \underline{\xi}(\mathcal{S}; m) . \tag{6.6}$$

Moreover, when ξ is stable on $\mathcal{S}$, we let $\xi_\infty(\mathcal{S})$ denote the value of the common limit and we call $\xi_\infty(\mathcal{S})$ the influence function *of ξ on $\mathcal{S}$.*

The locally finite sets that are of most interest to us are naturally the realizations of homogeneous Poisson processes, and one should note that if $\mathcal{P}_\tau$ is the homogeneous Poisson process with constant intensity τ on $\mathbb{R}^d$ then the sequence of random variables $\{\overline{\xi}(\mathcal{P}_\tau; m) : m = 1, 2, ...\}$ is nonincreasing and the sequence $\{\underline{\xi}(\mathcal{P}_\tau; m) : m = 1, 2, ...\}$ is nondecreasing. Thus, both sequences converge almost surely, and we see that ξ is almost surely stable on $\mathcal{P}_\tau$ if the two limits are equal with probability one. Many of the natural functionals $\xi(x; \mathcal{X})$ of Euclidean combinatorial optimization have this property, and there are some useful consequences of this fact.

A Weak Law of Large Numbers

Now consider a sequence $X_1, X_2, \ldots$ of independent d-dimensional random variables that have a common density f. Next, set $\mathcal{X}_n = \{X_1, \ldots, X_n\}$ and consider the scaled summand

$$\xi_n(x; \mathcal{X}) := \xi(n^{1/d}x; n^{1/d}\mathcal{X}_n) \tag{6.7}$$

together with the corresponding scaled sums

$$\tilde{H}_\xi(\mathcal{X}_n) := \sum_{x \in \mathcal{X}_n} \xi_n(x; \mathcal{X}_n) = \sum_{x \in \mathcal{X}_n} \xi(n^{1/d}x; n^{1/d}\mathcal{X}_n) \, .$$

When ξ is stable for almost every realization of any homogeneous Poisson process, these sums typically satisfy a weak of large numbers; in fact, this notion of stability was introduced in Penrose and Yukich [55] for the purpose of framing such weak laws. The next theorem is typical of the results obtained there.

Theorem 6.3. *For each constant $0 < \tau < \infty$ we suppose that ξ is almost surely stable for the homogenous Poisson process $\mathcal{P}_\tau$ with intensity τ, and we let $\xi_\infty(\mathcal{P}_\tau)$ denote the corresponding influence function. If the sequence*

$$\left\{ \xi(n^{1/d}X_1; n^{1/d}\mathcal{X}_n) : 1 \leq n < \infty \right\} \quad \text{is uniformly integrable} \,, \tag{6.8}$$

where observations in the sample $\mathcal{X}_n = \{X_1, X_2, \ldots, X_n\}$ are i.i.d. with density f on $\mathbb{R}^d$, then the influence function satisfies

$$I[\tau] = E[\xi_\infty(\mathcal{P}_\tau)] < \infty \quad \text{for each } 0 < \tau < \infty \text{ such that } f(\tau) > 0 \,, \tag{6.9}$$

and the normalized sums satisfy

$$n^{-1}\tilde{H}_\xi(\mathcal{X}_n) \to \int_{\mathbb{R}^d} I[f(x)]f(x)dx \quad \text{in} L^1 \, . \tag{6.10}$$

Consequences of Homogeneity

Many of the functionals addressed by Theorem 6.3 have the property that there is a constant $\gamma > 0$ such that

$$\xi(a\,x; a\mathcal{X}) = a^\gamma \xi(x; \mathcal{X}) \quad \text{for all } a \in \mathbb{R}^+ \,,$$

and, in this case the righthand side of the limit (6.10) is more simply written as

$$E[\xi_\infty(\mathcal{P}_1)] \int_{\mathbb{R}^d} f(x)^{(d-\gamma)/d}dx \, . \tag{6.11}$$

Moreover, when ξ is *scale invariant*, or homogeneous of order 0, the limit (6.11) simply boils down to $E[\xi_\infty(\mathcal{P}_1)]$, and the limit given by Theorem 6.3 does not depend on the density of underlying point set. Formula (6.11) also makes it clear that although ξ_∞ may at first seem subtle to define, it has an inevitable place in the limit theory of the normalized sums $\tilde{H}_\xi$.

Confirmation of Stability

Theorem 6.3 applies to many problems of computational and it provides limit laws for the minimal spanning tree, the k-nearest neighbors graph ([34], [20]), the Voronoi graph [32], the Delaunay graph, and the sphere of influence graphs ([27], [33]). Nevertheless, before Theorem 6.3 can be applied, one needs to prove the associated summands $\xi(x, \mathcal{X})$ are indeed stable on $\mathcal{P}_\tau$ for all $0 \leq \tau < \infty$.

In some cases, stability may be proved easily, but, in other cases, the proof may be quite challenging. For example, even the stability of the MST is not easy without the help of some powerful tools. For example, Penrose and Yukich [55] prove stability MST with help from the idea of a *blocking set*, which is a set of points that effectively isolates the MST in a neighborhood of the origin from any influence by points outside of a certain large disk. This basic structure was introduce by Kesten and Lee [40] (cf. Lemmas 3 and 5), and it provides one of the keys to the of the central limit theory for the MST as further developed by Lee ([43], [44]).

A Cox Process Coupling

Finally, we should note that the proof of Theorem 6.3 given in Penrose and Yukich [55] calls on an interesting coupling argument that is likely to have further applications. The key observation is that one may simulate samples that closely approximate the samples from a general density f with help from a Cox process that has a random *but conditionally uniform* intensity. This approximation provides one with a new way to convert results for uniform samples to results for nonuniform samples, and, in the right circumstances, it seems to have advantages over older methods where one first approximates f by a locally flat density and then creates a coupling. Nevertheless, each method ultimately relies on some sort of smoothness of the sums $\widetilde{H}_\xi$.

6.3 Euclidean Methods and Some Observations in Passing

There are so many connections between percolation theory, minimum spanning trees, and the objective method that we cannot survey them in detail, but we do want to review a few developments that seem particularly informative. The first of these is a recent observation of Bezuidenhout, Grimmett, and Löffler [16] that draws a simple connection between MSTs and the notion of free energy that originates in statistical mechanics.

MSTs, Free Energy, and Percolation Connections

To begin, if V is a subset of $\mathbb{R}^d$ and $\alpha > 0$, we let $\Gamma(\alpha, V)$ denote the graph with vertex set V and edge set consisting of pairs (x, y) of elements of V such that $|x - y| \leq \alpha$. Next, we let $\mathcal{P}$ denotes the homogeneous Poisson process on $\mathbb{R}^d$ with unit intensity, and take $S(\alpha)$ to be the size of the connected component of $\Gamma(\alpha, \mathcal{P} \cup \{0\})$ that contains 0, the origin of $\mathbb{R}^d$, and, finally, we define $\kappa(\alpha)$, the *free energy of the Poisson process*, by setting

$$\kappa(\alpha) = E\big[1/S(\alpha)\big] .$$

The free energy turns out to have a remarkably simple relationship to the empirical distribution of the edge lengths of the minimal spanning tree T_n of the set $V_n = \mathcal{P} \cap [-n, n]^d$. Specifically, if we consider the empirical distribution of the set of edge lengths of T_n,

$$F_n(\alpha) = (|V_n| - 1)^{-1} \sum_{e \in T_n} \mathbf{1}(|e| \le \alpha) ,$$

then we have the succinct limit

$$\lim_{n \to \infty} F_n(\alpha) = 1 - \kappa(\alpha) \quad \text{a.s. and in } L^1 . \tag{6.12}$$

This engaging identity may be explained in just a few lines. The key observation is that if $\nu(\alpha, n)$ denotes the number of connected components of the graph $\Gamma(\alpha, \mathcal{P} \cap [-n, n]^d)$, then by counting the missing edges we have

$$F_n(\alpha) = \frac{|V_n| - \nu(\alpha, n)}{|V_n| - 1} . \tag{6.13}$$

To finish off, one now just needs to note that the Ergodic Theorem for the Poisson process can be used to show

$$\lim_{n \to \infty} \nu(\alpha, n)/|V_n| = E\big[1/S(\alpha)\big] := \kappa(\alpha) \quad \text{a.s. and in } L^1 ,$$

so the limit (6.12) follows immediately from the identity (6.13).

Naturally, if one is to make good use of this observation more work is needed, and the real effort in Bezuidenhout, Grimmett, and Löffler [16] goes into the development of the analytical properties of the free energy and its analogs for Bernoulli site percolation. In that setting, they show that the free energy defect $H(\alpha) = 1 - \kappa(\alpha)$ is indeed a distribution function, and they also show that one has the convergence of the corresponding mth moments for all $0 \le m < \infty$,

$$\int_0^\infty \alpha^m dF_n(\alpha) \to \int_0^\infty \alpha^m dH(\alpha) \quad \text{a.s. and in } L^1 .$$

A Second Percolation Connection – Forest vs. Tree

When the minimal spanning forest for the Poisson process was introduce in Aldous and Steele [6], there was a strong suspicion that the minimal spanning forest might actually be a tree – at least in low dimensions. This conjecture was confirmed in dimension $d = 2$ by Alexander [8], Example 2.8, but, the question remains open for dimension $d \geq 3$.

The suspicion now is that the MSF of the Poisson process is not a tree if the dimension is sufficiently high, and Theorem 2.5, part (ii), of Alexander [8] provides some indirect evidence for this conjecture. Because of results of Hara and Slade ([28], [29]) that show one does not have percolation at the critical point for dimension $d > 19$ for Bernoulli bond percolation, Alexander's theorem implies that in high dimensions a related MSF does not have any doubly infinite paths. If one relies on the analogies between Bernoulli bond percolation, Bernoulli site percolation, and continuum percolation, Alexander's result suggests in turn that if the MSF of the Poisson process is a tree, then in high dimensions that tree must consist of a sequence of finite trees that are connected to one singly infinite path. Finally, for such a structure to cover all of the point of a Poisson process, this path would then be forced to wander around space in a way that does not seem feasible.

Inclusion Criteria and Computational Complexity

Almost all of the probability problems of Euclidean combinatorial optimization may be framed in terms of a graph G_n that depends on a random sample $\{X_1, X_2, ..., X_n\}$ in $\mathbb{R}^d$, and for many of these problems the subadditive method introduced by Beardwood, Halton and Hammersley [14] provides a competing technology to the objective method. Although the objective method has intuitive and analytical advantages in almost every case where it applies, there are still instances where the only concrete results have been found with the subadditive method.

For example, subadditive methods work quite naturally for traveling salesman problem (TSP), but when one tries to apply the objective method to the TSP one some deep and interesting questions emerge. Specifically, one needs to define an analog T of the traveling salesman tour for the Poisson process in $\mathbb{R}^d$, and in practical terms this means one needs to specify an inclusion criterion that tells one when an edge (x, y) is in T. We were able to provide such an inclusion criterion for the maximal matching problem, and classical results for the MST suggested the defining criterion for the minimal spanning forest. On the other hand, the theory of computational complexity suggest that one is unlikely to find an inclusion criterion that will provide a suitable analog for the TSP of the Poisson process. The basic intuition is that no problem that is NP-complete for finite sets of points can be expected to have an inclusion criterion that provides a suitable limit object on the Poisson process.

Further Comparison to Subadditive Methods

In addition to the examples suggested by computational complexity, there are other cases where subadditive methods seem to have an advantage over the objective method. One such situation is provided by the asymptotic analysis of heuristic methods, and the greedy matching problem [11] is perhaps the most natural example.

Nevertheless, there are surely cases where subadditive methods have been used in the past, but where the objective now provides an easier or more informative approach. Confirmed cases on this list include the limit theory for MST with power weighted edges [60] and MST vertex degrees [63]; and, with some systematic effort, the list can probably be extended to include the theory of Euclidean semi-matchings [61], optimal cost triangulations [58] and the K-median problem [35]. Moreover, there are many cases where the objective method quickly gives one the essential limit theory, yet subadditive methods appear to be awkward to apply. Here the list can be made as long as one likes, but one should certainly include the limit theory for Voronoi regions [32] and the sphere of influence graphs ([27], [33]).

6.4 Recurrence of Random Walks in Limits of Planar Graphs

Most of the local weak limits that we have surveyed here have their origins in combinatorial optimization, but the objective method and the theory of local weak convergence can also be applied to problems that are more purely probabilistic. One striking example is given by the recurrence theory for random walks on planar graphs recently given by Benjamini and Schramm [15].

Theorem 6.4. *Suppose that G is a $\mathcal{G}_\star$-valued random variable that is a weak limit in $\mathcal{G}_\star$ of a sequence $\{G_n[X_n] : n = 1, 2, ...\}$ of uniformly rooted random graphs where*

(1) for each $n \geq 1$, the graph G_n is planar with probability one, and
(2) $\max\{\deg(v) : v \in G_n\} \leq M < \infty$ for all $n \geq 1$,

then with probability one, the random walk on the graph $G(\omega)$ is recurrent.

This theorem is clearly part of the theory of local weak convergence, but the theorem and its proof differ greatly the other results in this survey. Even a sketch of the proof of Theorem 6.4 would take us far off our path, but we should at least note that one key to the proof is the remarkable Circle Packing Theorem of Koebe [41]. This theorem says that if a finite graph $G = (V, E)$ is planar, then there exist a collection C of circles in the plane with disjoint interiors and a one-to-one correspondence $\phi : V \to C$ such that $(x, y) \in E$ if and only if the circles $\phi(x)$ and $\phi(y)$ make contact.

From the perspective of this survey, there are two attractive features of Theorem 6.4. First, its very difference serves as a reminder that there must be other, as yet unimagined, areas where the theory of local weak convergence can play a natural role. Second, Theorem 6.4 suggest a useful generic question: What are the properties of a sequence of random rooted finite graphs that are preserved under local weak convergence? In many cases this question is bound to yield only pedestrian answers, but the Benjamini-Schramm Theorem shows that there are indeed cases when the generic question can lead one to remarkable results.

7 Limitations, Challenges, and Perspectives

This treatment of the objective method is part survey and part tutorial. The basic intention has been to make the objective method more accessible, while also illustrating its effectiveness on problems with some subtlety and substance. Any survey has an element of advocacy, and here we certainly hoped to show the objective method in a positive light. Nevertheless, one should be aware of the method's limitations.

Some Intrinsic Limitations

First, the objective method deals with reasonably complex entities such as the Skeleton Tree $PGW^\infty(1)$, the PWIT, or – de minimus – the Poisson process in $\mathbb{R}^d$. In many cases, one can make immediate and effective use of these off-the-shelf structures, but in more original problems, one is forced to invent (or discover) new objects. Any such object begins life as the weak limit in $\mathcal{G}_\star$ of a sequence of random finite graphs with uniformly distributed random roots, but, if it is to be used with effect, one almost always needs an independent characterization. This highly constructive feature of the objective method means that it seldom provides short proofs. Ironically, this remains so even when the result that one wants to prove is "conceptually clear" from the point of view offered by the limiting object.

Second, when one focuses directly on a limit object, one tends to lose almost all trace of how one gets to that limit. As a consequence, the objective method is typically ill suited for obtaining information about rates of convergence. To be sure, there are special circumstances where rates may be found; for example, the recursion used in our analysis of the maximum partial matching problem provides at least some rate information. Nevertheless, this is an exception that reflects the hybrid nature of one particular problem.

Third, the theory of local weak convergence has a substantial limitation on the type of information that it can provide. Basically, the method only addresses those problems that may be reduced to a calculation in a neighborhood of a randomly chosen root. For example, we are able to address total-sum-of-length problems only because of a lucky piece of arithmetic; the sum of the lengths of the edges in a graph with n vertices is just n times the expected value of half of the sum of the lengths of the edges incident to a randomly chosen root.

Many similar sounding problems are not so lucky. For example, consider the problem of calculating the expected length of the longest path in the MST of a uniform random sample of size n from the unit square. This problem sounds quite similar to the total length question, but, on reflection, one finds that it cannot be reduced to a calculation that just depends on a neighborhood of a random root; thus, it falls outside of the scope of the objective method. For the same reason, the objective method cannot help us show that the variance of A_n is asymptotic to σ^2/n, nor can it help with more general measures of concentration. Local weak convergence is always a prisoner that cannot escape the neighborhoods of a random root.

A Current Challenge: The Largest Independent Set

The limitations on the objective method are substantial, but one still finds a wide range of challenges for which the objective method offers the most likely path for progress. Here we will just discuss one problem that we find especially attractive.

An *independent set S* in a graph G is collection of vertices such that no two elements of S are joined by an edge of G, and the *independence number $\alpha(G)$* is the maximum cardinality of an independent set in G. The independence number is one of the most studied quantities of graph theory, and it also leads one to an interesting test of potential of the objective method.

For any integer $k \geq 3$ there is a standard and model for a *random k-regular graph* with n vertices, and the properties of these graphs a studied in detail in the monographs of Bollobás [17] and Janson, et.al. [36]. Now, if one starts to think about the independence number of a random k-regular graph $G_{k,n}$, one almost immediately comes to conjecture that there exists a constant $\alpha_k > 0$ such that

$$\lim_{n \to \infty} n^{-1} E[\alpha(G_{k,n})] = \alpha_k \ . \tag{7.1}$$

The origins of this conjecture are lost in the mists of time, and no doubt it has occurred independently to many people. From the perspective of this survey, one of its interesting features is that it offers several parallels to the random assignment problem as that problem was understood circa 1986.

For example, no one has yet proved that indicated limit (7.1) actually exists – yet no one doubts that it does. Moreover, some concrete limits have already been place on the possible size of α_k. Specifically, Frieze and Suen [24] studied the greedy algorithm for constructing an independent set for $G_{3,n}$, and they have proved

$$\liminf_{n \to \infty} n^{-1} E[\alpha(G_{3,n})] \geq 6 \log(3/2) - 2 = 0.432 \cdots \tag{7.2}$$

Finally, the methods of statistical mechanics have already found a role. In particular, Hartmann and Weigt [31] have used the replica method to study the independence number in a model that is closely related to $G_{k,n}$.

When we view conjecture (7.1) from the perspective of the objective method, we do not need long to hit on a promising plan. In fact, we can take one step almost automatically, since it is reasonably obvious that $G_{k,n}$ converges in $\mathcal{G}_*$ to the randomly rooted infinite k-regular tree $\mathcal{T}$. Thus, the intrigue only begins once one starts to look for a large involution invariant independent subset S of $\mathcal{T}$, and, remarkably enough, an attractive candidate comes to mind almost immediately.

To define our candidate S, we first flip a coin. If it comes up heads, we take the root r to be an element of S, and then we add to S every other vertex on the paths that descend from r, but if the coin comes up tails, we take all of the children of r to be in S and then add to S every other vertex on the paths that descend from these children. The set S is clearly an independent set in $\mathcal{T}$, and one can check that S is involution invariant. No checking is needed to see that S has density one-half, and, since no independent set can have density larger than one-half, the principles developed in Section 3 might seem to suggest that we are well on our way to proving that $\alpha_r = 1/2$ for all r.

Unfortunately, we meet a bump in the road. Our candidate S cannot be realized as the weak limit of a sequence of independent subsets of $G_{n,k}$, even though S does satisfy the necessary condition of involution invariance. The problem with S is that it exhibits a extreme form of long-range dependence, and, in this particular case, one can exploit that dependence to show that S is not the weak limit of sequence independent subsets of $G_{n,k}$ as $n \to \infty$.

Thus, we come back to the challenge of defining an independent subset $\mathcal{T}$ that can be realized as the limit of independent subsets of the sequence $G_{n,k}$, $n = 1, 2, \ldots$. One natural thought is that it may be possible to characterize α_k as the maximum density of an involution invariant random independent set in $\mathcal{T}$ that does not exhibit some particular type of long-range dependence, but, so far, there are only the faintest hints how one might specify the precise form of dependence that one should rule out.

Some Final Perspectives

One can never know how a mathematical theory will evolve, and even well-informed guesses are still just guesses. Nevertheless, the objective method and the theory of local weak convergence have a certain inevitability to them that invites speculation.

At the most concrete level, the objective method helps one to decouple two basic tasks: the proof of the existence of a limit and the determination of a limit constant. This contrasts with analytic methods where techniques based on recursions and generating functions typically address these two tasks simultaneously. It also contrasts with subadditive methods where one may prove the existence of a limit and then be left without any serious indication of its numerical value.

At a more abstract level, the objective method helps one focus on the most essential elements of a complex limit theorem. Just as one finds in the traditional theory of weak convergence that Brownian motion appears as the limit of many different processes, one finds in the theory of local weak convergence that basic objects such as the skeleton tree and the PWIT appear as the limit of many different processes. These new structures do not have the amazing universality of Brownian motion, yet they do have *some* measure of universality. Moreover, even though these semi-discrete structures are simpler than Brownian motion, they still have substantial internal structure that can support concrete computation; the calculation in Lemma 4.3 showing that $\zeta(3)$

is the average length per vertex of the PWIT provides one example. Finally, the list of objects is surely far from exhausted, and perhaps, it has only been begun.

One surely expects to see new local weak limits with all the charm of the skeleton tree or the PWIT. Recent support for this view is offered by Angel and Schramm (2002) which proves local weak convergence of uniform random triangulations of the sphere with n vertices to a limit random triangulation of the plane and by Gamarnik (2002) where arguments motivated by the PWIT-based analysis of the random assignment are applied to a linear programming relaxation of random K-SAT.

Finally, we should note that the objective method is tightly bound with the theory of recursive distributional equations. For example, the fixed point equations (3.10) and (5.18) were at the heart of our analysis of the minimum cost perfect matching problems, and this situation seems to be typical. In fact, equations that characterize an unknown distribution by equality in distributional with a function of a collection of independent copies of itself arise in a wide variety of probabilistic settings. Aldous and Bandyopadhyay (2002) provide a survey of such equations, both with and without direct connections to the objective method.

Acknowledgement. The authors are pleased to thank Kenneth Alexander, Harry Kesten, Sungchul Lee, Mathew Penrose, Gordon Slade, Z. Su, and Joseph Yukich for kindly providing preprints, references, and comments. The research of David Aldous is supported by the NSF under Grant DMS-9970901.

References

1. Aldous, D.J. (1990): A random tree model associated with random graphs. Random Structures Algorithms, **1**, 383–402.
2. Aldous, D.J. (1991): Asymptotic fringe distributions for general families of random trees. Ann. Appl. Probab. **1**, 228–266.
3. Aldous, D.J. (1992): Asymptotics in the random assignment problem. Probab. Th. Rel. Fields, **93**, 507–534.
4. Aldous, D.J. (2001): The $\zeta(2)$ limit in the random assignment problem. Random Structures Algorithms, **18**, 381–418.
5. Aldous, D.J. and Bandyopadhyay, A. (2002): A Survey of Max-type Recursive Distributional Equations. Technical Report, U.C. Berkeley.
6. Aldous, D.J. and Steele, J.M. (1992): Asymptotics of Euclidean Minimal Spanning Trees on Random Samples. Probab. Th. Rel. Fields, **92** 247–258.
7. Aldous, D.J. and Steele, J.M. (2002): The asymptotic essential uniqueness property for minimal spanning trees on random points, manuscript in preparation.
8. Alexander, K.S. (1995) Percolation and minimal spanning forest in infinite graphs. Ann. Probab. **23**, 87–104.
9. Alm, S.E. and Sorkin, G.B. (2002): Exact expectations and distributions for the random assigmnent problem. Combinatorics, Probability, and Computing **11**, 217–248.

70   David Aldous and J. Michael Steele

10. Angel, O. and Schramm, O. (2002): Uniform Infinite Planar Triangulations, arXiv:math.PR/0207153.
11. Avis, D., Davis, B., and Steele, J.M. (1988): Probabilistic analysis of a greedy heuristic for Euclidean matching. Probability in the Engineering and Information Sciences, **2**, 143–156.
12. Avram, F. and Bertsimas, D. (1992): The minimum spanning tree constant in geometric probability and under the independent model: a unified approach. Annals of Applied Probability, **2**, 113–130.
13. Balas, E. and Toth, P. (1985): Branch and bound methods. In The Traveling Salesman Problem: A Guided Tour of Combinatorial Optimzation. Lawler, E.L, Lenstra, J.K, Rinnooy Kan, A.H.G., and Smoys, D.B. (eds), Wiley, NY.
14. Beardwood, J., Halton, J.H., and Hammersley, J.M. (1959): The shortest path through many points. Proceedings of the Cambridge Philosphical Society, **55**, 299–327.
15. Benjamini, I. and Schramm, O. (2001): Recurrence of distributional limits of finite planar graphs. Electronic Joural of Probability, **6**, Paper No. 23, 1–13.
16. Bezuidenhout, C., Grimmett, G., and Löffler, A. (1998): Percolation and minimal spanning trees. J. Statist. Phys., **92**, 1–34.
17. Bollobás, B. (1985): Random Graphs. Academic Press, London.
18. Devroye, L. (1998): Branching processes and their application in the analysis of tree structures and tree algorithms. In M. Habib, editor, Probabilistic Methods for Algorithmic Discrete Mathematics, Springer-Verlag.
19. Dyer, M. E., Frieze, A. M., and McDiarmid, C. (1986): On linear programs with random costs. Math. Programming, **35**, 3–16.
20. Eppstein, D., Paterson, M.S. and Yao, F.F. (1997): On nearest-neighbor graphs. Discrete Comput. Geom., **17**, 263–282.
21. Frieze, A.M. (1985): On the value of a random minimum spanning tree problem. Discrete Appl. Math., **10**, 47–56.
22. Frieze, A.M. and McDiarmid, C.J.H. (1989): On random minimum lenght spanning trees. Combinatorica, **9**, 363–374.
23. Frieze, A. and Sorkin, G.B. (2001): The probabilistic relationship between the assignment problem and asymmetric traveling salesman problems. Proceedings of SODA, ACM Publishers, 652–660.
24. Frieze, A. and Suen, S. (1994): On the independence number of random cubic graphs. Random Structures and Algorithms, **5**, 640–664.
25. Gamarnik, D. (2002): Linear Phase Transition in Random Linear Constraint Satisfaction Problem. Technical Report: IBM T.J. Watson Research Center.
26. Grimmett, G.R. (1980): Random labelled trees and their branching networks. J. Austral. Math. Soc. (Ser. A), **30**, 229–237.
27. Füredi, Z. (1995): The expected size of a random sphere of influence graph, *Intuitive Geometry*, Bolyai Mathematical Society, **6**, 319–326.
28. Hara, T. and Slade, G. (1990): Mean-field critical behaviour for percolation in high dimensions. Commun. Math. Phys., **128**, 333–391.
29. Hara, T. and Slade, G. (1994): Mean-field behaviour and the lace expansion. Probability and Phase Transition (G. Grimmett, ed.), Kluwer, Dordrecht.
30. Harris, T.H. (1989): The Theory of Branching Processes. Dover Publications, New York.
31. Hartmann, A.K. and Weigt, M. (2001): Statistical mechanics perspective on the phase transition in vertex covering of finite-connectivity random graphs. Theoretical Computer Science, **265**, 199–225.

32. Hayen, A. and Quine, M.P. (2000): The proportion of triangles in a Poisson-Voronoi tessellation of the plane. Adv. in Appl. Probab., **32**, 67–74.
33. Hitczenko, P., Janson, S., and Yukich, J.E. (1999): On the variance of the random sphere of influence graph. Random Structures Algorithms, **14**, 139–152.
34. Henze, N. (1987): On the fraction of random points with specified nearest neighbor interrelations and degree of attraction. Adv. Appl. Prob., **19**, 873–895.
35. Hochbaum, D. and Steele, J.M. (1982): Steinhaus' geometric location problem for random samples in the plane. Advances in Applied Probability, **14**, 55–67.
36. Janson, S., Luczak, T., and Ruciński (2000): Random Graphs. Wiley Interscience Publishers, New York.
37. Karp, R.M. (1979): A patching algorithm for the non-symmetric traveling salesman problem. SIAM Journal on Computing, **8**, 561–573.
38. Karp, R.M. (1987): An upper bound on the expected cost of an optimal assignment. In: Discrete Algorithms and Complexity: Proceedings of the Japan-U. S. Joint Seminar, Academic Press, New York.
39. Karp, R.M. and Steele, J.M. (1985): Probabilistic analysis of heuristics. In The Traveling Salesman Problem: A Guided Tour of Combinatorial Optimzation. Lawler, E.L, Lenstra, J.K, Rinnooy Kan, A.H.G., and Smoys, D.B. (eds), Wiley, NY, 181–205.
40. Kesten, H. and Lee, S. (1996): The central limit theorem for weighted minimal spanning trees on random points. Annals of Applied Probability, **6**, 495–527.
41. Koebe, P. (1936): Kontaktprobleme der Konformen Abbildung. Ber. Sächs. Akad. Wiss. Leipzig, Math.-Phys. Kl. **88**, 141–164.
42. Kurtzberg, J. M. (1962): On approximation methods for the assignment problem. J. Assoc. Comput. Mach., **9**, 419–439.
43. Lee, S. (1997): The central limit theorem for Euclidean minimal spanning trees I. Annals of Applied Probability, **7**, 996–1020.
44. Lee, S. (1999): The central limit theorem for Euclidean minimal spanning trees II., Advances in Applied Probability, **31**, 969–984.
45. Lee, S. and Su, Z. (2002): On the fluctutation in the random assigmnent problem. Commun. Korean. Math. Soc., **17**, 321–330.
46. Linusson, S. and Wästlund, J. (2003): A Proof of Parisi's Conjecture on the Random Assignment Problem. Unpublished.
47. Lovász, L. and Plummer, M.D. (1986): Matching Theory. Annals of Discrete Mathematics, vol. 29. North-Holland Publishers. Amsterdam.
48. Mahmoud, H.M. (1992): Evolution of Random Search Trees, Wiley, New York, 1992.
49. Matoušek, J. and Neštřil, Y. (1998): Discrete Mathematics, Oxford Universtity Press, Oxford.
50. Meyn, S.P. and Tweedie, R.L. (1993): Markov Chains and Stochastic Stability, Springer-Verlag, New York.
51. Mézard, M. and Parisi, G. (1987): On the solution of the random link matching problem. J. Physique, **48**, 1451–1459.
52. Nair, C. and Prabhakar, B. and Sharma, M. (2003): A Proof of Parisi's Conjecture for the Finite Random Assignment Problem. Unpublished.
53. Parisi, G. (1998): A conjecture on random bipartite matching. ArXiv Condmat **13**, 277–303.
54. Penrose, M.D. (1996): The random minimal spanning tree in high dimensions. Ann of Probability, **24** 1903–1925.

55. Penrose, M.D. and Yukich, J.E. (2002): Weak laws of large numbers geometric probability, Annals of Applied Probability, **13**, 277–303.
56. Propp, J. and Wilson D. (1998): Coupling from the past: a user's guide, In D. Aldous and J. Propp, editors, Microsurveys in Discrete Probability, number 41 in DIMACS Ser. Discrete Math. Theoret. Comp. Sci., pages 181–192.
57. Steele, J.M. (1981): Subadditive Euclidean functionals and non-linear growth in geometric probability. Annals of Probability, **9**, 365–376.
58. Steele, J.M. (1982): Optimal triangulation of random samples in the plane. *Annals of Probability*, **10**, 548–553.
59. Steele, J.M. (1987): On Frieze's $\zeta(3)$ limit for the lenths of minimal spanning trees, Discrete Applied Mathematics, **18**, 99–103.
60. Steele, J.M. (1988): Growth rates of Euclidean minimal spanning trees with power weighted edges. *Annals of Probability*, **16**, 1767–1787, 1988.
61. Steele, J.M. (1992): Euclidean semi-matchings of random samples. *Mathematical Programming*, **53**, 127–146, 1992.
62. Steele, J.M. (1997): Probability Theory and Combinatorial Optimization, NSF-CBMS Volume 69. Society for Industrial and Applied Mathematics, Philadelphia.
63. Steele, J.M., Shepp, L.A. J.M. Eddy, W. (1987): On the number of leaves of a Euclidean minimal spanning tree. J. Appl. Probab., **24**, 809–826.
64. Talagrand, M. (1995): Concentration of measure and isoperimetric inequalities in product spaces. Publ. Math. IHES, **81**, 73–205.
65. Vitter, J.S. and Flajolet, P. (1990): Analysis of algorithms and data structures. In *Handbook of Theoretical Computer Science*, volume A: Algorithms and Complexity (Chapter 9), North-Holland, 431–524.
66. Walkup, D. W. (1979): On the expected value of a random assigmnent problem. SIAM J. Comput., **8**, 440–442.
67. Yukich, J.E. (1998): Probability Theory of Classical Euclidean Optimization Problems, Lecture Notes in Mathematics, **1675**, Springer-Verlag, New York.

The Random-Cluster Model

Geoffrey Grimmett

Abstract. The class of random-cluster models is a unification of a variety of stochastic processes of significance for probability and statistical physics, including percolation, Ising, and Potts models; in addition, their study has impact on the theory of certain random combinatorial structures, and of electrical networks. Much (but not all) of the physical theory of Ising/Potts models is best implemented in the context of the random-cluster representation. This systematic summary of random-cluster models includes accounts of the fundamental methods and inequalities, the uniqueness and specification of infinite-volume measures, the existence and nature of the phase transition, and the structure of the subcritical and supercritical phases. The theory for two-dimensional lattices is better developed than for three and more dimensions. There is a rich collection of open problems, including some of substantial significance for the general area of disordered systems, and these are highlighted when encountered. Amongst the major open questions, there is the problem of ascertaining the exact nature of the phase transition for general values of the cluster-weighting factor q, and the problem of proving that the critical random-cluster model in two dimensions, with $1 \leq q \leq 4$, converges when re-scaled to a stochastic Löwner evolution (SLE). Overall the emphasis is upon the random-cluster model for its own sake, rather than upon its applications to Ising and Potts systems.

1 Introduction

During a classical period, probabilists studied the behaviour of *independent* random variables. The emergent theory is rich, and is linked through theory and application to areas of pure/applied mathematics and to other sciences. It is however unable to answer important questions from a variety of sources concerning large families of *dependent* random variables. Dependence comes in many forms, and one of the targets of modern probability theory has been to derive robust techniques for studying it. The voice of statistical physics has been especially loud in the call for rigour in this general area. In a typical scenario, we are provided with an infinity of random variables, indexed by the vertices of some graph such as the cubic lattice, and which have some dependence structure governed by the geometry of the graph. Thus mathematicians and physicists have had further cause to relate probability and geometry. One major outcome of the synthesis of ideas from physics and probability is the theory of Gibbs states, [59], which is now established as a significant branch of probability theory.

A classic example of a Gibbs state is the (Lenz–)Ising model [89] for a ferromagnet. When formulated on the bounded region B of the square lattice $\mathbb{L}^2$, a random variable σ_x taking values -1 and $+1$ is assigned to each vertex x of B, and the probability of the configuration σ is proportional to $\exp(-\beta H(\sigma))$, where $\beta > 0$ and the 'energy' $H(\sigma)$ is the negative of the sum of $\sigma_x \sigma_y$ over all neighbouring pairs x, y of B. This 'starter model' has proved extraordinarily successful in generating beautiful and relevant mathematics, and has been useful and provocative in the mathematical theory of phase transitions and cooperative phenomena (see, for example, [50]).

There are many possible generalisations of the Ising model in which the σ_x may take a general number q of values, rather than $q = 2$ only. One such generalisation, the so-called Potts model [126], has attracted especial interest amongst physicists, and has displayed a complex and varied structure; for example, when q is large, it enjoys a discontinuous phase transition, in contrast to the continuous transition believed to take place for small q. Ising/Potts models are the first of three principal ingredients in the story of random-cluster models. Note that they are 'vertex models' in the sense that they involve random variables σ_x indexed by the vertices x of the underlying graph.

The '(bond) percolation model' was inspired by problems of physical type, and emerged from the mathematics literature of the 1950s [29, 150]. In this model for a porous medium, each edge of a graph is declared 'open' (to the passage of fluid) with probability p, and 'closed' otherwise, different edges having independent states. The problem is to determine the typical large-scale properties of connected components of open edges, as the parameter p varies. Percolation theory is now a mature part of probability, at the core of the study of random media and interacting systems, and it is the second ingredient in the story of random-cluster models. Note that bond percolation is an 'edge model', in that the random variables are indexed by the set of edges of the underlying graph. [There is a variant termed 'site percolation' in which the vertices are open/closed at random rather than the edges.]

The third and final ingredient preceded the first two, and is the theory of electrical networks. Dating back at least to the 1847 paper [102] of Kirchhoff, this sets down a method for calculating macroscopic properties of an electrical network in terms of its local structure. In particular, it explains the relevance of counts of certain types of spanning trees of the graph. In the modern vernacular, an electrical network on a graph G may be studied via the properties of a 'uniformly random spanning tree' on G (see [16]).

These three ingredients seemed fairly distinct until Fortuin and Kasteleyn discovered, around 1970, [53, 54, 55, 94], that each features in a certain way within a family of probability measures of 'edge models', parameterised by two quantities, $p \in [0, 1]$ and $q \in (0, \infty)$. [In actuality, electrical networks arise as a weak limit of such measures.] These models they termed 'random-cluster models', and they developed the basic theory – correlation inequalities and the like – in a series of papers published thereafter. The true power of random-cluster models as a mechanism for studying Ising/Potts models has emerged only gradually over the intervening thirty years.

We note in passing that the genesis of the random-cluster model lay in Kasteleyn's observation that each of the three ingredients above satisfies certain series/parallel laws: any two edges in series (or parallel) may be replaced by a single edge in such a way that, if the interaction function is adapted accordingly, then the distributions of large-scale properties remain unchanged.

The family of random-cluster measures (that is, probability measures which govern random-cluster models) is not an extension of the Potts measures. The relationship is more sophisticated, and is such that *correlations* for Potts models correspond to *connections* in random-cluster models. Thus the *correlation structure* of a Potts model may be studied via the *stochastic geometry* of a corresponding random-cluster model. The intuition behind this geometrical study comes often from percolation, of which the random-cluster model is indeed an extension.

It turns out that, in many situations involving ferromagnetic Ising/Potts models, the best way forward is via the random-cluster model. As examples of this we mention the existence of discontinuous phase transitions [105], exact computations in two dimensions [14], the verification of the Wulff construction for Ising droplets [36], and the Dobrushin theory of interfaces and non-translation-invariant measures [62]. As a major exception to the mantra 'everything worth doing for Ising/Potts is done best via random-cluster', we remind the reader of the so-called random-current expansion for the Ising model, wielded with effect in [1, 2, 5] and elsewhere. The random-current method appears to be Ising-specific, and has enabled a deep analysis of the Ising model unparalleled in more general Potts systems. (See Section 5.3)

The primary target of this review is to summarise and promote the theory of random-cluster models for its own sake. In doing so, we encounter many results having direct impact on Ising/Potts systems, but we shall not stress such connections. Some of the theory has been discovered several times by apparently independent teams; whilst making a serious attempt to list key references, we apologise for unwitting omissions of which there will certainly be a few. The large number of references to work of the author is attributable in part to the fact that he is acquainted with these contributions.

It is a lesson in humility to return to the original Fortuin–Kasteleyn papers [53, 55], and especially [54], where so much of the basic theory was first presented. These authors may not have followed the slickest of routes, but they understood rather well the object of their study. Amongst the many papers of general significance since, we highlight: [4], which brought the topic back to the fore; [47], where the coupling between Potts and random-cluster models was so beautifully managed; [69], where the random-cluster model was studied systematically on infinite grids; [81], which links the theory to several other problems of interest in statistical mechanics; and [61], where random-cluster models are placed in the perspective of stochastic geometry as a tool for studying phase transitions.

This review is restricted mostly to core material for random-cluster models on the nearest-neighbour cubic lattice in a general number d of dimensions. Only in passing do we mention such subjects as extensions to long-range systems [4], mean-field behaviour in high dimensions [99], and mixing properties [7]. Neither do we stress the impact that graphical methods of the random-cluster type have had on a variety of other disordered systems, such as the Ashkin–Teller model [12, 81, 122, 132, 148], the Widom–Rowlinson model [39, 40, 60, 81, 146], or on methods for simulating disordered physical systems [41, 42, 141, 149].

Random-cluster methods may be adapted to systems with random interactions [3, 72, 81], and even to non-ferromagnetic systems of Edwards–Anderson spin-glass type [46, 117, 118] where, for example, they have been used to prove that, for a given set $\{J_e\}$ of positive or negative interactions, uniqueness of the infinite-volume Gibbs measure for the ferromagnetic system having interactions $\{|J_e|\}$ implies uniqueness for the original system.

Amongst earlier papers on random-cluster models, the following include a degree of review material: [4, 23, 61, 66, 81, 113].

Notwithstanding the fairly mature theory which has evolved, there remain many open problems including some of substantial significance for the general area. Many of these are marked in the text with the acronym **OP**.

2 Potts and random-cluster processes

We write $\mu(f)$ for the expectation of a random variable f under a probability measure μ.

2.1 Random-cluster measures

Let $G = (V, E)$ be a finite graph. An edge e having endvertices x and y is written as $e = \langle x, y \rangle$. A random-cluster measure on G is a member of a certain class of probability measures on the set of subsets of the edge set E. We take as state space the set $\Omega = \{0, 1\}^E$, members of which are vectors $\omega = (\omega(e) : e \in E)$. We speak of the edge e as being *open* (in ω) if $\omega(e) = 1$, and as being *closed* if $\omega(e) = 0$. For $\omega \in \Omega$, let $\eta(\omega) = \{e \in E : \omega(e) = 1\}$ denote the set of open edges, and let $k(\omega)$ be the number of connected components (or 'open clusters') of the graph $(V, \eta(\omega))$. Note that $k(\omega)$ includes a count of isolated vertices, that is, of vertices incident to no open edge. We assign to Ω the σ-field $\mathcal{F}$ of all its subsets.

A *random-cluster measure* on G has two parameters satisfying $0 \leq p \leq 1$ and $q > 0$, and is the measure $\phi_{p,q}$ on the measurable pair $(\Omega, \mathcal{F})$ given by

$$\phi_{p,q}(\omega) = \frac{1}{Z} \left\{ \prod_{e \in E} p^{\omega(e)}(1-p)^{1-\omega(e)} \right\} q^{k(\omega)}, \qquad \omega \in \Omega,$$

where the 'partition function', or 'normalising constant', Z is given by

$$Z = \sum_{\omega \in \Omega} \left\{ \prod_{e \in E} p^{\omega(e)} (1-p)^{1-\omega(e)} \right\} q^{k(\omega)} .$$

This measure differs from product measure only through the inclusion of the term $q^{k(\omega)}$. Note the difference between the cases $q \le 1$ and $q \ge 1$: the former favours fewer clusters, whereas the latter favours many clusters. When $q = 1$, edges are open/closed independently of one another. This very special case has been studied in detail under the titles *percolation* and *random graphs*; see [25, 71, 90]. Perhaps the most important values of q are the integers, since the random-cluster model with $q \in \{2, 3, \dots\}$ corresponds, in a way sketched in the next two sections, to the Potts model with q local states. The bulk of this review is devoted to the theory of random-cluster measures when $q \ge 1$. The case $q < 1$ seems to be harder mathematically and less important physically. There is some interest in the limit as $q \downarrow 0$; see Sections 2.4 and 3.6.

We shall sometimes write $\phi_{G,p,q}$ for $\phi_{p,q}$, when the choice of graph G is to be stressed. Samples from random-cluster measures on $\mathbb{Z}^2$ are presented in Fig. 2.1.

2.2 Ising and Potts models

In a famous experiment, a piece of iron is exposed to a magnetic field. The field is increased from zero to a maximum, and then diminishes to zero. If the temperature is sufficiently low, the iron retains some 'residual magnetisation', otherwise it does not. There is a critical temperature for this phenomenon, often called the *Curie point*. The famous (Lenz–)Ising model for such ferromagnetism, [89], may be summarised as follows. One supposes that particles are positioned at the points of some lattice in Euclidean space. Each particle may be in either of two states, representing the physical states of 'spin up' and 'spin down'. Spin-values are chosen at random according to a certain probability measure, known as a *Gibbs state*, which is governed by interactions between neighbouring particles. This measure may be described as follows.

Let $G = (V, E)$ be a finite graph. We think of each vertex $v \in V$ as being occupied by a particle having a random spin. Since spins are assumed to come in two basic types, we take as sample space the set $\Sigma = \{-1, +1\}^V$. The appropriate probability mass function $\lambda_{\beta, J, h}$ on Σ has three parameters satisfying $0 \le \beta, J < \infty$ and $h \in \mathbb{R}$, and is given by

$$\lambda_{\beta, J, h}(\sigma) = \frac{1}{Z_{\mathrm{I}}} e^{-\beta H(\sigma)} , \qquad \sigma \in \Sigma ,$$

where the partition function Z_{I} and the 'Hamiltonian' $H : \Sigma \to \mathbb{R}$ are given by

$$Z_{\mathrm{I}} = \sum_{\sigma \in \Sigma} e^{-\beta H(\sigma)} , \qquad H(\sigma) = - \sum_{e = \langle x, y \rangle \in E} J \sigma_x \sigma_y - h \sum_{x \in V} \sigma_x .$$

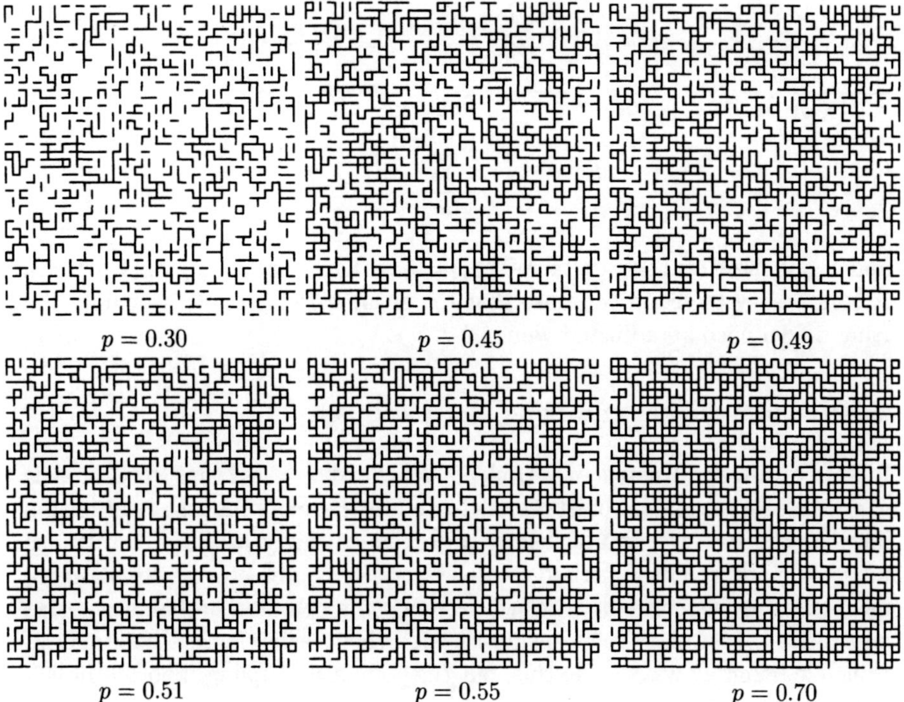

Fig. 2.1. Samples from the random-cluster measure with $q = 1$ on the box $[0, 40]^2$ of the square lattice. We have set $q = 1$ for ease of programming, the measure being of product form in this case. The critical value is $p_c(1) = \frac{1}{2}$. Samples with more general values of q may be obtained by the method of 'coupling from the past', as described in Section 8.2.

The physical interpretation of β is as the reciprocal $1/T$ of temperature, of J as the strength of interaction between neighbours, and of h as the external field. For reasons of simplicity, we shall consider only the case of zero external field, and we assume henceforth that $h = 0$. Each edge has equal interaction strength J in the above formulation. Since β and J occur only as a product βJ, the measure $\lambda_{\beta, J, 0}$ has effectively only a single parameter. In a more complicated measure not studied here, different edges e are permitted to have different interaction strengths J_e.

As pointed out by Baxter, [14], the Ising model permits an infinity of generalisations. Of these, the extension to so-called 'Potts models' has proved especially fruitful. Whereas the Ising model permits only two possible spin-values at each vertex, the Potts model [126] permits a general number $q \in \{2, 3, \ldots\}$, and is given as follows.

Let q be an integer satisfying $q \geq 2$, and take as sample space $\Sigma = \{1, 2, \ldots, q\}^V$. Thus each vertex of G may be in any of q states. The relevant probability measure is now given by

$$\pi_{\beta,J,q}(\sigma) = \frac{1}{Z_{\mathrm{P}}} e^{-\beta H'(\sigma)}, \qquad \sigma \in \Sigma,$$

where Z_{P} is the appropriate normalising constant,

$$H'(\sigma) = - \sum_{e=\langle x,y \rangle} J\delta_{\sigma_x,\sigma_y},$$

and $\delta_{u,v}$ is the Kronecker delta. When $q=2$, we have that $\delta_{\sigma_x,\sigma_y} = \frac{1}{2}(1+\sigma_x\sigma_y)$. It is now easy to see in this case that the ensuing Potts model is simply the Ising model with an adjusted value of J.

2.3 Random-cluster and Ising–Potts coupled

It was Fortuin and Kasteleyn [53, 54, 55, 94] who discovered that Potts models may be recast as random-cluster models, and furthermore that the relationship between the two systems facilitates an extended study of phase transitions in Potts models. Their methods were essentially combinatorial. In the more modern game, we construct the two systems on a common probability space, and then observe their relationship through their realisations. There may in principle be many ways to do this, but the standard coupling reported in [47] is of special value.

Let $q \in \{2, 3, \ldots\}$, $0 \le p \le 1$, and let $G = (V, E)$ be a finite graph, as before. We consider the product sample space $\Sigma \times \Omega$ where $\Sigma = \{1, 2, \ldots, q\}^V$ and $\Omega = \{0, 1\}^E$ as above. We now define a probability mass function μ on $\Sigma \times \Omega$ by

$$\mu(\sigma,\omega) \propto \prod_{e \in E} \left\{ (1-p)\delta_{\omega(e),0} + p\delta_{\omega(e),1}\delta_e(\sigma) \right\}, \qquad (\sigma,\omega) \in \Sigma \times \Omega,$$

where $\delta_e(\sigma) = \delta_{\sigma_x,\sigma_y}$ for $e = \langle x, y \rangle \in E$. Elementary calculations reveal the following facts.

(a) *Marginal on Σ.* The marginal measure $\mu_1(\sigma) = \sum_{\omega \in \Omega} \mu(\sigma,\omega)$ is given by

$$\mu_1(\sigma) \propto \exp\left\{ \beta \sum_e J\delta_e(\sigma) \right\}$$

where $p = 1 - e^{-\beta J}$. This is the Potts measure.

(b) *Marginal on Ω.* The second marginal of μ is

$$\mu_2(\omega) = \sum_{\sigma \in \Sigma} \mu(\sigma,\omega) \propto \left\{ \prod_e p^{\omega(e)}(1-p)^{1-\omega(e)} \right\} q^{k(\omega)}.$$

This is the random-cluster measure.

(c) *The conditional measures.* Given ω, the conditional measure on Σ is obtained by putting (uniformly) random spins on entire clusters of ω (of which there are $k(\omega)$). These spins are constant on given clusters, and are independent between clusters. Given σ, the conditional measure on Ω is obtained by setting $\omega(e) = 0$ if $\delta_e(\sigma) = 0$, and otherwise $\omega(e) = 1$ with probability p (independently of other edges).

In conclusion, the measure μ is a coupling of a Potts measure $\pi_{\beta,J,q}$ on V, together with the random-cluster measure $\phi_{p,q}$ on Ω. The parameters of these measures are related by the equation $p = 1 - e^{-\beta J}$. Since $0 \le p \le 1$, this is possible only if $\beta J \ge 0$.

This special coupling may be used in a particularly simple way to show that correlations in Potts models correspond to open connections in random-cluster models. When extended to infinite graphs, this implies as discussed in Section 4 that the phase transition of a Potts model corresponds to the creation of an infinite open cluster in the random-cluster model. Thus arguments of stochastic geometry, and particularly those developed for the percolation model, may be harnessed directly in order to understand the correlation structure of the Potts system. The basic step is as follows.

We write $\{x \leftrightarrow y\}$ for the set of all $\omega \in \Omega$ for which there exists an open path joining vertex x to vertex y. The complement of the event $\{x \leftrightarrow y\}$ is denoted $\{x \nleftrightarrow y\}$.

The 'two-point correlation function' of the Potts measure $\pi_{\beta,J,q}$ on the finite graph $G = (V, E)$ is defined to be the function $\tau_{\beta,J,q}$ given by

$$\tau_{\beta,J,q}(x, y) = \pi_{\beta,J,q}(\sigma_x = \sigma_y) - \frac{1}{q}, \qquad x, y \in V.$$

The term q^{-1} is the probability that two independent and uniformly distributed spins are equal. The 'two-point connectivity function' of the random-cluster measure $\phi_{p,q}$ is defined as the function $\phi_{p,q}(x \leftrightarrow y)$ for $x, y \in V$, that is, the probability that x and y are joined by a path of open edges. It turns out that these 'two-point functions' are (except for a constant factor) the same.

Theorem 2.1 (Correlation/connection [94]). *If $q \in \{2, 3, \ldots\}$ and $p = 1 - e^{-\beta J}$ satisfies $0 \le p \le 1$, then*

$$\tau_{\beta,J,q}(x, y) = (1 - q^{-1})\phi_{p,q}(x \leftrightarrow y), \qquad x, y \in V.$$

Proof. The indicator function of an event A is denoted 1_A. We have that

$$\tau_{\beta,J,q}(x, y) = \sum_{\sigma,\omega} \left\{ 1_{\{\sigma_x = \sigma_y\}}(\sigma) - q^{-1} \right\} \mu(\sigma, \omega)$$

$$= \sum_{\omega} \phi_{p,q}(\omega) \sum_{\sigma} \mu(\sigma \mid \omega) \left\{ 1_{\{\sigma_x = \sigma_y\}}(\sigma) - q^{-1} \right\}$$

$$= \sum_{\omega} \phi_{p,q}(\omega) \left\{ (1 - q^{-1}) 1_{\{x \leftrightarrow y\}}(\omega) + 0 \cdot 1_{\{x \nleftrightarrow y\}}(\omega) \right\}$$

$$= (1 - q^{-1})\phi_{p,q}(x \leftrightarrow y),$$

where μ is the above coupling of the Potts and random-cluster measures. $\quad\square$

The theorem may be generalised as follows. Suppose we are studying the Potts model, and are interested in some 'observable' $f : \Sigma \to \mathbb{R}$. The mean value of $f(\sigma)$ satisfies

$$\pi_{\beta,J,q}(f) = \sum_{\sigma} f(\sigma)\pi_{\beta,J,q}(\sigma) = \sum_{\sigma,\omega} f(\sigma)\mu(\sigma,\omega)$$
$$= \sum_{\omega} F(\omega)\phi_{p,q}(\omega) = \phi_{p,q}(F)$$

where $F : \Omega \to \mathbb{R}$ is given by

$$F(\omega) = \mu(f \mid \omega) = \sum_{\sigma} f(\sigma)\mu(\sigma \mid \omega).$$

The above theorem is obtained in the case $f(\sigma) = \delta_{\sigma_x,\sigma_y} - q^{-1}$, where $x, y \in V$.

The Potts models considered above have zero external field. Some complications arise when an external field is added; see the discussions in [8, 23].

2.4 The limit as $q \downarrow 0$

Let $G = (V, E)$ be a finite connected graph, and let $\phi_{p,q}$ be the random-cluster measure on the associated sample space $\Omega = \{0,1\}^E$. We consider first the weak limit of $\phi_{p,q}$ as $q \downarrow 0$ for fixed $p \in (0,1)$. This limit may be ascertained by observing that the dominant terms in the partition function

$$Z(p,q) = \sum_{\omega \in \Omega} p^{|\eta(\omega)|}(1 - p)^{|E \setminus \eta(\omega)|} q^{k(\omega)}$$

are those for which $k(\omega)$ is a minimum, that is, those with $k(\omega) = 1$. It follows that $\lim_{q\downarrow 0} \phi_{p,q}$ is precisely the product measure $\phi_{p,1}$ (that is, percolation with intensity p) conditioned on the resulting graph $(V, \eta(\omega))$ being connected. A more interesting limit arises if we allow p to converge to 0 with q, as follows.

The random-cluster model originated in a systematic study by Fortuin and Kasteleyn of systems of a certain type which satisfy certain parallel and series laws. Electrical networks are the best known such systems – two parallel (respectively, series) connections of resistances r_1 and r_2 may be replaced by a single connection with resistance $(r_1^{-1} + r_2^{-1})^{-1}$ (respectively, $r_1 + r_2$). Fortuin and Kasteleyn [55] realised that the electrical-network theory of a graph G is related to the limit as $q \downarrow 0$ of the random-cluster model on G. Their argument may be expanded as follows.

Suppose $p = p_q$ is related to q in such a way that $p \to 0$ and $q/p \to 0$ as $q \to 0$. We may write $Z(p,q)$ as

$$Z(p,q) = (1 - p)^{|E|} \sum_{\omega \in \Omega} \left(\frac{p}{1-p}\right)^{|\eta(\omega)|+k(\omega)} \left(\frac{q(1-p)}{p}\right)^{k(\omega)}.$$

Note that $p/(1-p) \to 0$ and $q(1-p)/p \to 0$ as $q \to 0$. Now $k(\omega) \geq 1$ and $|\eta(\omega)|+k(\omega) \geq |V|$ for all $\omega \in \Omega$; these two inequalities are satisfied simultaneously with equality if and only if $\eta(\omega)$ is a spanning tree of G. It follows that, in the limit as $q \to 0$, the 'mass' is concentrated on such configurations, and it is easily seen that the limit mass is uniformly distributed. That is, $\lim_{q \downarrow 0} \phi_{p,q}$ is a probability measure which selects, uniformly at random, a spanning tree of G; in other words, the limit measure is $\phi_{\frac{1}{2},1}$ conditioned on the resulting graph being a spanning tree.

The link to the theory of electrical networks is now provided by Kirchhoff's theorem [102], which expresses effective resistances in terms of counts of spanning trees. See also [79].

The theory of random spanning trees is beautiful in its own right (see [16]), and is linked in an important way to the emerging field of stochastic growth processes of 'stochastic Löwner evolution' (SLE) type (see [111, 130]), to which we return in Section 6.4. Another limit emerges if $p = q$ and $q \downarrow 0$, namely uniform measure on the set of forests of G. More generally, take $p = \alpha q$ where $\alpha \in (0, \infty)$ is constant, and take the limit as $q \downarrow 0$. The limit measure is the percolation measure $\phi_{\beta,1}$ conditioned on the non-existence of open circuits, where $\beta = \alpha/(1+\alpha)$. If $p/q \to 0$ as $p, q \to 0$, the limit measure is concentrated on the empty set of edges.

2.5 Rank-generating functions

The partition functions of Potts and random-cluster measures are particular evaluations of rank-generating functions, defined as follows. The *rank-generating function* of the simple graph $G = (V, E)$ is the function

$$W_G(u,v) = \sum_{E' \subseteq E} u^{r(G')} v^{c(G')}, \qquad u, v \in \mathbb{R},$$

where $r(G') = |V| - k(G')$ is the *rank* of the graph $G' = (V, E')$, and $c(G') = |E'| - |V| + k(G')$ is its *co-rank*; here, $k(G')$ denotes the number of components of the graph G'. The rank-generating function has various useful properties, and occurs in several contexts in graph theory; see [20, 142]. It crops up in other forms also. For example, the function

$$T_G(u,v) = (u-1)^{|V|-1} W_G\big((u-1)^{-1}, v-1\big)$$

is known as the *dichromatic* (or *Tutte*) *polynomial*, [142]. The partition function $Z = Z_G$ of the random-cluster measure on G with parameters p, q is easily seen to satisfy

$$Z_G = q^{|V|}(1-p)^{|E|} W_G\left(\frac{p}{q(1-p)}, \frac{p}{1-p}\right),$$

a relationship which provides a link with other classical quantities associated with a graph. See [20, 21, 53, 144] also.

3 Infinite-volume random-cluster measures

It is in the infinite-volume limit that random-cluster measures exhibit phase transitions. There are two ways of constructing random-cluster measures on infinite graphs, namely by taking weak limits as a finite domain approaches the infinite system, and by studying measures on the infinite graph having the 'correct' conditional versions. Such matters are discussed in this section, which begins with a summary of certain valuable properties of random-cluster measures on finite graphs.

3.1 Stochastic ordering

The stochastic ordering of measures provides a technique fundamental to the study of random-cluster measures. Let $G = (V, E)$ be a finite or countably infinite graph as above; let $\Omega = \{0, 1\}^E$, and let $\mathcal{F}$ be the σ-field of Ω generated by the finite-dimensional cylinders. Note first that Ω is a partially ordered set with partial order $\omega_1 \leq \omega_2$ if $\omega_1(e) \leq \omega_2(e)$ for all e. A random variable $f : \Omega \to \mathbb{R}$ is called *increasing* if $f(\omega_1) \leq f(\omega_2)$ whenever $\omega_1 \leq \omega_2$. An event $A \in \mathcal{F}$ is called *increasing* if its indicator function 1_A is increasing. The word 'decreasing' should be interpreted in the natural way. Given two probability measures μ_1, μ_2 on Ω, we write $\mu_1 \leq_{\mathrm{st}} \mu_2$, and say that μ_1 is stochastically smaller than μ_2, if $\mu_1(f) \leq \mu_2(f)$ for all bounded increasing random variables f on Ω.

We return now to the case when G is a *finite* graph. Let μ_1, μ_2 be probability measures on Ω, and assume for the moment that the μ_i are strictly positive in the sense that $\mu_i(\omega) > 0$ for all $\omega \in \Omega$. An important sufficient condition for the inequality $\mu_1 \leq_{\mathrm{st}} \mu_2$ was found by Holley [88], namely that

$$\mu_2(\omega_1 \vee \omega_2)\mu_1(\omega_1 \wedge \omega_2) \geq \mu_1(\omega_1)\mu_2(\omega_2) \qquad \text{for all } \omega_1, \omega_2 \in \Omega,$$

where $\omega_1 \vee \omega_2$ and $\omega_1 \wedge \omega_2$ are the maximum and minimum configurations given respectively as $\max\{\omega_1(e), \omega_2(e)\}$ and $\min\{\omega_1(e), \omega_2(e)\}$, for $e \in E$. A probability measure μ on Ω is said to have the *FKG lattice property* if

$$\mu(\omega_1 \vee \omega_2)\mu(\omega_1 \wedge \omega_2) \geq \mu(\omega_1)\mu(\omega_2) \qquad \text{for all } \omega_1, \omega_2 \in \Omega,$$

and it is a consequence of Holley's argument that any strictly positive measure with the FKG lattice property satisfies the so-called FKG inequality. This amounts to the following for random-cluster measures.

Theorem 3.1 (FKG inequality [54, 56]). *Suppose that* $0 \leq p \leq 1$ *and* $q \geq 1$. *If* f *and* g *are increasing functions on* Ω, *then* $\phi_{p,q}(fg) \geq \phi_{p,q}(f)\phi_{p,q}(g)$.

Specialising to indicator functions, we obtain that

$$\phi_{p,q}(A \cap B) \geq \phi_{p,q}(A)\phi_{p,q}(B) \qquad \text{for increasing events } A, B,$$

whenever $q \geq 1$. It is not difficult to see that the FKG inequality does not generally hold when $0 < q < 1$.

Holley's theorem leads easily to the following comparison inequalities, which were first proved by Fortuin.

Theorem 3.2 (Comparison inequalities [54]). *It is the case that*

$$\phi_{p',q'} \leq \phi_{p,q} \qquad if \quad q' \geq q,\, q' \geq 1,\, and\, p' \leq p,$$

$$\phi_{p',q'} \geq \phi_{p,q} \qquad if \quad q' \geq q,\, q' \geq 1,\, and\, \frac{p'}{q'(1-p')} \geq \frac{p}{q(1-p)}\,.$$

3.2 A differential formula

One way of estimating the probability of an event A is via an estimate of its derivative $d\phi_{p,q}(A)/dp$. When $q = 1$, there is a formula for this derivative which has proved very useful, and which is commonly attributed to Russo, see [13, 71, 131]. This formula may be generalised to random-cluster measures as follows. The proof is an exercise in the differentiation of summations.

Theorem 3.3 ([19]). *Let $0 < p < 1$, $q > 0$, and let $\phi_{p,q}$ be the corresponding random-cluster measure on a finite graph $G = (V, E)$. Then*

$$\frac{d}{dp}\phi_{p,q}(A) = \frac{1}{p(1-p)}\big\{\phi_{p,q}(|\eta|1_A) - \phi_{p,q}(|\eta|)\phi_{p,q}(A)\big\}$$

for any event A, where $|\eta| = |\eta(\omega)| = \sum_{e\in E}\omega(e)$ is the number of open edges of the configuration ω.

3.3 Conditional probabilities

Whether or not an edge e is open depends on the configuration on $E \setminus \{e\}$, and a further important property of random-cluster measures summarises the nature of this dependence.

For $e \in E$, we denote by $G \setminus e$ (respectively, $G.e$) the graph obtained from G by deleting (respectively, contracting) e. We write $\Omega_e = \{0,1\}^{E\setminus\{e\}}$; for $\omega \in \Omega$ we define $\omega_e \in \Omega_e$ by $\omega_e(f) = \omega(f)$ for $f \neq e$. For $e = \langle x, y\rangle$, we write K_e for the event that x and y are joined by an open path not using e.

Theorem 3.4 ([54]). *Let $e \in E$. We have that*

$$\phi_{G,p,q}\big(\omega \,\big|\, \omega(e) = j\big) = \begin{cases} \phi_{G\setminus e,p,q}(\omega_e) & if\ j = 0,\\[2mm] \phi_{G.e,p,q}(\omega_e) & if\ j = 1, \end{cases}$$

and

$$\phi_{G,p,q}\big(\omega(e) = 1 \,\big|\, \omega_e\big) = \begin{cases} p & if\ \omega_e \in K_e,\\[2mm] \dfrac{p}{p + (1-p)q} & if\ \omega_e \notin K_e. \end{cases}$$

That is to say, the effect of conditioning on the absence or presence of an edge e is to replace the measure $\phi_{G,p,q}$ by the random-cluster measure on the respective graph $G \setminus e$ or $G.e$. Secondly, the conditional probability that e is open, given the configuration elsewhere, depends only on whether or not K_e occurs, and is then given by the stated formula. The proof is elementary. The final equation of the theorem leads to properties of random-cluster measures referred to elsewhere as 'insertion tolerance' and the 'finite-energy property'.

3.4 Infinite-volume weak limits

In studying random-cluster measures on infinite graphs, we restrict ourselves to the case of the hypercubic lattice in d dimensions, where $d \geq 2$; similar observations are valid in greater generality. Let $d \geq 2$, and let $\mathbb{Z}^d$ be the set of all d-vectors of integers; for $x \in \mathbb{Z}^d$, we normally write $x = (x_1, x_2, \ldots, x_d)$. For $x, y \in \mathbb{Z}^d$, let

$$\|x - y\| = \sum_{i=1}^{d} |x_i - y_i| \, .$$

We place an edge $\langle x, y \rangle$ between x and y if and only if $\|x - y\| = 1$; the set of such edges is denoted by $\mathbb{E}^d$, and we write $\mathbb{L}^d = (\mathbb{Z}^d, \mathbb{E}^d)$ for the ensuing lattice. For any subset S of $\mathbb{Z}^d$, we write ∂S for its boundary, that is,

$$\partial S = \{ s \in S : \langle s, t \rangle \in \mathbb{E}^d \text{ for some } t \notin S \} \, .$$

Let $\Omega = \{0, 1\}^{\mathbb{E}^d}$, and let $\mathcal{F}$ be the σ-field of subsets of Ω generated by the finite-dimensional cylinders. The letter Λ is used to denote a finite box of $\mathbb{Z}^d$, which is to say that $\Lambda = \prod_{i=1}^{d} [x_i, y_i]$ for some $x, y \in \mathbb{Z}^d$; we interpret $[x_i, y_i]$ as the set $\{x_i, x_i + 1, x_i + 2, \ldots, y_i\}$. The set Λ generates a subgraph of $\mathbb{L}^d$ having vertex set Λ and edge set $\mathbb{E}_\Lambda$ containing all $\langle x, y \rangle$ with $x, y \in \Lambda$.

We are interested in the 'thermodynamic limit' (as $\Lambda \uparrow \mathbb{Z}^d$) of the random-cluster measure on the finite box Λ. In order to describe such weak limits, we shall need to introduce the notion of a 'boundary condition'.

For $\xi \in \Omega$, we write Ω_Λ^ξ for the (finite) subset of Ω containing all configurations ω satisfying $\omega(e) = \xi(e)$ for $e \in \mathbb{E}^d \setminus \mathbb{E}_\Lambda$; these are the configurations which 'agree with ξ off Λ'. For $\xi \in \Omega$ and values of p, q satisfying $0 \leq p \leq 1$, $q > 0$, we define $\phi_{\Lambda,p,q}^\xi$ to be the random-cluster measure on the finite graph $(\Lambda, \mathbb{E}_\Lambda)$ 'with boundary condition ξ'; this is the equivalent of a 'specification' for Gibbs states. More precisely, let $\phi_{\Lambda,p,q}^\xi$ be the probability measure on the pair $(\Omega, \mathcal{F})$ given by

$$\phi_{\Lambda,p,q}^\xi(\omega) = \begin{cases} \dfrac{1}{Z_{\Lambda,p,q}^\xi} \left\{ \displaystyle\prod_{e \in \mathbb{E}_\Lambda} p^{\omega(e)} (1-p)^{1-\omega(e)} \right\} q^{k(\omega, \Lambda)} & \text{if } \omega \in \Omega_\Lambda^\xi, \\ 0 & \text{otherwise,} \end{cases}$$

where $k(\omega, \Lambda)$ is the number of components of the graph $(\mathbb{Z}^d, \eta(\omega))$ which intersect Λ, and where $Z_{\Lambda,p,q}^\xi$ is the appropriate normalising constant

$$Z^\xi_{\Lambda,p,q} = \sum_{\omega \in \Omega^\xi_\Lambda} \left\{ \prod_{e \in \mathbb{E}_\Lambda} p^{\omega(e)}(1-p)^{1-\omega(e)} \right\} q^{k(\omega,\Lambda)} .$$

Note that $\phi^\xi_{\Lambda,p,q}(\Omega^\xi_\Lambda) = 1$.

Definition 3.5. *Let $0 \le p \le 1$ and $q > 0$. A probability measure ϕ on $(\Omega, \mathcal{F})$ is called a* limit random-cluster measure *with parameters p and q if there exist $\xi \in \Omega$ and a sequence $\Lambda = (\Lambda_n : n \ge 1)$ of boxes satisfying $\Lambda_n \to \mathbb{Z}^d$ as $n \to \infty$ such that*

$$\phi^\xi_{\Lambda_n,p,q} \Rightarrow \phi \qquad as\ n \to \infty .$$

The set of all such measures ϕ is denoted by $\mathcal{W}_{p,q}$, and the closed convex hull of $\mathcal{W}_{p,q}$ is denoted $\overline{\mathrm{co}\,\mathcal{W}_{p,q}}$.

In writing $\Lambda_n \to \mathbb{Z}^d$ we mean that, for all m, $\Lambda_n \supseteq [-m, m]^d$ for all large n. The arrow '$\Rightarrow$' denotes weak convergence.

It might seem reasonable to define a limit random-cluster measure to be any weak limit of the form $\lim_{n\to\infty} \phi^{\xi_n}_{\Lambda_n,p,q}$ for some sequence $(\xi_n : n \ge 1)$ of members of Ω and some sequence $\Lambda = (\Lambda_n : n \ge 1)$ of boxes satisfying $\Lambda_n \to \mathbb{Z}^d$. It may however be shown that this adds no extra generality to the class as defined above, [69]. The dependence of the limit measure ϕ on the choice of sequence (Λ_n) can be subtle, especially when $q < 1$ (**OP**).

It is standard that $\mathcal{W}_{p,q} \neq \varnothing$ for all $0 \le p \le 1, q > 0$, and one way of seeing this is as follows. The sample space Ω is the product of discrete spaces, and is therefore compact. It follows that any class of probability measures on Ω is tight, and hence relatively compact (see the account of Prohorov's theorem in [22]), which is to say that any infinite sequence of probability measures contains a weakly convergent subsequence.

When does the limit $\lim_{n\to\infty} \phi^\xi_{\Lambda_n,p,q}$ exist, and when does it depend on the choice of boundary condition ξ? The FKG inequality provides a route to a partial answer to this important question. Suppose for the moment that $q \ge 1$. Two extremal boundary conditions of special importance are provided by the configurations 0 and 1, comprising 'all edges closed' and 'all edges open' respectively. One speaks of configurations in Ω^0_Λ as having 'free' boundary conditions, and configurations in Ω^1_Λ as having 'wired' boundary conditions.

Theorem 3.6 (Thermodynamic limit [4, 27, 54, 66, 69]). *Suppose $0 \le p \le 1$ and $q \ge 1$.*

(a) *Let $\Lambda = (\Lambda_n : n \ge 1)$ be a sequence of boxes satisfying $\Lambda_n \to \mathbb{Z}^d$ as $n \to \infty$. The weak limits*

$$\phi^b_{p,q} = \lim_{n\to\infty} \phi^b_{\Lambda_n,p,q}, \qquad for\ b = 0, 1,$$

exist and are independent of the choice of Λ.

(b) We have that each $\phi_{p,q}^b$ is translation-invariant, and

$$\phi_{p,q}^0 \leq_{\text{st}} \phi \leq_{\text{st}} \phi_{p,q}^1 \qquad \text{for all } \phi \in \mathcal{W}_{p,q}.$$

(c) For $b = 0,1$, the measure $\phi_{p,q}^b$ is ergodic, in that any translation-invariant random variable is $\phi_{p,q}^b$-a.s. constant.

The FKG inequality underlies all parts of Theorem 3.6. The claim (c) of ergodicity has until recently been considered slightly subtle (see the discussion after the forthcoming Theorem 3.9) but an easy proof may be found in [113].

It follows from the inequality of part (b) that $|\mathcal{W}_{p,q}| = 1$ if and only if $\phi_{p,q}^0 = \phi_{p,q}^1$. It is an important open problem to determine for which p, q this holds, and we shall return to this question in Section 5 (**OP**). For the moment, we note one sufficient condition for uniqueness, proved using a certain convexity property of the logarithm of a partition function Z.

Theorem 3.7 ([67, 69]). *Let $q \geq 1$. There exists a subset $\mathcal{D}_q$ of $[0,1]$, at most countably infinite in size, such that $\phi_{p,q}^0 = \phi_{p,q}^1$, and hence $|\mathcal{W}_{p,q}| = 1$, if $p \notin \mathcal{D}_q$.*

It is believed but not proved (**OP**) that: for any given $q \geq 1$, $\mathcal{D}_q$ either is empty or consists of a singleton (the critical point, to be defined in Section 4), the former occurring if and only if q is sufficiently small.

3.5 Random-cluster measures on infinite graphs

One may define a class of measures on the infinite lattice without having recourse to weak limits. The following definition of a random-cluster measure is based upon the Dobrushin–Lanford–Ruelle (DLR) definition of a Gibbs state, [44, 59, 106]. It was introduced in [66, 67], and discussed further in [27, 69]. For any box Λ, we write $\mathcal{T}_\Lambda$ for the σ-field generated by the set $\{\omega(e) : e \in \mathbb{E}^d \setminus \mathbb{E}_\Lambda\}$ of states of edges having at least one endvertex outside Λ.

Definition 3.8. *Let $0 \leq p \leq 1$ and $q > 0$. A probability measure ϕ on $(\Omega, \mathcal{F})$ is called a* random-cluster measure *with parameters p and q if*

for all $A \in \mathcal{F}$ and all finite boxes Λ, $\phi(A \mid \mathcal{T}_\Lambda)(\xi) = \phi_{\Lambda,p,q}^\xi(A)$ for ϕ-a.e. ξ.

The set of such measures is denoted $\mathcal{R}_{p,q}$.

The condition of this definition amounts to the following. Suppose we are given that the configuration off the finite box Λ is that of ξ. Then, for almost every $\xi \in \Omega$, the (conditional) measure on Λ is simply the random-cluster measure with boundary condition ξ. No further generality is gained by replacing the finite box Λ by a general finite subset of $\mathbb{Z}^d$.

Some information about the structure of $\mathcal{R}_{p,q}$, and its relationship to $\mathcal{W}_{p,q}$, is provided in [69]. For example, for all p, q, $\mathcal{R}_{p,q}$ is non-empty and convex. We have no proof that $\mathcal{W}_{p,q} \subseteq \mathcal{R}_{p,q}$, but we state one theorem in this direction. For $\omega \in \Omega$, let $I(\omega)$ be the number of infinite open clusters of ω. We say that a probability measure ϕ on $(\Omega, \mathcal{F})$ has the $0/1$-*infinite-cluster property* if $\phi(I \in \{0, 1\}) = 1$.

Theorem 3.9 ([69, 70, 73]). *Let* $0 \le p \le 1$ *and* $q > 0$. *If* $\phi \in \overline{\mathrm{co}\,\mathcal{W}_{p,q}}$ *and* ϕ *has the* $0/1$-*infinite-cluster property, then* $\phi \in \mathcal{R}_{p,q}$.

Since, [30], any translation-invariant probability measure satisfying a finite-energy property (see the discussion after Theorem 3.4) necessarily has the $0/1$-infinite-cluster property, we have that all translation-invariant members of $\overline{\mathrm{co}\,\mathcal{W}_{p,q}}$ lie in $\mathcal{R}_{p,q}$. Suppose for the moment that $q \ge 1$. By Theorem 3.6(b), the weak limits $\phi_{p,q}^b$, $b = 0, 1$, are translation-invariant, and therefore they belong to $\mathcal{R}_{p,q}$. It is not difficult to see, by the FKG inequality, that

$$\phi_{p,q}^0 \le_{\mathrm{st}} \phi \le_{\mathrm{st}} \phi_{p,q}^1 \qquad \text{for all } \phi \in \mathcal{R}_{p,q}\,, \tag{$*$}$$

and it follows that $|\mathcal{R}_{p,q}| = 1$ if and only if $\phi_{p,q}^0 = \phi_{p,q}^1$. The claim of ergodicity in Theorem 3.6(c) is one consequence of the extremality $(*)$ of the $\phi_{p,q}^b$ within the class $\mathcal{R}_{p,q}$ (see also [113, page 1113]).

It may be seen by an averaging argument, [69], that $\overline{\mathrm{co}\,\mathcal{W}_{p,q}}$ necessarily contains at least one translation-invariant measure, for all $p \in [0, 1]$ and $q \in (0, \infty)$. Therefore, $\mathcal{R}_{p,q}$ is non-empty for all p and q.

We note that Theorem 3.9, and particularly the $0/1$-infinite-cluster property, is linked to the property of so-called 'almost sure quasilocality', a matter discussed in [121].

3.6 The case $q < 1$

The FKG inequality, a keystone of many arguments when $q \ge 1$, is not valid when $q < 1$. Consequently, many fundamental questions are unanswered to date, and the theory of random-cluster models on a finite graph $G = (V, E)$ remains obscure when $q < 1$. The intuition is that certain positive correlations should be replaced by negative correlations; however, the theory of negative correlation is more problematic than that of positive correlation (see [120]). We return to this point later in this subsection.

As referred to above, there is an existence proof of infinite-volume weak limits and random-cluster measures for all $q > 0$. On the other hand, no constructive proof is known of the existence of such measures when $q < 1$ **(OP)**. More specifically, the existence of the weak limits $\lim_{\Lambda \uparrow \mathbb{Z}^d} \phi_{\Lambda,p,q}^b$, $b = 0, 1$, is not known when $q < 1$. The best that can be shown currently is that the two limits exist and are equal when p is either sufficiently small or sufficiently large, [73]. This may be achieved by comparison with percolation models having different values of p, very much as in [69] (the claim for small p may also be shown by the arguments of [49, 51]).

The theory of percolation gives a clue to a possible way forward. When $q = 1$, the FKG inequality is complemented by the so-called 'disjoint-occurrence' (or 'BK') inequality. This latter inequality is said to be valid for a measure μ if $\mu(A \circ B) \le \mu(A)\mu(B)$ for all increasing events A, B, where $A \circ B$ is the event that A and B occur disjointly (see [18, 71] for a discussion of this and the more general 'Reimer inequality' [129]). The disjoint-occurrence inequality has been established for classes of measures which are only slightly more general than product measures, and it is an interesting open question whether it is valid for a wider class of measures of importance (**OP**). It has been asked whether the disjoint-occurrence inequality could be valid for random-cluster measures with $q < 1$ (**OP**). A positive answer would aid progress substantially towards an understanding of limit random-cluster measures.

We illustrate this discussion about disjoint-occurrence with the following test question (**OP**): is it generally the case that the random-cluster measure $\phi_{p,q}$ on G satisfies

$$\phi_{p,q}(\text{edges } e \text{ and } f \text{ are open}) \le \phi_{p,q}(e \text{ is open})\phi_{p,q}(f \text{ is open}) \qquad (*)$$

for $e \ne f$ and $q < 1$? (See [120].) This equation would be a very special instance of the disjoint-occurrence inequality. A further restriction arises if we take the limit as $q \downarrow 0$; recall the discussion of Section 2.4. This leads to certain open questions of a purely graph-theoretic type, which combinatorial theorists might elevate to the status of conjectures. The first such question is the following. Let $K(e_1, e_2, \dots)$ be the number of subsets F of the edge set E, containing $e_1, e_2, \dots$, such that the graph (V, F) is connected. Is it the case that (**OP**)

$$K(e, f)K(\varnothing) \le K(e)K(f) \quad \text{if } e \ne f? \qquad (**)$$

(See [93].) In the second such question, we ask if the same inequality is valid with $K(e_1, e_2, \dots)$ redefined as the number of subsets F containing $e_1, e_2, \dots$ such that (V, F) is a forest (**OP**). These two questions are dual to one another in the sense that the first holds for a planar graph G if and only if the second holds for its planar dual. Explicit computations have confirmed the forest conjecture for all graphs G having eight or fewer vertices, [78].

In the 'intermediate regime', with $K(e_1, e_2, \dots)$ redefined as the number of spanning trees (that is, connected forests) of G containing $e_1, e_2, \dots$, the corresponding inequality is indeed valid. An extra ingredient in this case is the link to electrical networks, and particularly the variational principle known as the Thomson or Dirichlet principle (see [45]). Further results and references are provided in [16]. Substantially more is known for spanning trees, namely a general result concerning the 'negative association' of the uniform measure on the set of spanning trees of G, [48].

We note a more general version of conjecture $(**)$, namely

$$K_\alpha(e, f)K_\alpha(\varnothing) \le K_\alpha(e)K_\alpha(f) \quad \text{for } e \ne f, \ 0 < \alpha < \infty,$$

where

$$K_\alpha(e_1, e_2, \dots) = \sum_{\substack{F \subseteq E \\ F \supseteq \{e_1, e_2, \dots\} \\ (V, F) \text{ connected}}} \alpha^{|F|}.$$

This is equivalent to $(*)$ in the limit as $q \downarrow 0$, where $\alpha = p/(1-p)$.

By other means one may establish a certain non-trivial monotonicity when $q < 1$, but by a more complicated reasoning than before involving a property of convexity of the logarithm of the partition function. Namely, the mean number of open edges is non-decreasing in p, for $0 < q < \infty$, [69].

4 Phase transition, the big picture

Phase transition in a Potts model corresponds to the creation of an infinite open cluster in the corresponding random-cluster model. There are rich predictions concerning the nature of such a phase transition, but these have been proved only in part. This section is a summary of the expected properties of the phase diagram for different dimensions d and cluster-weighting factors q. The corresponding rigorous theory is described in Sections 5 and 6.

4.1 Infinite open clusters

We assume henceforth that $q \geq 1$, and we concentrate here on the extremal random-cluster measures $\phi_{p,q}^0$ and $\phi_{p,q}^1$. The phase transition of a random-cluster measure is marked by the onset of an infinite open cluster. We write $\{0 \leftrightarrow \infty\}$ for the event that the origin is the endvertex of some infinite open path, and we define the $\phi_{p,q}^b$ *percolation probability* by

$$\theta^b(p, q) = \phi_{p,q}^b(0 \leftrightarrow \infty), \qquad b = 0, 1.$$

It is almost immediate by a stochastic-ordering argument that $\theta^b(p, q)$ is non-decreasing in p, and therefore

$$\theta^b(p, q) \begin{cases} = 0 & \text{if } p < p_c^b(q), \\ > 0 & \text{if } p > p_c^b(q), \end{cases} \qquad b = 0, 1,$$

for critical points $p_c^b(q)$ given by

$$p_c^b(q) = \sup\{p : \theta^b(p, q) = 0\}, \qquad b = 0, 1.$$

It is an easy exercise to show that the number I of infinite open clusters satisfies:

$$\phi_{p,q}^b(I \geq 1) = \begin{cases} 0 & \text{if } \theta^b(p, q) = 0, \\ 1 & \text{if } \theta^b(p, q) > 0. \end{cases}$$

We shall see in Section 5.2 that any infinite open cluster is $\phi^b_{p,q}$-a.s. unique whenever it exists.

We have by Theorem 3.7 that $\phi^0_{p,q} = \phi^1_{p,q}$ for almost every p, whence $\theta^0(p,q) = \theta^1(p,q)$ for almost every p, and therefore $p^0_{\mathrm{c}}(q) = p^1_{\mathrm{c}}(q)$. Henceforth we use the abbreviated notation $p_{\mathrm{c}}(q) = p^0_{\mathrm{c}}(q) = p^1_{\mathrm{c}}(q)$, and we refer to $p_{\mathrm{c}}(q)$ as the *critical point* of the corresponding random-cluster measures. The non-triviality of $p_{\mathrm{c}}(q)$ may be proved by comparisons of random-cluster measures with product measures via Theorem 3.2. Recall the fact, [71, Chapter 1], that $0 < p_{\mathrm{c}}(1) < 1$ if $d \geq 2$.

Theorem 4.1 ([4]). *We have for $q \geq 1$ that*

$$p_{\mathrm{c}}(1) \leq p_{\mathrm{c}}(q) \leq \frac{q p_{\mathrm{c}}(1)}{1 + (q-1)p_{\mathrm{c}}(1)}.$$

When q is an integer satisfying $q \geq 2$, the phase transition of the random-cluster model corresponds in a special way to that of the Potts model with the same value of q. An indicator of phase transition in the Potts model is the 'magnetisation', defined as follows. Consider a Potts measure π^1_Λ on Λ having parameters β, J, q, and with '1' boundary conditions, which is to say that all vertices on the boundary $\partial\Lambda$ are constrained to have spin value 1. Let $\tau_\Lambda = \pi^1_\Lambda(\sigma_0 = 1) - q^{-1}$, a quantity which represents the net effect of this boundary condition on the spin at the origin. The corresponding random-cluster measure ϕ^1_Λ has parameters $p = 1 - e^{-\beta J}$ and q, and has wired boundary condition. We apply Theorem 2.1 to the graph obtained from Λ by identifying all vertices in $\partial\Lambda$, and we find that

$$\tau_\Lambda = (1 - q^{-1})\phi^1_\Lambda(0 \leftrightarrow \partial\Lambda).$$

The limit function $\tau = \lim_{\Lambda\uparrow\mathbb{Z}^d}\tau_\Lambda$ is called the magnetisation, it is a non-decreasing function of βJ and satisfies

$$\tau \begin{cases} = 0 & \text{if } \beta J \text{ is small}, \\ > 0 & \text{otherwise}. \end{cases}$$

It is not hard to show, [4], that $\phi^1_\Lambda(0 \leftrightarrow \partial\Lambda) \to \phi^1(0 \leftrightarrow \infty)$ as $\Lambda \uparrow \mathbb{Z}^d$, whence $\tau = (1-q^{-1})\theta^1(p,q)$ where $p = 1-e^{-\beta J}$. Therefore there is long-range order in the Potts model (that is, $\tau > 0$) if and only if the origin lies in an infinite open cluster with strictly positive $\phi^1_{p,q}$-probability. In particular, $p_{\mathrm{c}}(q) = 1 - e^{-\beta_{\mathrm{c}} J}$ where β_{c} is the critical value of β for the Potts model in question.

4.2 First- and second-order phase transition

There is a rich physical theory of phase transitions in percolation, Ising, and Potts models, some of which has been made rigorous in the context of the random-cluster model. There follows a broad sketch of the big picture, a full rigorous verification of which is far from complete. Rigorous mathematical progress is described in Section 5.

I. The subcritical phase, $p < p_c(q)$

It is standard, [4], that

$$\phi^0_{p,q} = \phi^1_{p,q} \qquad \text{if } \theta^1(p,q) = 0,$$

implying that there exists a unique random-cluster measure whenever $\theta^1(p,q) = 0$. In particular, $|\mathcal{W}_{p,q}| = |\mathcal{R}_{p,q}| = 1$ when $0 \le p < p_c(q)$. Assume for the moment that $p < p_c(q)$, and denote the unique random-cluster measure by $\phi_{p,q}$. By the definition of the critical point, all open clusters are $\phi_{p,q}$-a.s. finite. It is believed that they have a tail which decays exponentially, in that there exist $\gamma = \gamma(p,q) > 0$ and $\eta = \eta(p,q) > 0$ such that

$$\phi_{p,q}(|C| = n) = e^{-\gamma n(1+o(1))}, \quad \phi_{p,q}(\mathrm{rad}(C) = n) = e^{-\eta n(1+o(1))}, \quad \text{as } n \to \infty,$$

where C denotes the open cluster containing the origin, and its *radius* $\mathrm{rad}(C)$ is defined as $\sup\{\|x\| : x \in C\}$. Such exponential decay would be the starting point for a complete exploration of the subcritical phase. More detailed asymptotics should then emerge, including the Ornstein–Zernike decay of the connectivity functions:

$$\phi_{p,q}(0 \leftrightarrow e_n) \sim \frac{c(p,q)}{n^{(d-1)/2}} e^{-n/\xi(p,q)} \qquad \text{as } n \to \infty,$$

where $e_n = (n, 0, 0, \ldots, 0)$ and $\xi(p,q)$ is termed the 'correlation length'.

II. The supercritical phase, $p > p_c(q)$

This phase is characterised by the existence of one or more infinite open clusters (exactly one, in fact, for translation-invariant measures at least, see Section 5.2). It is believed that, as in the subcritical phase, we have that $\phi^0_{p,q} = \phi^1_{p,q}$ when $p > p_c(q)$; this remains unproven in general. Thus the first main problem is to prove that there is a unique random-cluster measure when $p > p_c(q)$.

The theory of percolation, [71], suggests a route towards understanding the geometry of the supercritical phase, namely by developing a rigorous block renormalisation argument. This should permit the use of theory developed when p is close to 1 in order to understand the model when p is close to $p_c(q)$. In particular, one expects an exponential estimate for the decay of the probabilities of long-range connections within finite open clusters, and a Wulff construction for the shape of such clusters.

III. Near the critical point, $p \simeq p_c(q)$

The main open problem is to understand the way in which the nature of the phase transition depends on the value of q. It is believed that the transition is continuous and governed by critical exponents and scaling theory when q is small, and is discontinuous when q is large. Presumably there exists a threshold for q which separates the so-called 'second-order' (or continuous) transition from the so-called 'first-order' (or discontinuous) transition. More specifically, it is believed that there exists $Q = Q(d)$ satisfying

$$Q(d) = \begin{cases} 4 & \text{if } d = 2\,, \\ 2 & \text{if } d \geq 6\,, \end{cases}$$

such that the following hold.

(i) *Assume that $q < Q$.*
  - For any p, there exists a unique random-cluster measure, denoted $\phi_{p,q}$. In particular $\phi^0_{p_c(q),q} = \phi^1_{p_c(q),q}$.
  - $\theta(p,q) = \phi_{p,q}(0 \leftrightarrow \infty)$ is a continuous function of p. There is no percolation at the critical point, in the sense that $\theta(p_c(q), q) = 0$.
  - The edge-density $h(p,q) = \phi_{p,q}(e \text{ is open})$, viewed as a function of p, is continuous at the critical point $p = p_c(q)$. [The letter e denotes a typical edge of the lattice.]
  - These functions and others have power-law singularities at $p_c(q)$, and the associated critical exponents satisfy the scaling relations (see [71, Chapter 9]).
  - When d is large (how large depends on the value of q), these critical exponents take on their 'mean-field' values, and depend no further on the value of d.
  - There is no 'mass gap', in the sense that the correlation length $\xi(p,q)$ satisfies $\lim_{p \uparrow p_c(q)} \xi(p,q) = \infty$.
  - Universality reigns, in that the critical exponents depend on the number d of dimensions but not on the choice of lattice. For example, the exponents associated with the square lattice are expected to be the same as those for the triangular lattice.
  - Assume $d = 2$ and $1 \leq q < 4$. The process with $p = p_c(q)$ converges as the lattice spacing shrinks to zero, the limit process when suitably defined being a stochastic Löwner evolution SLE_κ having parameter κ satisfying $\cos(4\pi/\kappa) = -\tfrac{1}{2}\sqrt{q}$, $\kappa \in (4, 8)$ (see Section 6.4 and [130]).

(ii) *Assume that $q > Q$.*
  - There exists a unique random-cluster measure if and only if $p \neq p_c(q)$. When $d = 2$ and $p = p_c(q)$, there are exactly two extremal members of $\mathcal{R}_{p,q}$, namely the free and the wired measures $\phi^b_{p,q}$, $b = 0, 1$. When $d \geq 3$ and $p = p_c(q)$ there exist other extremal members of $\mathcal{R}_{p,q}$ including a variety of non-translation-invariant measures.
  - We have that $\theta^0(p_c(q), q) = 0$ but $\theta^1(p_c(q), q) > 0$.
  - The edge-density $h(p,q)$ is a discontinuous function of p at the critical point $p_c(q)$.
  - There is a 'mass gap' in the sense that the correlation length $\xi(p,q)$ satisfies $\lim_{p \uparrow p_c(q)} \xi(p,q) < \infty$.

5 General results in d (≥ 2) dimensions

The properties of the random-cluster model depend pivotally on whether the process is subcritical ($p < p_c(q)$), supercritical ($p > p_c(q)$), or critical ($p \simeq p_c(q)$). We consider these situations in turn, in each case identifying major results and open problems. There is a bulk of information available for certain values of q, namely when $q = 1, 2$ and q is sufficiently large. In addition, the case $d = 2$ is special, and we shall return to this in Section 6. We assume throughout this section that $q \geq 1$.

Little is known in general about the numerical values of $p_c(q)$. For example, it is known that $p_c(q)$ is Lipschitz-continuous and strictly increasing when $d \geq 2$, [68], and there is a striking conjecture (**OP**) that $p_c(q) = \sqrt{q}/(1 + \sqrt{q})$ when $d = 2$ (see Section 6.2). Some concrete inequalities involving the $p_c(q)$ are implied by the comparison inequalities of Theorem 3.2.

5.1 The subcritical phase, $p < p_c(q)$

As remarked in Section 4.2, there is a unique random-cluster measure when $p < p_c(q)$, and we shall denote this by $\phi_{p,q}$.

The key theorem for understanding the subcritical phase of percolation states that long-range connections have exponentially decaying probabilities. Such a result is believed to hold for all random-cluster models with $q \geq 1$, but no proof has been found (**OP**) which is valid for all $q \geq 1$ and all $p < p_c(q)$. The full result is known only when $q = 1$, $q = 2$, or q is sufficiently large, and the three sets of arguments for these cases are somewhat different from one another. As for results valid for all q (≥ 1), the best that is currently known is that the connectivity function decays exponentially for sufficiently small p (this follows by Theorem 3.2 and the corresponding $q = 1$ result), and that it decays *exponentially* whenever it decays at a sufficient *polynomial* rate. We describe the last result next.

As a preliminary we introduce another definition of a critical point. Let $B(n)$ be the cube $[-n, n]^d$. We write

$$Y(p, q) = \limsup_{n \to \infty} \left\{ n^{d-1} \phi_{p,q}\big(0 \leftrightarrow \partial B(n)\big) \right\}$$

and $p_g(q) = \sup\{p : Y(p, q) < \infty\}$. Evidently $0 < p_g(q) \leq p_c(q)$, and it is believed that $p_g(q) = p_c(q)$ for all $q \geq 1$ (**OP**).

Theorem 5.1 ([76]). *Let* $q \geq 1$, $d \geq 2$, *and* $0 \leq p < p_g(q)$. *There exists* $\gamma = \gamma(p, q)$ *satisfying* $\gamma > 0$ *such that*

$$\phi_{p,q}(0 \leftrightarrow \partial B(n)) \leq e^{-\gamma n} \qquad \textit{for all large } n.$$

The spirit of the theorem is close to that of Hammersley [84] and Simon–Lieb [112, 136] who proved exponential estimates when $q = 1, 2$ subject to a hypothesis of finite susceptibility (that is, under the hypothesis that $\sum_x \phi_{p,q}(0 \leftrightarrow x) < \infty$). The latter assumption is slightly stronger than the assumption of the above theorem when $d = 2$.

Connectivity functions are expected to decay exponentially with a correction term of power order. More specifically, it is expected as reported in Section 4.2 that

$$\phi_{p,q}(0 \leftrightarrow x) \sim \frac{c}{|x|^{(d-1)/2}} \exp(-|x|/\xi) \qquad \text{as } |x| \to \infty,$$

for constants $c(p,q)$ and $\xi(p,q)$, and for some suitable norm $|\cdot|$ on $\mathbb{Z}^d$. Such 'Ornstein–Zernike' decay is a characteristic of many systems in their disordered phases. No proof is known (**OP**), except in the special cases when $q = 1$ and $q = 2$, [32, 33]. In [9] may be found a weaker result which bounds the fluctuations by a power-law when $d = 2$, under the assumption that the function does indeed decay exponentially.

5.2 The supercritical phase, $p > p_{\mathrm{c}}(q)$

We assume as usual that $q \geq 1$, and we begin with a discussion of the number of infinite clusters. For $\omega \in \Omega$, let $I(\omega)$ be the number of infinite open clusters.

Suppose that $\phi_{p,q}$ is a translation-invariant member of $\mathcal{R}_{p,q}$. If in addition $\phi_{p,q}$ is ergodic, then, by a well known theorem of Burton and Keane [30],

$$\text{either} \quad \phi_{p,q}(I = 0) = 1 \quad \text{or} \quad \phi_{p,q}(I = 1) = 1\,;$$

that is to say, the infinite open cluster is almost surely unique whenever it exists. It is noted in [30] that methods of ergodic decomposition enable the extension of such results to translation-invariant measures which are not necessarily ergodic. That is, under the assumption of translation-invariance alone,

$$\phi_{p,q}(I \in \{0,1\}) = 1\,,$$

which is to say that translation-invariant random-cluster measures have the 0/1-infinite-cluster property. A further comment on the use of ergodic decomposition in this context is to be found in [31, 61].

In two dimensions, the supercritical process is best studied via the subcritical process which arises as its graphical dual (see Section 6). There are two general approaches to the supercritical phase in a general number d (≥ 3) of dimensions. The less powerful is to derive results for large p by comparison with percolation, the theory of which is relatively complete. Without an extra ingredient, such an approach will not reveal the structure of the supercritical phase all the way down to the critical value $p_{\mathrm{c}}(q)$. As an example, we present one theorem concerning the uniqueness of random-cluster measures.

Theorem 5.2 ([69]). *If $d \geq 2$ and $q \geq 1$, there exists $p' = p'(d,q) < 1$ such that $\phi_{p,q}^0 = \phi_{p,q}^1$ whenever $p > p'$.*

It is an important open problem to prove that $\phi^0_{p,q} = \phi^1_{p,q}$ for all $p > p_c(q)$, or equivalently that there exists a unique random-cluster measure throughout the phase (**OP**).

A more powerful approach, sometimes used in conjunction with the comparison argument summarised above, is the 'block argument' laid out in [36, 125]. One may think of block arguments as a form of rigorous renormalisation. One divides space into blocks, constructs events of an appropriate nature on such blocks, having large probabilities, and then allows these events to combine across space. There have been substantial successes using this technique, of which the most striking is the resolution, subject to certain side conditions, of the so-called Wulff construction for the asymptotic shape of large Ising droplets.

Rather than discussing the physical background of the Wulff construction, we mention instead its impact on random-cluster models. Let $B(n) = [-n, n]^d$, and consider the wired random-cluster measure $\phi^1_{B(n),p,q}$ with $p > p_c(q)$. The larger is an open cluster, the more likely it is to be joined to the boundary $\partial B(n)$. Suppose that we condition on the event that there exists in $B(n)$ an open cluster C which does not touch $\partial B(n)$ and which has volume of the order of the volume n^d of the box. What can be said about the shape of C? Since $p > p_c(q)$, there is little cost in having large *volume*, and the price is spent around its *boundary*. Indeed, the price may be expressed as a surface integral of an appropriate function termed 'surface tension'. This 'surface tension' may be specified as the exponential rate of decay of a certain probability. The Wulff prediction for the shape of C is that, when re-scaled in the limit of large n, its shape converges to the solution of a certain variational problem, that is, the limit shape is obtained by minimising a certain surface integral subject to a condition on its volume.

No proof of this general picture for random-cluster models has appeared in the literature, although it is believed that the methods of [36, 37, 125] enable such a proof. The authors of [36] have instead concentrated on using random-cluster technology to solve the corresponding question for the asymptotic shape of large droplets in the Ising model. The outcome is an important 'large deviation' theorem which utilises block arguments and yields a full solution to the Ising problem whenever the corresponding random-cluster model (which has $q = 2$) has parameter p satisfying $p > \widehat{p}_c(2)$ and $\phi^0_{p,2} = \phi^1_{p,2}$. Here, $\widehat{p}_c(2)$ is the limit of a certain decreasing sequence of critical points defined on slabs in $\mathbb{Z}^d$, and is conjectured (**OP**) to be equal to the critical point $p_c(2)$. [Closely related results have been obtained in [24]. Fluctuations in droplet shape for *two-dimensional* random-cluster models have been studied in [10, 11].]

The 'slab critical point' $\widehat{p}_c(q)$ may be defined for any random-cluster model as follows. Fix $q \geq 1$, and let $d \geq 3$. Let $S(n, L) = [-n, n]^{d-1} \times [-L, L]$. Let $\psi^{n,L}_{p,q}$ be the random-cluster measure on $S(n, L)$ with parameters p, q (and with free boundary conditions). We denote by $\Pi(p, L)$ the property that:

there exists $\alpha > 0$ such that, for all $x \in S(n, L)$ and all n, $\psi_{p,q}^{n,L}(0 \leftrightarrow x) > \alpha$.

It is not hard to see that $\Pi(p, L) \Rightarrow \Pi(p', L')$ if $p \leq p'$ and $L \leq L'$. It is thus natural to define the quantities

$$\widehat{p}_{\mathrm{c}}(q, L) = \inf\{p : \Pi(p, L) \text{ occurs}\}, \qquad \widehat{p}_{\mathrm{c}}(q) = \lim_{L \to \infty} \widehat{p}_{\mathrm{c}}(q, L),$$

and it is clear that $\widehat{p}_{\mathrm{c}}(q) \geq p_{\mathrm{c}}(q)$.

Conjecture 5.3 ([125]). Let $q \geq 1$ and $d \geq 3$. We have that $\widehat{p}_{\mathrm{c}}(q) = p_{\mathrm{c}}(q)$.

Subject to a verification of this conjecture, and of a positive answer to the question of the uniqueness of random-cluster measures when $p > p_{\mathrm{c}}(q)$, the block arguments of [36, 125] may be expected to result in a fairly complete picture of the supercritical phase of random-cluster models with $q \geq 1$; see [37] also.

The case $q = 1$ is special, percolation enjoys a spatial independence not shared with general random-cluster models. This additional property has been used in the formulation of a type of 'dynamic renormalisation', which has in turn yielded a proof that $\widehat{p}_{\mathrm{c}}(1) = p_{\mathrm{c}}(1)$ for percolation in three or more dimensions, [71, Chapter 7], [74]. Such arguments do not to date have a random-cluster counterpart.

As a further application of a block argument we note the following bound, [125], for the tail of the size of the open cluster C at the origin,

$$\phi_{p,q}^{b}(|C| = n) \leq \exp\left(-\alpha n^{(d-1)/d}\right) \qquad \text{for all } n,$$

for some $\alpha = \alpha(p, q) > 0$, and valid for $d \geq 3$, $b = 0, 1$, and p sufficiently close to 1. The complementary inequality

$$\phi_{p,q}^{b}(|C| = n) \geq \exp\left(-\alpha' n^{(d-1)/d}\right) \qquad \text{for all } n,$$

may be obtained for large p as done in the case of percolation, [71, Section 8.6].

5.3 Near the critical point, $p \simeq p_{\mathrm{c}}(q)$

Surprisingly little is known about random-cluster measures near the critical point, except in the cases $q = 1, 2$ and q large. In each such case, there are special arguments which are apparently not suitable for generalisation. We summarise such results as follows.

I. Percolation, $q = 1$

There is a full theory of the subcritical and supercritical phases of percolation, [71]. The behaviour when $p \simeq p_{\mathrm{c}}(1)$ has been the subject of deep study, and many beautiful results are known. Nevertheless, the picture is incomplete.

For example, it is believed but not proved that $\theta(p_c(1), 1) = 0$ for all $d \geq 2$, but this is known only when $d = 2$ (because of special properties of two dimensions explored for $\mathbb{L}^2$ in Section 6) and when d is large ($d \geq 19$ suffices) using a method termed the 'lace expansion'. The lace expansion explains also the values of some critical exponents when d is large; see, for example, [85, 86].

Great progress has been made in recent years towards understanding the phase transition when $d = 2$. The idea is to work at the critical point $p = p_c(1)$, and to observe the process over an increasing sequence of regions of $\mathbb{Z}^2$. It is believed that the process, re-scaled as the regions become larger, converges in a certain manner to a stochastic process generated in a prescribed way by a differential equation, known as a Löwner equation, which is driven in a certain way by a Brownian motion. Stochastic processes which arise in this way have been termed *stochastic Löwner evolutions* by Schramm, [135], and denoted SLE_κ, where κ is the variance parameter of the Brownian motion. It is believed that the space of stochastic Löwner evolutions is a canonical family of processes which arise as scaling limits of discrete processes such as critical percolation, critical random-cluster models with $q \leq 4$, self-avoiding walks, loop-erased random walk, and uniform spanning trees. Full proofs are not yet known (**OP**). We expand on this very important development in Section 6.4.

II. Ising model, $q = 2$

Integer values of q are special, and the value $q = 2$ particularly so because of certain transformations which permit the passage to a model which might be termed a 'Poisson graph'. Let $G = (V, E)$ be a finite graph and let $0 < \lambda < \infty$. Suppose that $\pi = \{\pi(e) : e \in E\}$ is a family of independent random variables each having the Poisson distribution with parameter λ. We now construct a random graph $G_\pi = (V, E_\pi)$ having vertex set V and, for each $e \in E$, having exactly $\pi(e)$ edges in parallel joining the endvertices of the edge e [the original edge e is itself removed]. We call G_π a *Poisson graph with intensity* λ, and write $\mathbb{P}_\lambda$ and $\mathbb{E}_\lambda$ for the appropriate probability measure and expectation operator.

We introduce next the concept of a flow on an oriented graph. Let $q \in \{2, 3, \ldots\}$ and let $G' = (V', E')$ be a finite oriented graph. Let $f : E' \to \{0, 1, 2, \ldots, q-1\}$. For $x \in V'$, the *total flow into* x is the sum of $\pm f(e')$ over all edges e' incident to x, with $+1$ when e' is oriented towards x and -1 otherwise. The function f is called a *mod-q flow* if the total flow into x is zero (modulo q) for all $x \in V'$. The mod-q flow f is called *non-zero* if $f(e') \neq 0$ for every $e' \in E'$. We write $F_q(G')$ for the number of non-zero mod-q flows on G'. It is a remarkable fact, [142], that $F_q(G')$ does not depend on the orientations of edges in E', and thus one may define $F_q(G')$ unambiguously for any *unoriented* graph G'.

We return now to the Poisson graph G_π. For $x, y \in V$, $x \neq y$, we denote by $G_\pi^{x,y}$ the graph obtained from G_π by adding an edge with endvertices x, y. [If x and y are already adjacent in G_π, we add exactly one further edge between them.] Connection probabilities and flows are related by the following theorem, which may be proved using properties of Tutte polynomials (see [142] and Section 2.5).

Let $G = (V, E)$ be a finite graph, and write $\phi_{G,p,q}$ for the random-cluster measure on G with parameters p, q.

Theorem 5.4 ([63, 73]). *Let $q \in \{2, 3, \dots\}$ and $0 \le p = 1 - e^{-\lambda q} < 1$. We have that*

$$(q-1)\phi_{G,p,q}(x \leftrightarrow y) = \frac{\mathbb{E}_\lambda(F_q(G_\pi^{x,y}))}{\mathbb{E}_\lambda(F_q(G_\pi))} \qquad \text{for all } x, y \in V, \ x \neq y.$$

This formula takes on an especially simple form when $q = 2$, since non-zero mod-2 flows necessarily take only the value 1. It follows that, for any graph G', $\mathbb{E}_\lambda(F_2(G'))$ equals the $\mathbb{P}_\lambda$-probability that the degree of every vertex of G' is even, [1]. Observations of this sort have led when $q = 2$ to the so-called 'random-current' expansion for Ising models, thereby after some work [1, 2, 5] leading to proofs amongst other things of the following, expressed here in the language of random-cluster measures.

(i)   When $q = 2$ and $p < p_c(q)$, we have exponential decay of the radius distribution,

$$\phi_{p,2}(\mathrm{rad}(C) = n) \le e^{-\eta n} \qquad \text{for all } n,$$

where $\eta = \eta(p) > 0$; exponential decay of the two-point connectivity function follows.

(ii) When $q = 2$ and $d \neq 3$, there is a unique random-cluster measure $\phi_{p,2}$ for all p, in that $|\mathcal{R}_{p,q}| = 1$.

(iii) The phase transition is continuous when $q = 2$ and $d \neq 3$. In particular, $\theta^0(p_c(2), 2) = \theta^1(p_c(2), 2) = 0$, and the edge-density $h(p, 2) = \phi_{p,2}(e$ is open$)$ is a continuous function of p at the critical point $p_c(2)$.

(iv) When $d \ge 4$, some (at least) critical exponents take their mean-field values, and depend no further on the value of d.

Note that the nature of the phase transition in three dimensions remains curiously undecided (**OP**).

III. The case of large q

It is not known whether the phase transition is continuous for all small q (**OP**). The situation for large q is much better understood owing to a method known as Pirogov–Sinai theory [123, 124] which may be adapted in a convenient manner to random-cluster measures. The required computation, which may be found in [105], has its roots in an earlier paper [103] dealing with Potts models. A feature of such arguments is that they are valid 'all the way to the critical point' (rather than for 'small p' or 'large p' only), so long as q is sufficiently large. One obtains thereby a variety of conclusions including the following.

(i)   The edge-densities $h^b(p,q) = \phi^b_{p,q}(e$ is open), $b = 0, 1$, are discontinuous functions of p at the critical point.

(ii)   The percolation probabilities satisfy $\theta^0(p_{\mathrm{c}}(q), q) = 0$, $\theta^1(p_{\mathrm{c}}(q), q) > 0$.

(iii) There is a multiplicity of random-cluster measures when $p = p_{\mathrm{c}}(q)$, in that $\phi^0_{p_{\mathrm{c}}(q),q} \neq \phi^1_{p_{\mathrm{c}}(q),q}$.

(iv)   If $p < p_{\mathrm{c}}(q)$, there is exponential decay and a mass gap, in that the unique random-cluster measure satisfies

$$\phi_{p,q}(0 \leftrightarrow e_n) = e^{-(1+\mathrm{o}(1))n/\xi} \qquad \text{as } n \to \infty,$$

where $e_n = (n, 0, 0, \dots, 0)$ and the correlation length $\xi = \xi(p,q)$ is such that $\lim_{p \uparrow p_{\mathrm{c}}(q)} \xi(p,q) = \psi(q) < \infty$.

(v)   If $d = 3$ and $p = p_{\mathrm{c}}(q)$, there exists a non-translation-invariant random-cluster measure, [38, 116].

It is not especially fruitful to seek numerical estimates on the required size $Q(d)$ of q for the above conclusions to be valid. Such estimates may be computed, but turn out to be fairly distant from those anticipated, namely $Q(2) = 4$, $Q(d) = 2$ for $d \geq 6$.

The proofs of the above facts are rather complicated and will not be explained here. Proofs are much easier and not entirely dissimilar when $d = 2$, and a very short sketch of such a proof is provided in Section 6.3.

6 In two dimensions

The duality theory of planar graphs provides a technique for studying random-cluster models in two dimensions. We shall see in Section 6.1 that, for a dual pair (G, G^{d}) of planar graphs, the measures $\phi_{G,p,q}$ and $\phi_{G^{\mathrm{d}},p^{\mathrm{d}},q}$ are dual measures in a certain geometrical sense, where p, p^{d} are related by $p^{\mathrm{d}}/(1 - p^{\mathrm{d}}) = q(1 - p)/p$. Such a duality permits an analysis by which many results for $\mathbb{L}^2$ may be derived. Of particular interest is the value of p for which $p = p^{\mathrm{d}}$. This 'self-dual point' is easily found to be $p = p_{\mathrm{sd}}(q)$ where

$$p_{\mathrm{sd}}(q) = \frac{\sqrt{q}}{1 + \sqrt{q}},$$

and it is conjectured that $p_{\mathrm{c}}(q) = p_{\mathrm{sd}}(q)$ for $q \geq 1$.

6.1 Graphical duality

Let $G = (V, E)$ be a simple planar graph imbedded in $\mathbb{R}^2$. We obtain its dual graph $G^{\mathrm{d}} = (V^{\mathrm{d}}, E^{\mathrm{d}})$ as follows (the roman letter 'd' denotes 'dual' rather than number of dimensions). We place a dual vertex within each face of G, including the infinite face of G if G is finite. For each $e \in E$ we place a dual edge $e^{\mathrm{d}} = \langle x^{\mathrm{d}}, y^{\mathrm{d}} \rangle$ joining the two dual vertices lying in the two faces of G

abutting e; if these two faces are the same, then $x^{\mathrm{d}} = y^{\mathrm{d}}$ and e^{d} is a loop. Thus E^{d} is in one–one correspondence to E. It is easy to see that the dual of $\mathbb{L}^2$ is isomorphic to $\mathbb{L}^2$. What is the relevance of graphical duality to random-cluster measures on G?

Suppose that G is finite. Any configuration $\omega \in \Omega\ (= \{0,1\}^E)$ gives rise to a dual configuration ω^{d} lying in the space $\Omega^{\mathrm{d}} = \{0,1\}^{E^{\mathrm{d}}}$ defined by $\omega^{\mathrm{d}}(e^{\mathrm{d}}) = 1 - \omega(e)$. As before, to each configuration ω^{d} corresponds the set $\eta(\omega^{\mathrm{d}}) = \{e^{\mathrm{d}} \in E^{\mathrm{d}} : \omega^{\mathrm{d}}(e^{\mathrm{d}}) = 1\}$ of its 'open edges'. Let $f(\omega)$ be the number of faces of the graph $(V, \eta(\omega))$, including the infinite face. By drawing a picture, one may easily be convinced (see Fig. 6.1) that the faces of $(V, \eta(\omega))$ are in one–one correspondence with the components of $(V^{\mathrm{d}}, \eta(\omega^{\mathrm{d}}))$, and therefore $f(\omega) = k(\omega^{\mathrm{d}})$, in the obvious notation. We shall make use of Euler's formula (see [147]),

$$k(\omega) = |V| - |\eta(\omega)| + f(\omega) - 1\,, \qquad \omega \in \Omega\,.$$

The random-cluster measure on G is given by

$$\phi_{G,p,q}(\omega) \propto \left(\frac{p}{1-p}\right)^{|\eta(\omega)|} q^{k(\omega)}\,, \qquad \omega \in \Omega\,.$$

Using Euler's formula and the equality $f(\omega) = k(\omega^{\mathrm{d}})$, we find that

$$\phi_{G,p,q}(\omega) = \phi_{G^{\mathrm{d}},p^{\mathrm{d}},q}(\omega^{\mathrm{d}}) \qquad \text{for } \omega \in \Omega\,,$$

where the dual parameter p^{d} is given according to

$$\frac{p^{\mathrm{d}}}{1-p^{\mathrm{d}}} = \frac{q(1-p)}{p}\,.$$

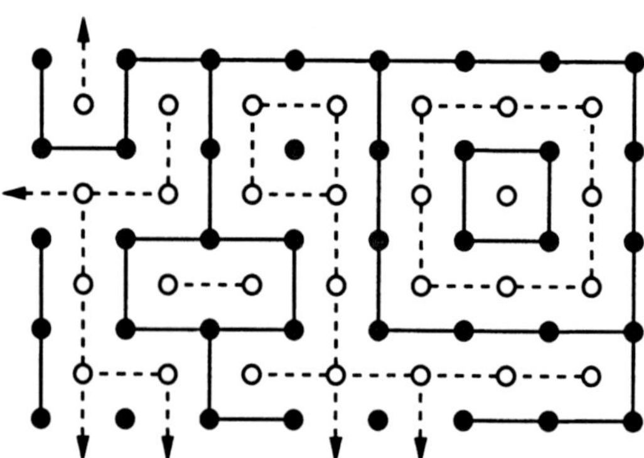

Fig. 6.1. A primal configuration ω (with solid lines and vertices) and its dual configuration ω^{d} (with dashed lines and hollow vertices). The arrows join the given vertices of the dual to a dual vertex in the infinite face. Note that each face of the primal graph (including the 'infinite face') corresponds to a unique component of the dual graph.

The unique fixed point of the mapping $p \mapsto p^{\mathrm{d}}$ is given by $p = p_{\mathrm{sd}}(q)$ where $p_{\mathrm{sd}}(q) = \sqrt{q}/(1 + \sqrt{q})$. We note at this point that

$$\phi_{G,p_{\mathrm{sd}}(q),q}(\omega) \propto q^{\frac{1}{2}|\eta(\omega)|+k(\omega)} \propto q^{\frac{1}{2}(k(\omega^{\mathrm{d}})+k(\omega))},$$

by Euler's formula. This representation of the random-cluster measure at the 'self-dual point' $p_{\mathrm{sd}}(q)$ highlights the duality of measures.

Turning to the square lattice, let $\Lambda_n = [0,n]^2$, whose dual graph Λ_n^{d} may be obtained from $[-1,n]^2 + (\frac{1}{2},\frac{1}{2})$ by identifying all boundary vertices. By the above,

$$\phi^0_{\Lambda_n,p,q}(\omega) = \phi^1_{\Lambda_n^{\mathrm{d}},p^{\mathrm{d}},q}(\omega^{\mathrm{d}})$$

for configurations ω on Λ_n (and with a small 'fix' on the boundary of Λ_n^{d}). Letting $n \to \infty$, we obtain that $\phi^0_{p,q}(A) = \phi^1_{p^{\mathrm{d}},q}(A^{\mathrm{d}})$ for all cylinder events A, where $A^{\mathrm{d}} = \{\omega^{\mathrm{d}} : \omega \in A\}$.

6.2 Value of the critical point

Consider the random-cluster process on the two-dimensional lattice $\mathbb{L}^2 = (\mathbb{Z}^2, \mathbb{E}^2)$, with parameters p and q satisfying $q \geq 1$. The following remarkable conjecture is widely believed (**OP**).

Conjecture 6.1. Let $q \geq 1$. The critical value $p_{\mathrm{c}}(q)$ of $\mathbb{L}^2$ is given by

$$p_{\mathrm{c}}(q) = \frac{\sqrt{q}}{1 + \sqrt{q}} \qquad \text{for } q \geq 1.$$

This conjecture is known to hold when $q = 1$, $q = 2$, and for $q \geq 25.72$. The $q = 1$ case was answered by Kesten [97] in his famous proof that the critical probability of bond percolation on $\mathbb{L}^2$ is $\frac{1}{2}$. For $q = 2$, the value of $p_{\mathrm{c}}(2)$ given above agrees with the celebrated calculation by Onsager [119] of the critical temperature of the Ising model on $\mathbb{Z}^2$, and is implied by probabilistic results in the modern vernacular of [2]. The formula for $p_{\mathrm{c}}(q)$ has been established rigorously in [104, 105] for sufficiently large (real) values of q, specifically $q \geq 25.72$ (see also [70]).

Conjecture 6.1 arises in a natural manner from the observation that $\mathbb{L}^2$ is a self-dual graph, and $p = p_{\mathrm{sd}}(q) = \sqrt{q}/(1 + \sqrt{q})$ is the self-dual point of a random-cluster measure on $\mathbb{L}^2$ with parameters p, q.

Several other remarkable conjectures about the phase transition in $\mathbb{L}^2$ may be found in the physics literature (see [14]), as consequences of 'exact' but non-rigorous arguments involving ice-type models. These include exact formulae for the asymptotic behaviour of the partition function $\lim_{\Lambda \uparrow \mathbb{Z}^2} \{Z_{\Lambda,p,q}\}^{1/|\Lambda|}$, and also for the edge-densities $h^b(p,q) = \phi^b_{p,q}(e \text{ is open})$, $b = 0,1$, at the self-dual point $p = p_{\mathrm{sd}}(q)$.

Progress towards a rigorous verification of the conjecture may be summarised briefly as follows. Using an argument, [152], taken from percolation using the uniqueness of infinite open clusters, we obtain by duality that

$\theta^0\big(p_{\mathrm{sd}}(q), q\big) = 0$ (see [69, 143]), whence the critical value of the square lattice satisfies $p_{\mathrm{c}}(q) \geq p_{\mathrm{sd}}(q)$ for $q \geq 1$. The complementary inequality $p_{\mathrm{c}}(q) \leq p_{\mathrm{sd}}(q)$ has eluded mathematicians despite progress by physicists, [87].

Suppose on the contrary that $p_{\mathrm{c}}(q) > p_{\mathrm{sd}}(q)$, so that $p_{\mathrm{c}}(q)^{\mathrm{d}} < p_{\mathrm{sd}}(q)$. For $p \in (p_{\mathrm{c}}(q)^{\mathrm{d}}, p_{\mathrm{c}}(q))$ we have also that $p^{\mathrm{d}} \in (p_{\mathrm{c}}(q)^{\mathrm{d}}, p_{\mathrm{c}}(q))$. Therefore, for $p \in (p_{\mathrm{c}}(q)^{\mathrm{d}}, p_{\mathrm{c}}(q))$, both primal and dual processes comprise, almost surely, the union of *finite* open clusters. This contradicts the intuitive picture, supported for $p \neq p_{\mathrm{c}}(q)$ by our knowledge of percolation, of finite clusters of one process floating in an infinite ocean of the other process.

Exact values for the critical points of the triangular and hexagonal lattices may be conjectured similarly, using graphical duality together with the star–triangle transformation [14, 73, 101].

Rigorous numerical upper bounds of impressive accuracy have been achieved for the square lattice and other two-dimensional lattices via an extension of the basic model to a larger class termed in [8] the 'asymmetric random-cluster model'. The bound in question for $\mathbb{L}^2$ is

$$p_{\mathrm{c}}(q) \leq \frac{\sqrt{q}}{\sqrt{1 - q^{-1}} + \sqrt{q}}, \qquad q \geq 1.$$

For example, when $q = 10$, we have that $0.760 \leq p_{\mathrm{c}}(10) \leq 0.769$, to be compared with the conjecture that $p_{\mathrm{c}}(10) = \sqrt{10}/(1 + \sqrt{10}) \simeq 0.760$. A valuable consequence of the comparison methods developed in [8] is the exponential decay of connectivity functions when $q > 2$ and p is such that

$$p < p_{\mathrm{sd}}(q - 1) = \frac{\sqrt{q - 1}}{1 + \sqrt{q - 1}}.$$

6.3 First-order phase transition

There is a special argument discovered first for Potts models, [104], which may be used to show first-order phase transition when q is sufficiently large.

Let a_n be the number of self-avoiding walks on $\mathbb{L}^2$ beginning at the origin. It is standard, [114], that $a_n^{1/n} \to \mu$ as $n \to \infty$, for some constant μ called the *connective constant* of the lattice. Let $Q = \big\{\frac{1}{2}\big(\mu + \sqrt{\mu^2 - 4}\big)\big\}^4$. We have that $2.620 < \mu < 2.696$ (see [137]), whence $21.61 < Q < 25.72$. We set

$$\psi(q) = \frac{1}{24} \log \left\{ \frac{(1 + \sqrt{q})^4}{q\mu^4} \right\},$$

noting that $\psi(q) > 0$ if and only if $q > Q$. We write $B(n) = [-n, n]^2$.

Theorem 6.2 ([70, 104]). *If $d = 2$ and $q > Q$ then the following hold.*

(a) *The critical point is given by $p_{\mathrm{c}}(q) = \sqrt{q}/(1 + \sqrt{q})$.*
(b) *We have that $\theta^1(p_{\mathrm{c}}(q), q) > 0$.*
(c) *For any $\psi < \psi(q)$ and all large n, $\phi^0_{p_{\mathrm{c}}(q), q}\big(0 \leftrightarrow \partial B(n)\big) \leq e^{-n\psi}$. Hence, in particular, $\theta^0(p_{\mathrm{c}}(q), q) = 0$.*

The idea of the proof is as follows. There is a partial order on circuits of $\mathbb{L}^2$ given by $\Gamma \leq \Gamma'$ if the bounded component of $\mathbb{R}^2 \setminus \Gamma$ is a subset of the bounded component of $\mathbb{R}^2 \setminus \Gamma'$. We work at the self-dual point $p = p_{\mathrm{sd}}(q)$, and with the box $B(n)$ with wired boundary conditions. An 'outer contour' is defined to be a circuit Γ of the dual graph $B(n)^{\mathrm{d}}$ all of whose edges are open in the dual (that is, they traverse closed edges in the primal graph $B(n)$), and which is maximal with this property. Using self-duality, one may show that

$$\phi^1_{B(n),p_{\mathrm{sd}}(q),q}(\Gamma \text{ is an outer circuit}) \leq \frac{1}{q}\left(\frac{q}{(1+\sqrt{q})^4}\right)^{|\Gamma|/4},$$

for any given circuit Γ of $B(n)^{\mathrm{d}}$. Combined with a circuit-counting argument of Peierls-type involving the connective constant, this estimate implies after a little work the claims of Theorem 6.2. The idea of the proof appeared in [104] in the context of Potts models, and the random-cluster formulation may be found in [70].

We stress that corresponding conclusions may be obtained for general d (≥ 2) when q is sufficiently large ($q > Q(d)$ for suitable $Q(d)$), as shown in [105] using so-called Pirogov–Sinai theory. Whereas, in the case $d = 2$, the above duality provides an especially simple proof, the proof for general d utilises nested sequences of surfaces of $\mathbb{R}^d$ and requires a control of the effective boundary conditions within the surfaces.

6.4 SLE limit when $q \leq 4$

Many exact calculations are 'known' for critical processes in two dimensions, but the physical arguments involved have sometimes appeared in varying degrees magical or revelationary to mathematicians. The new technology of stochastic Löwner evolutions (SLE), discovered by Schramm [135] and mentioned in Section 5.3, threatens to provide a rigorous underpinning of many such arguments in a manner most consonant with modern probability theory. Roughly speaking, the theory of SLE informs us of the correct weak limit of a critical process in the limit of large spatial scales, and in addition provides a mechanism for performing calculations for the limit process.

Let $\mathbb{H} = (-\infty, \infty) \times (0, \infty)$ be the upper half-plane of $\mathbb{R}^2$, with closure $\overline{\mathbb{H}}$. We view $\mathbb{H}$ and $\overline{\mathbb{H}}$ as subsets of the complex plane. Consider the ordinary differential equation

$$\frac{d}{dt}g_t(z) = \frac{2}{g_t(z) - B_{\kappa t}}, \qquad z \in \overline{\mathbb{H}} \setminus \{0\},$$

subject to the boundary condition $g_0(z) = z$, where $t \in [0, \infty)$, κ is a positive constant, and $(B_t : t \geq 0)$ is a standard Brownian motion. The solution exists when $g_t(z)$ is bounded away from $B_{\kappa t}$. More specifically, for $z \in \overline{\mathbb{H}}$, let τ_z be the infimum of all times τ such that 0 is a limit point of $g_s(z) - B_{\kappa s}$ in the limit as $s \uparrow \tau$. We let

$$H_t = \{z \in \mathbb{H} : \tau_z > t\}, \qquad K_t = \{z \in \overline{\mathbb{H}} : \tau_z \leq t\},$$

so that H_t is open, and K_t is compact. It may now be seen that g_t is a conformal homeomorphism from H_t to $\mathbb{H}$.

We call $(g_t : t \geq 0)$ a *stochastic Löwner evolution* (SLE) with parameter κ, written SLE_κ, and we call the K_t the *hulls* of the process. There is good reason to believe that the family $K = (K_t : t \geq 0)$ provides the correct scaling limit of a variety of random spatial processes, the value of κ being chosen according to the process in question. General properties of SLE_κ, viewed as a function of κ, have been studied in [130, 145], and a beautiful theory has emerged. For example, the hulls K form almost surely a simple path if and only if $\kappa \leq 4$. If $\kappa > 8$, then SLE_κ generates almost surely a space-filling curve.

Schramm [135] has identified the relevant value of κ for several different processes, and has indicated that percolation has scaling limit SLE_6, but full rigorous proofs are incomplete. In the case of percolation, Smirnov [138, 139] has proved the very remarkable result that, *for site percolation on the triangular lattice*, the scaling limit exists and is SLE_6 (this last statement is illustrated and partly explained in Fig. 6.2), but the existence of the limit is open for other lattices and for bond percolation.

It is possible to perform calculations on stochastic Löwner evolutions, and in particular to confirm, [110, 140], the values of many critical exponents associated with percolation (or, at least, site percolation on the triangular lattice). The consequences are in agreement with predictions of mathematical physicists previously considered near-miraculous (see [71, Chapter 9]). In addition, SLE_6 satisfies the appropriate version of Cardy's formula, [34, 107].

The technology of SLE is a major piece of contemporary mathematics which promises to explain phase transitions in an important class of two-dimensional disordered systems, and to help bridge the gap between probability theory and conformal field theory. It has already provided complete explanations of conjectures, by mathematicians and physicists, associated with two-dimensional Brownian motions and specifically their intersection exponents and fractionality of frontier, [108, 109].

Extra work is needed in order to prove the validity of the limiting operation for other percolation models and random processes. In another remarkable recent paper [111], Lawler, Schramm, and Werner have verified the existence of the scaling limit for loop-erased random walk and for the uniform spanning tree Peano curve, and have shown them to be SLE_2 and SLE_8 respectively. It is believed that self-avoiding walk on $\mathbb{L}^2$, [114], has scaling limit $\mathrm{SLE}_{8/3}$.

We turn now to the random-cluster model on $\mathbb{L}^2$ with parameters p and q. For $1 \leq q < 4$, it is believed that the percolation probability $\theta(p, q)$, viewed as a function of p, is continuous at the critical point $p_c(q)$ (**OP**), and furthermore that $p_c(q) = \sqrt{q}/(1 + \sqrt{q})$. It seems likely that, when re-scaled in the manner similar to that of percolation (illustrated in Fig. 6.2), the exploration process of the model converges to a limit process of SLE type. It then

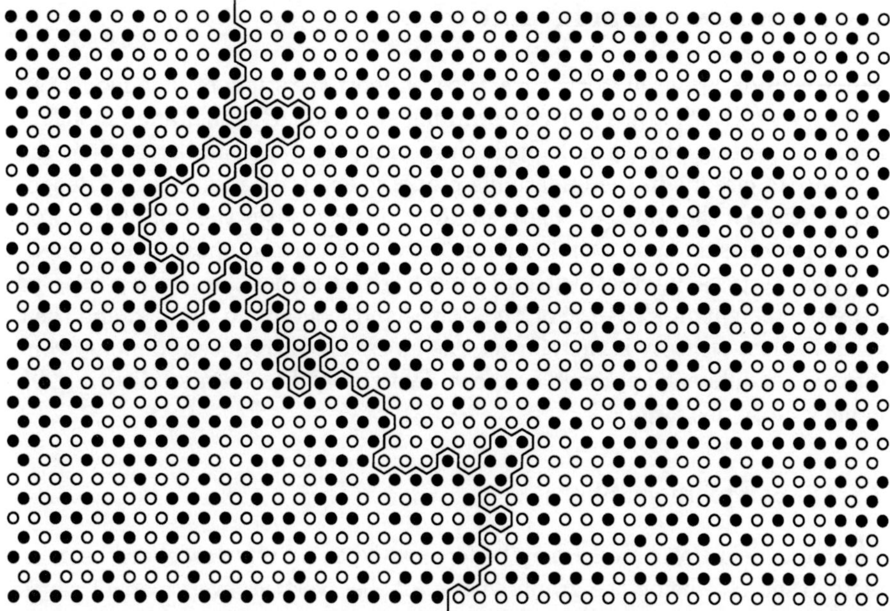

Fig. 6.2. Site percolation on the triangular lattice with p equal to the critical point $\frac{1}{2}$, and with a mixed boundary condition along the lower side. The interface traces the boundary between the white and the black clusters touching the boundary, and is termed the 'exploration process'. In the limit of small lattice-spacing, the interface converges in a certain manner to the graph of a function which satisfies the Löwner differential equation driven by a Brownian motion with variance parameter $\kappa = 6$.

remains only to specify the parameter κ of the limit in terms of q. It has been conjectured in [130] that κ satisfies $\cos(4\pi/\kappa) = -\frac{1}{2}\sqrt{q}$, $\kappa \in (4,8)$. This value is consistent with Smirnov's theorem [139], and also with the finding of [111] that the scaling limit of the uniform spanning tree Peano curve is SLE_8, on recalling that the uniform spanning tree measure is obtainable as a limit of the random-cluster measure as $p, q \downarrow 0$ (recall Section 2.4).

There are uncertainties over how this programme will develop. For a start, the theory of random-cluster models is not as complete as that of percolation and of the uniform spanning tree. Secondly, the existence of spatial limits is currently known only in certain special cases. The programme is however ambitious and full of promise, and should ultimately yield a full picture of the critical behaviour – including values of exponents – of random-cluster models, and hence of Ising/Potts models, with $q \leq 4$.

7 On complete graphs and trees

While considerations of 'real space–time' support the study of such models on lattices such as $\mathbb{L}^d$, it has proved rewarding also to analyse the random-cluster model on certain other graphs including complete graphs and trees. It is the presence of circuits in the underlying graph which is the root cause of dependence between the states of edges, and for this reason it is the complete graph which provides an appropriate setting for what is termed 'mean-field theory', in which vertices 'interact' with *all* other vertices rather than with a selected subset of 'neighbours'. Trees, on the other hand, contain no circuits, and their random-cluster theory is therefore sterile unless one introduces boundary conditions. [A different approach to mean-field theory has been studied in [99], namely on $\mathbb{L}^d$ for large d.]

7.1 On complete graphs

The mean-field Potts model may be formulated as a Potts model on the complete graph K_n, being the graph with n labelled vertices every pair of which is joined by an edge. The study of such a process dates back at least to 1954, [100], and has been continued over the last fifty years [26, 99, 151]. The model is exactly soluble in the sense that quantities of interest may be calculated exactly and rigorously. It is therefore not surprising that the corresponding random-cluster models (for real q) have 'exact solutions' also, [26].

Consider the random-cluster measure $\psi_{n,\lambda,q} = \phi_{K_n,\lambda/n,q}$ on the complete graph K_n, having parameters $p = \lambda/n$ and q; this is the appropriate scaling to allow an interesting limit as $n \to \infty$. In the case $q = 1$, this measure is product measure, and therefore the ensuing graph is an Erdős–Rényi random graph [25, 90]. The overall picture for general values of q is rather richer than for the case $q = 1$, and many exact calculations may be performed rigorously. It turns out that the phase transition is of first-order if and only if $q > 2$, and the behaviour of the system depends on how λ compares with a 'critical value' $\lambda_c(q)$ taking the value

$$\lambda_c(q) = \begin{cases} q & \text{if } 0 < q \le 2, \\ 2\left(\dfrac{q-1}{q-2}\right)\log(q-1) & \text{if } q > 2. \end{cases}$$

From the detailed picture described in [26] the following information may be extracted. The given properties occur with $\psi_{n,\lambda,q}$-probability tending to 1 as $n \to \infty$.

I. Subcritical case, when $\lambda < \lambda_c(q)$

The largest component of the graph is of order $\log n$.

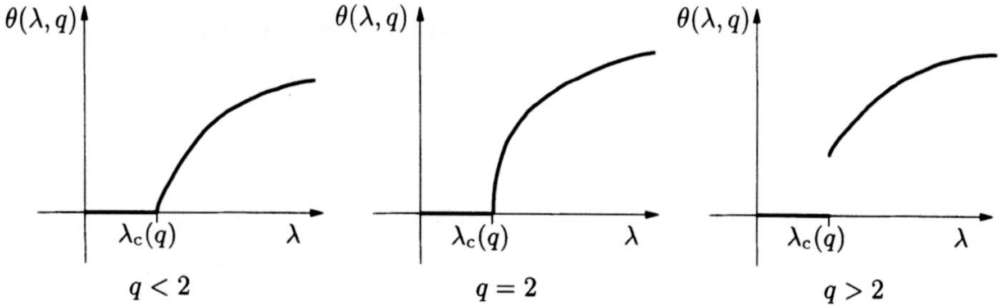

Fig. 7.1. The function $\theta(\lambda, q)$ for the three cases $q < 2$, $q = 2$, $q > 2$.

II. Supercritical case, when $\lambda > \lambda_c(q)$

There is a 'giant component' having order $\theta(\lambda, q)n$ where θ is defined to be the largest root of the equation

$$e^{\lambda\theta} = \frac{1 + (q-1)\theta}{1 - \theta}.$$

III. Critical case, when $\lambda = \lambda_c(q)$, $0 < q \le 2$

The largest component has order $n^{2/3}$.

IV. Critical case, when $\lambda = \lambda_c(q)$, $q > 2$

The largest component is either of order $\log n$ or of order $\theta(\lambda, q)n$, where θ is given as in case II above.

The dichotomy between first- and second-order phase transition is seen by studying the function $\theta(\lambda, q)$, sketched in Fig. 7.1. When $0 < q \le 2$, the function $\theta(\lambda, q)$ descends continuously to 0 as $\lambda \downarrow \lambda_c(q)$. On the other hand, this limit is strictly positive when $q > 2$.

The above results are obtained via a relationship between the model for general q and the model for the special case $q = 1$. The latter system has been analysed extensively, [25, 90]. We illustrate the argument in the case $q \ge 1$; a similar approach is valid when $q < 1$. Consider the open clusters $C_1, C_2, \ldots, C_m$ of a sample from the random-cluster measure $\phi_{K_n, p, q}$. We colour each such cluster *red* with probability ϱ, and *white* otherwise, different clusters receiving independent colours. We delete all vertices in white clusters, and let H denote the remaining graph, comprising a certain random number N of vertices (from the red clusters) together with certain open edges joining pairs of them. It may be seen that, conditional on the value of N, the measure governing H is the random-cluster measure with parameters p and $q\varrho$. We choose $\varrho = 1/q$ to obtain an Erdős–Rényi random graph on a *random* set of vertices. This is the observation which permits the full analysis to proceed.

One consequence of this study is an explicit identification of the exponential asymptotics of the partition function $Z_{K_n, \lambda/n, q}$, namely of the limit function

$$f(\lambda, q) = \lim_{n \to \infty} \left\{ \frac{1}{n} \log Z_{K_n, \lambda/n, q} \right\}.$$

This provides information via the Gärtner–Ellis theorem, [43], concerning the large-deviation theory of the number of clusters in such systems.

7.2 On trees and non-amenable graphs

Whereas physical considerations support the study of interacting systems on finite-dimensional lattices, mathematicians have been drawn also to the study of general graphs, thus enabling a clearer elucidation of the mathematical structure of such systems (see the discussion in [134]). A subject of special focus has been the class of graphs for which the ratio of surface to volume of finite boxes does not approach zero in the limit as the size of the box tends to infinity. A prime example of such a graph is an infinite regular tree with vertex degree at least three. We make the distinction more concrete as follows. Let $G = (V, E)$ be an infinite connected graph with finite vertex degrees. For $W \subseteq V$, we define its *boundary* ∂W to be the set of all $w \in W$ having some neighbour v not belonging to W. The countably infinite graph G is called *amenable* if its 'Cheeger constant'

$$\kappa(G) = \inf \left\{ \frac{|\partial W|}{|W|} : W \subseteq V, \; 0 < |W| < \infty \right\}$$

satisfies $\kappa(G) = 0$; G is called *non-amenable* if $\kappa(G) > 0$. It is easily seen that $\mathbb{L}^d$ is amenable, whereas an infinite regular tree with degree at least three is non-amenable.

The role of amenability in probability theory has been evident since the work of Kesten [95, 96] concerning random walks on a general graph G. More relevant to this review are [17, 75], which consider the number of infinite clusters in the bond percolation model on G. Suppose G is a quasi-transitive graph (that is, its vertex set has only finitely many orbits under its automorphism group). Suppose in addition that G is amenable. Consider bond percolation on G with density p. It may be proved as in [30, 58] that the number I of infinite open clusters satisfies

$$\text{either} \quad P_p(I = 0) = 1 \quad \text{or} \quad P_p(I = 1) = 1.$$

That is, if an infinite open cluster exists, then it is almost surely unique. Under similar assumptions on a non-amenable graph G, it is believed but not yet proved in full generality that there exists an interval of values of p for which $P_p(I = \infty) = 1$; see, for example, the discussion in [113]. A corresponding question for random-cluster models is to ascertain for which graphs G and

values of q there is non-uniqueness of random-cluster measures for an interval
of values of p. [Recall Theorem 3.7, easily extended to more general amenable
graphs, which states that, for $q \geq 1$, there is a unique random-cluster measure
on $\mathbb{L}^d$ for all except at most countably many values of p.] See [82, 92] and especially [134, Section 6.1], for recent accounts of this and associated questions,
and [80] for an analysis of random-cluster measures on regular trees.

8 Time-evolutions of random-cluster models

Let μ be a probability measure on a space $(\Omega, \mathcal{F})$. We may study stochastic processes taking values in Ω which converge weakly to μ in the limit of
large times. There are a multiplicity of reasons for and benefits in studying
time-evolutions. First, physical systems generally have dynamics as well as
equilibria. Secondly, new questions of interest arise, such as that of the estimation of a relaxation time. Thirdly, the dynamics thus introduced can yield
a new technique for studying the limit measure μ.

When studying a physical system, it is often acceptable to restrict oneself
to dynamics which are reversible in time. In Section 8.1, we describe a natural
reversible dynamic for a random-cluster model, akin to the Glauber dynamics
of the Ising model. This dynamic permits an extension which couples together
the random-cluster measures on a given graph as p and q range over their
possible values.

The problem commonly arises in statistics, computer science, and statistical physics of how to obtain a sample from a system governed by a probability measure with complex structure. In Section 8.2 we summarise the
Propp–Wilson 'coupling from the past' approach, [128], to this problem in
the context of the random-cluster measure.

Since Potts models may be obtained from random-cluster models, there is
an interplay between the dynamics for these two systems. A famous instance
of this relationship is the so-called Swendsen–Wang dynamic [141], which is
described in Section 8.3.

We assume throughout this section that $G = (V, E)$ is a finite connected
graph, and that $\phi_{p,q}$ is the random-cluster measure on $\Omega = \{0,1\}^E$ with
$0 < p < 1$ and $q > 0$.

8.1 Reversible dynamics

We shall consider transitions from a configuration ω to configurations which
differ from ω on one edge only. Thus we introduce the following notation. For
$\omega \in \Omega$ and $e \in E$, let ω^e and ω_e be the configurations obtained by 'switching
e on' and 'switching e off', respectively, that is

$$\omega^e(f) = \begin{cases} 1 & \text{if } f = e, \\ \omega(f) & \text{if } f \neq e, \end{cases} \qquad \omega_e(f) = \begin{cases} 0 & \text{if } f = e, \\ \omega(f) & \text{if } f \neq e. \end{cases}$$

Let $(X_t : t \geq 0)$ be a Markov chain, [77], on the state space Ω with generator $Q = \{q_{\omega,\omega'} : \omega, \omega' \in \Omega\}$ satisfying

$$q_{\omega_e,\omega^e} = p, \quad q_{\omega^e,\omega_e} = (1-p)q^{D(e,\omega_e)}, \qquad \text{for } \omega \in \Omega,\ e \in E,$$

where $D(e,\xi)$ is the indicator function of the event that the endpoints of e are joined by no open path of ξ. This specifies the rate at which single edges are acquired or lost. We set $q_{\omega,\xi} = 0$ if ω and ξ differ on two or more edges, and we choose the diagonal elements $q_{\omega,\omega}$ in such a way that Q, when viewed as a matrix, has row sums zero, that is,

$$q_{\omega,\omega} = -\sum_{\xi:\xi\neq\omega} q_{\omega,\xi}\,.$$

It is elementary that the 'detailed balance equations'

$$\phi_{p,q}(\omega)q_{\omega,\omega'} = \phi_{p,q}(\omega')q_{\omega',\omega}\,, \quad \omega, \omega' \in \Omega,$$

hold, whence X is reversible with respect to $\phi_{p,q}$. It follows by the irreducibility of the chain that $X_t \Rightarrow \phi_{p,q}$ as $t \to \infty$ (where '$\Rightarrow$' denotes weak convergence). There are of course many Markov chains with generators satisfying the above detailed balance equations, the important quantity is the ratio $q_{\omega,\omega'}/q_{\omega',\omega}$.

Two extensions of this dynamical structure which have proved useful are as follows. The evolution may be specified in terms of a 'graphical representation' constructed via a family of independent Poisson processes. This allows a natural coupling of the measures $\phi_{p,q}$ for different p and q. Such couplings are monotone in p when $q \geq 1$. One may similarly couple the unconditional measure $\phi_{p,q}(\cdot)$ and the conditioned measure $\phi_{p,q}(\cdot \mid A)$. Such couplings permit probabilistic interpretations of differences of the form $\phi_{p',q}(B \mid A) - \phi_{p,q}(B)$ when $q \geq 1$, $p \leq p'$, and A and B are increasing, and this can be useful in particular calculations (see [19, 68, 69]).

We turn now to the thermodynamic limit, and the question of the structure of a Markovian random-cluster process on an infinite connected graph. In the case $q \geq 1$, the above couplings are monotone in the choice of the underlying graph G. Therefore there exist 'limit dynamics' as G passes through an increasing sequence of finite graphs. Boundary conditions may be introduced, and one may obtain thereby a certain Markov process $\zeta = (\zeta_t : t \geq 0)$ on the state space $[0,1]^{\mathbb{E}}$, where $\mathbb{E}$ is the limiting (infinite) edge set. This process, which does not generally have the Feller property, generates a pair of 'level-set processes' taking values in $\{0,1\}^{\mathbb{E}}$, defined for $0 \leq p \leq 1$ by

$$\zeta_t^{p,-}(e) = 1_{\{\zeta_t(e)>1-p\}}\,, \quad \zeta_t^{p,+}(e) = 1_{\{\zeta_t(e)\geq 1-p\}}\,, \qquad e \in \mathbb{E},$$

where, as before, 1_A denotes the indicator function of an event A. These two processes are Markovian and are reversible with respect to the infinite-volume free and wired random-cluster measures, respectively. See [69].

Note that the generator of the Markov chain given above depends on the random variable $D(e, \omega_e)$, and that this random variable is a 'non-local' function of the configuration ω in the sense that there is no absolute bound on the distance from e of edges whose states may be relevant to its value. It is this feature of non-locality which leads to interesting complications linked in part to the 0/1-infinite-cluster property introduced before Theorem 3.9. Further discussion may be found in [69, 121].

8.2 Coupling from the past

In running Monte Carlo experiments, one requires the ability to sample from the probability measure $\phi_{p,q}$. The Markov chain X_t of Section 8.1 certainly converges weakly to $\phi_{p,q}$ as $t \to \infty$, but this is not as good as having in the hand a sample with the *exact* distribution. Random-cluster measures are well suited to the Propp–Wilson approach to sampling termed 'coupling from the past', [128], and we sketch this here. Some illustrations may be found in [91].

First we provide ourselves with a discrete-time reversible Markov chain $(Z_n : n \geq 0)$ on the state space Ω having invariant measure $\phi_{p,q}$. The so-called *heat-bath algorithm* provides a suitable example of such a chain, and proceeds as follows. At each stage, we pick a random edge e, chosen uniformly from E and independently of all earlier choices, and we make e open with the correct conditional probability, given the configuration on the other edges. The corresponding transition matrix is given by $\Pi = \{\pi_{\omega,\omega'} : \omega, \omega' \in \Omega\}$ where

$$\pi_{\omega_e,\omega^e} = \frac{1}{|E|} \cdot \frac{\phi_{p,q}(\omega^e)}{\phi_{p,q}(\omega^e) + \phi_{p,q}(\omega_e)},$$

$$\pi_{\omega^e,\omega_e} = \frac{1}{|E|} \cdot \frac{\phi_{p,q}(\omega_e)}{\phi_{p,q}(\omega^e) + \phi_{p,q}(\omega_e)}.$$

A neat way to do this is as follows. Suppose that $Z_n = \omega$. Let e_n be a random edge of E, and let U_n be uniformly distributed on the interval $[0, 1]$, these variables being chosen independently of all earlier choices. We obtain Z_{n+1} from ω by retaining the states of all edges except possibly that of e_n. We set

$$Z_{n+1}(e_n) = 0 \quad \text{if and only if} \quad U_n \leq \frac{\phi_{p,q}(\omega_e)}{\phi_{p,q}(\omega^e) + \phi_{p,q}(\omega_e)}.$$

Thus the evolution of the chain is determined by the sequences e_n, U_n, and the initial state Z_0. One may make this construction explicit by writing $Z_{n+1} = \psi(Z_n, e_n, U_n)$ $(= \psi(\omega, e_n, U_n))$ for some function $\psi : \Omega \times E \times [0, 1] \to \Omega$. It is easily seen by the Holley condition of Section 3.1 that, if $q \geq 1$, and for every e and u, the function $\psi(\cdot, e, u)$ is non-decreasing in its first argument. It follows that the coupling is 'monotone' in the sense that, if $\omega \leq \omega'$, then chain starting at ω lies at all times beneath the chain starting at ω' (using the partial order on Ω).

We let $W = (W(\omega) : \omega \in \Omega)$ be a vector of random variables such that $W(\omega)$ has the distribution of Z_1 conditional on $Z_0 = \omega$. Following the scheme described above, we may take $W(\omega) = \psi(\omega, e, U)$ where e and U are chosen at random. Let W_{-m}, $m \geq 1$, be independent random vectors distributed as W, that is, $W_{-m}(\cdot) = \psi(\cdot, e_m, U_m)$ where the set $\{(e_m, U_m) : m \geq 1\}$ comprises independent pairs of independent random variables, each e_i being uniform on E, and each U_i being uniform on $[0, 1]$. We now construct a sequence Y_{-n}, $n \geq 1$, of random maps from Ω to Ω by the following inductive procedure. First, for $\omega \in \Omega$, we set $Y_{-1}(\omega) = W_{-1}(\omega)$. Having found $Y_{-1}, Y_{-2}, \ldots, Y_{-m}$, we define $Y_{-m-1}(\omega) = Y_{-m}(W_{-m-1}(\omega))$. That is, $Y_{-m-1}(\omega)$ is obtained from ω by passing in one step to $W_{-m-1}(\omega)$, and then applying Y_{-m} to this new state. The exact dependence structure of this scheme is an important ingredient of what follows.

We stop this process at the earliest time m at which 'coalescence has occurred', that is, at the moment M given by $M = \min\{m : Y_{-m}(\cdot)$ is the constant function$\}$. It is a theorem, [128], that M is $\phi_{p,q}$-a.s. finite and, for any ω, the random output $Y_{-M}(\omega)$ is governed exactly by the probability measure $\phi_{p,q}$.

This procedure looks unwieldy, since Ω is typically rather large, but the reality is simpler when $q \geq 1$. By the monotonicity of the above coupling when $q \geq 1$, it suffices to follow the trajectories of the 'smallest' and 'largest' configurations, namely those beginning, respectively, with every edge closed and with every edge open. The processes starting at intermediate configurations remain sandwiched between the extremal processes, for all times t. Thus one may define M by $M = \min\{m : Y_{-m}(0) = Y_{-m}(1)\}$, where 0 and 1 denote the vectors of zeros and ones as before.

8.3 Swendsen–Wang dynamics

It is a major target of statistical physics to understand the time-evolution of disordered systems, and a prime example lies in the study of the Ising model. A multiplicity of types of dynamics have been proposed. The majority of these share a quality of 'locality' in the sense that the evolution involves changes to the states of vertices in close proximity to one another, perhaps single spin-flips, or spin-exchanges. The state space is generally large, of size 2^N where N is the number of vertices, and the Hamiltonian has complicated structure. When subjected to 'local dynamics', the process may approach equilibrium very slowly (see [115, 133] for accounts of recent work of relevance). 'Non-local dynamics', on the other hand, have the potential to approach equilibrium faster, since they permit large jumps around the state space, relatively unconstrained by neighbourly relations. The random-cluster model has played a role in the development of a simple but attractive such system, namely that proposed by Swendsen and Wang [141] and described as follows for the Potts model with q states.

As usual, $G = (V, E)$ is a finite graph, typically a large box in $\mathbb{Z}^d$, and $\Sigma = \{1, 2, \ldots, q\}^V$ is the state space of a Potts model on G. We write $\Omega = \{0, 1\}^E$. Suppose that, at some time n, we have obtained a configuration σ_n ($\in \Sigma$). We construct σ_{n+1} as follows. Let $p = 1 - e^{-\beta J}$ where $0 < \beta J < \infty$.

I.   We let $\omega_n \in \Omega$ be given as follows. For $e = \langle x, y \rangle \in E$,

$$\text{if } \sigma_n(x) \neq \sigma_n(y), \text{ let } \omega_n(e) = 0\,,$$

$$\text{if } \sigma_n(x) = \sigma_n(y), \text{ let } \omega_n(e) = \begin{cases} 1 & \text{with probability } p\,, \\ 0 & \text{otherwise}\,, \end{cases}$$

different edges receiving independent states. The edge configuration ω_n is carried forward to the next stage.

II. To each cluster C of the graph $(V, \eta(\omega_n))$ we assign an integer chosen uniformly at random from the set $\{1, 2, \ldots, q\}$, different clusters receiving independent labels. We let $\sigma_{n+1}(x)$ be the value thus assigned to the cluster containing the vertex x.

It may be checked that the Markov chain $(\sigma_n : n \geq 0)$ has as unique invariant measure the Potts measure on Σ with parameters β and J. (Recall paragraph (c) of Section 2.3.)

The Swendsen–Wang algorithm leads to samples which generally converge to equilibrium faster than those defined via local dynamics. This is especially evident in the 'high β' (or 'low temperature') phase, for the following reason. Consider for example the simulation of an Ising model on a finite box with free boundary conditions, and suppose that the initial state is $+1$ at all vertices. If β is large, then local dynamics result in samples which remain close to the '+ phase' for a very long time. Only after a long wait will the process achieve an average magnetisation close to 0. Swendsen–Wang dynamics, on the other hand, can achieve large jumps in average magnetisation even in a single step, since the spin allocated to a given large cluster of the corresponding random-cluster model is equally likely to be either of the two possibilities. A rigorous analysis of rates of convergence is however incomplete. It turns out that, *at* the critical point, Swendsen–Wang dynamics approach equilibrium only slowly, [28]. A further discussion is available in [61].

Algorithms of Swendsen–Wang type have been described for other statistical mechanical models having graphical representations of random-cluster-type; see [41, 42]. Related work may be found in [149].

Acknowledgements. GRG recalls John Hammersley passing to him in 1971 a copy of Fortuin's thesis [52] in which much of the basic theory is developed. Piet Kasteleyn kindly filled out the origins of random-cluster models in two letters addressed to GRG in November 1992. The author acknowledges the opportunity given by the Landau Center of the Hebrew University, Jerusalem, to deliver a course of lectures on the random-cluster model during July 2001. Harry Kesten kindly criticised a draft of the work. Thanks are due to Malwina Luczak for her contributions to discussions on certain topics in this paper, and to Ágoston Pisztora for reading and commenting on parts of it. The further suggestions of Christian Borgs, Olle Häggström, Russell Lyons, Roberto Schonmann, Oded Schramm, and Alan Sokal have been appreciated. The paper was completed during a programme at the Isaac Newton Institute (Cambridge).

References

1. Aizenman, M., Geometric analysis of ϕ^4 fields and Ising models, Communications in Mathematical Physics **86**, 1–48 (1982)
2. Aizenman, M., Barsky, D. J., Fernández, R., The phase transition in a general class of Ising-type models is sharp, Communications in Mathematical Physics **47**, 343–374 (1987)
3. Aizenman, M., Chayes, J. T., Chayes, L., Newman, C. M., The phase boundary in dilute and random Ising and Potts ferromagnets, Journal of Physics A: Mathematical and General **20**, L313–L318 (1987)
4. Aizenman, M., Chayes, J. T., Chayes, L., Newman, C. M., Discontinuity of the magnetization in one-dimensional $1/|x - y|^2$ Ising and Potts models, Journal of Statistical Physics **50**, 1–40 (1988)
5. Aizenman, M., Fernández, R., On the critical behavior of the magnetization in high-dimensional Ising models, Journal of Statistical Physics **44**, 393–454 (1986)
6. Alexander, K., Simultaneous uniqueness of infinite clusters in stationary random labeled graphs, Communications in Mathematical Physics **168**, 39–55 (1995)
7. Alexander, K., Weak mixing in lattice models, Probability Theory and Related Fields **110**, 441–471 (1998)
8. Alexander, K., The asymmetric random cluster model and comparison of Ising and Potts models, Probability Theory and Related Fields **120**, 395–444 (2001)
9. Alexander, K., Power-law corrections to exponential decay of connectivities and correlations in lattice models, Annals of Probability **29**, 92–122 (2001)
10. Alexander, K., Cube-root boundary fluctuations for droplets in random cluster models, Communications in Mathematical Physics **224**, 733–781 (2001)
11. Alexander, K., The single-droplet theorem for random-cluster models, In: *In and Out of Equilibrium*, ed. Sidoravicius, V., Birkhäuser, Boston, 47–73 (2002)
12. Ashkin, J., Teller, E., Statistics of two-dimensional lattices with four components, The Physical Review **64**, 178–184 (1943)
13. Barlow, R. N., Proschan, F., *Mathematical Theory of Reliability*, Wiley, New York, (1965)
14. Baxter, R. J., *Exactly Solved Models in Statistical Mechanics*, Academic Press, London, (1982)
15. Beijeren, H. van, Interface sharpness in the Ising system, Communications in Mathematical Physics **40**, 1–6 (1975)
16. Benjamini, I., Lyons, R., Peres, Y., Schramm, O., Uniform spanning forests, Annals of Probability **29**, 1–65 (2001)
17. Benjamini, I., Schramm, O., Percolation beyond $\mathbb{Z}^d$, many questions and a few answers, Electronic Communications in Probability **1**, 71–82 (1996)

18. Berg, J. van den, Kesten, H. , Inequalities with applications to percolation and reliability, Journal of Applied Probability **22**, 556–569 (1985)

19. Bezuidenhout, C. E., Grimmett, G. R., Kesten, H., Strict inequality for critical values of Potts models and random-cluster processes, Communications in Mathematical Physics **158**, 1–16 (1993)

20. Biggs, N. L., *Algebraic Graph Theory*, Cambridge University Press, Cambridge, (1984)

21. Biggs, N. L., *Interaction Models*, Cambridge University Press, LMS Lecture Note Series no. 30, Cambridge, (1977)

22. Billingsley, P., *Convergence of Probability Measures*, Wiley, New York, (1968)

23. Biskup, M., Borgs, C., Chayes, J. T., Kotecký, R., Gibbs states of graphical representations of the Potts model with external fields. Probabilistic techniques in equilibrium and nonequilibrium statistical physics, Journal of Mathematical Physics **41**, 1170–1210 (2000)

24. Bodineau, T., The Wulff construction in three and more dimensions, Communications in Mathematical Physics **207**, 197–229 (1999)

25. Bollobás, B., *Random Graphs*, Academic Press, London, (1985)

26. Bollobás, B., Grimmett, G. R., Janson, S., The random-cluster process on the complete graph, Probability Theory and Related Fields **104**, 283–317 (1996)

27. Borgs, C., Chayes, J. T., The covariance matrix of the Potts model: A random-cluster analysis, Journal of Statistical Physics **82**, 1235–1297 (1996)

28. Borgs, C., Chayes, J. T., Frieze, A. M., Kim, J. H., Tetali, E., Vigoda, E., Vu, V. V., Torpid mixing of some MCMC algorithms in statistical physics, Proceedings of the 40th IEEE Symposium on the Foundations of Computer Science 218–229 (1999)

29. Broadbent, S.R., Hammersley, J. M., Percolation processes I. Crystals and mazes, Proceedings of the Cambridge Philosophical Society **53**, 629–641 (1957)

30. Burton, R. M., Keane, M., Density and uniqueness in percolation, Communications in Mathematical Physics **121**, 501–505 (1989)

31. Burton, R. M., Keane, M., Topological and metric properties of infinite clusters in stationary two-dimensional site percolation, Israel Journal of Mathematics **76**, 299–316 (1991)

32. Campanino, M., Chayes, J. T., Chayes, L., Gaussian fluctuations of connectivities in the subcritical regime of percolation, Probability Theory and Related Fields **88**, 269–341 (1991)

33. Campanino, M., Ioffe, D., Velenik, Y., Ornstein–Zernike theory for the finite range Ising models above T_c, Probability Theory and Related Fields **125**, 305–349 (2003)

34. Cardy, J., Critical percolation in finite geometries, Journal of Physics A: Mathematical and General **25**, L201 (1992)

35. Cerf, R., Kenyon, R., The low-temperature expansion of the Wulff crystal in the 3D Ising model, Communications in Mathematical Physics **222**, 147–179 (2001)

36. Cerf, R., Pisztora, Á., On the Wulff crystal in the Ising model, Annals of Probability **28**, 947–1017 (2000)

37. Cerf, R., Pisztora, Á., Phase coexistence in Ising, Potts and percolation models, Annales de l'Institut Henri Poincaré, Probabilités et Statistiques **37**, 643–724 (2001)

38. Černý, J., Kotecký, R., Interfaces for random cluster models, Journal of Statistical Physics **111**, 73–106 (2003)

39. Chayes, J. T., Chayes, L., Kotecký, R., The analysis of the Widom–Rowlinson model by stochastic geometric methods, Communications in Mathematical Physics **172**, 551–569 (1995)

40. Chayes, L., Kotecký, R., Intermediate phase for a classical continuum model, Physical Review B **54**, 9221–9224 (1996)

41. Chayes, L., Machta, J., Graphical representations and cluster algorithms, Part I: discrete spin systems, Physica A **239**, 542–601 (1997)

42. Chayes, L., Machta, J., Graphical representations and cluster algorithms, II, Physica A **254**, 477–516 (1998)

43. Dembo, A., Zeitouni, O., *Large deviations techniques and applications*, 2nd edition, Springer, New York, (1998)

44. Dobrushin, R. L., Gibbsian random fields for lattice systems and pairwise interactions, Functional Analysis and its Applications (in translation) **2**, 292–301 (1968)

45. Doyle, P. G., Snell, J. L., *Random Walks and Electric Networks*, Carus Mathematical Monographs 22, Mathematical Association of America, Washington, DC, (1984)

46. Edwards, S. F., Anderson, P. W., Theory of spin glasses, Journal of Physics F: Metal Physics **5**, 965–974 (1975)

47. Edwards, R. G., Sokal, A. D., Generalization of the Fortuin–Kasteleyn–Swendsen–Wang representation and Monte Carlo algorithm, The Physical Review D **38**, 2009–2012 (1988)

48. Feder, T., Mihail, M., Balanced matroids, Proceedings of the 24th ACM Symposium on the Theory of Computing 26–38 (1992)

49. Fernández, R., Ferrari, P. A., Garcia, N. L., Loss network representation for Peierls contours, Annals of Probability **29**, 902–937 (2001)

50. Fernández, R., Fröhlich, J., Sokal, A. D., *Random Walks, Critical Phenomena, and Triviality in Quantum Field Theory*, Springer, Berlin (1992)

51. Ferrari, P. A., Fernández, R., Garcia, N. L., Perfect simulation for interacting point processes, loss networks and Ising models, Stochastic Processes and their Applications **102**, 63–88 (2002)

52. Fortuin, C. M., *On the random-cluster model*, Doctoral thesis, University of Leiden, (1971)

53. Fortuin, C. M., On the random-cluster model. II. The percolation model, Physica **58**, 393–418 (1972)

54. Fortuin, C. M., On the random-cluster model. III. The simple random-cluster process, Physica **59**, 545–570 (1972)

55. Fortuin, C. M., Kasteleyn, P. W., On the random-cluster model. I. Introduction and relation to other models, Physica **57**, 536–564 (1972)

56. Fortuin, C. M., Kasteleyn, P. W., Ginibre, J., Correlation inequalities on some partially ordered sets, Communications in Mathematical Physics **22**, 89–103 (1971)

57. Gallavotti, G., Miracle-Solé, S., Equilibrium states of the Ising model in the two-phase region, Physical Review B **5**, 2555–2559 (1972)

58. Gandolfi, A., Keane, M., Newman, C. M., Uniqueness of the infinite component in a random graph with applications to percolation and spin glasses, Probability Theory and Related Fields **92**, 511–527 (1992)

59. Georgii, H.-O., *Gibbs measures and phase transitions*, Walter de Gruyter, Berlin, (1988)

60. Georgii, H.-O., Häggström, O., Phase transition in continuum Potts models, Communications in Mathematical Physics **181**, 507–528 (1996)

61. Georgii, H.-O., Häggström, O., Maes, C., The random geometry of equilibrium phases, ed. Domb, C., Lebowitz, J. L., In: *Phase Transitions and Critical Phenomena*, **18**, 1–142, Academic Press, London, (2000)

62. Gielis, G., Grimmett, G. R., Rigidity of the interface in percolation and random-cluster models, Journal of Statistical Physics **109**, 1–37 (2002)

63. Grimmett, G. R., Unpublished (1991)

64. Grimmett, G. R., Differential inequalities for Potts and random-cluster processes, In: *Cellular Automata and Cooperative Systems*, ed. N. Boccara et al., Kluwer, Dordrecht, 227–236 (1993)

65. Grimmett, G. R., Potts models and random-cluster processes with many-body interactions, Journal of Statistical Physics **75**, 67–121 (1994)

66. Grimmett, G. R., The random-cluster model, In: *Probability, Statistics and Optimisation*, Wiley, ed. F. P. Kelly, Chichester, 49–63 (1994)

67. Grimmett, G. R., Percolative problems, In: *Probability and Phase Transition*, ed. G. R. Grimmett, Kluwer, Dordrecht, 69–86 (1994)

68. Grimmett, G. R., Comparison and disjoint-occurrence inequalities for random-cluster models, Journal of Statistical Physics **78**, 1311–1324 (1995)

69. Grimmett, G. R., The stochastic random-cluster process and the uniqueness of random-cluster measures, Annals of Probability **23**, 1461–1510 (1995)

70. Grimmett, G. R., Percolation and disordered systems, In: *Ecole d'Eté de Probabilités de Saint Flour XXVI–1996*, ed. P. Bernard, Lecture Notes in Mathematics no. 1665, Springer, Berlin, 153–300 (1997)

71. Grimmett, G. R., *Percolation*, 2nd edition, Springer, Berlin, (1999)

72. Grimmett, G. R., Inequalities and entanglements for percolation and random-cluster models, In: *Perplexing Problems in Probability; Festschrift in Honor of Harry Kesten*, ed. M. Bramson, R. Durrett, Birkhäuser, Boston, 91–105 (1999)

73. Grimmett, G. R., *The Random-Cluster Model*, in preparation, (2002)

74. Grimmett, G. R., Marstrand, J. M., The supercritical phase of percolation is well behaved, Proceedings of the Royal Society (London), Series A **430**, 439–457 (1990)

75. Grimmett, G. R., Newman, C. M., Percolation in $\infty + 1$ dimensions, In: *Disorder in Physical Systems*, ed. G. R. Grimmett, D. J. A. Welsh, Oxford University Press, Oxford, 219–240 (1990)

76. Grimmett, G. R., Piza, M. S. T., Decay of correlations in subcritical Potts and random-cluster models, Communications in Mathematical Physics **189**, 465–480 (1997)

77. Grimmett, G. R., Stirzaker, D. R., *Probability and Random Processes*, 3rd edition, Oxford University Press, Oxford, (2001)

78. Grimmett, G. R., Winkler, S., Negative association in uniform forests and connected graphs, (to appear)(2003)

79. Häggström, O., Random-cluster measures and uniform spanning trees, Stochastic Processes and their Applications **59**, 267–275 (1995)

80. Häggström, O., The random-cluster model on a homogeneous tree, Probability Theory and Related Fields **104**, 231–253 (1996)

81. Häggström, O., Random-cluster representations in the study of phase transitions, Markov Processes and Related Fields **4**, 275–321 (1998)

82. Häggström, O., Jonasson, J., Lyons, R., Explicit isoperimetric constants and phase transitions in the random-cluster model, Annals of Probability **30**, 443–473 (2002)

83. Häggström, O., Jonasson, J., Lyons, R., Coupling and Bernoullicity in random-cluster and Potts models, Bernoulli **8**, 275–294 (2002)

84. Hammersley, J. M., Percolation processes. Lower bounds for the critical probability, Annals of Mathematical Statistics **28**, 790–795 (1957)

85. Hara, T., Slade, G., Mean-field critical behaviour for percolation in high dimensions, Communications in Mathematical Physics **128**, 333–391 (1990)

86. Hara, T., Slade, G., The scaling limit of the incipient infinite cluster in high-dimensional percolation. II. Integrated super-Brownian excursion, Journal of Mathematical Physics **41**, 1244–1293 (2000)

87. Hintermann, D., Kunz, H., Wu, F. Y., Exact results for the Potts model in two dimensions, Journal of Statistical Physics **19**, 623–632 (1978)

88. Holley, R., Remarks on the FKG inequalities, Communications in Mathematical Physics **36**, 227–231 (1974)

89. Ising, E., Beitrag zur Theorie des Ferromagnetismus, Zeitschrift für Physik **31**, 253–258 (1925)

90. Janson, S., Łuczak, T., Ruciński, A., *Random Graphs*, Wiley, New York, (2000)

91. Jerrum, M., Mathematical foundations of the Markov chain Monte Carlo method, In: *Probabilistic Methods for Algorithmic Discrete Mathematics*, ed. Habib, M., McDiarmid, C., Ramirez-Alfonsin, J., Reed, B., Springer, Berlin, (1998)

92. Jonasson, J., The random cluster model on a general graph and a phase transition characterization of nonamenability, Stochastic Processes and their Applications **79**, 335–354 (1999)

93. Kahn, J., A normal law for matchings, Combinatorica **20**, 339–391 (2000)

94. Kasteleyn, P. W., Fortuin, C. M., Phase transitions in lattice systems with random local properties, Journal of the Physical Society of Japan, Supplement **26**, 11–14 (1969)

95. Kesten, H., Symmetric random walks on groups, Transactions of the American Mathematical Society **92**, 336–354 (1959)

96. Kesten, H., Full Banach mean values on countable groups, Mathematica Scandinavica **7**, 146–156 (1959)

97. Kesten, H., The critical probability of bond percolation on the square lattice equals $\frac{1}{2}$, Communications in Mathematical Physics **74**, 41–59 (1980)

98. Kesten, H., *Percolation Theory for Mathematicians*, Birkhäuser, Boston, (1982)

99. Kesten, H., Schonmann, R. H., Behavior in large dimensions of the Potts and Heisenberg models, Reviews in Mathematical Physics **1**, 147–182 (1990)

100. Kihara, T., Midzuno, Y., Shizume, J., Statistics of two-dimensional lattices with many components, Journal of the Physical Society of Japan **9**, 681–687 (1954)

101. Kim, D., Joseph, R. I., Exact transition temperatures for the Potts model with q states per site for the triangular and honeycomb lattices, Journal of Physics C: Solid State Physics **7**, L167–L169 (1974)

102. Kirchhoff, G., Über die Auflösung der Gleichungen, auf welche man bei der Untersuchung der linearen Verteilung galvanischer Ströme geführt wird, Annalen der Physik und Chemie **72**, 497–508 (1847)

103. Kotecký, R., Shlosman, S., First order phase transitions in large entropy lattice systems, Communications in Mathematical Physics **83**, 493–515 (1982)

104. Laanait, L., Messager, A., Ruiz, J., Phase coexistence and surface tensions for the Potts model, Communications in Mathematical Physics **105**, 527–545 (1986)

105. Laanait, L., Messager, A., Miracle-Solé, S., Ruiz, J., Shlosman, S., Interfaces in the Potts model I: Pirogov–Sinai theory of the Fortuin–Kasteleyn representation, Communications in Mathematical Physics **140**, 81–91 (1991)

106. Lanford, O. E., Ruelle, D., Observables at infinity and states with short range correlations in statistical mechanics, Communications in Mathematical Physics **13**, 194–215 (1969)

107. Langlands, R., Pouliot, P., Saint-Aubin, Y., Conformal invariance in two-dimensional percolation, Bulletin of the American Mathematical Society **30**, 1–61 (1994)

108. Lawler, G. F., Schramm, O., Werner, W., The dimension of the planar Brownian frontier is 4/3, Mathematics Research Letters **8**, 401–411 (2001)

109. Lawler, G. F., Schramm, O., Werner, W., Values of Brownian intersection exponents III: Two-sided exponents, Annales de l'Institut Henri Poincaré, Probabilités et Statistiques **38**, 109–123 (2002)

110. Lawler, G. F., Schramm, O., Werner, W., One-arm exponent for critical 2D percolation, Electronic Journal of Probability **7**, 1–13 (2002)

111. Lawler, G. F., Schramm, O., Werner, W., Conformal invariance of planar loop-erased random walks and uniform spanning trees, (to appear)(2001)

112. Lieb, E. H., A refinement of Simon's correlation inequality, Communications in Mathematical Physics **77**, 127–135 (1980)

113. Lyons, R., Phase transitions on nonamenable graphs, Journal of Mathematical Physics **41**, 1099–1126 (2001)

114. Madras, N., Slade, G., *The Self-Avoiding Walk*, Birkhäuser, Boston, (1993)

115. Martinelli, F., Lectures on Glauber dynamics for discrete spin models, In: *Ecole d'Eté de Probabilités de Saint Flour XXVII–1997*, ed. P. Bernard, Lecture Notes in Mathematics no. 1717, Springer, Berlin, 93–191 (1999)

116. Messager, A., Miracle-Solé, S., Ruiz, J., Shlosman, S., Interfaces in the Potts model. II. Antonov's rule and rigidity of the order disorder interface, Communications in Mathematical Physics **140**, 275–290 (1991)

117. Newman, C. M., Disordered Ising systems and random cluster representations, In: *Probability and Phase Transition*, ed. G. R. Grimmett, Kluwer, Dordrecht, 247–260 (1994)

118. Newman, C. M., *Topics in Disordered Systems*, Birkhäuser, Boston, (1997)

119. Onsager, L., Crystal statistics, I. A two-dimensional model with an order-disorder transition, The Physical Review **65**, 117–149 (1944)

120. Pemantle, R., Towards a theory of negative dependence, Journal of Mathematical Physics **41**, 1371–1390 (2000)

121. Pfister, C.-E., Vande Velde, K., Almost sure quasilocality in the random cluster model, Journal of Statistical Physics **79**, 765–774 (1995)

122. Pfister, C.-E., Velenik, Y., Random-cluster representation for the Ashkin–Teller model, Journal of Statistical Physics **88**, 1295–1331 (1997)

123. Pirogov, S. A., Sinai, Ya. G., Phase diagrams of classical lattice systems, Theoretical and Mathematical Physics **25**, 1185–1192 (1975)

124. Pirogov, S. A., Sinai, Ya. G., Phase diagrams of classical lattice systems, continuation, Theoretical and Mathematical Physics **26**, 39–49 (1976)

125. Pisztora, Á., Surface order large deviations for Ising, Potts and percolation models, Probability Theory and Related Fields **104**, 427–466 (1996)

126. Potts, R. B., Some generalized order–disorder transformations, Proceedings of the Cambridge Philosophical Society **48**, 106–109 (1952)

127. Preston, C. J., *Gibbs States on Countable Sets*, Cambridge University Press, Cambridge, (1974)

128. Propp, J. G., Wilson, D. B., Exact sampling with coupled Markov chains and applications to statistical mechanics, Random Structures and Algorithms **9**, 223–252 (1996)

129. Reimer, D., Proof of the van den Berg–Kesten conjecture, Combinatorics, Probability, Computing **9**, 27–32 (2000)

130. Rohde, S., Schramm, O., Basic properties of SLE, (to appear)(2001)

131. Russo, L., On the critical percolation probabilities, Zeitschrift für Wahrscheinlichkeitstheorie und Verwandte Gebiete **56**, 229–237 (1981)

132. Salas, J., Sokal, A. D., Dynamic critical behavior of a Swendsen–Wang-type algorithm for the Ashkin–Teller model, Journal of Statistical Physics **85**, 297–361 (1996)

133. Schonmann, R. H., Metastability and the Ising model, Proceedings of the International Congress of Mathematicians, Berlin 1998, ed. G. Fischer, U. Rehmann, Documenta Mathematica, Extra volume **III**, 173–181 (1998)

134. Schonmann, R. H., Multiplicity of phase transitions and mean-field criticality on highly non-amenable graphs, Communications in Mathematical Physics **219**, 271–322 (2001)

135. Schramm, O., Scaling limits of loop-erased walks and uniform spanning trees, Israel Journal of Mathematics **118**, 221–288 (2000)

136. Simon, B., Correlation inequalities and the decay of correlations in ferromagnets, Communications in Mathematical Physics **77**, 111–126 (1980)

137. Slade, G., Bounds on the self-avoiding walk connective constant, Journal of Fourier Analysis and its Applications, Special Issue: Proceedings of the Conference in Honor of Jean-Pierre Kahane, 1993, 525–533 (1995)

138. Smirnov, S., Critical percolation in the plane: conformal invariance, Cardy's formula, scaling limits, Comptes Rendus des Séances de l'Académie des Sciences. Série I. Mathématique **333**, 239–244 (2001)

139. Smirnov, S., Critical percolation in the plane. I. Conformal invariance and Cardy's formula. II. Continuum scaling limit, preprint (2001)

140. Smirnov, S., Werner, W., Critical exponents for two-dimensional percolation, Mathematics Research Letters **8**, 729–744 (2001)

141. Swendsen, R. H., Wang, J. S., Nonuniversal critical dynamics in Monte Carlo simulations, Physical Review Letters **58**, 86–88 (1987)

142. Tutte, W. T., *Graph Theory*, Addison-Wesley, Menlo Park, California, (1984)

143. Welsh, D. J. A., Percolation in the random-cluster process, Journal of Physics A: Mathematical and General **26**, 2471–2483 (1993)

144. Welsh, D. J. A., Merino, C., The Potts model and the Tutte polynomial, Journal of Mathematical Physics **41**, 1127–1152 (2000)

145. Werner, W., Random planar curves and Schramm–Loewner evolutions, In: *Ecole d'Eté de Probabilités de Saint Flour*, (to appear)(2003)

146. Widom, B., Rowlinson, J. S., New model for the study of liquid–vapor phase transition, Journal of Chemical Physics **52**, 1670–1684 (1970)

147. Wilson, R. J., *Introduction to Graph Theory*, Longman, London, (1979)

148. Wiseman, S., Domany, E., Cluster method for the Ashkin–Teller model, Physical Review E **48**, 4080–4090 (1993)

149. Wolff, U., Collective Monte Carlo updating for spin systems, Physical Review Letters **62**, 361–364 (1989)

150. Wood, De Volson, Problem 5, American Mathematical Monthly **1**, 99, 211–212 (1894)

151. Wu, F. Y., The Potts model, Reviews in Modern Physics **54**, 235–268 (1982)

152. Zhang, Y., Unpublished, see [71], page 289, (1988)

Models of First-Passage Percolation

C. Douglas Howard*

* Research supported by NSF Grant DMS-02-03943.

1 Introduction

1.1 The Basic Model and Some Fundamental Questions

First-passage percolation (FPP) was introduced by Hammersley and Welsh in 1965 (see [26]) as a model of fluid flow through a randomly porous material. Envision a fluid injected into the material at a fixed site: as time elapses, the portion of the material that is wet expands in a manner that is a complicated function of the material's random structure. In the standard FPP model, the spatial randomness of the material is represented by a family of *non-negative* i.i.d. random variables indexed by the nearest neighbor edges of the $\mathbf{Z}^d$ lattice. (We take $d \geq 2$ throughout this chapter.) If edge e has endpoints $\mathbf{u}, \mathbf{v} \in \mathbf{Z}^d$ (so $|\mathbf{u} - \mathbf{v}| = 1$, where $|\cdot|$ denotes the usual Euclidean norm) then the associated quantity $\tau(e)$ represents the time it takes fluid to flow from site $\mathbf{u}$ to site $\mathbf{v}$, or the reverse, along the edge e. If the sequence of edges $r = (e_1, \ldots, e_n)$ forms a path from $\mathbf{u} \in \mathbf{Z}^d$ to $\mathbf{v} \in \mathbf{Z}^d$, then $T(r) \equiv \sum_i \tau(e_i)$ represents the time it takes fluid to flow from $\mathbf{u}$ to $\mathbf{v}$ along the path r. For any $\mathbf{u}, \mathbf{v} \in \mathbf{Z}^d$, we further define the *passage time* from $\mathbf{u}$ to $\mathbf{v}$ as

$$T(\mathbf{u}, \mathbf{v}) \equiv \inf\{T(r) : \text{the edges in } r \text{ form a path from } \mathbf{u} \text{ to } \mathbf{v}\}. \qquad (1.1)$$

If $\mathbf{u} = \mathbf{v}$ in (1.1), we take $T(\mathbf{u}, \mathbf{v}) = 0$. With the origin $\mathbf{0}$ representing the fluid injection site, $T(\mathbf{0}, \mathbf{v})$ is the time at which the site $\mathbf{v}$ is first wetted by fluid flowing along some path from $\mathbf{0}$ to $\mathbf{v}$ (assuming the infimum in (1.1) is attained).

In [26], Hammersley and Welsh study, among other things, the asymptotic behavior as $n \to \infty$ of $a_{0,n} \equiv T(\mathbf{0}, n\hat{\mathbf{e}}_1)$ and other similar quantities, where $\hat{\mathbf{e}}_i$ denotes the unit vector in the ith coordinate direction. More generally, with $a_{m,n} \equiv T(m\hat{\mathbf{e}}_1, n\hat{\mathbf{e}}_1)$, a powerful subadditivity emerges:

$$a_{0,n} \leq a_{0,m} + a_{m,n} \quad \text{for } 0 \leq m \leq n. \qquad (1.2)$$

This holds for the simple reason that a path from $\mathbf{0}$ to $m\hat{\mathbf{e}}_1$ concatenated with one from $m\hat{\mathbf{e}}_1$ to $n\hat{\mathbf{e}}_1$ produces a path from $\mathbf{0}$ to $n\hat{\mathbf{e}}_1$. From (1.2) and the invariance of the joint distribution of the $\tau(e)$'s under lattice translations it follows that $g_n \equiv Ea_{0,n}$ satisfies $g_{m+n} \leq g_m + g_n$ for all $m, n \geq 0$. Hence, from basic properties of subadditive sequences, one may define the *time constant* μ as

$$\mu = \mu(F, d) \equiv \lim_{n \to \infty} \frac{Ea_{0,n}}{n} = \inf_{n > 0} \frac{Ea_{0,n}}{n}, \qquad (1.3)$$

where $F(x) \equiv P[\tau \leq x]$ is the distribution of the edge variables. Central questions in the early development of FPP were:

- Under what conditions and in what sense does $a_{0,n}/n$ converge to μ?
- For what edge distributions F is $\mu(F) = 0$?

- Is the infimum in (1.1) always attained? That is, do *minimizing paths*, also called *routes*, exist for all edge distributions F? A minimizing path r satisfies $T(\mathbf{u}, \mathbf{v}) = T(r)$.
- Where minimizing paths exist, how many edges do they typically have as a function of the distance between their endpoints?

Some very basic questions remain unanswered or have only partial answers to date. For example:

- Is $Ea_{0,n}$ monotonically increasing in n?
- Is $a_{0,n}$ stochastically increasing in n?
- What can we say about $\mathrm{Var}\, a_{0,n}$ for large n?

It is convenient to extend the definition of $T(\cdot, \cdot)$ to all of $\mathbf{R}^d \times \mathbf{R}^d$ as follows: for $\mathbf{u}, \mathbf{v} \in \mathbf{R}^d$, $T(\mathbf{u}, \mathbf{v}) \equiv T(\mathbf{u}^*, \mathbf{v}^*)$ where $\mathbf{u}^*$ (resp. $\mathbf{v}^*$) is the site in $\mathbf{Z}^d$ that is closest to $\mathbf{u}$ (resp. $\mathbf{v}$), with some fixed rule for breaking ties. Another object of study in FPP is the *wet region* at time t:

$$\mathcal{W}_t \equiv \{\mathbf{x} \in \mathbf{R}^d : T(\mathbf{0}, \mathbf{x}) \leq t\}. \tag{1.4}$$

Plainly $\mathcal{W}_t$ is a stochastic subset of $\mathbf{R}^d$ that is growing with time. Many questions fundamental to FPP concern properties of $\mathcal{W}_t$. Some examples:

- Does $\mathcal{W}_t$, when scaled properly, converge in some sense to an asymptotic shape?
- When it does, what can we say about the shape?
- What can we say about the roughness of the interface between the wet region and the dry region as time elapses?
- How do the answers to these questions depend on the distribution F of the $\tau(e)$'s?

An alternative visualization of FPP comes from the observation that a configuration of edge values induces a random geometry on $\mathbf{Z}^d$ in the sense that $T(\cdot, \cdot)$ satisfies the triangle inequality: $T(\mathbf{u}, \mathbf{w}) \leq T(\mathbf{u}, \mathbf{v}) + T(\mathbf{v}, \mathbf{w})$ (see (1.6) below). $T(\cdot, \cdot)$ is therefore a metric on $\mathbf{Z}^d$ if $F(0) = 0$, and a pseudometric on $\mathbf{Z}^d$ if $F(0) > 0$. Similarly, $T(\cdot, \cdot)$ is a pseudometric on $\mathbf{R}^d$. In this spirit, we will sometimes use the term *geodesic* (or finite geodesic), rather than route, to describe minimizing paths. A semi-infinite path $r = (e_k : k \geq 0)$ will be called a *semi-infinite geodesic* if every finite portion of r is minimizing: $T(\tilde{r}) = T(\mathbf{u}, \mathbf{v})$ for all $\tilde{r} = (e_k : 0 \leq m \leq k \leq n)$ where $\mathbf{u}$ and $\mathbf{v}$ are the endpoints of the path $\tilde{r}$. A doubly infinite path $r = (e_k : k \in \mathbf{Z})$ is called a *doubly infinite geodesic* if it satisfies the same property (without the restriction $0 \leq m$). Concerning geodesics:

- How far do finite geodesics typically wander from the straight line segment connecting their endpoints?
- Do semi-infinite geodesics exist?
- Do doubly infinite geodesics exist?

- Where they do exist, what can we say about the straightness (in the sense of Euclidean geometry) of infinite geodesics?

To date, none of these questions concerning geodesics has a completely dispositive answer.

FPP is a rather mature subject that has developed a substantial literature. This survey is necessarily incomplete. For the following important topics, for example, we provide only a partial list of references: FPP on trees [6, 8, 52, 57]; higher-dimensional analogs of FPP [12, 38, 39]; Ising FPP [15, 22, 27]; FPP on finite graphs [34]; and reverse shape models [23]. Additionally, with a few exceptions, we report only on subcritical FPP, i.e., on the case where $\mu > 0$. The case $\mu = 0$ is considerably different in flavor (see [13, 14, 43, 44, 82, 83, 84]). Beginning in Section 3, we outline, with varying precision, proofs of some of the more recent major developments in FPP. We prove special cases where doing so illustrates the main ideas with greater ease.

1.2 Notation

We will be working with FPP on other graphs in addition to the $\mathbf{Z}^d$ lattice. Our general setting is a graph whose vertex set $\mathcal{V}$ is a locally finite subset of $\mathbf{R}^d$. The edge set $\mathcal{E}$ will be some subset of $\{\{\mathbf{u}, \mathbf{v}\} : \mathbf{u}, \mathbf{v} \in \mathcal{V}, \mathbf{u} \neq \mathbf{v}\}$. We informally identify the edge $\{\mathbf{u}, \mathbf{v}\}$ with the line segment $\overline{\mathbf{u}\mathbf{v}}$, but caution that some of the graphs we will work with in dimension 2 are not planar.

In defining paths, it is more convenient to think of vertices rather than edges. For distinct $\mathbf{u}, \mathbf{v} \in \mathcal{V}$, a *path* from $\mathbf{u}$ to $\mathbf{v}$ will mean a sequence of vertices $r = (\mathbf{u} = \mathbf{v}_0, \mathbf{v}_1, \ldots, \mathbf{v}_n = \mathbf{v})$ such that each $e_i \equiv \{\mathbf{v}_{i-1}, \mathbf{v}_i\} \in \mathcal{E}$. With some abuse of notation, we refer to $\{e_i\}$ as the edges of r and write $e \in r$ if $e \in \{e_i\}$. The length of r, denoted $|r|$, is the number of edges in $\{e_i\}$. The path r will be called *self-avoiding* if the vertices $\mathbf{v}_i$ are distinct. For vertices $\mathbf{u}$, $\mathbf{v}$, $\mathcal{R}(\mathbf{u}, \mathbf{v})$ will denote all paths from $\mathbf{u}$ to $\mathbf{v}$; $\mathcal{R}^{\text{s.a.}}(\mathbf{u}, \mathbf{v})$ will denote the subset that are self-avoiding. For general $\mathbf{u}, \mathbf{v} \in \mathbf{R}^d$, we will define $\mathcal{R}(\mathbf{u}, \mathbf{v}) \equiv \mathcal{R}(\mathbf{u}^*, \mathbf{v}^*)$ and $\mathcal{R}^{\text{s.a.}}(\mathbf{u}, \mathbf{v}) \equiv \mathcal{R}^{\text{s.a.}}(\mathbf{u}^*, \mathbf{v}^*)$, where $\mathbf{u}^*$ (resp. $\mathbf{v}^*$) is the vertex in $\mathcal{V}$ that is closest to $\mathbf{u}$ (resp. $\mathbf{v}$) with some deterministic rule for breaking ties. We will use $\mathcal{R}^{\text{s.a.}}(\mathbf{u}, \cdot)$ to denote all self-avoiding paths starting at $\mathbf{u}$, i.e.,

$$\mathcal{R}^{\text{s.a.}}(\mathbf{u}, \cdot) \equiv \bigcup_{\mathbf{v} \in \mathcal{V}} \mathcal{R}^{\text{s.a.}}(\mathbf{u}, \mathbf{v}).$$

For $\mathbf{u}, \mathbf{v} \in \mathcal{V}$ with $m \equiv \mathbf{u} \cdot \hat{\mathbf{e}}_1 < \mathbf{v} \cdot \hat{\mathbf{e}}_1 \equiv n$, $\mathcal{R}^{\text{cyl}}(\mathbf{u}, \mathbf{v})$ will denote the *cylinder paths* from $\mathbf{u}$ to $\mathbf{v}$: those paths $r \in \mathcal{R}(\mathbf{u}, \mathbf{v})$ such that $m < \mathbf{x} \cdot \hat{\mathbf{e}}_1 \leq n$ for all vertices $\mathbf{x} \in r$ except $\mathbf{v}_0$. Sometimes, when the edge set is ambiguous, we include it as an argument: $\mathcal{R}(\mathbf{u}, \mathbf{v}, \mathcal{E})$, for example.

The edge variables, which will be denoted by τ or $\tau(e)$, are always nonnegative random variables but are not always i.i.d., as they are in the standard model discussed above. With $T(r) \equiv \sum_{e \in r} \tau(e)$, the passage time from $\mathbf{u}$ to $\mathbf{v}$ is defined by

$$T(\mathbf{u},\mathbf{v}) \equiv \begin{cases} 0 & \text{if } \mathbf{u}^* = \mathbf{v}^* \\ \inf\{T(r) : r \in \mathcal{R}(\mathbf{u},\mathbf{v})\} & \text{if } \mathbf{u}^* \neq \mathbf{v}^*. \end{cases} \tag{1.5}$$

We note that $\mathcal{R}(\mathbf{u},\mathbf{v})$ may be replaced with $\mathcal{R}^{\text{s.a.}}(\mathbf{u},\mathbf{v})$ in (1.5). With T defined in this more general context, definition (1.4) for the wet region $\mathcal{W}_t$ continues to apply. Also, the *triangle inequality* remains valid:

$$\begin{aligned} T(\mathbf{u},\mathbf{w}) &= \inf\{T(r'') : r'' \in \mathcal{R}(\mathbf{u},\mathbf{w})\} \\ &\leq \inf\{T(r'') : r'' \in \mathcal{R}(\mathbf{u},\mathbf{w}),\ \mathbf{v} \in r''\} \\ &= \inf\{T(r) : r \in \mathcal{R}(\mathbf{u},\mathbf{v})\} + \inf\{T(r') : r' \in \mathcal{R}(\mathbf{v},\mathbf{w})\} \\ &= T(\mathbf{u},\mathbf{v}) + T(\mathbf{v},\mathbf{w}). \end{aligned} \tag{1.6}$$

$M(\mathbf{u},\mathbf{v})$ will denote the set of all vertices on minimizing paths from $\mathbf{u}$ to $\mathbf{v}$:

$$M(\mathbf{u},\mathbf{v}) \equiv \bigcup_{r:T(r)=T(\mathbf{u},\mathbf{v})} \{\text{vertices on } r\}. \tag{1.7}$$

When the minimizing path is unique, we will also think of $M(\mathbf{u},\mathbf{v})$ as a path, so, for example, $T(\mathbf{u},\mathbf{v}) = T(M(\mathbf{u},\mathbf{v}))$.

An FPP model is completely specified once the graph $(\mathcal{V},\mathcal{E})$ and the joint distribution of $(\tau(e) : e \in \mathcal{E})$ are determined. We will refer to $\mathbf{Z}^d$ lattice FPP with i.i.d. edge variables satisfying $\mu > 0$ as *Standard FPP*.

Points in $\mathbf{R}^d$ (and therefore $\mathbf{Z}^d$) are written in bold, e.g., $\mathbf{u}$, $\mathbf{v}$, $\mathbf{x}$; real numbers are denoted by u, v, x, etc. The origin will be denoted by $\mathbf{0}$. For $\mathbf{x} \in \mathbf{R}^d$ and $\varrho \geq 0$, $B(\mathbf{x},\varrho)$ will denote the ball $\{\mathbf{y} \in \mathbf{R}^d : |\mathbf{x} - \mathbf{y}| \leq \varrho\}$. In certain inequalities, C_0 will denote a positive constant, thought of as small, which can be replaced with any smaller constant without destroying the validity of the inequality. Similarly, C_1 will denote a (large) finite constant which can be replaced with similar impunity by any larger constant. The values of C_0 and C_1 in each setting will depend only the specification of the FPP model in question, i.e., on $(\mathcal{V},\mathcal{E})$ and the distribution of $(\tau(e) : e \in \mathcal{E})$.

2 The Time Constant

2.1 The Fundamental Processes of Hammersley and Welsh

In addition to the process $(a_{m,n} : 0 \leq m < n)$, Hammersley and Welsh studied three other related quantities. There, the setting was the $\mathbf{Z}^2$ lattice. Here, we present their definitions for arbitrary dimension, and include the formal definition of $a_{m,n}$:

$$a_{m,n} \equiv \inf\{T(r) : r \in \mathcal{R}(m\hat{\mathbf{e}}_1, n\hat{\mathbf{e}}_1)\}, \tag{2.1}$$

$$b_{m,n} \equiv \inf\{T(r) : r \in \mathcal{R}(m\hat{\mathbf{e}}_1, \mathbf{v}),\ \mathbf{v} \in \mathbf{Z}^d \text{ with } \mathbf{v} \cdot \hat{\mathbf{e}}_1 = n\}, \tag{2.2}$$

$$t_{m,n} \equiv \inf\{T(r) : r \in \mathcal{R}^{\text{cyl}}(m\hat{\mathbf{e}}_1, n\hat{\mathbf{e}}_1)\}, \text{ and} \qquad (2.3)$$

$$s_{m,n} \equiv \inf\{T(r) : r \in \mathcal{R}^{\text{cyl}}(m\hat{\mathbf{e}}_1, \mathbf{v}), \mathbf{v} \in \mathbf{Z}^d \text{ with } \mathbf{v} \cdot \hat{\mathbf{e}}_1 = n\}. \qquad (2.4)$$

In terms of the fluid analogy, $b_{0,n}$ represents the time at which some site on $\{\mathbf{v} \in \mathbf{Z}^d : \mathbf{v} \cdot \hat{\mathbf{e}}_1 = n\}$ (a line for $d = 2$, a plane for $d = 3$, and a hyper-plane for higher d) is first wetted by fluid forced in at $\mathbf{0}$. Clearly, then $b_{0,n} \leq a_{0,n}$. The process $t_{m,n}$ is very similar to $a_{m,n}$, except that the infimum is restricted to paths from $m\hat{\mathbf{e}}_1$ to $n\hat{\mathbf{e}}_1$ with vertices, except for the first vertex, that lie in the cylinder of sites whose first coordinate is between $m+1$ and n. It is immediate from this restriction that $t_{m,n} \geq a_{m,n}$. The process $s_{m,n}$ is analogous to $b_{m,n}$ with a similar restriction on paths, so $s_{m,n} \geq b_{m,n}$. The s and t processes were introduced as approximations to a and b enjoying properties making them easier to study. For example, $t_{0,m}$ and $t_{m,n}$ are independent, while $a_{0,m}$ and $a_{m,n}$ are not.

Regarding these processes, Hammersley and Welsh showed, for $\theta = a, t$, or s, that

$$\lim_{n \to \infty} \frac{\theta_{0,n}}{n} = \mu \text{ in probability}, \qquad (2.5)$$

$$P[\limsup_{n \to \infty} \frac{\theta_{0,n}}{n} = \mu] = 1, \text{ and} \qquad (2.6)$$

$$\lim_{n \to \infty} \frac{\operatorname{Var} \theta_{0,n}}{n^2} = 0, \qquad (2.7)$$

where (2.5) and (2.6) assume that $E\tau < \infty$, and (2.7) additionally assumes $E\tau^2 < \infty$. They conjectured that (2.5) holds also with $\theta = b$.

A crucial advance in the theory of subadditive processes came in 1968 with Kingman's *subadditive ergodic theorem*, which in fact was motivated by its application to FPP (see [45, 46, 47]). Kingman's theorem was improved upon by Liggett in 1985 [51], who showed that slightly relaxed hypotheses suffice. (See [51] also for an example of a process requiring the weakened hypotheses.) While Kingman's version was perfectly adequate for FPP, we state Liggett's version here.

Theorem 2.1. *Suppose $(X_{m,n} : 0 \leq m < n)$ is a family of random variables satisfying:*

$$X_{0,n} \leq X_{0,m} + X_{m,n} \text{ whenever } 0 < m < n \qquad (2.8)$$

the joint distribution of $(X_{m,m+k} : k \geq 1)$ does not depend on m $\qquad (2.9)$

for each $k \geq 1$, $(X_{nk,(n+1)k} : n \geq 1)$ is a stationary sequence $\qquad (2.10)$

$$EX_{0,1}^+ < \infty \text{ and } EX_{0,n} \geq -cn \text{ for some constant } c. \qquad (2.11)$$

Then

$$\lim_{n \to \infty} \frac{EX_{0,n}}{n} \text{ exists and equals } \gamma \equiv \inf_n \frac{EX_{0,n}}{n}, \qquad (2.12)$$

$$X \equiv \lim_{n \to \infty} \frac{X_{0,n}}{n} \text{ exists a.s. and in } L^1, \text{ and} \qquad (2.13)$$

$$EX = \gamma. \qquad (2.14)$$

If the stationary sequences in (2.10) are ergodic, then $X = \gamma$ a.s.

Kingman's version of the theorem required, in place of (2.8), (2.9) and (2.10), that $X_{l,n} \leq X_{l,m} + X_{m,n}$ for all $0 \leq l < m < n$ and that the entire joint distribution of $(X_{m+k,n+k} : 0 \leq m < n)$ does not depend on k.

From Theorem 2.1 it follows immediately, for $\theta = a$ or t, that

$$\lim_{n \to \infty} \frac{\theta_{0,n}}{n} = \mu \text{ a.s. and in } L^1, \tag{2.15}$$

provided

$$E\theta_{0,1} < \infty. \tag{2.16}$$

Clearly $E\tau < \infty$ implies (2.16), for $a_{0,1}$ and $t_{0,1}$ are bounded above by the edge variable corresponding to the edge $\{0, \hat{e}_1\}$. Work done by Reh, Smythe, and Wierman in the late 1970's progressively refined how these four processes converged (a.s., in probability, etc.) in dimension 2 and under what hypotheses for the distribution F of the $\tau(e)$. (See [59, 64, 66, 76, 77, 78]. See also [67] for a full accounting of the state of FPP in 1978.) We summarize results of [59] **(A)** and [77] **(B)** in the next theorem.

Theorem 2.2. (A) *Let* $Y = \min(\tau_1, \tau_2, \tau_3, \tau_4)$ *and* $\widetilde{Y} = \min(\tau_1, \tau_2, \tau_3)$, *where the* τ_i's *are independent and distributed according to* F. *If* $EY < \infty$ *then: convergence (to* μ*) for* $a_{0,n}/n$ *is almost sure and in* L^1*; convergence for* $b_{0,n}/n$ *is in* L^1*; convergence for* $s_{0,n}/n$ *and* $t_{0,n}/n$ *is in probability. If* $EY^2 < \infty$ *then convergence for* $b_{0,n}/n$ *and* $s_{0,n}/n$ *is almost sure. If* $E\widetilde{Y} < \infty$ *then convergence for* $t_{0,n}/n$ *is almost sure. If* $EY = \infty$ *then* $\limsup_n a_{0,n}/n = \infty$ *almost surely.* **(B)** *If* $E\tau_1^\varepsilon < \infty$ *for any* $\varepsilon > 0$ *then convergence for* $b_{0,n}/n$ *and* $s_{0,n}/n$ *is almost sure and convergence for* $a_{0,n}/n$ *and* $t_{0,n}/n$ *is in probability.*

The relevance of Y and $\widetilde{Y}$ will become apparent in Section 3.1.

2.2 About μ

It is natural to ask if $\mu(F)$ can be computed explicitly when F is specified. Unfortunately, we are far from being able to do that for any non-trivial F. Some early work (in 2 dimensions) of Reh, Smythe, and Wierman [59, 66, 78] concerned which F's have $\mu(F) = 0$, an issue definitively resolved by Kesten [40] for all d in 1986.

Theorem 2.3. *Let* $p_c(d)$ *denote the critical probability for d-dimensional bond percolation. Then for d-dimensional lattice FPP*

$$\mu > 0 \text{ if and only if } F(0) < p_c(d). \tag{2.17}$$

Other good general information about μ is scarce. Let $\lambda = \lambda(F) \equiv \inf\{x \geq 0 : F(x) > 0\}$. If $0 < F(\lambda) < 1$ then

$$\lambda < \mu, \tag{2.18}$$

and, if F is not concentrated at one point,

$$\mu \; < \; E\tau. \tag{2.19}$$

Note that trivially $\lambda \leq \mu$; $\mu \leq E\tau$ follows from subadditivity. The bound (2.19) is due to Hammersley and Welsh [26]; the bound (2.18) is due to Kesten [40]. See [33, 65, 67, 76] for additional bounds in special cases. More recently, Sidoravicius, Vares, and Surgailis [63] have proven the lower bound

$$\mu \; \geq 1 - \sqrt{2(1 - p^4)}$$

in dimension 2 for the case of 0–1 valued edge variables with $P[\tau = 1] = p$. This bound is meaningful for $2^{-1/4}(\approx 0.84) < p < 1$.

It is also known that $\mu(F)$ varies continuously with F. Specifically, if the distributions F_n (supported on $[0, \infty)$) converge weakly to F then $\mu(F_n) \to \mu(F)$ as $n \to \infty$. This was shown in [16] with the hypothesis that the F_n are all stochastically bounded by a common distribution with finite mean. In [18] that hypothesis was removed.

Van den Berg and Kesten, in [10] give a criterion insuring that $\mu(F) < \mu(\widetilde{F})$ for two edge distributions F and $\widetilde{F}$. Specifically, if X and $\widetilde{X}$ are distributed according to F and $\widetilde{F}$ respectively, they call $\widetilde{F}$ *more variable* than F if $E\psi(\widetilde{X}) \leq E\psi(X)$ for all increasing concave $\psi : \mathbf{R} \to \mathbf{R}$ for which $E|\psi(\widetilde{X})|$ and $E|\psi(X)|$ are finite. (Note that automatically $\widetilde{F}$ is more variable than F if X stochastically dominates $\widetilde{X}$, i.e., $F(x) \leq \widetilde{F}(x)$ for all $x \in \mathbf{R}$.) Then, for Standard FPP, we have the following theorem.

Theorem 2.4. *Let F and $\widetilde{F}$ be two edge distributions with finite mean such that $\widetilde{F}$ is more variable than F. Then $\mu(\widetilde{F}) \leq \mu(F)$. If, in addition, F satisfies*

$$\lambda(F) = 0 \quad and \quad F(0) < p_c(d), \; or \tag{2.20}$$
$$\lambda(F) > 0 \quad and \quad F(\lambda) < p_c^{dir}(d), \tag{2.21}$$

and $\widetilde{F} \neq F$, then $\mu(\widetilde{F}) < \mu(F)$.

Here, $p_c^{\mathrm{dir}}(d)$ denotes the critical probability for directed bond percolation on the $\mathbf{Z}^d$ lattice.

As remarked in [10] (see Remark 2.15), this theorem has an interesting corollary. If τ is distributed according to F, let F_x denote the distribution of $\tau \wedge x$. If the distribution F is not bounded ($F(x) < 1$ for all x) but F has finite mean, then $\mu(F_x) < \mu(F)$ provided F also satisfies (2.20) or (2.21). Roughly speaking, this implies that arbitrarily large edge values appear with positive density along minimizing paths from $\mathbf{0}$ to $n\hat{\mathbf{e}}_1$ as $n \to \infty$.

Kesten, in [40], has also studied the asymptotics of $\mu(F, d)$ for fixed F as a function of dimension d. Roughly, he shows that if F has finite mean, $F(0) = 0$, and F has a density function $f(x)$ that is sufficiently flat and bounded away from 0 as $x \downarrow 0$, then

$$C_0 \frac{\log d}{d} \leq \mu(F, d) \leq C_1 \frac{\log d}{d}.$$

See also [19] for the case of exponential edge variables.

2.3 Minimizing Paths

Another matter studied early in the development of FPP concerned the existence of routes, or paths that achieve the infima in (2.1) – (2.4). If the distribution of the edge variables satisfies $\lambda > 0$, then the a.s. existence of routes is easy to see. For example, on the event $\{a_{0,n} \leq A\}$, any path r with more than $(A+1)/\lambda$ edges will have $T(r) \geq a_{0,n} + 1$. The problem therefore reduces to finding a *minimum* over the finitely many paths with $(A+1)/\lambda$ or fewer edges. The conclusion follows by letting $A \uparrow \infty$. More generally, for subcritical FPP where $F(0) < p_c(d)$, one may choose $\varepsilon > 0$ with $F(\varepsilon) < p_c(d)$. The conclusion follows from the non-percolation of edges with edge values of ε or less. In dimension 2, it is known that routes exist for *all* edge distributions. See [26, 68, 78] for details. For critical FPP (where $F(0) = p_c(d)$) and supercritical FPP (where $F(0) > p_c(d)$), the issue has not been generally resolved for $d > 2$ – but see [84] for a special case.

In the subcritical regime, where routes exist, let N_n^θ denote the number of edges on the shortest route realizing $\theta_{0,n}$ for $\theta = a, b, t,$ or s. It is natural to ask about the $n \to \infty$ asymptotics of N_n^θ, and early work addressed this question. For any $x \in \mathbf{R}$, let $F \oplus x$ denote the distribution of $\tau + x$. Note that negative edge values may occur under the $F \oplus x$ distribution for $x < 0$. In [68], Smythe and Wierman show that $\mu(F \oplus x) \geq 0$ for x in an open interval containing 0 provided $F(0) < 1/L$, where, with $\mathcal{E} = \{\mathbf{Z}^2 \text{ nearest neighbor edges}\}$,

$$L \equiv \lim_{n \to \infty} [\mathrm{Card}\{r \in \mathcal{R}^{\mathrm{s.a.}}(\mathbf{0}, \cdot, \mathcal{E}) : |r| = n\}]^{1/n}. \tag{2.22}$$

L is the *connectivity constant* of the $\mathbf{Z}^2$ lattice. It is known from [26] that $\mu(F \oplus x)$ is concave and increasing in x where $\mu(F \oplus x) > -\infty$. It follows that $\mu(F \oplus x)$ has left and right derivatives, $\mu^-(x)$ and $\mu^+(x)$ respectively, on this open interval. Also, $\mu^-(x) \geq \mu^+(x)$ with $\mu^-(x) = \mu^+(x)$ except for possibly countably many values of x. Smythe and Wierman show the following.

Theorem 2.5. *Consider 2 dimensional Standard FPP with* $F(0) < 1/L$. *Then, almost surely, for* $\theta = a$ *and* $\theta = t$,

$$\mu^+(0) \leq \liminf_{n \to \infty} \frac{N_n^\theta}{n} \leq \limsup_{n \to \infty} \frac{N_n^\theta}{n} \leq \mu^-(0). \tag{2.23}$$

This result has intuitive appeal, as the following very informal heuristic makes clear. Suppose r^* is a route for $a_{0,n}$ (for $x = 0$), and suppose further that $\mu^-(0) = \mu^+(0) \equiv \mu^\pm(0)$. Then:

$$N_n^a = \frac{d}{dx} T(r^*)\Big|_{x=0} \approx \frac{d}{dx} n\mu(F \oplus x)\Big|_{x=0} = n\mu^\pm. \tag{2.24}$$

See also [73] for similar results for a different FPP model. Kesten [37] shows for subcritical FPP (i.e., under the weaker condition $F(0) < p_c(d)$) that $\limsup_{n\to\infty} N_n^a/n < \infty$.

In view of (2.23), it seemed for many years that establishing $\mu^-(0) = \mu^+(0)$ was a promising strategy to establish that N_n^a/n converges almost surely to some $C \in (0, \infty)$. However, Steele and Zhang [69] have recently shown in 2 dimensions that $\mu^+(0) < \mu^-(0)$ for 0–1 Bernoulli edge variables with $P[\tau = 0]$ sufficiently close to $p_c(2) = 1/2$.

3 Asymptotic Shape and Shape Fluctuations

3.1 Shape Theorems for Standard FPP

Many interesting questions emerge when μ is thought of as a function of direction. As discussed below, for any unit vector $\hat{\mathbf{x}}$,

$$\mu(\hat{\mathbf{x}}) \equiv \lim_{n\to\infty} \frac{ET(\mathbf{0}, n\hat{\mathbf{x}})}{n} \tag{3.1}$$

exists and varies continuously with $\hat{\mathbf{x}}$. Note that subadditivity together with invariance of the distribution of the passage times with respect to lattice symmetries implies that $\mu(\hat{\mathbf{x}}) = 0$ in all directions if $\mu(\hat{\mathbf{e}}_1) = 0$. Assuming $F(0) < p_c(d)$ so $\mu(\hat{\mathbf{e}}_1) > 0$, roughly speaking $\mu(\hat{\mathbf{x}})^{-1}$ is the distance the wet region spreads in the $\hat{\mathbf{x}}$ direction per unit of time. With linear growth in all directions, it is natural to ask how closely $t^{-1}\mathcal{W}_t$ resembles $\widetilde{W}$ where

$$\widetilde{W} \equiv \{\mathbf{x} \in \mathbf{R}^d : |\mathbf{x}| \le \mu(\mathbf{x}/|\mathbf{x}|)^{-1}\}. \tag{3.2}$$

If, on the other hand, $F(0) \ge p_c(d)$ so $\mu(\hat{\mathbf{e}}_1) = 0$, then, roughly speaking, growth is super-linear in all directions and one would expect $t^{-1}\mathcal{W}_t$ to eventually cover any bounded set.

In their celebrated *shape theorem*, Cox and Durrett [17] provide necessary and sufficient conditions yielding an FPP analog of a strong law of large numbers. For $\mathbf{z} \in \mathbf{Z}^d$, let $Y(\mathbf{z})$ denote the minimum of the $\tau(e)$ over the $2d$ edges e incident to $\mathbf{z}$. If $EY^d = \infty$ (where $Y \equiv Y(\mathbf{0})$) then

$$\sum_{\mathbf{z}\in(2\mathbf{Z})^d} P[Y(\mathbf{z}) > K|\mathbf{z}|] = \infty$$

for any $K < \infty$. Since $T(\mathbf{0}, \mathbf{z}) \ge Y(\mathbf{z})$ and the random variables $(Y(\mathbf{z}) : \mathbf{z} \in (2\mathbf{Z})^d)$ are independent, it follows from the Borel-Cantelli Lemma that

$$\frac{T(\mathbf{0}, \mathbf{z})}{|\mathbf{z}|} > K \text{ for infinitely many } \mathbf{z} \in \mathbf{Z}^d, \text{ a.s.}$$

Without this moment condition, while the leading edge of the wet region is growing linearly (or faster if $\mu(\hat{\mathbf{e}}_1) = 0$), there are pockets in the interior that remain dry for prolonged periods of time. On the other hand, if

$$EY^d < \infty, \tag{3.3}$$

we have the next theorem concerning the asymptotic shape of the wet region as time elapses.

Theorem 3.1. *Suppose $\mu(\hat{\mathbf{e}}_1) > 0$. Then*

$$\text{for all } \varepsilon \in (0,1), \; (1-\varepsilon)\widetilde{W} \subset t^{-1}W_t \subset (1+\varepsilon)\widetilde{W} \text{ for all large } t, \text{ a.s.} \tag{3.4}$$

if and only if (3.3) holds. If $\mu(\hat{\mathbf{e}}_1) = 0$ then

$$\text{for all compact } K \subset \mathbf{R}^d, \; K \subset t^{-1}W_t \text{ for all large } t, \text{ a.s.}$$

if and only if (3.3) holds.

We outline here a proof (different in some details from Cox and Durrett's) of the more interesting $\mu(\hat{\mathbf{e}}_1) > 0$ part of this theorem, the general structure of which is applicable to other shape theorems (i.e., for other variations of FPP). We have already seen the necessity of $EY^d < \infty$ for almost sure convergence to an asymptotic shape. Suppose, then, that $EY^d < \infty$.

The basic idea is to first establish radial convergence, i.e., that $\lim_{n\to\infty} T(\mathbf{0}, n\hat{\mathbf{x}})/n$ exists for a set of directions $\hat{\mathbf{x}}$ that are dense in the unit sphere, and then to patch this together to obtain

$$\limsup_{\substack{\mathbf{x}\to\infty \\ \mathbf{x}\in\mathbf{Z}^d}} \left| \frac{T(\mathbf{0},\mathbf{x})}{|\mathbf{x}|} - \mu(\mathbf{x}/|\mathbf{x}|) \right| = 0. \tag{3.5}$$

The shape theorem follows, loosely speaking, from an "inversion" of this.

To establish radial convergence, note that, for $\mathbf{z} \in \mathbf{Z}^d$, the family of random variables $(X_{m,n} \equiv T(m\mathbf{z}, n\mathbf{z}) : 0 \le m < n)$ satisfies the conditions of Theorem 2.1. That (2.8) – (2.10) hold is straightforward; also we may take $c = 0$ in (2.11). Now $EY^d < \infty$ (in fact $EY < \infty$) implies that $ET(\mathbf{0},\mathbf{z}) < \infty$ for all $\mathbf{z} \in \mathbf{Z}^d$. By subadditivity and lattice invariance, $ET(\mathbf{0},\mathbf{z}) \le \|\mathbf{z}\|_1 ET(\mathbf{0},\hat{\mathbf{e}}_1)$, so it suffices to prove this for $\mathbf{z} = \hat{\mathbf{e}}_1$. Note that there are $2d$ edge-disjoint paths from $\mathbf{0}$ to $\hat{\mathbf{e}}_1$, the longest of which can be taken to have nine edges. See Figure 1 for the construction with $d = 2$. (For higher d, the additional paths are of the form $(\mathbf{0}, \hat{\mathbf{e}}_k, \hat{\mathbf{e}}_1 + \hat{\mathbf{e}}_k, \hat{\mathbf{e}}_1)$ for $2 < k \le d$.) Let T^* denote the passage time for the nine-edge path. Then

$$P[T(\mathbf{0},\hat{\mathbf{e}}_1) > s] \;\le\; P[T^* > s]^{2d} \;\le\; 9^{2d} P[\tau > s/9]^{2d} \;=\; 9^{2d} P[Y > s/9],$$

where τ is a generic edge variable. The second inequality holds since $T^* > s$ implies that one of the edges e on T^* has $\tau(e) > s/9$. That $ET(\mathbf{0},\hat{\mathbf{e}}_1) < \infty$ follows immediately from this. From Theorem 2.1, $\lim_{n\to\infty} T(\mathbf{0}, n\mathbf{z})/n$ exists almost surely.

Let $\hat{\mathbf{z}} = \mathbf{z}/|\mathbf{z}|$ and put

$$\mu(\hat{\mathbf{z}}) \;\equiv\; \lim_{n\to\infty} \frac{T(\mathbf{0}, n\mathbf{z})}{n|\mathbf{z}|}. \tag{3.6}$$

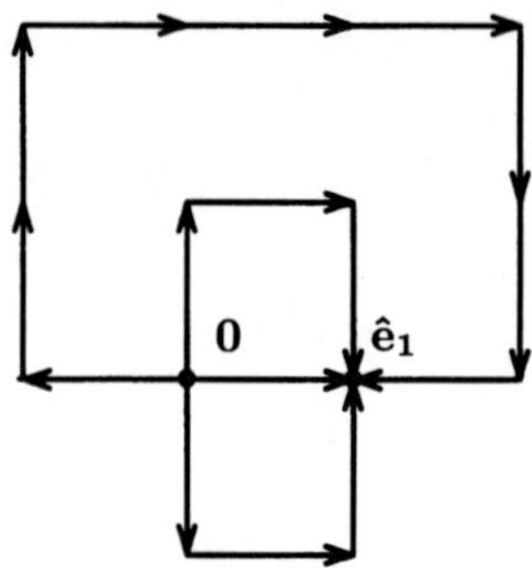

Fig. 1. Four edge-disjoint paths from $\mathbf{0}$ to $\hat{\mathbf{e}}_1$.

This makes sense as the right side of (3.6) is independent of the choice of $\mathbf{z}$, provided $\mathbf{z}/|\mathbf{z}| = \hat{\mathbf{z}}$. We claim that we have also $\lim_{n\to\infty} T(\mathbf{0}, n\hat{\mathbf{z}})/n = \mu(\hat{\mathbf{z}})$. To see this, write $n\hat{\mathbf{z}} = m_n \mathbf{z} + \mathbf{v}_n$, for integer m_n with $||\mathbf{v}_n||_1 \leq ||\mathbf{z}||_1$. Then, by the triangle inequality, $|T(\mathbf{0}, n\hat{\mathbf{z}}) - T(\mathbf{0}, m_n\mathbf{z})| \leq T(n\hat{\mathbf{z}}, m_n\mathbf{z})$. Also, for any $\delta > 0$, $\sum_n P[T(n\hat{\mathbf{z}}, m_n\mathbf{z}) > n\delta] < \infty$, since $T(n\hat{\mathbf{z}}, m_n\mathbf{z})$ is stochastically bounded by the sum of $||\mathbf{z}||_1$ random variables of finite mean. So, almost surely, for large n, $T(\mathbf{0}, n\hat{\mathbf{z}}) = T(\mathbf{0}, m_n\mathbf{z}) + n\Delta_n$ where $|\Delta_n| \leq \delta$. Note also that $n = m_n|\mathbf{z}| + c_n$ where $|c_n| \leq ||\mathbf{z}||_1$, so, for large n,

$$\frac{T(\mathbf{0}, n\hat{\mathbf{z}})}{n} = \frac{T(\mathbf{0}, m_n\mathbf{z}) + n\Delta_n}{n} = \frac{T(\mathbf{0}, m_n\mathbf{z})}{m_n|\mathbf{z}| + c_n} + \Delta_n,$$

where $|\Delta_n| \leq \delta$ and $|c_n| \leq ||\mathbf{z}||_1$. The conclusion follows since δ was arbitrary and $m_n \to \infty$ as $n \to \infty$. We now have almost sure radial convergence for the countable dense set of directions $U \equiv \{\mathbf{z}/|\mathbf{z}| : \mathbf{z} \in \mathbf{Z}^d\}$.

By Theorem 2.1, for directions $\hat{\mathbf{z}} \in U$, we have also that $\lim_{n\to\infty} ET(\mathbf{0}, n\hat{\mathbf{z}})/n = \mu(\hat{\mathbf{z}})$. If $\hat{\mathbf{x}}$ and $\hat{\mathbf{y}}$ are any unit vectors, then

$$\left| \frac{ET(\mathbf{0}, n\hat{\mathbf{x}})}{n} - \frac{ET(\mathbf{0}, n\hat{\mathbf{y}})}{n} \right| \leq \left| \frac{ET(n\hat{\mathbf{x}}, n\hat{\mathbf{y}})}{n} \right|$$

$$\leq \frac{C_1 ||n\hat{\mathbf{x}} - n\hat{\mathbf{y}}||_1}{n} \leq C_1 |\hat{\mathbf{x}} - \hat{\mathbf{y}}|,$$

yielding that

$$\limsup_{n\to\infty} \left| \frac{ET(\mathbf{0}, n\hat{\mathbf{x}})}{n} - \frac{ET(\mathbf{0}, n\hat{\mathbf{y}})}{n} \right| \leq C_1 |\hat{\mathbf{x}} - \hat{\mathbf{y}}|,$$

and, for directions $\hat{\mathbf{z}} \in U$, that

$$\limsup_{n\to\infty} \left| \frac{ET(\mathbf{0}, n\hat{\mathbf{x}})}{n} - \mu(\hat{\mathbf{z}}) \right| \leq C_1 |\hat{\mathbf{x}} - \hat{\mathbf{z}}|.$$

From this it follows that

$$\mu(\hat{\mathbf{x}}) \equiv \lim_{n\to\infty} \frac{ET(\mathbf{0}, n\hat{\mathbf{x}})}{n} \text{ exists and equals } \lim_{\substack{\hat{\mathbf{z}}\to\hat{\mathbf{x}} \\ \hat{\mathbf{z}}\in U}} \mu(\hat{\mathbf{z}}).$$

If $\hat{\mathbf{x}}$ and $\hat{\mathbf{y}}$ are any unit vectors, we have by this reasoning that $|\mu(\hat{\mathbf{x}}) - \mu(\hat{\mathbf{y}})| \leq C_1|\hat{\mathbf{x}} - \hat{\mathbf{y}}|$. Note that we have not yet proved almost sure radial convergence for directions other than those in U.

To patch things together, fix any $\varepsilon \in (0,1)$ and choose finitely many unit vectors $\hat{\mathbf{u}}_1, \ldots, \hat{\mathbf{u}}_m \in U$ such that

$$\bigcup_{a \geq 0} \bigcup_{j=1}^{m} B(a\hat{\mathbf{u}}_j, a\varepsilon) = \mathbf{R}^d.$$

For any $\mathbf{x} \in \mathbf{R}^d$, with some abuse of notation choose $\hat{\mathbf{u}}_{\mathbf{x}} \in \{\hat{\mathbf{u}}_1, \ldots, \hat{\mathbf{u}}_m\}$ so that for some $a \geq 0$, $\mathbf{x} \in B(a\hat{\mathbf{u}}_{\mathbf{x}}, a\varepsilon)$ and let $a_{\mathbf{x}} \equiv \inf\{a : \mathbf{x} \in B(a\hat{\mathbf{u}}_{\mathbf{x}}, a\varepsilon)\}$ and $\mathbf{u}_{\mathbf{x}} \equiv a_{\mathbf{x}}\hat{\mathbf{u}}_{\mathbf{x}}$, so $|\mathbf{u}_{\mathbf{x}} - \mathbf{x}| = a_{\mathbf{x}}\varepsilon$ and $a_{\mathbf{x}}(1 - \varepsilon) \leq |\mathbf{x}| \leq a_{\mathbf{x}}(1 + \varepsilon)$. Then, with $\hat{\mathbf{x}} = \mathbf{x}/|\mathbf{x}|$,

$$\begin{aligned}
|T(\mathbf{0}, \mathbf{x}) - |\mathbf{x}|\mu(\hat{\mathbf{x}})| \leq\ & |T(\mathbf{0}, \mathbf{x}) - T(\mathbf{0}, \mathbf{u}_{\mathbf{x}})| \\
& + |T(\mathbf{0}, \mathbf{u}_{\mathbf{x}}) - a_{\mathbf{x}}\mu(\hat{\mathbf{u}}_{\mathbf{x}})| \\
& + |a_{\mathbf{x}}\mu(\hat{\mathbf{u}}_{\mathbf{x}}) - |\mathbf{x}|\mu(\hat{\mathbf{u}}_{\mathbf{x}})| \\
& + ||\mathbf{x}|\mu(\hat{\mathbf{u}}_{\mathbf{x}}) - |\mathbf{x}|\mu(\hat{\mathbf{x}})|.
\end{aligned} \tag{3.7}$$

The second term in the right side of (3.7) is almost surely bounded by $\varepsilon|\mathbf{u}_{\mathbf{x}}|$ and, therefore, by $\frac{\varepsilon}{1-\varepsilon}|\mathbf{x}|$ for $|\mathbf{u}_{\mathbf{x}}|$, hence for $|\mathbf{x}|$, sufficiently large by almost sure radial convergence for the m unit vectors chosen from U. The third term is bounded by $a_{\mathbf{x}}\varepsilon\bar{\mu} \leq \frac{\varepsilon\bar{\mu}}{1-\varepsilon}|\mathbf{x}|$ where $\bar{\mu} = \sup_{\hat{\mathbf{x}}} \mu(\hat{\mathbf{x}}) < \infty$. The last term is bounded by $C_1|\hat{\mathbf{u}}_{\mathbf{x}} - \hat{\mathbf{x}}||\mathbf{x}|$, which in turn is bounded by $\frac{C_1\varepsilon}{1-\varepsilon}|\mathbf{x}|$.

The moment condition is needed to get that $|T(\mathbf{0}, \mathbf{x}) - T(\mathbf{0}, \mathbf{u}_{\mathbf{x}})|$ is bounded by $K\varepsilon|\mathbf{x}|$ for sufficiently large $|\mathbf{x}|$, almost surely, for some K that is independent of ε. This will yield that

$$\limsup_{\mathbf{x} \to \infty} \left| \frac{T(\mathbf{0}, \mathbf{x})}{|\mathbf{x}|} - \mu(\hat{\mathbf{x}}) \right| \leq \frac{C_1\varepsilon}{1 - \varepsilon} \text{ a.s.}$$

for some $C_1 < \infty$ that is independent of ε. By letting $\varepsilon \downarrow 0$ through, say, the rationals, one obtains (3.5).

Now $|T(\mathbf{0}, \mathbf{x}) - T(\mathbf{0}, \mathbf{u}_{\mathbf{x}})| \leq |T(\mathbf{x}, \mathbf{u}_{\mathbf{x}})|$, $T(\mathbf{x}, \mathbf{u}_{\mathbf{x}}) \overset{\mathrm{d}}{=} T(\mathbf{0}, \mathbf{x} - \mathbf{u}_{\mathbf{x}})$, and $|\mathbf{x} - \mathbf{u}_{\mathbf{x}}|$ is of order $\varepsilon|\mathbf{x}|$, so it suffices to show that

$$\sum_{\mathbf{x} \in \mathbf{Z}^d} P[T(\mathbf{0}, \mathbf{x} - \mathbf{u}_{\mathbf{x}}) \geq K|\mathbf{x} - \mathbf{u}_{\mathbf{x}}|] < \infty.$$

A simple geometric argument shows that $\mathrm{Card}\{\mathbf{x} \in \mathbf{Z}^d : (\mathbf{x} - \mathbf{u}_{\mathbf{x}})^* = \mathbf{z}\} \leq C_1(\varepsilon) < \infty$ for all $\mathbf{z} \in \mathbf{Z}^d$ (here $(\mathbf{x} - \mathbf{u}_{\mathbf{x}})^*$ is the element of $\mathbf{Z}^d$ closest to $\mathbf{x} - \mathbf{u}_{\mathbf{x}}$). It therefore suffices to show that

$$\sum_{\mathbf{x} \in \mathbf{Z}^d} P[T(\mathbf{0}, \mathbf{x}) \geq K|\mathbf{x}|] < \infty. \tag{3.8}$$

Cox and Durrett use the moment condition $EY^d < \infty$ to establish this – we omit the details of the argument. Clearly a stronger moment condition, a finite exponential moment for example, would easily yield (3.8) for some $K < \infty$.

Cox and Durrett ([17], for $d = 2$) and, later, Kesten ([40], for arbitrary dimension) found a clever means of obtaining a shape theorem that requires no moment condition. The basic idea is to allow, when computing $T(\mathbf{x}, \mathbf{y})$, free passage from $\mathbf{x}$ to sites in some configuration-dependent set $\Delta(\mathbf{x})$ surrounding $\mathbf{x}$ and, similarly, from $\mathbf{y}$ to points in the corresponding $\Delta(\mathbf{y})$. They define

$$\hat{T}(\mathbf{x}, \mathbf{y}) \;=\; \inf\{T(\mathbf{u}, \mathbf{v}) : \mathbf{u} \in \Delta(\mathbf{x}), \mathbf{v} \in \Delta(\mathbf{y})\}.$$

As constructed in [17, 40], $\Delta(\mathbf{x})$ contains only points "near" $\mathbf{x}$ in the sense that Card $\Delta(\mathbf{x})$ has finite moments of all order. If $\mu(\hat{\mathbf{e}}_1) > 0$, there is a convex $\widehat{W}$ of non-empty interior that is invariant under lattice symmetries such that

$$\text{for all } \varepsilon \in (0, 1), \ (1 - \varepsilon)\widehat{W} \subset t^{-1}\widehat{W}_t \subset (1 + \varepsilon)\widehat{W} \text{ for all large } t, \text{ a.s.,}$$

where $\widehat{W}_t = \{\mathbf{x} : \hat{T}(\mathbf{0}, \mathbf{x}) \leq t\}$. If $\mu(\hat{\mathbf{e}}_1) = 0$, every compact set will a.s. be eventually covered by $t^{-1}\widehat{W}_t$.

Boivin [11] generalizes the shape theorem in a different direction. Boivin shows that (3.5) will hold provided the edge times satisfy a certain integrability condition if the sequence of random variables

$$\{\tau(\{\mathbf{x}, \mathbf{x} + \hat{\mathbf{e}}_i\}) : 1 \leq i \leq d, \ \mathbf{x} \in \mathbf{Z}^d\}$$

is stationary and ergodic. Boivin's integrability condition is satisfied by edge variables with finite moment of order $d + \delta$ for some $\delta > 0$.

3.2 About the Asymptotic Shape for Lattice FPP

Standard FPP. For Standard FPP (in contrast to other FPP models discussed below), very little is known about the asymptotic shape $\widetilde{W}$. Clearly $\widetilde{W}$ is invariant with respect to the lattice symmetries, i.e., coordinate permutations and multiplication of any of the coordinate values by -1. Additionally, one sees that $\widetilde{W}$ is convex through an application of the triangle inequality (1.6). By convexity, $\widetilde{W}$ must contain the diamond $D_0 = \{\mathbf{x} \in \mathbf{R}^d : ||\mathbf{x}||_1 \leq \mu(\hat{\mathbf{e}}_1)^{-1}\}$. When $\lambda = \lambda(F) > 0$, one obtains that $\widetilde{W}$ is contained in the diamond $D_1 = \{\mathbf{x} \in \mathbf{R}^d : ||\mathbf{x}||_1 \leq \lambda^{-1}\}$ by considering only directed paths (paths where the individual coordinate values either monotonically increase or monotonically decrease). Finally, it is a direct consequence of convexity and symmetry that $\widetilde{W}$ is contained in the box $B = \{\mathbf{x} \in \mathbf{R}^d : ||\mathbf{x}||_\infty \leq \mu(\hat{\mathbf{e}}_1)^{-1}\}$. See Figure 2 for the picture in 2 dimensions, which is drawn for $\lambda > \mu(\hat{\mathbf{e}}_1)/2$ so that $B \not\subset D_1$. Durrett and Liggett [20] have shown that if $\lambda > 0$ and $F(\lambda)$ is sufficiently large (but still less than 1) then $\widetilde{W}$ actually reaches out to ∂D_1 covering four intervals on ∂D_1 that contain, respectively, the four points $(\pm(2\lambda)^{-1}, \pm(2\lambda)^{-1})$.

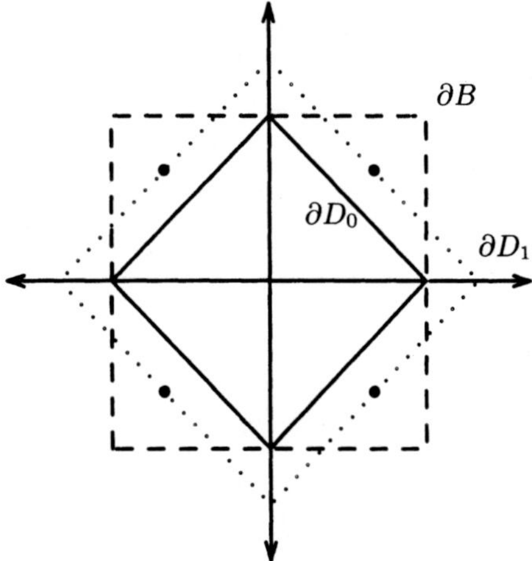

Fig. 2. $\widetilde{W}$ is convex and symmetric. It contains D_0 and is contained in $D_1 \cap B$. The four points $(\pm(2\lambda)^{-1}, \pm(2\lambda)^{-1})$ are highlighted.

Early Monte Carlo simulation results (see [21, 60]) suggested that, for $d = 2$, the asymptotic shape could be a Euclidean ball. This now seems unlikely in view of Kesten's result that in high d under mild conditions on F the asymptotic shape is not a ball (see [40], Corollary 8.4).

Little else is rigorously known about the asymptotic shape. In fact, the general lack of information about $\widetilde{W}$ is a technical stumbling block for completing rigorous proofs of a number of important results in lattice FPP (see Sections 4 and 5 below). Specifically, it would be very useful to have information about the curvature of the boundary of $\widetilde{W}$. We have seen that we can construct examples where, in 2 dimensions, $\partial\widetilde{W}$ contains straight line segments. It is not expected that these flat portions of $\partial\widetilde{W}$ are typical, but nothing along these lines is rigorously known. Returning to general $d \geq 2$, we follow Newman and Piza [55] and call direction $\hat{\mathbf{x}}$ a *direction of curvature* for $\widetilde{W}$ if, for $\mathbf{z} = \hat{\mathbf{x}}/\mu(\hat{\mathbf{x}}) \in \partial\widetilde{W}$, there is a (Euclidean) ball D such that

$$\mathbf{z} \in \partial D \text{ and } \widetilde{W} \subset D. \tag{3.9}$$

It turns out that quite a number of things about minimizing paths and fluctuation exponents are known to hold in directions of curvature. Unfortunately, all that is known is that there are directions of curvature – though possibly only finitely many. For example, take $D^* = B(\mathbf{0}, \varrho^*)$, where $\varrho^* = \inf\{\varrho > 0 : \partial B(\mathbf{0}, \varrho) \cap \widetilde{W} = \emptyset\}$. Then $\mathbf{z}/|\mathbf{z}|$ is a direction of curvature for all $\mathbf{z} \in \partial D^* \cap \widetilde{W}$. Yet no specific direction, $\hat{\mathbf{e}}_1$ for example, is known to be a direction of curvature for the asymptotic shape corresponding to any distribution F.

Other Lattice Models. There is a modified lattice FPP model, due to Seppäläinen [62], where we have complete information about the asymptotic shape. The setting here is *directed FPP* on the first quadrant of the $\mathbf{Z}^2$ lattice, $\mathbf{Z}_+^2 \equiv \{(x_1, x_2) \in \mathbf{Z}^2 : x_1 \geq 0, x_2 \geq 0\}$ with nearest neighbor edges. If e is a vertical edge, then $\tau(e) = \tau_0$, a fixed positive constant. If e is a horizontal edge, then $\tau(e)$ is Bernoulli with $P[\tau(e) = \lambda] = p$ and $P[\tau(e) = \kappa] = q \equiv 1 - p$ where $0 \leq \lambda < \kappa$. The horizontal edge variables are independent. Additionally, only directed paths (paths with monotonically increasing $\hat{\mathbf{e}}_1$ and $\hat{\mathbf{e}}_2$ components) are considered:

$$T^*(\mathbf{0}, \mathbf{v}) = \inf\{T(r) : r \text{ is a directed lattice path from } \mathbf{0} \text{ to } \mathbf{v}\},$$

and, for unit vector $\hat{\mathbf{x}} = (x_1, x_2)$,

$$\mu^*(\hat{\mathbf{x}}) \equiv \lim_{n \to \infty} \frac{ET^*(\mathbf{0}, n\hat{\mathbf{x}})}{n}.$$

With this set up, Seppäläinen shows that

$$\mu^*(\hat{\mathbf{x}}) = \begin{cases} \lambda x_1 + \tau_0 x_2 & \text{if } px_2 > qx_1 \\ \lambda x_1 + \tau_0 x_2 + (\kappa - \lambda)(\sqrt{qx_1} - \sqrt{px_2})^2 & \text{if } px_2 \leq qx_1. \end{cases}$$

The situation for the stationary case of Boivin is quite interesting. Here, the edge variables are drawn from up to d different distributions – one for each coordinate direction. It is therefore typically the case that $\mu(\hat{\mathbf{e}}_i) \neq \mu(\hat{\mathbf{e}}_j)$ for $i \neq j$, and one may even have $\mu(\hat{\mathbf{e}}_i) = 0$ and $\mu(\hat{\mathbf{e}}_j) > 0$: there is not the dichotomy of linear growth in all directions or super-linear growth in all directions. In [24], Häggström and Meester study the case of linear growth in all directions (i.e., all $\mu(\hat{\mathbf{e}}_i) > 0$) and ask what asymptotic shapes are possible. Convexity of $\widetilde{W}$ is assured in the stationary case just as in the i.i.d. case by (1.6). However, $\widetilde{W}$ is not necessarily invariant under coordinate permutations nor under multiplication of any individual coordinate value by -1. What remains is a much weaker symmetry. It follows from

$$T(\mathbf{0}, \mathbf{x}) = T(\mathbf{x}, \mathbf{0}) \stackrel{\mathrm{d}}{=} T(\mathbf{0}, -\mathbf{x})$$

that $\mu(\hat{\mathbf{x}}) = \mu(-\hat{\mathbf{x}})$ for all directions $\hat{\mathbf{x}}$ and therefore that $\widetilde{W} = -\widetilde{W}$. Remarkably, Häggström and Meester show that if W is any compact and convex set with nonempty interior such that $W = -W$, then W can arise as the asymptotic shape for the right edge distributions $F_1, \ldots, F_d$ for stationary first-passage percolation.

3.3 FPP Based on Poisson Point Processes

Vahidi-Asl and Wierman [71, 72] studied FPP on two dual planar graphs induced by a homogeneous Poisson point process of, say, unit mean density. Let $Q \subset \mathbf{R}^d$ denote a realization of a d-dimensional homogeneous Poisson process of unit density, so Q is infinite but locally finite. We refer to elements of Q as *particles*, or Poisson particles. For $\mathbf{x} \in \mathbf{R}^d$, let $Q(\mathbf{x})$ denote the particle that is closest to $\mathbf{x}$, with some fixed rule for breaking ties. Then, for $\mathbf{q} \in Q$,

$$V(\mathbf{q}) \;\equiv\; \{\mathbf{x} \in \mathbf{R}^d : Q(\mathbf{x}) = \mathbf{q}\}^{\circ} \tag{3.10}$$

is the Voronoi region associated with Q at $\mathbf{q}$. It consists of all points that are strictly closer to $\mathbf{q}$ than to any other Poisson particle. Specializing now to 2 dimensions, $\partial V(\mathbf{q})$ is a convex polygon surrounding $\mathbf{q}$. The *Voronoi graph* $\mathcal{V}_2$ is the graph whose edges (resp. vertices) are collectively the edges (resp. vertices) of the polygons $\partial V(\mathbf{q})$ as $\mathbf{q}$ ranges over all of Q. Formally, the *Delaunay graph* $\mathcal{D}_2$ is dual to the Voronoi graph : its vertex set is Q and $\mathbf{q}, \mathbf{q}' \in Q$ have an edge between them if $\partial V(\mathbf{q})$ and $\partial V(\mathbf{q}')$ share a common line segment. (See Figure 3 for a patch of $\mathcal{V}_2$ and $\mathcal{D}_2$.)

FPP on the graphs $\mathcal{V}_2$ and $\mathcal{D}_2$ is entirely analogous to FPP on the $\mathbf{Z}^2$ lattice, as described in Section 1.2. In the models of Vahidi-Asl and Wierman, the edge variables are taken to be i.i.d. (and independent of the particle configuration) with common distribution F. Note that the distribution of the process $T(\cdot, \cdot)$ is determined *jointly* by the particle configuration (through $\mathcal{V}_2$ or $\mathcal{D}_2$) and the edge variable configuration. That is, for these models the path set $\mathcal{R}(\mathbf{u}, \mathbf{v})$ in (1.5) is random and is determined by the Poisson particle configuration. For any direction $\hat{\mathbf{x}}$, the process $(T(m\hat{\mathbf{x}}, n\hat{\mathbf{x}}) : 0 \le m < n)$ is easily seen to satisfy the hypotheses of Theorem 2.1, provided $ET(\mathbf{0}, \hat{\mathbf{e}}_1) < \infty$ for F, insuring that

$$\mu(\hat{\mathbf{x}}) \;\equiv\; \lim_{n \to \infty} \frac{T(\mathbf{0}, n\hat{\mathbf{x}})}{n}$$

exists for all $\hat{\mathbf{x}}$. A fundamental advantage of Poisson-based FPP models is that $\mu(\hat{\mathbf{x}})$ is independent of $\hat{\mathbf{x}}$. This follows immediately from the fact that the distribution of Q is invariant with respect to all rigid motions of $\mathbf{R}^2$ – rotation in particular. We refer to the common value as μ. This implies, of

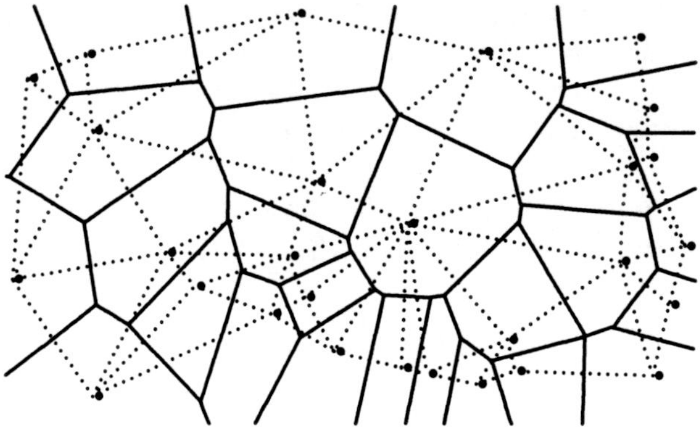

Fig. 3. The Voronoi graph (*solid lines*) and the Delaunay graph (*dotted lines*) for a particular Poisson particle configuration (shown as •'s).

course, that if an asymptotic shape exists for these models, it must be the Euclidean ball $B = B(\mathbf{0}, 1/\mu)$.

Now each vertex of the graph $\mathcal{V}_2$ has degree exactly 3 while the Delaunay graph is actually a triangulation of $\mathbf{R}^2$, so each vertex of $\mathcal{D}_2$ has degree at least 3. In this model, therefore, the natural condition to insure that sites not get left "dry" for too long is $EY^2 < \infty$, where here $Y \equiv \min(\tau_1, \tau_2, \tau_3)$ with the τ_i independent and distributed according to F. Indeed, Vahisi-Asl and Wierman show that, for FPP on either $\mathcal{V}_2$ or $\mathcal{D}_2$, if $\mu > 0$ then

$$\text{for all } \varepsilon \in (0,1), \ (1-\varepsilon)B \subset t^{-1}\mathcal{W}_t \subset (1+\varepsilon)B \text{ for all large } t, \text{ a.s.} \quad (3.11)$$

if and only if $EY^2 < \infty$. (The μ's and Y's are naturally different in the two settings.)

Serafini [61] extends this result to the Delaunay graph $\mathcal{D}_d$ in arbitrary dimension d. The vertex set of $\mathcal{D}_d$ is again Q, the set of Poisson particles. In $\mathcal{D}_d$, there is an edge corresponding to a pair of particles $\{\mathbf{q}, \mathbf{q}'\}$ if and only if $\partial V(\mathbf{q})$ and $\partial V(\mathbf{q}')$ share a common $d-1$ - dimensional face. The required moment condition here is that $EY^d < \infty$, where $Y = \min\{\tau_1, \ldots, \tau_{d+1}\}$ and the τ_i's are independent and distributed according to F. On $\mathcal{V}_2$ and all the $\mathcal{D}_d$, $\mu > 0$ provided $F(0)$, the atom at 0, is sufficiently small.

Note that the shape theorem for these graphs is non-trivial even when the $\tau(e)$'s are constant random variables. On the Delaunay graph with all $\tau(e) = 1$, for example, $T(\mathbf{u}, \mathbf{v})$ counts the minimum number of Voronoi regions that one must travel through to get from $\mathbf{u}$ to $\mathbf{v}$, counting, say, $\mathbf{v}$'s region but not $\mathbf{u}$'s.

Howard and Newman have studied a different family of Poisson-based FPP models called *Euclidean FPP* – so named because of the invariance of the distribution of Q with respect to all rigid motions of $\mathbf{R}^d$. Here, FPP takes place on the complete graph $\mathcal{C}(Q)$ with vertex set Q. For any edge $e = \{\mathbf{q}, \mathbf{q}'\}$, they put $\tau(e) = \phi(|\mathbf{q} - \mathbf{q}'|)$, where $\phi : \mathbf{R}^+ \to \mathbf{R}^+$ satisfies $\phi(0) = 0$, $\phi(1) = 1$, and ϕ is strictly convex. These conditions imply that $\phi(x)$ is continuous and strictly increasing. Note that in Euclidean FPP, the edge variables are deterministic given the particle configuration Q. Howard and Newman have restricted their attention to ϕ's of the form $\phi(x) = x^\alpha$, for some $\alpha > 1$, but their results should extend to a broader class of functions. Henceforth we make the same restriction.

The strict convexity of $\phi(x)$ implies that long jumps are discouraged on minimizing paths. In fact, the edge $\{\mathbf{q}, \mathbf{q}'\}$ can (possibly) belong to a minimizing path r only if the region

$$R_\alpha(\mathbf{q}, \mathbf{q}') \equiv \{\mathbf{x} \in \mathbf{R}^d : |\mathbf{q} - \mathbf{x}|^\alpha + |\mathbf{x} - \mathbf{q}'|^\alpha < |\mathbf{q} - \mathbf{q}'|^\alpha\}$$

$$\text{is devoid of Poisson particles.} \quad (3.12)$$

(If $\widetilde{\mathbf{q}} \in R_\alpha(\mathbf{q}, \mathbf{q}')$, then the path $\widetilde{r}$ where $(\mathbf{q}, \mathbf{q}')$ in r is replaced with $(\mathbf{q}, \widetilde{\mathbf{q}}, \mathbf{q}')$ would have $T(\widetilde{r}) < T(r)$.) It follows that the complete graph may be replaced with the graph $\mathcal{C}_\alpha(Q)$ that has vertex set Q and edge set

$$\mathcal{C}_\alpha^{\mathrm{edge}}(Q) \equiv \{\{\mathbf{q}, \mathbf{q}'\} : \mathbf{q}, \mathbf{q}' \in Q,\ R_\alpha(\mathbf{q}, \mathbf{q}') \cap Q = \emptyset\}. \qquad (3.13)$$

The graph $\mathcal{C}_\alpha(Q)$ almost surely has finite degree at each vertex. For fixed Q, the edge set of $\mathcal{C}_\alpha(Q)$ is decreasing in α and, in dimension 2, for $1 < \alpha < 2$, $\mathcal{C}_\alpha(Q)$ is almost surely not planar, but for $\alpha \geq 2$ it almost surely is. For any $\mathbf{u}, \mathbf{v} \in \mathbf{R}^d$, (1.5) produces

$$
\begin{aligned}
T(\mathbf{u}, \mathbf{v}) &= \inf\left\{ \sum_{i=0}^{k-1} |\mathbf{q}_i - \mathbf{q}_{i+1}|^\alpha : k \geq 2,\ \mathbf{q}_i \in Q,\ \mathbf{q}_0 = Q(\mathbf{u}),\ \mathbf{q}_k = Q(\mathbf{v}) \right\} \\
&= \inf\left\{ \sum_{i=0}^{k-1} |\mathbf{q}_i - \mathbf{q}_{i+1}|^\alpha : k \geq 2,\ \{\mathbf{q}_i, \mathbf{q}_{i+1}\} \in \mathcal{C}_\alpha^{\mathrm{edge}}(Q), \right. \\
&\qquad\qquad\qquad\qquad\qquad\qquad \left. \mathbf{q}_0 = Q(\mathbf{u}),\ \mathbf{q}_k = Q(\mathbf{v}) \right\},
\end{aligned}
\qquad (3.14)
$$

where the second equality holds when $Q(\mathbf{u}) \neq Q(\mathbf{v})$ ($T(\mathbf{u}, \mathbf{v}) = 0$ otherwise).

In [30], Howard and Newman show that, for all $\alpha > 1$, $\mu = \mu(\alpha, d)$ satisfies $0 < \mu < \infty$ and that (3.11) holds for $B = B(\mathbf{0}, 1/\mu)$ and with $\mathcal{W}_t$ as defined in (1.4). (Here, "almost surely" is with respect to the measure on particle configurations.)

3.4 Upper Bounds on Shape Fluctuations

More recent work of Alexander and Kesten (for Standard FPP) and Howard and Newman (for Euclidean FPP) has focused on replacing ε in the shape theorem (see (3.4)) with $\varepsilon(t)$, where $\varepsilon(t) \to 0$ as $t \to \infty$. In [41] Kesten shows that for Standard FPP where the edge distribution satisfies the exponential moment condition

$$M_\tau(\gamma) \equiv Ee^{\gamma\tau} < \infty \text{ for some } \gamma > 0 \qquad (3.15)$$

one has, for some finite A,

$$(t - At^\kappa \log t)\widetilde{\mathcal{W}} \subset \mathcal{W}_t \subset (t + At^{1/2} \log t)\widetilde{\mathcal{W}} \text{ for all large } t,\ \text{a.s.} \qquad (3.16)$$

where $\kappa = (2d + 3)/(2d + 4)$. A key ingredient for obtaining (3.16) is the moderate deviation estimate

$$P[|T(\mathbf{0}, \mathbf{x}) - ET(\mathbf{0}, \mathbf{x})| \geq x|\mathbf{x}|^{1/2}] \leq C_1 e^{-C_0 x} \text{ for } x \leq C_0|\mathbf{x}|. \qquad (3.17)$$

Unfortunately, (3.17) bounds the probability of moderate deviations about the *mean*, rather than about $|\mathbf{x}|\mu(\hat{\mathbf{x}})$. The second key ingredient is therefore the bound

$$|\mathbf{x}|\mu(\hat{\mathbf{x}}) \leq ET(\mathbf{0}, \mathbf{x}) \leq |\mathbf{x}|\mu(\hat{\mathbf{x}}) + C_1|\mathbf{x}|^\kappa \log |\mathbf{x}|, \qquad (3.18)$$

where $\hat{\mathbf{x}} = \mathbf{x}/|\mathbf{x}|$, proved by Kesten in [41] for the same value of κ.

Alexander, in [1] for $\hat{\mathbf{x}} = \hat{\mathbf{e}}_1$, and later in [2] for arbitrary directions, improves this by showing that we may take $\kappa = 1/2$ in (3.18) and therefore also in (3.16). With the *longitudinal fluctuation exponent* χ_1 defined as

$$\chi_1 \equiv \inf\{\kappa : (t - t^\kappa)\widetilde{W} \subset W_t \subset (t + t^\kappa)\widetilde{W} \text{ for all large } t, \text{ a.s.}\}, \qquad (3.19)$$

these results may be summarized as follows.

Theorem 3.2. *For Standard FPP with edge variables satisfying the moment condition (3.15), $\chi_1 \leq 1/2$.*

Below, we follow Alexander in [2] to prove (3.16) (with $\kappa = 1/2$) given (3.17) and (3.18) (again with $\kappa = 1/2$). Later, we outline proofs of (3.17) and a version of (3.18).

Proof of (3.16) using (3.17) and (3.18). Let $A, t > 0$ and suppose

$$\text{there exists } \mathbf{x} \in W_t \cap \mathbf{Z}^d \text{ with } \mathbf{x} \notin (t + At^{1/2}\log t)\widetilde{W}. \qquad (3.20)$$

Let $\hat{\mathbf{x}} = \mathbf{x}/|\mathbf{x}|$. Then $T(\mathbf{0}, \mathbf{x}) \leq t$ but $ET(\mathbf{0}, \mathbf{x}) \geq |\mathbf{x}|\mu(\hat{\mathbf{x}}) > t + At^{1/2}\log t$, so

$$ET(\mathbf{0}, \mathbf{x}) - T(\mathbf{0}, \mathbf{x}) \geq |\mathbf{x}|\mu(\hat{\mathbf{x}}) - t \geq At^{1/2}\log t. \qquad (3.21)$$

From Section 3.2, we know that $\inf_{\hat{\mathbf{x}}} \mu(\hat{\mathbf{x}}) > 0$, so if $|\mathbf{x}|\mu(\hat{\mathbf{x}}) \leq 2t$ then $t \geq C_0|\mathbf{x}|$ and (3.21) yields, for large $|\mathbf{x}|$, that

$$ET(\mathbf{0}, \mathbf{x}) - T(\mathbf{0}, \mathbf{x}) \geq AC_0|\mathbf{x}|^{1/2}\log|\mathbf{x}|.$$

On the other hand, if $|\mathbf{x}|\mu(\hat{\mathbf{x}}) > 2t$, then the first inequality in (3.21) yields, for large $|\mathbf{x}|$, that

$$ET(\mathbf{0}, \mathbf{x}) - T(\mathbf{0}, \mathbf{x}) \geq |\mathbf{x}|\frac{\mu(\hat{\mathbf{x}})}{2} \geq AC_0|\mathbf{x}|^{1/2}\log|\mathbf{x}|.$$

Thus if (3.20) occurs for arbitrarily large t then

$$ET(\mathbf{0}, \mathbf{x}) - T(\mathbf{0}, \mathbf{x}) \geq AC_0|\mathbf{x}|^{1/2}\log|\mathbf{x}| \quad \text{for infinitely many } \mathbf{x} \in \mathbf{Z}^d.$$

But, for large $|\mathbf{x}|$, (3.17) gives that

$$P[ET(\mathbf{0}, \mathbf{x}) - T(\mathbf{0}, \mathbf{x}) \geq AC_0|\mathbf{x}|^{1/2}\log|\mathbf{x}|] \leq C_1|\mathbf{x}|^{-AC_0^2},$$

which is summable over $\mathbf{x} \in \mathbf{Z}^d$ if we choose A sufficiently large. It follows from the Borel-Cantelli lemma that (3.20) occurs for only finitely many $\mathbf{x} \in \mathbf{Z}^d$ yielding the second inclusion in (3.16).

For the first inclusion, suppose $A > 0$ and that for arbitrarily large t there exists $\mathbf{x} \in \mathbf{Z}^d$ with $\mathbf{x} \in (t - At^{1/2}\log t)\widetilde{W}$ but with $\mathbf{x} \notin W_t$. Then $C_0|\mathbf{x}| \leq t$, $T(\mathbf{0}, \mathbf{x}) > t$, $|\mathbf{x}|\mu(\hat{\mathbf{x}}) \leq t - At^{1/2}\log t$, and, for large t, $|\mathbf{x}| > 1$. So, by (3.18), for large t,

$$ET(\mathbf{0}, \mathbf{x}) \;\leq\; t - At^{1/2}\log t + C_1|\mathbf{x}|^{1/2}\log|\mathbf{x}|$$
$$\leq\; t - AC_0|\mathbf{x}|^{1/2}\log(C_0|\mathbf{x}|) + C_1|\mathbf{x}|^{1/2}\log|\mathbf{x}|$$
$$\leq\; t - (AC_0 - C_1)|\mathbf{x}|^{1/2}\log|\mathbf{x}|$$

and

$$T(\mathbf{0}, \mathbf{x}) - ET(\mathbf{0}, \mathbf{x}) \;\geq\; (AC_0 - C_1)|\mathbf{x}|^{1/2}\log|\mathbf{x}|.$$

But it follows from (3.17), as above, that if A is sufficiently large this almost surely will happen for only finitely many $\mathbf{x} \in \mathbf{Z}^d$.

Proof of (3.17). We outline here Kesten's proof of (3.17) and along the way prove another of the fluctuation results in [41], namely that

$$\operatorname{Var} T(\mathbf{0}, \mathbf{x}) \;\leq\; C_1|\mathbf{x}| \quad \text{provided the edge variables satisfy } E\tau^2 < \infty. \quad (3.22)$$

(This has recently been improved somewhat for Bernoulli edge times by Benjamini, Kalai, and Schramm [7] to $\operatorname{Var} T(\mathbf{0}, \mathbf{x}) \leq C_1|\mathbf{x}|/\log|\mathbf{x}|$.) Later, we outline a proof of (3.18) in the context of Euclidean FPP, where the model's isotropy makes the argument considerably easier.

Toward (3.17) and (3.22), let $(e_i : i \geq 1)$ be any enumeration of the $\mathbf{Z}^d$ lattice edge set and put $\tau_i = \tau(e_i)$. Let $\mathcal{F}_m = \sigma(\tau_1, \ldots, \tau_m)$ with $\mathcal{F}_0 \equiv \{\emptyset, \Omega\}$, and with $T \equiv T(\mathbf{0}, \mathbf{x})$, express $T - ET$ as a sum of martingale increments as follows:

$$T - ET \;=\; \sum_{m=1}^{\infty} E[T|\mathcal{F}_m] - E[T|\mathcal{F}_{m-1}] \;=\; \sum_{m=1}^{\infty} \Delta_m, \quad (3.23)$$

where $\Delta_m \equiv E[T|\mathcal{F}_m] - E[T|\mathcal{F}_{m-1}]$. Put $\widetilde{\mathcal{F}}_m = \sigma(\tau_1, \ldots, \tau_{m-1}, \tau_{m+1}, \tau_{m+2}, \ldots)$ and define $\widetilde{\Delta}_m = T - E[T|\widetilde{\mathcal{F}}_m]$. Then one verifies that $E[E[T|\widetilde{\mathcal{F}}_m]|\mathcal{F}_m] = E[T|\mathcal{F}_{m-1}]$, yielding that $E[\widetilde{\Delta}_m|\mathcal{F}_m] = \Delta_m$ and hence that

$$\operatorname{Var} T \;=\; \sum_{m=1}^{\infty} E\Delta_m^2 \;\leq\; \sum_{m=1}^{\infty} E\widetilde{\Delta}_m^2. \quad (3.24)$$

Let $M^*(\mathbf{0}, \mathbf{x})$ be a minimizing route from $\mathbf{0}$ to $\mathbf{x}$ chosen according to some fixed rule if the route is not unique, and put $R_m \equiv \{e_m \in M^*(\mathbf{0}, \mathbf{x})\}$. Choose $\widetilde{t}$ so that $\widetilde{p} \equiv P[\tau \leq \widetilde{t}] > 0$, and let $\widetilde{T}_m$ denote the minimal passage time from $\mathbf{0}$ to $\mathbf{x}$ when the edge variable configuration is altered so that $\tau_m = \widetilde{t}$. Also, let $\widetilde{R}_m$ be the event that e_m is on the mimimizing path (again chosen according to some rule when not unique) that realizes $\widetilde{T}_m$. Note that

$$\widetilde{p}P[\widetilde{R}_m] \;\leq\; \widetilde{p}P[R_m|\tau_m \leq \widetilde{t}] \;\leq\; P[R_m]. \quad (3.25)$$

In general, $|T - \widetilde{T}_m| \leq |\tau_m - \widetilde{t}| \leq \tau_m + \widetilde{t}$ (since this bound holds for passage times for individual paths), while on $(R_m \cup \widetilde{R}_m)^c$, $T - \widetilde{T}_m = 0$. On $R_m \setminus \widetilde{R}_m$, we have $\tau_m \leq \widetilde{t}$, so $|T - \widetilde{T}_m| \leq \widetilde{t}$. It follows that

$$(T - \widetilde{T}_m)^2 \; \leq \; (\tau_m + \widetilde{t})^2 I_{\widetilde{R}_m} + \widetilde{t}^{\,2} I_{R_m}. \qquad (3.26)$$

Now $\widetilde{T}_m \in \widetilde{\mathcal{F}}_m$ and $\widetilde{R}_m \in \widetilde{\mathcal{F}}_m$ so, in particular, τ_m and $\widetilde{R}_m$ are independent. In general, if X and Y are L^2 random variables with Y measurable with respect to some σ-field $\mathcal{G}$, then

$$E[(X - E[X|\mathcal{G}])^2|\mathcal{G}] \; \leq \; E[(X - Y)^2|\mathcal{G}], \qquad (3.27)$$

so, also, $E(X - E[X|\mathcal{G}])^2 \leq E(X - Y)^2$. It follows that

$$\begin{aligned}
E\widetilde{\Delta}_m^2 \; = \; E[(T - E[T|\widetilde{\mathcal{F}}_m])^2] \; &\leq \; E[(T - \widetilde{T}_m)^2] \\
&\leq \; E[(\tau_m + \widetilde{t})^2 I_{\widetilde{R}_m} + \widetilde{t}^{\,2} I_{R_m}] \\
&\leq \; E[(\tau_m + \widetilde{t})^2] E I_{\widetilde{R}_m} + \widetilde{t}^{\,2} E I_{R_m} \\
&\leq \; BEI_{R_m}, \qquad (3.28)
\end{aligned}$$

where $B \equiv \widetilde{p}^{\,-1} E[(\tau + \widetilde{t})^2] + \widetilde{t}^{\,2}$. Setting this into (3.24) yields that $\operatorname{Var} T \leq BE|M^*(\mathbf{0}, \mathbf{x})|$, where $|M^*(\mathbf{0}, \mathbf{x})|$ is the number of edges along $M^*(\mathbf{0}, \mathbf{x})$. To bound $E|M^*(\mathbf{0}, \mathbf{x})|$ note that

$$E|M^*(\mathbf{0}, \mathbf{x})| \; = \; |\mathbf{x}| \int_0^\infty P[|M^*(\mathbf{0}, \mathbf{x})| \geq x|\mathbf{x}|] \, dx$$

and, for any $a, y > 0$,

$$\begin{aligned}
P[|M^*(\mathbf{0}, \mathbf{x})| \geq y] \; \leq \; &P[T(\mathbf{0}, \mathbf{x}) \geq ay] \\
&+ \; P[\exists\, r \in \mathcal{R}^{\text{s.a.}}(\mathbf{0}, \cdot) : |r| \geq y, \, T(r) < ay]. \quad (3.29)
\end{aligned}$$

For this application, we take $y = x|\mathbf{x}|$ in (3.29). Now

$$|\mathbf{x}| \int_0^\infty P[T(\mathbf{0}, \mathbf{x}) \geq ax|\mathbf{x}|] \, dx \; = \; a^{-1} ET(\mathbf{0}, \mathbf{x}) \; \leq \; a^{-1} C_1 |\mathbf{x}|,$$

where the inequality is easily obtained by considering the passage time along any directed path from $\mathbf{0}$ to $\mathbf{x}$. In [40] Kesten shows that, for a sufficiently small,

$$P[\exists\, r \in \mathcal{R}^{\text{s.a.}}(\mathbf{0}, \cdot) : |r| \geq y, \, T(r) < ay] \; \leq \; C_1 \exp(-C_0 y), \qquad (3.30)$$

so, for $|\mathbf{x}| \geq 1$,

$$|\mathbf{x}| \int_0^\infty P[\exists\, r \in \mathcal{R}^{\text{s.a.}}(\mathbf{0}, \cdot) : |r| \geq x|\mathbf{x}|, \, T(r) < ax|\mathbf{x}|] \, dx \; \leq \; C_1 |\mathbf{x}|,$$

yielding that $\operatorname{Var} T \leq BE|M^*(\mathbf{0}, \mathbf{x})| \leq C_1 |\mathbf{x}|$.

Kesten shows that (3.30) holds provided $F(0) < p_c(d)$. If the atom at 0 is sufficiently small, (3.30) follows from an easy Peierls argument, which we include for completeness. Let L_d denote the connectivity constant for the $\mathbf{Z}^d$ lattice; so if $L > L_d$, then for large n there are fewer than L^n self-avoiding lattice paths starting at $\mathbf{0}$ of length n. Suppose $F(0) < 1/L_d$ and choose $L > L_d$ and $x_0 > 0$ so that $q \equiv F(x_0)$ satisfies $F(0) < q < 1/L$. Then for large n,

$$P[\exists \; r \in \mathcal{R}^{\text{s.a.}}(\mathbf{0}, \cdot) : |r| \geq n, \; T(r) < an] \; \leq \; L^n P[S_n \leq an],$$

where S_n is distributed as the sum of n independent edge variables. Letting $p = 1 - q$, clearly if $B_n \sim \text{Binomial}(n, p)$ then S_n stochastically dominates $x_0 B_n$ so that

$$L^n P[S_n \leq an] \; \leq \; L^n P[B_n \leq an/x_0] \; \leq \; [Le^{a\theta/x_0}(pe^{-\theta} + q)]^n,$$

where the second inequality holds for any $\theta > 0$ by standard large deviation techniques. One may then choose $\theta > 0$ large so that $pe^{-\theta} + q < L^{-1}$, and then $a > 0$ small so that $e^{a\theta/x_0}(pe^{-\theta} + q) < L^{-1}$. This choice of a yields (3.30) for appropriate C_0 and C_1.

The proof of the much harder (3.17) rests on moderate deviation result for martingales with bounded differences. The version below, found in [32], when taken with $\gamma = 1$ is (roughly) equivalent to Kesten's Theorem 3 in [41]. Howard and Newman's proof closely parallel's Kesten's $\gamma = 1$ case – the details are omitted.

Theorem 3.3. *Let $(M_m : m \geq 0)$, $M_0 \equiv 0$, be a martingale with respect to the filtration $\mathcal{F}_m \uparrow \mathcal{F}$. Put $\Delta_m = M_m - M_{m-1}$ and suppose $(U_m : m \geq 1)$ is a sequence of $\mathcal{F}$-measurable positive random variables satisfying $E[\Delta_m^2|\mathcal{F}_{m-1}] \leq E[U_m|\mathcal{F}_{m-1}]$. With $S = \sum_{m=1}^{\infty} U_m$, suppose further that for finite constants $C_1 > 0$, $0 < \gamma \leq 1$, $c \geq 1$, and $x_0 \geq c^2$ we have $|\Delta_m| \leq c$ and*

$$P[S > x] \; \leq \; C_1 \exp(-x^\gamma), \quad \text{when } x \geq x_0. \tag{3.31}$$

Then $\lim_{m \to \infty} M_m = M$ exists and is finite almost surely and there are constants (not depending on c and x_0) $C_2 = C_2(C_1, \gamma) < \infty$ and $C_3 = C_3(\gamma) > 0$ such that

$$P[|M| \geq x\sqrt{x_0}] \; \leq \; C_2 \exp(-C_3 x) \text{ when } x \leq x_0^\gamma.$$

We apply this here to $M_m \equiv E[T|\mathcal{F}_m] - ET$, with $U_m \equiv BI_{R_m}$. We verify that $E[\Delta_m^2|\mathcal{F}_{m-1}] \leq E[U_m|\mathcal{F}_{m-1}]$ as follows:

$$
\begin{aligned}
E[\Delta_m^2|\mathcal{F}_{m-1}] &= E[(E[\tilde{\Delta}_m|\mathcal{F}_m])^2 \, |\mathcal{F}_{m-1}] \\
&\leq E[E[\tilde{\Delta}_m^2|\mathcal{F}_m] \, |\mathcal{F}_{m-1}] \\
&= E[\tilde{\Delta}_m^2|\mathcal{F}_{m-1}] \\
&= E[(T - E[T|\tilde{\mathcal{F}}_m])^2|\mathcal{F}_{m-1}] \\
&= E[E[(T - E[T|\tilde{\mathcal{F}}_m])^2|\tilde{\mathcal{F}}_m]|\mathcal{F}_{m-1}] \\
&\leq E[E[(T - \tilde{T}_m)^2|\tilde{\mathcal{F}}_m]|\mathcal{F}_{m-1}] \quad \text{by (3.27)} \\
&= E[(T - \tilde{T}_m)^2|\mathcal{F}_{m-1}] \\
&\leq BE[I_{R_m}|\mathcal{F}_{m-1}],
\end{aligned}
$$

where the last inequality follows along the lines of (3.28), but using that $\tilde{R}_m$ and τ_m remain independent even conditioned on the values of $\tau_1, \ldots, \tau_{m-1}$. To verify that (3.31) holds for our choice of U_m, note that $S = B\sum_m I_{R_m} = B|M^*(\mathbf{0}, \mathbf{x})|$, so

$$P[S > x] \;=\; P[|M^*(\mathbf{0}, \mathbf{x})| > x/B].$$

We use (3.29) again with a chosen as before so that (3.30) holds. It will then suffice to provide an exponential bound for $P[T(\mathbf{0}, \mathbf{x}) \geq ax/B]$. Using the hypothesis that the τ's have finite exponential moment, choose $\theta > 0$ so that $M_\tau(\theta) < \infty$. Letting r^* denote any direct path (with $\|\mathbf{x}\|_1$ edges), we have:

$$
\begin{aligned}
P[T(\mathbf{0}, \mathbf{x}) \geq ax/B] \;&\leq\; P[T(r^*) \geq ax/B] \\
&=\; P[\exp(\theta T(r^*)) \geq \exp(\theta ax/B] \\
&\leq\; \exp(-\theta ax/B) M_{T(r^*)}(\theta) \\
&=\; \exp(-\theta ax/(2B)) \exp(-\theta ax/(2B)) M_\tau(\theta)^{\|\mathbf{x}\|_1} \\
&\leq\; \exp(-\theta ax/(2B)),
\end{aligned}
$$

where the last inequality holds provided $\exp(-\theta ax/(2B)) M_\tau(\theta)^{\|\mathbf{x}\|_1} \leq 1$. This latter condition is equivalent to

$$x \;\geq\; A\|\mathbf{x}\|_1, \tag{3.32}$$

where

$$A \;\equiv\; \frac{2B \log M_\tau(\theta)}{\theta a}. \tag{3.33}$$

We therefore take x_0 to be the right hand side of (3.32), and note that for any c, $x_0 \geq c^2$ (as is required in Theorem 3.3) for sufficiently large $|\mathbf{x}|$.

The one hypotheses of Theorem 3.3 that is *not* satisfied is the bound on the martingale differences: $|\Delta_m| \leq c$. This will be trivially satisfied if we replace the original FPP edge variables τ_i with truncated variables $\hat{\tau}_i \equiv \tau_i \wedge c$. Letting $\hat{T}(\mathbf{0}, \mathbf{x})$ denote passage time with the $\hat{\tau}$'s as edge variables, Theorem 3.3 yields that

$$P[|\hat{T}(\mathbf{0}, \mathbf{x}) - E\hat{T}(\mathbf{0}, \mathbf{x})| \geq x\sqrt{A\|\mathbf{x}\|_1}] \;\leq\; C_2 \exp(-C_3 x), \text{ provided } x \leq A\|\mathbf{x}\|_1.$$

The key point here is that the constants A, C_2 and C_3 do not depend on c. The proof of (3.17) is then completed by showing that a "large" difference between T and $\hat{T}$ occurs with small probability for appropriate c. The exponentially small tail of the τ's is used to obtain this. (See [41], Lemma 1, for details.)

For Euclidean FPP, Howard and Newman [31, 32] have proved similar results. In this setting, $\operatorname{Var} T(\mathbf{0}, \mathbf{x}) \leq C_1|\mathbf{x}|$ holds and, for $\kappa_1 \equiv \min(1, d/\alpha)$ and $\kappa_2 \equiv 1/(4\alpha + 3)$, the following version of (3.17) holds:

$$P[|T(\mathbf{0}, \mathbf{x}) - ET(\mathbf{0}, \mathbf{x})| > x\sqrt{|\mathbf{x}|}] \;\leq\; C_1 \exp(-C_0 x^{\kappa_1}) \text{ for } x \leq C_0|\mathbf{x}|^{\kappa_2}. \tag{3.34}$$

The exponents κ_1 and κ_2 (which are both ≤ 1) arise from an application of Theorem 3.3 with $\gamma = \kappa_2$. This occurs because the edge variables in Euclidean

FPP do not satisfy (3.15) when the parameter α is larger than the dimension d. To see why this is true, pick a generic vertex of Q, say, $\mathbf{q} \equiv Q(0)$, and let $\tau_{\min}$ denote the value of the smallest edge variable over those edges in $C_\alpha^{\mathrm{edge}}(Q)$ that are incident to $\mathbf{q}$. With $\mathbf{q}'$ denoting the particle in Q that is *second* closest to $\mathbf{0}$, we have

$$
\begin{aligned}
P[\tau_{\min} > x] \;&\geq\; P\big[\,\big|\,|\mathbf{q}| - |\mathbf{q}'|\,\big|^\alpha > x\big] \\
&\geq\; P[|\mathbf{q}| \leq 1 \cap |\mathbf{q}'| > x^{1/\alpha} + 1] \\
&=\; P[|\mathbf{q}'| > x^{1/\alpha} + 1 \,\big|\, |\mathbf{q}| \leq 1] P[|\mathbf{q}| \leq 1] \\
&=\; P[(B(\mathbf{0}, x^{1/\alpha} + 1) \setminus B(\mathbf{0}, |\mathbf{q}|)) \cap Q = \emptyset \,\big|\, |\mathbf{q}| \leq 1] P[|\mathbf{q}| \leq 1] \\
&\geq\; P[B(\mathbf{0}, x^{1/\alpha} + 1) \cap Q = \emptyset] P[|\mathbf{q}| \leq 1] \\
&\geq\; C_0 \exp(-C_0 x^{d/\alpha}),
\end{aligned}
$$

where the first inequality follows from (3.12).

The moderate deviation estimate (3.34) together with the Euclidean FPP analog of (3.18), namely

$$
\mu|\mathbf{x}| \;\leq\; ET(\mathbf{0}, \mathbf{x}) \;\leq\; \mu|\mathbf{x}| + |\mathbf{x}|^{1/2}(\log|\mathbf{x}|)^{1/\kappa_1}, \tag{3.35}
$$

are sufficient to obtain that

$$
(t - t^{1/2}(\log t)^{2/\kappa_1})\widetilde{W} \subset W_t \subset (t + t^{1/2}(\log t)^{2/\kappa_1})\widetilde{W} \text{ for large } t, \text{ a.s.,} \tag{3.36}
$$

where here $\widetilde{W} = B(\mathbf{0}, \mu^{-1})$. Stated in terms of χ_1, we have the following theorem.

Theorem 3.4. *For Euclidean FPP in dimension d with $\alpha > 1$, $\chi_1 \leq 1/2$.*

Proof of a version of (3.18). As promised, we conclude this section by outlining a proof of (3.35), the Euclidean version of (3.18), where in this setting the model's isotropy considerably simplifies the argument. Toward this end, define

$$
\begin{aligned}
T_n \;&\equiv\; T(\mathbf{0}, n\hat{\mathbf{e}}_1), \tag{3.37} \\
T_n^* \;&\equiv\; \inf\{T(\mathbf{0}, \mathbf{q}) : \mathbf{q} \in Q, |\mathbf{q}| \geq n\}, \text{ and} \\
T_n^{**} \;&\equiv\; \inf\{T(2n\hat{\mathbf{e}}_1, \mathbf{q}) : \mathbf{q} \in Q, |\mathbf{q} - 2n\hat{\mathbf{e}}_1| \geq n\}.
\end{aligned}
$$

Typically, the (a.s. unique) minimizing path $M(\mathbf{0}, 2n\hat{\mathbf{e}}_1)$ will touch a particle $\mathbf{q} \in Q$ that is outside of $B(\mathbf{0}, n) \cup B(2n\hat{\mathbf{e}}_1, n)$. In this case, we will clearly have $T_{2n} \geq T_n^* + T_n^{**}$. It is possible, however, that $M(\mathbf{0}, 2n\hat{\mathbf{e}}_1)$ travels directly from some $\mathbf{q} \in B(\mathbf{0}, n)$ to some $\mathbf{q}' \in B(2n\hat{\mathbf{e}}_1, n)$. In this latter case, one has that

$$
T_{2n} \;\geq\; T_n^* + T_n^{**} - |\mathbf{q} - \mathbf{q}'|^\alpha.
$$

Setting the random variable A_n equal to 0 in the first case and $|\mathbf{q} - \mathbf{q}'|^\alpha$ in the second case one has:

$$T_{2n} \geq T_n^* + T_n^{**} - A_n.$$

Using (3.12) to show that long (i.e., length exceeding n^γ for any particular $\gamma < 1$) edges in $\mathcal{C}_\alpha^{\text{edge}}(Q)$ occur within a distance n of $\mathbf{0}$ with small probability, one sees that $EA_n \leq n^{1/2}$. Since clearly $T_n^* \overset{\mathrm{d}}{=} T_n^{**}$, one obtains that

$$ET_{2n} \geq 2ET_n^* - n^{1/2}.$$

The strategy then is to use (3.34) to show that $|ET_n^* - ET_n| \leq C_1 n^{1/2} (\log n)^{1/\kappa_1}$ (this is where isotropy is useful), yielding the following weak superadditivity:

$$ET_{2n} \geq 2ET_n - C_1 n^{1/2} (\log n)^{1/\kappa_1}.$$

We then apply the following proposition, taken directly from [32].

Proposition 3.5. *Suppose the functions $a : \mathbf{R}^+ \to \mathbf{R}$ and $g : \mathbf{R}^+ \to \mathbf{R}^+$ satisfy the following conditions: $a(n)/n \to \nu \in \mathbf{R}$, $g(n)/n \to 0$ as $n \to \infty$, $a(2n) \geq 2a(n) - g(n)$, and $\psi \equiv \limsup_{n\to\infty} g(2n)/g(n) < 2$. Then, for any $c > 1/(2 - \psi)$, $a(n) \leq \nu n + cg(n)$ for all large n.*

Based on general subadditivity considerations, we have that

$$0 < \mu \equiv \inf_{n>0} \frac{ET_n}{n} < \infty \quad \text{and} \quad \lim_{n\to\infty} \frac{T_n}{n} = \mu \ (\text{a.s. and in } L^1).$$

Taking $a(n) = ET_n$ and $g(n) = c_1 n^{1/2} (\log n)^{1/\kappa_1}$ (so that $\limsup_n g(2n)/g(n) = \sqrt{2} < 2$), we get that, for appropriate C_1,

$$\mu n \leq ET_n \leq \mu n + C_1 n^{1/2} (\log n)^{1/\kappa_1} \text{ for large } n.$$

Regarding the proposition, it is easily verified that, for $c > 1/(2 - \psi)$, $\tilde{a}(n) \equiv a(n) - cg(n)$ satisfies $\tilde{a}(2n) \geq 2\tilde{a}(n)$ for all large n. Iterating this n times yields $\tilde{a}(2^n n) \geq 2^n \tilde{a}(n)$ or $\tilde{a}(2^n n)/(2^n n) \geq \tilde{a}(n)/n$. Under our hypotheses on a and g, $\tilde{a}(x)/x \to \nu$ as $x \to \infty$, so letting $n \to \infty$ shows that $\tilde{a}(n)/n \leq \nu$ for all large n.

While isotropy has considerably simplified matters here, arguments of this sort are generally based on some sort of superadditivity.

3.5 Some Related Longitudinal Fluctuation Exponents

The exponent χ_1, as defined in (3.19), is one of many ways of measuring longitudinal fluctuations. This particular exponent measures fluctuations in all directions simultaneously, but analogous exponents can be defined for each direction:

$$\chi_1(\hat{\mathbf{x}}) \equiv \inf\{\kappa : T(\mathbf{0}, n\hat{\mathbf{x}}) \leq t \text{ for } n \leq \mu(\hat{\mathbf{x}})^{-1}(t - t^\kappa), \text{ and}$$
$$T(\mathbf{0}, n\hat{\mathbf{x}}) \geq t \text{ for } n \geq \mu(\hat{\mathbf{x}})^{-1}(t + t^\kappa)$$
$$\text{for all large } t, \text{ a.s.}\}. \tag{3.38}$$

For Euclidean FPP models, $\chi_1(\hat{\mathbf{x}})$ is clearly independent of $\hat{\mathbf{x}}$. For Standard FPP, direction-independence is still believed to hold but this is not known rigorously.

In addition to these "almost sure" definitions, one may measure longitudinal fluctuations by computing variance, as in:

$$\chi_2(\hat{\mathbf{x}}) \equiv \inf\left\{\kappa : \limsup_{n \to \infty} \frac{\operatorname{Var} T(\mathbf{0}, n\hat{\mathbf{x}})}{n^{2\kappa}} < \infty\right\}, \tag{3.39}$$

with

$$\chi_2 \equiv \inf\left\{\kappa : \limsup_{|\mathbf{x}| \to \infty} \frac{\operatorname{Var} T(\mathbf{0}, |\mathbf{x}|)}{|\mathbf{x}|^{2\kappa}} < \infty\right\}. \tag{3.40}$$

Again, $\chi_2(\hat{\mathbf{x}})$ is independent of $\hat{\mathbf{x}}$ for Euclidean FPP and believed, but not known, to be independent for Standard FPP as well. It is also generally believed that $\chi_1 = \chi_2$ but, again, this is not known. In this terminology, (3.22) may be restated as $\chi_2 \leq 1/2$.

3.6 Monotonicity

Returning to Standard FPP models, the passage times $a_{0,n}$ will not be monotonically increasing in n, yet it is natural to expect that $E a_{0,n}$ is increasing, at least for large values of n. (For small values of n funny things can happen. See, e.g., [9, 36].) Now (3.18), with $\kappa = 1/2$, implies that $E a_{0,n+\Delta n} > E a_{0,n}$ for large n whenever $\Delta n > n^{1/2} \log^{1+\varepsilon}(n)$. Little, however, is known about the case where Δn is smaller, in particular, when $\Delta n = 1$, although it seems natural to conjecture that $\lim_{n \to \infty} E[a_{0,n+1} - a_{0,n}]$ exists and is, therefore, equal to $\mu(\hat{\mathbf{e}}_1)$.

While conjectures about monotonicity in FPP date back to [26], only recently have there been some results in this direction. In [4], a clever deterministic *crossing inequality* (first appearing less generally in [3]) is used to obtain some restricted results in dimension 2. For example, with the half-plane $H = \{(x, y) : y \geq 0\}$ and, for $0 \leq m < n$,

$$a_{m,n}^H \equiv \inf\{T(r) : r \in \mathcal{R}((m,0), (n,0)) \text{ with } r \text{ contained in } H\},$$

one has that $E a_{0,n}^H$ is an increasing concave function of n. This implies that $\lim_{n \to \infty} E[a_{0,n+1}^H - a_{0,n}^H] = \mu$, since the time constant in the direction of the x-axis for FPP in the half-space equals its unrestricted counterpart [42]. Additionally, with n a non-negative integer, $C = \{(x, y) : 0 \leq x \leq n\}$, and

$$t(m) \equiv \inf\{T(r) : r \in \mathcal{R}((0,0), (n,m)) \text{ with } r \text{ contained in } C\},$$

one has that $Et(m)$ is an increasing convex function of m for $m \geq 0$.

The path restrictions to H and C, as well as the $d = 2$ restriction, are used in the arguments to insure that two particular minimizing paths cross. Consider $a_{0,n}^H$, for example. Let $r_0 = (\mathbf{u}_0, \ldots, \mathbf{u}_{i_0}, \ldots, \mathbf{u}_{k_0})$ realize $a_{0,n}^H$, i.e., $r_0 \in \mathcal{R}((0,0),(n,0))$, r_0 is contained in H, and $T(r_0) = a_{0,n}^H$. (See Figure 4.) Let $r_1 = (\mathbf{v}_0, \ldots, \mathbf{v}_{i_1}, \ldots, \mathbf{v}_{k_1})$ realize $a_{1,n+1}^H$. Here $\mathbf{u}_{i_0} = \mathbf{v}_{i_1}$ denotes the first (if there are more than one) vertex where r_0 and r_1 cross, which they must.

Put

$$r_+ = (\mathbf{u}_0, \ldots, \mathbf{u}_{i_0}, \mathbf{v}_{i_1+1}, \ldots, \mathbf{v}_{k_1}), \text{ and}$$
$$r_- = (\mathbf{v}_0, \ldots, \mathbf{v}_{i_1}, \mathbf{u}_{i_0+1}, \ldots, \mathbf{u}_{k_0}),$$

so $r_+ \in \mathcal{R}((0,0),(n+1,0))$ and $r_- \in \mathcal{R}((1,0),(n,0))$. Then

$$a_{0,n+1}^H + a_{1,n}^H \leq T(r_+) + T(r_-) = a_{0,n}^H + a_{1,n+1}^H.$$

Taking expectations, using translation invariance of the edge variables, and rearranging yields

$$Ea_{0,n+1}^H - Ea_{0,n}^H \leq Ea_{0,n}^H - Ea_{0,n-1}^H. \tag{3.41}$$

Concavity clearly follows from (3.41), but so does monotonicity. For if $Ea_{0,n^*+1}^H - Ea_{0,n^*}^H = -\delta < 0$, then (3.41) implies that

$$Ea_{0,n}^H \leq Ea_{0,n^*}^H - (n - n^*)\delta \text{ for } n > n^*,$$

and hence that $Ea_{0,n}^H$ is eventually negative, which is impossible. Additional monotonicity results in dimension 2 are discussed at the end of Section 5.2.

In Euclidean FPP, other tools are available that do not lead to dimension dependent arguments. Following [29], we will think of the mean particle density as a parameter $\lambda > 0$; we have been working with $\lambda = 1$. Let $E^\lambda(\cdot)$ denote expectation with respect to the measure making particle configurations homogeneous Poisson with density λ. Re-scaling length, i.e.,

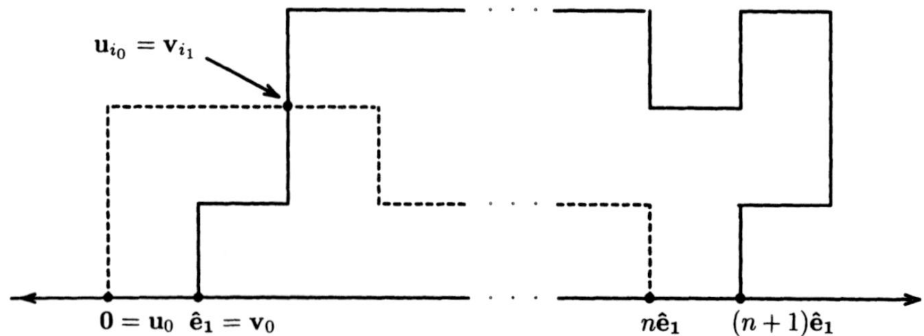

Fig. 4. The *dashed path* (r_0) realizes $a_{0,n}^H$; the *solid path* (r_1) realizes $a_{1,n+1}^H$.

changing Q to $n^{-1}Q \equiv \{n^{-1}\mathbf{q} : \mathbf{q} \in Q\}$, carries minimizing paths in Q to minimizing paths in $n^{-1}Q$. Furthermore, if Q is Poisson with mean density λ, then $n^{-1}Q$ is Poisson with mean density λn^d. It follows from this reasoning that

$$E^\lambda T(\mathbf{0}, n\hat{\mathbf{e}}_1) = n^\alpha \, E^{\lambda n^d} T(\mathbf{0}, \hat{\mathbf{e}}_1). \qquad (3.42)$$

Several applications of (3.42) then produces

$$\frac{d}{dn}E^1(\mathbf{0}, n\hat{\mathbf{e}}_1) = \alpha n^{\alpha-1}E^{n^d}T(\mathbf{0}, \hat{\mathbf{e}}_1) + n^\alpha \Big(\frac{d}{d\lambda}E^\lambda T(\mathbf{0}, \hat{\mathbf{e}}_1)\Big|_{\lambda=n^d}\Big)dn^{d-1}$$
$$= \alpha \frac{E^1 T(\mathbf{0}, n\hat{\mathbf{e}}_1)}{n} + d\frac{\frac{d}{d\lambda}E^\lambda T(\mathbf{0}, n\hat{\mathbf{e}}_1)\big|_{\lambda=1}}{n}. \qquad (3.43)$$

The first term $\to \alpha\mu > 0$ as $n \to \infty$, so the key is to understand $\frac{d}{d\lambda}E^\lambda T(\mathbf{0}, n\hat{\mathbf{e}}_1)|_{\lambda=1}$. This is a negative quantity because adding points to a particle configuration decreases passage time. Now (3.43) has conveniently turned differentiation with respect to n (distance from the origin), into differentiation with respect to particle density. As we will see, thinking of a density $1 - \Delta\lambda$ particle configuration as a density 1 particle configuration with some of its particles removed then leads to

$$\frac{d}{dn}E^1 T(\mathbf{0}, n\hat{\mathbf{e}}_1) = \alpha\frac{E^1 T(\mathbf{0}, n\hat{\mathbf{e}}_1)}{n} - d\frac{E^1 S(\mathbf{0}, n\hat{\mathbf{e}}_1)}{n}, \qquad (3.44)$$

where

$$S(\mathbf{0}, n\hat{\mathbf{e}}_1) = S(\mathbf{0}, n\hat{\mathbf{e}}_1; Q) \equiv \sum_{\mathbf{q} \in M(\mathbf{0}, n\hat{\mathbf{e}}_1)} [T(\mathbf{0}, n\hat{\mathbf{e}}_1; Q \setminus \mathbf{q}) - T(\mathbf{0}, n\hat{\mathbf{e}}_1; Q)].$$
$$(3.45)$$

Here, for $Q^* = Q$ and $Q^* = Q\setminus\mathbf{q}$, $T[\mathbf{0}, n\hat{\mathbf{e}}_1, Q^*]$ denotes the passage time from $\mathbf{0}$ to $n\hat{\mathbf{e}}_1$ in the particle configuration Q^*. One obtains by simple geometric arguments (see [29] for details) that

$$S(\mathbf{0}, n\hat{\mathbf{e}}_1) \leq C(\alpha, d)T(\mathbf{0}, n\hat{\mathbf{e}}_1),$$

with, for each d, $C(\alpha, d) \downarrow 0$ as $\alpha \downarrow 1$ giving, from (3.44), that $\frac{d}{dn}ET(\mathbf{0}, n\hat{\mathbf{e}}_1) > 0$ for large n, for $1 < \alpha < \alpha(d)$.

To see (3.44), let $\widetilde{Q} = Q \cap [-\frac{L}{2}, \frac{L}{2}]$ so that, for large L, we have $T(\mathbf{0}, n\hat{\mathbf{e}}_1) = T(\mathbf{0}, n\hat{\mathbf{e}}_1; Q) \approx T(\mathbf{0}, n\hat{\mathbf{e}}_1, \widetilde{Q})$ and

$$\frac{d}{d\lambda}ET^\lambda(\mathbf{0}, n\hat{\mathbf{e}}_1)|_{\lambda=1} \approx \frac{d}{d\lambda}ET^\lambda(\mathbf{0}, n\hat{\mathbf{e}}_1; \widetilde{Q})|_{\lambda=1}$$
$$\approx \frac{E^1 T(\mathbf{0}, n\hat{\mathbf{e}}_1; \widetilde{Q}) - E^{1-\Delta\lambda}T(\mathbf{0}, n\hat{\mathbf{e}}_1; \widetilde{Q})}{\Delta\lambda}. \qquad (3.46)$$

Here we use "$\approx$" to mean, in an informal sense, "approximately equals". Letting N denote the number of particles in $\widetilde{Q}$, we have

$$E^\lambda N = \begin{cases} L^d & \text{when } \lambda = 1 \\ L^d - 1 & \text{when } \lambda = 1 - L^{-d} \end{cases} \tag{3.47}$$

Taking $\Delta\lambda = L^{-d}$ in (3.46), one gets

$$\begin{aligned}
\frac{d}{d\lambda} E^\lambda T(\mathbf{0}, n\hat{\mathbf{e}}_1)\big|_{\lambda=1} &\approx L^d E^1\big[T(\mathbf{0}, n\hat{\mathbf{e}}_1; \widetilde{Q}) \big| N = L^d\big] \\
&\quad - L^d E^1\big[T(\mathbf{0}, n\hat{\mathbf{e}}_1; \widetilde{Q}) \big| N = L^d - 1\big] \\
&= E^1\big[\{\sum_{\mathbf{q}\in\widetilde{Q}} T(\mathbf{0}, n\hat{\mathbf{e}}_1; \widetilde{Q}) - T(\mathbf{0}, n\hat{\mathbf{e}}_1; \widetilde{Q} \setminus \mathbf{q})\} \big| N = L^d\big] \\
&\approx E^1 \sum_{\mathbf{q}\in Q} \big[T(\mathbf{0}, n\hat{\mathbf{e}}_1; Q) - T(\mathbf{0}, n\hat{\mathbf{e}}_1; Q \setminus \mathbf{q})\big].
\end{aligned}$$

Noting that $T(\mathbf{0}, n\hat{\mathbf{e}}_1; Q) = T(\mathbf{0}, n\hat{\mathbf{e}}_1; Q\setminus\mathbf{q})$ for $\mathbf{q} \notin M(\mathbf{0}, n\hat{\mathbf{e}}_1)$, one gets (3.44).

Expected passage time for Euclidean FPP eventually strictly increases with distance from the origin, at least for certain values of the model's parameter. (One expects this to hold for all $\alpha > 1$.) For Standard FPP, there are no spatially unrestricted monotonicity results; as discussed above all analogous lattice results are in dimension 2 and involve FPP on a half-plane or cylinder.

4 Transversal Fluctuations and the Divergence of Shape Fluctuations

4.1 Transversal Fluctuation Exponents

In Section 3 we discussed longitudinal fluctuations – deviations of $T(\mathbf{0}, \mathbf{x})$ about either its mean or $\mu(\hat{\mathbf{x}})|\mathbf{x}|$. Here we discuss the closely related issue of measuring how far minimizing paths from $\mathbf{0}$ to $\mathbf{x}$ typically wander from the straight line segment $\overline{\mathbf{0}\mathbf{x}}$. As with longitudinal fluctuations, there are many ways of measuring this. For any two subsets A and B of $\mathbf{R}^d$, we put

$$\text{Dist}(\mathbf{x}, B) \equiv \inf\{|\mathbf{x} - \mathbf{y}| : \mathbf{y} \in B\}, \text{ and} \tag{4.1}$$

$$d_{\max}(A, B) \equiv \sup\{\text{Dist}(\mathbf{x}, B) : \mathbf{x} \in A\}, \tag{4.2}$$

so $d_{\max}(A, B)$ is the maximal distance from points in A to the set B. We then define the *transversal fluctuation exponents*:

$$\xi(\hat{\mathbf{x}}) \equiv \inf\{\kappa : \lim_{n\to\infty} P[d_{\max}(M(\mathbf{0}, n\hat{\mathbf{x}}), \overline{\mathbf{0}\,n\hat{\mathbf{x}}}) \leq n^\kappa] = 1\}, \text{ and} \tag{4.3}$$

$$\xi \equiv \inf\{\kappa : \lim_{|\mathbf{x}|\to\infty} P[d_{\max}(M(\mathbf{0}, \mathbf{x}), \overline{\mathbf{0}\mathbf{x}}) \leq |\mathbf{x}|^\kappa] = 1\}. \tag{4.4}$$

Here again, $\xi(\hat{\mathbf{x}})$ is believed not to depend on $\hat{\mathbf{x}}$, but for Standard FPP this is not known. For Euclidean FPP, this is clearly the case.

It is generally believed that all reasonable definitions of χ and ξ should yield the same values and that these values are independent of direction, if applicable. Yet, as mentioned, there is next to nothing rigorously know about this. There is strong reason to believe that, in dimension 2, $\chi = 1/3$ and $\xi = 2/3$. In fact, as discussed in Section 4.5, Baik, Deift, and Johansson obtain precisely this (and much more) for a related growth model. Additionally, it is conjectured by Krug and Spohn [48] that $\chi = 2\xi - 1$ should hold in all dimensions. Theorem 4.1 below is verification of one of the inequalities. Note that this scaling relation would imply that $\xi \geq 1/2$. Predictions of how these exponents vary (if at all) with dimension are all over the board; see [55] with an accounting and references to their sources in the physics literature.

4.2 Upper Bounds on ξ

We begin with the following result of Newman and Piza [55] (for Standard FPP) and Howard and Newman [32] (for Euclidean FPP).

Theorem 4.1. *For Standard FPP satisfying (3.3) and for Euclidean FPP, $\xi(\hat{\mathbf{x}}) \leq (1 + \chi_1)/2$ for any direction of curvature $\hat{\mathbf{x}}$ (see (3.9)). In view of Theorems 3.2 and 3.4, $\xi(\hat{\mathbf{x}}) \leq 3/4$ for all directions of curvature.*

We remark that for Euclidean FPP all directions are directions of curvature, and, for Euclidean FPP, $\xi \leq (1 + \chi_1)/2 \leq 3/4$. It was precisely the restriction of Theorem 4.1 to directions of curvature combined with the lack of good information about $\widetilde{W}$ for Standard FPP that motivated the development of Euclidean FPP. In this setting, Howard and Newman use (3.34) and (3.35) to prove a stronger result giving a lower bound on the rate of convergence to 1 in (4.4). They show that, for any $\varepsilon > 0$, there are constants C_0 and C_1 such that

$$P[d_{\max}(M(\mathbf{0}, n\hat{\mathbf{x}}), \overline{\mathbf{0}\,n\hat{\mathbf{x}}}) \geq n^{\frac{3}{4}+\varepsilon}] \;\leq\; C_1 \exp(-C_0 n^{3\varepsilon\kappa_1/4}). \qquad (4.5)$$

The proof of Theorem 4.1 given in [55] (which we follow here) is quite geometrical, using only that FPP models obey a shape theorem. As such, it should be applicable to a wide assortment of random growth models. We prove the case where $\widetilde{W} = B(\mathbf{0}, \mu^{-1})$, i.e., where the asymptotic shape is a perfect Euclidean ball (Euclidean FPP, e.g.). This catches the essence of the argument in [55] while avoiding some messy details.

Fix any direction $\hat{\mathbf{x}}$ (here, automatically a direction of curvature), $\varepsilon > 0$, and $\kappa \in (\chi_1, 1)$ (we assume $\chi_1 < 1$, since the shape theorem gives immediately that $\xi(\hat{\mathbf{x}}) \leq 1$). With $\mathcal{W}_t(\mathbf{x}) \equiv \{\mathbf{y} \in \mathbf{R}^d : T(\mathbf{x}, \mathbf{y}) \leq t\}$, we clearly have

$$M(\mathbf{0}, n\hat{\mathbf{x}}) \;\subset\; \bigcup_{0 \leq t \leq T(\mathbf{0}, n\hat{\mathbf{x}})} [\mathcal{W}_t(\mathbf{0}) \cap \mathcal{W}_{T(\mathbf{0}, n\hat{\mathbf{x}})-t}(n\hat{\mathbf{x}})].$$

It follows from the definition of χ_1 that, except for an exceptional set of configurations E with $P[E] < \varepsilon$,

$$\mathcal{W}_t(\mathbf{0}) \subset (t + t^\kappa)\widetilde{W} \subset (t + C_1 n^\kappa)\widetilde{W}$$

for $t(\varepsilon) \leq t \leq \mu n + n^\kappa$. Noting also that

$$P[\mathcal{W}_{t(\varepsilon)}(\mathbf{0}) \subset C_1 n^\kappa \widetilde{W}] \geq 1 - \varepsilon$$

for n exceeding some $n(\varepsilon)$, we obtain that

$$\mathcal{W}_t(\mathbf{0}) \subset (t + C_1 n^\kappa)\widetilde{W} \text{ for } 0 \leq t \leq \mu n + n^\kappa \qquad (4.6)$$

occurs with probability approaching 1 as $n \to \infty$. Also, except for an exceptional set E' with $P[E'] < \varepsilon$, $T(\mathbf{0}, n\hat{\mathbf{x}}) \leq \mu n + n^\kappa$ for all $n \geq n(\varepsilon)$, for a possibly larger $n(\varepsilon)$. Combining this with the reasoning behind (4.6) (shifted to $n\hat{\mathbf{x}}$) yields that

$$\mathcal{W}_{T(\mathbf{0}, n\hat{\mathbf{x}}) - t}(n\hat{\mathbf{x}}) \subset \mathcal{W}_{\mu n + n^\kappa - t}(n\hat{\mathbf{x}}) \subset n\hat{\mathbf{x}} + (\mu n + C_1 n^\kappa - t)\widetilde{W}$$
$$\text{for } 0 \leq t \leq \mu n + n^\kappa$$

occurs with probability approaching 1 as $n \to \infty$. Hence, with probability approaching 1 as $n \to \infty$,

$$M(\mathbf{0}, n\hat{\mathbf{x}}) \subset H_n \equiv \bigcup_{0 \leq t \leq \mu n + n^\kappa} [(t + C_1 n^\kappa)\widetilde{W} \cap (n\hat{\mathbf{x}} + (\mu n + C_1 n^\kappa - t)\widetilde{W})]. \quad (4.7)$$

From here it is pure geometry to show that

$$d_{\max}(H_n, \overline{\mathbf{0}\, n\hat{\mathbf{x}}}) \leq C_1 n^{(1+\kappa)/2},$$

and therefore that, when (4.7) holds,

$$d_{\max}(M(\mathbf{0}, n\hat{\mathbf{x}}), \overline{\mathbf{0}\, n\hat{\mathbf{x}}}) \leq C_1 n^{(1+\kappa)/2}.$$

Figure 5 shows the situation in dimension 2 which, in our simplified case where $\widetilde{W} = B(\mathbf{0}, \mu^{-1})$, easily extends to higher d. The figure shows $\partial[(t + C_1 n^\kappa)\widetilde{W}]$ (a circle centered at $\mathbf{0}$ with radius $\mu^{-1}t + \mu^{-1}C_1 n^\kappa$) and $\partial[n\hat{\mathbf{x}} + (\mu n + C_1 n^\kappa - t)\widetilde{W}]$ (a circle centered at $n\hat{\mathbf{x}}$ with radius $n + \mu^{-1}C_1 n^\kappa - \mu^{-1}t$). Clearly $|\mathbf{x} - \mathbf{y}| = 2\mu^{-1}C_1 n^\kappa$, so the highlighted right triangle has a leg of length $a = \mu^{-1}t + O(n^\kappa)$ with hypotenuse also of length $c = \mu^{-1}t + O(n^\kappa)$. It follows directly from the Pythagorean Theorem that the other leg has length b with $b^2 = O(n^{1+\kappa}) + O(n^{2\kappa}) = O(n^{1+\kappa})$, with the second equality following from $\kappa < 1$. The theorem follows since clearly

$$d_{\max}((t + C_1 n^\kappa)\widetilde{W} \cap (n\hat{\mathbf{x}} + (\mu n + C_1 n^\kappa - t)\widetilde{W}), \overline{\mathbf{0}\, n\hat{\mathbf{x}}}) = b = O(n^{(1+\kappa)/2}),$$

and $\kappa \in (\chi_1, 1)$ was arbitrary.

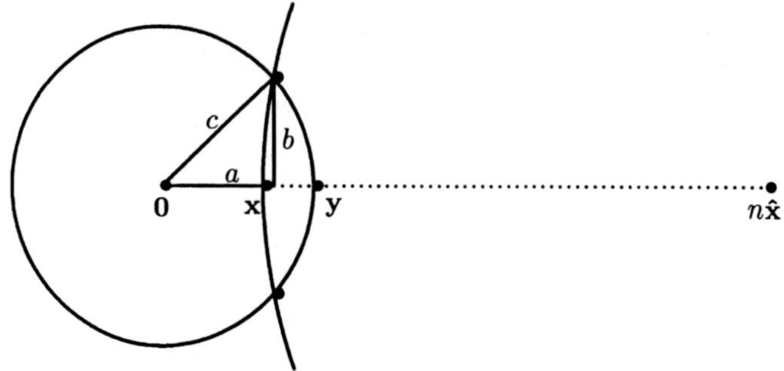

Fig. 5. The circle is $\partial[(t + C_1 n^\kappa)\widetilde{\mathcal{W}}] = \partial B(\mathbf{0}, \mu^{-1}t + \mu^{-1}C_1 n^k)$, the arc is part of $\partial[n\hat{\mathbf{x}} + (\mu n + C_1 n^\kappa - t)\widetilde{\mathcal{W}}] = \partial B(n\hat{\mathbf{x}}, n + \mu^{-1}C_1 n^\kappa - \mu^{-1}t)$.

4.3 Lower Bounds on χ

Theorem 4.1 shows that upper bounds on χ translate into upper bounds on ξ, so it is somewhat amusing that upper bounds on ξ translate into lower bounds on χ as shown by Wehr and Aizenman [74] (for other models), Newman and Piza [55] (for Standard FPP) and Howard [28] (for Euclidean FPP).

Theorem 4.2. *For Standard FPP where the edge variables satisfy (2.20) or (2.21), and $E\tau^2 < \infty$, and for Euclidean FPP, we have*

$$\chi_2(\hat{\mathbf{x}}) \ \geq \ \frac{1 - (d-1)\xi(\hat{\mathbf{x}})}{2}. \tag{4.8}$$

In 2 dimensional Standard FPP satisfying the above hypotheses, $\mathrm{Var}[T(\mathbf{0}, n\hat{\mathbf{x}})] \geq C_0 \log n$ *for all directions* $\hat{\mathbf{x}}$.

In dimension 2, the bound $\xi(\hat{\mathbf{x}}) \leq 3/4$ yields that $\chi_2(\hat{\mathbf{x}}) \geq 1/8$, for directions of curvature $\hat{\mathbf{x}}$. The $d = 2$ logarithmic lower bound improves Kesten's $\mathrm{Var}[T(\mathbf{0}, n\hat{\mathbf{x}})] \geq C_0$ (see [41]), but does not help in bounding χ from below. Pemantle and Peres [58] have independently (with different techniques) obtained the logarithmic bound for exponentially distributed edge times. If the scaling relation $\chi = 2\xi - 1$ holds, Theorem 4.2 is of no help in $d > 2$.

That upper bounds on ξ should somehow translate into lower bounds on χ_2, which measures variance of passage time, has a certain intuitive appeal. If $\xi(\hat{\mathbf{e}}_1)$ were very small (think of it as 0), $T(\mathbf{0}, n\hat{\mathbf{e}}_1)$ should behave like $T(r(\mathbf{0}, n\hat{\mathbf{e}}_1))$, where $r(\mathbf{0}, n\hat{\mathbf{e}}_1)$ is the direct path from $\mathbf{0}$ to $n\hat{\mathbf{e}}_1$. But $T(r(\mathbf{0}, n\hat{\mathbf{e}}_1))$ is the sum of n independent edge variables and its standard deviation grows like $n^{1/2}$ – the known upper bound for the rate of growth of $T(\mathbf{0}, n\hat{\mathbf{e}}_1)$'s standard deviation. Indeed, setting $\xi = 0$ into (4.8) produces a lower bound of $1/2$.

Below we follow Newman and Piza's proof of (4.8) for Standard FPP, but only for the case of Bernoulli edge variables: $P[\tau = 1] = p$ and $P[\tau = 0] = q \equiv 1 - p < p_c(d)$. Note that conditions (2.20) and (2.21) imply that $\mathrm{Var}\,\tau > 0$, and that in our simplified Bernoulli setting (2.21) is not relevant. The approach is to use the martingale representation (3.23) to get an upper bound on the variance of $T \equiv T(\mathbf{0}, n\hat{\mathbf{x}})$. Fix any $\kappa > \xi(\hat{\mathbf{x}})$ and let $\mathcal{E}_n$ denote all edges e satisfying $d_{\max}(e, \overline{\mathbf{0}\,n\hat{\mathbf{x}}}) < n^{\kappa}$, so

$$|\mathcal{E}_n| \;\leq\; C_1 n^{1+(d-1)\kappa} \tag{4.9}$$

For $\delta = 0$ or $\delta = 1$, let T_m^{δ} denote the passage time from $\mathbf{0}$ to $n\hat{\mathbf{x}}$ when the configuration has been (possibly) altered so that $\tau_m = \delta$. Put $J_m \equiv T_m^1 - T_m^0$, so $\{J_m = 1\}$ is the event that the value of τ_m in the configuration "matters" when computing T. Note that

$$T \;=\; T_m^1 \tau_m + T_m^0 (1 - \tau_m) \;=\; T_m^0 + J_m \tau_m.$$

Also, T_m^0, T_m^1 and J_m are $\widetilde{\mathcal{F}}_m$-measurable and τ_m and $\widetilde{\mathcal{F}}_m$ are independent so $E[T_m^0 | \mathcal{F}_m] = E[T_m^0 | \mathcal{F}_{m-1}]$, $E[J_m \tau_m | \mathcal{F}_m] = \tau_m E[J_m | \mathcal{F}_{m-1}]$, and

$$\Delta_m \;=\; (\tau_m - p) E[J_m | \mathcal{F}_{m-1}],$$

where Δ_m is defined as in (3.23). It follows that

$$E\Delta_m^2 \;=\; pq E\big(E[J_m | \mathcal{F}_{m-1}]^2\big) \;\geq\; pq (EJ_m)^2.$$

Setting this into (3.24) gives

$$\operatorname{Var} T \;=\; \sum_{m=1}^{\infty} E\Delta_m^2 \;\geq\; pq \sum_{m:e_m \in \mathcal{E}_n} (EJ_m)^2$$

$$\geq\; \frac{pq}{|\mathcal{E}_n|} \Big(\sum_{m:e_m \in \mathcal{E}_n} EJ_m \Big)^2, \tag{4.10}$$

where the final inequality follows from the Cauchy-Schwarz inequality. If r is any minimizing path for T, then if $e_m \in r$ has $\tau_m = \tau(e_m) = 1$, then also $J_m = 1$ since changing τ_m to 0 would reduce T by 1. Letting $A_n \equiv \{d_{\max}(M(\mathbf{0}, n\hat{\mathbf{x}}), \overline{\mathbf{0}\,n\hat{\mathbf{x}}}) < n^{\kappa}\}$, we obtain

$$\sum_{m:e_m \in \mathcal{E}_n} EJ_m \;\geq\; P[A_n] E\Big[\sum_{m:e_m \in \mathcal{E}_n} J_m | A_n \Big] \;\geq\; P[A_n] E[T | A_n] \;\geq\; C_0 n. \tag{4.11}$$

The last inequality follows from the shape theorem and the fact that $\lim_n P[A_n] = 1$ by the definition of $\xi(\hat{\mathbf{x}})$. Setting (4.11) and (4.9) into (4.10) gives that $\operatorname{Var} T \geq C_0 n^{1-(d-1)\kappa}$. Since $\kappa > \xi(\hat{\mathbf{x}})$, (4.8) follows.

4.4 Lower Bounds on ξ

It is somewhat less surprising that lower bounds on χ translate into lower bounds on ξ. This observation is mined by Licea, Newman, and Piza [50]. Their first bound, for which we offer a hueristic argument for Bernoulli 0–1 valued edge times, is stated in the theorem below. The basic argument works in the Euclidean FPP setting as well [28].

Theorem 4.3. *For Standard FPP satisfying the hypotheses of Theorem 4.2 and for Euclidean FPP, $\xi \geq 1/(d+1)$.*

We work below in the $\hat{\mathbf{e}}_1$ direction, the generalization to arbitrary direction is easy (and unnecessary for Euclidean FPP). Choose $\kappa > \xi(\hat{\mathbf{e}}_1)$ and, as before, let $T \equiv T(\mathbf{0}, n\hat{\mathbf{e}}_1)$ and also put $\widetilde{T} \equiv T(3n^\kappa\hat{\mathbf{e}}_2, n\hat{\mathbf{e}}_1 + 3n^\kappa\hat{\mathbf{e}}_2)$. Additionally, let T^* denote the minimal passage time from $\mathbf{0}$ to $n\hat{\mathbf{e}}_1$ along paths r satisfying $d_{\max}(r, \overline{\mathbf{0}\,n\hat{\mathbf{e}}_1}) < n^\kappa$, with $\widetilde{T}^*$ denoting the minimal passage time from $3n^\kappa\hat{\mathbf{e}}_2$ to $n\hat{\mathbf{e}}_1 + 3n^\kappa\hat{\mathbf{e}}_2$ along paths r satisfying $d_{\max}(r, \overline{3n^\kappa\hat{\mathbf{e}}_2,\ n\hat{\mathbf{e}}_1 + 3n^\kappa\hat{\mathbf{e}}_2}) < n^\kappa$. Then T^* and $\widetilde{T}^*$ are independent and it follows from the arguments of Theorem 4.2 that

$$\mathrm{Var}(T^* - \widetilde{T}^*) \geq C_0 n^{1-(d-1)\kappa}.$$

Let

$$A_n \equiv \{d_{\max}(M(\mathbf{0}, n\hat{\mathbf{e}}_1), \overline{\mathbf{0}\,n\hat{\mathbf{e}}_1}) < n^\kappa\}, \text{ and}$$
$$\widetilde{A}_n \equiv \{d_{\max}(M(3n^\kappa\hat{\mathbf{e}}_2, n\hat{\mathbf{e}}_1 + 3n^\kappa\hat{\mathbf{e}}_2), \overline{3n^\kappa\hat{\mathbf{e}}_2,\ n\hat{\mathbf{e}}_1 + 3n^\kappa\hat{\mathbf{e}}_2}) < n^\kappa\},$$

so, on $A_n \cap \widetilde{A}_n$, $T = T^*$ and $\widetilde{T} = \widetilde{T}^*$. The idea is to use the fact that $P[A_n \cap \widetilde{A}_n] \to 1$ as $n \to \infty$ to show that we also have

$$\mathrm{Var}(T - \widetilde{T}) \geq C_0 n^{1-(d-1)\kappa}.$$

On the other hand

$$|T - \widetilde{T}| \leq T(\mathbf{0}, 3n^\kappa\hat{\mathbf{e}}_2) + T(n\hat{\mathbf{e}}_1 + 3n^\kappa\hat{\mathbf{e}}_2, n\hat{\mathbf{e}}_1) \leq C_1 n^\kappa,$$

so

$$\mathrm{Var}(T - \widetilde{T}) \leq C_1 n^{2\kappa}.$$

These bounds on $\mathrm{Var}(T - \widetilde{T})$ can hold simultaneously for large n only if $\kappa \geq 1/(d+1)$. But $\kappa > \xi(\hat{\mathbf{e}}_1)$ was arbitrary so $\xi(\hat{\mathbf{e}}_1) \geq 1/(d+1)$.

Point-to-Plane Definitions of ξ. We have been discussing up to now point-to-point fluctuation exponents which measure the fluctuations of minimizing paths with two fixed endpoints. *Point-to-plane* definitions allow one endpoint of the path to become "unstuck."

We present here a result of Serafini [61] in the context of FPP on the Delaunay graph, where isotropy simplifies the presentation. Serafini's proof, which we do not present, is an adaptation of the methodology of Licea, Newman, Piza, who obtain a host of similar results for various point-to-plane definitions of ξ in the context of Standard FPP. Significantly, one such result has $\xi \geq 3/5$ in dimension 2 for a suitable definition of ξ. Heuristically, in dimension 2 any strictly superdiffusive bound on transversal fluctuations (for the right definition of ξ) should lead to the non-existence of doubly-infinite geodesics (see [54], p. 9). Additionally, in dimension 2, the non-existence of such geodesics is equivalent to the non-existence of non-constant ground states for an associated disordered ferromagnetic Ising model (see, e.g., [54], Propositions 1.1 and 1.2). Unfortunately, the definition yielding the 3/5 bound is not sufficiently strong to give non-existence of doubly-infinite geodesics.

Returning to the setting of the Delaunay graph, for $n > 0$ define the half-space $H_n \equiv \{\mathbf{x} \in \mathbf{R}^d : \mathbf{x} \cdot \hat{\mathbf{e}}_1 \geq n\}$ and put

$$T(\mathbf{0}, H_n) \equiv \inf\{T(\mathbf{0}, \mathbf{q}) : \mathbf{q} \in H_n\}.$$

Then any path r with $T(r) = T(\mathbf{0}, H_n)$ has one endpoint at $Q(\mathbf{0})$ (the Poisson particle closest to $\mathbf{0}$) and the other just to the "right" of the hyper-plane $\mathbf{x} \cdot \hat{\mathbf{e}}_1 = n$. One may define ξ in terms of $d_{\max}(r, L)$, where $L \equiv \{a\hat{\mathbf{e}}_1 : a \in \mathbf{R}\}$ is the first coordinate axis. Already it is, in principle, possible that the freeing of the second endpoint will produce sufficient additional freedom to generate a larger exponent. (This is not believed to be the case.) Serafini's definition allows for additional wiggle room by considering "almost minimizing" paths. Put

$$A(n, \varepsilon) \equiv \{\mathbf{q} \in H_n : T(\mathbf{0}, \mathbf{q}) \leq T(\mathbf{0}, H_n) + \varepsilon\},$$

and define

$$\xi_\varepsilon \equiv \sup\{\kappa \geq 0 : \liminf_{n \to \infty} P[d_{\max}(A(n, \varepsilon), L) \leq n^\kappa] < 1\},$$

and finally $\xi^S \equiv \inf_{\varepsilon > 0} \xi_\varepsilon$. In [61], Serafini shows that for any $d \geq 2$, $\xi^S \geq 1/2$ provided the edge variables have finite expectation.

4.5 Fluctuations for Other Related Models

In [79], the numerical bounds of Theorems 4.2 and 4.3 are proven for analogous quantities in a setting other than FPP, again using the basic methodology of [50, 55]. In this model, Brownian paths starting at $\mathbf{0}$ are conditioned to hit a ball of radius 1 centered at $n\hat{\mathbf{e}}_1$ and are further weighted so that they tend to avoid rotationally invariant "soft obstacles" centered at Poisson particles. (See [70] for more on these models.) There is no specific minimizing path as in FPP models, rather a measure on Brownian paths that is relatively concentrated on paths having little interaction with the obstacles. Here, n^ξ is the minimal order of magnitude of the diameter of the cylinder about $\overline{\mathbf{0}\, n\hat{\mathbf{e}}_1}$ on which the resulting path measure is asymptotically supported as $n \to \infty$, while n^χ is the order of magnitude of the variance of a normalizing partition function viewed as a function of the Poisson realization. A version of $\xi \geq 3/5$ in dimension 2 is also proved in [80] for this model for a point-to-plane definition of ξ.

In [81] progress is made toward a version of $\chi = 2\xi - 1$ for the Brownian path model. The precise statement in [81] involves two inequalities using different definitions of χ which, if equal (an open question), would yield the scaling relation.

More recently, in another Poisson-based non-FPP model, exact results have been obtained for $d = 2$. In this model, one considers paths of straight line segments starting at $\mathbf{0}$ moving only in the up/right direction and ending at (n, n) with Poisson particles as the interim segment endpoints. The exponents ξ and χ concern, respectively, the path that *maximizes* the number of Poisson particles touched and the number of touched particles. Here, it is known that $\xi = 2/3$ [35] and $\chi = 1/3$ [5].

5 Infinite Geodesics and Spanning Trees

5.1 Semi-Infinite Geodesics and Spanning Trees

For Standard FPP and for the Poisson models of Vahidi-Asl, Wierman, and Serafini, $M(\mathbf{x}, \mathbf{y})$ will consist of a single minimizing path, degenerate if $\mathbf{x}^* = \mathbf{y}^*$ (for the Poisson models, if $Q(\mathbf{x}) = Q(\mathbf{y})$), provided F has no atoms:

$$P[\tau = x] = 0 \text{ for all } x \in \mathbf{R}. \tag{5.1}$$

For the remainder of this chapter we assume that (5.1) holds. For the Euclidean models of Howard and Newman, minimizing paths are always unique by virtue of the continuity of the Poisson point process (see [32], Proposition 1.1). For any of the models we have discussed, we may therefore define, for each vertex $\mathbf{u} \in \mathcal{V}$, the graph $\mathcal{T}(\mathbf{u})$ whose vertex set is $\mathcal{V}$ and whose edge set consists of $\cup_{\mathbf{v} \in \mathcal{V}}\{$edges of $M(\mathbf{u}, \mathbf{v})\}$. The following theorem is an easy consequence of route-uniqueness.

Theorem 5.1. *For Standard or Poisson FPP where (5.1) holds or for Euclidean FPP one has, almost surely: for every $\mathbf{u} \in \mathcal{V}$, $\mathcal{T}(\mathbf{u})$ is a spanning tree on $\mathcal{V}$ with every vertex having finite degree; there is at least one semi-infinite geodesic starting from every $\mathbf{u}$.*

We remark that for Euclidean FPP, one proves the finite degree statement using (3.12). Recall that a semi-infinite (resp. doubly infinite) path $r = (\mathbf{v}_i)$ is a semi-infinite (resp. doubly-infinite) geodesic if, for each $i < j$ the minimizing path $M(\mathbf{v}_i, \mathbf{v}_j)$ is $(\mathbf{v}_i, \mathbf{v}_{i+1}, \ldots, \mathbf{v}_j)$. The semi-infinite geodesic may be constructed inductively as follows. Take $M_0 = (\mathbf{v}_0 \equiv \mathbf{u})$ and suppose $M_n = (\mathbf{v}_0, \mathbf{v}_1, \ldots, \mathbf{v}_n)$ has already been constructed. Choose $e_{n+1} = \{\mathbf{v}_n, \mathbf{v}_{n+1}\}$ incident to $\mathbf{v}_n$ so that the path $M_{n+1} = (\mathbf{v}_0, \mathbf{v}_1, \ldots, \mathbf{v}_n, \mathbf{v}_{n+1})$ can be extended to arbitrarily long paths in $\mathcal{T}(\mathbf{u})$. This is possible and proceeds indefinitely since the vertex set is infinite and each vertex has finite degree. Then $M_\infty \equiv (\mathbf{v}_0, \mathbf{v}_1, \ldots)$ is easily seen to be a semi-infinite geodesic.

With the exception of this one guaranteed semi-infinite geodesic, little else is known unconditionally about their existence/abundance in Standard FPP. The strongest unconditional statement that can be made at present is due to Häggström and Pemantle [25], which specializes to the case of dimension 2 with exponential edge variables. In this setting, they show that, with positive probability, any particular site (e.g. $\mathbf{0}$) has at least two distinct semi-infinite geodesics. The existence of two semi-infinite geodesics in this FPP model turns out to be equivalent to the simultaneous survival of two types of infection in a generalized Richardson [60] growth model.

For Euclidean models, much more is known about infinite geodesics. For the remainder of this Section 5 we confine our attention, except where otherwise noted, to Euclidean FPP. If the vertices along a semi-infinite geodesic M satisfy

$$\lim_{\substack{\mathbf{v} \in M \\ |\mathbf{v}| \to \infty}} \frac{\mathbf{v}}{|\mathbf{v}|} = \hat{\mathbf{x}}, \tag{5.2}$$

we say that M has *asymptotic direction* $\hat{\mathbf{x}}$; we call M an $\hat{\mathbf{x}}$-*geodesic*. Howard and Newman [32] show the following.

Theorem 5.2. *For Euclidean FPP for all $d \geq 2$ and $\alpha \in (1, \infty)$, almost surely: (i) every semi-infinite geodesic has an asymptotic direction; (ii) for every vertex $\mathbf{q} \in \mathcal{V} = Q$ and every unit vector $\hat{\mathbf{x}}$, there is at least one $\hat{\mathbf{x}}$-geodesic starting at $\mathbf{q}$; (iii) for every $\mathbf{q} \in Q$, the set $U(\mathbf{q})$ of unit vectors $\hat{\mathbf{x}}$ such that there is more than one $\hat{\mathbf{x}}$-geodesic starting at $\mathbf{q}$ is dense in the unit sphere S^{d-1}.*

We remark that it is sufficient to prove the theorem for $\mathbf{q} = Q(\mathbf{0})$, the particle closest to $\mathbf{0}$. See Figure 6 for a simulated realization of Euclidean FPP in dimension 2.

The key to obtaining Theorem 5.2 is (4.5), which, recall, is known to hold for all directions $\hat{\mathbf{x}}$ because $\widetilde{\mathcal{W}}$ is a Euclidean ball for Euclidean FPP. We remark that a similar estimate should hold for the Poisson models of Vahidi-Asl, Wierman, and Serafini, but this has not yet been verified. Such an estimate would lead to the validity of Theorem 5.2 for these models as well. Indeed, the theorem would hold for Standard FPP (see [53]) if a technical hypothesis of "uniform curvature" for $\widetilde{\mathcal{W}}$ could be verified in this context. This hypothesis would be satisfied if, in (3.9), the (finite) radius of the sphere D could be taken to be independent of $\mathbf{z}$.

By virtue of (4.5), geodesics in Euclidean FPP satisfy a straightness property that is somewhat stronger than the statement $\xi \leq 3/4$, with ξ as in (4.4). Specifically, for any of the FPP models under discussion here, we put

$$\xi_2 \equiv \inf\{\kappa : \text{the number of } \mathbf{v} \in \mathcal{V} \text{ with}$$

$$d_{\max}(M(\mathbf{0}, \mathbf{v}), \overline{\mathbf{0}\,\mathbf{v}}) \geq |\mathbf{v}|^{\kappa} \text{ is a.s. finite}\}.$$

Then, for Euclidean FPP, an application of the Borel-Cantelli lemma, together with (4.5), easily yields $\xi_2 \leq 3/4$.

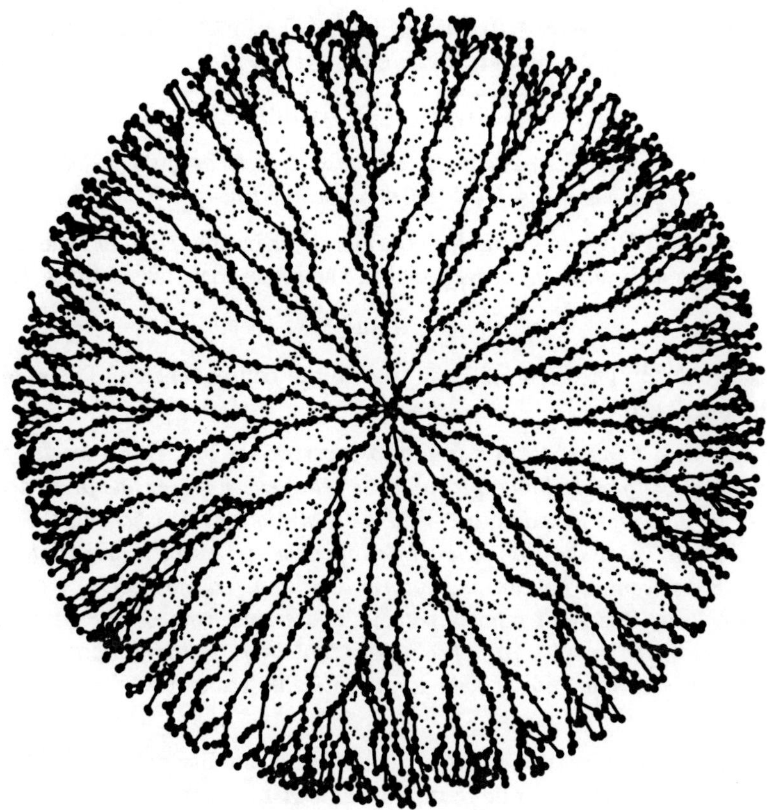

Fig. 6. Illustrated above is a simulated realization of Euclidean FPP in dimension 2 showing geodesics to particles near the boundary of a disk of radius 40. The particle density is 1 per unit area and $\alpha = 1.25$.

This provides a key bound on the wiggle-room of extensions of finite geodesics. Let $M^{\mathrm{out}}(\mathbf{q})$ denote all $\mathbf{q}' \in Q$ such that $\mathbf{q} \in M(\mathbf{0}, \mathbf{q}')$, that is, the finite geodesic from $\mathbf{0}$ to $\mathbf{q}$ extends to one from $\mathbf{0}$ to $\mathbf{q}'$. Similarly, for a tree $\mathcal{T}$ with vertices $\mathbf{u}$ and $\mathbf{v}$, let $\mathcal{T}^{\mathrm{out}}(\mathbf{u}, \mathbf{v})$ denote all vertices $\mathbf{v}'$ such that the path in $\mathcal{T}$ from $\mathbf{u}$ to $\mathbf{v}'$ goes through $\mathbf{v}$. With the cone

$$C(\mathbf{x}, \varepsilon) \equiv \{\mathbf{y} \in \mathbf{R}^d : \theta(\mathbf{x}, \mathbf{y}) \leq \varepsilon\},$$

where $\theta(\mathbf{x}, \mathbf{y})$ is the angle (in $[0, \pi]$) between $\mathbf{x}$ and $\mathbf{y}$, and h a positive function on $(0, \infty)$, we say that such a tree $\mathcal{T}$ is *h-straight* at $\mathbf{u}$ if for all but finitely many vertices $\mathbf{v}$ of $\mathcal{T}$,

$$\mathcal{T}^{\mathrm{out}}(\mathbf{u}, \mathbf{v}) \subset \mathbf{u} + C(\mathbf{v} - \mathbf{u}, h(|\mathbf{v} - \mathbf{u}|)). \tag{5.3}$$

The following is a consequence of $\xi_2 \leq 3/4$ together with a purely geometric argument (see [32], Lemma 2.7).

Theorem 5.3. *Choose $\delta \in (0, \frac{1}{4})$ and let $h_\delta(n) = n^{-\delta}$. Then for Euclidean FPP, almost surely, for all but finitely many $\mathbf{q} \in Q$,*

$$M^{out}(\mathbf{q}) \subset C(\mathbf{q}, h_\delta(|\mathbf{q}|)).$$

Equivalently, the tree $\mathcal{T}(\mathbf{0})$ is almost surely h_δ-straight at $Q(\mathbf{0})$. It follows that, almost surely, for every $\mathbf{q} \in Q$, $\mathcal{T}(\mathbf{q})$ is h_δ-straight at $\mathbf{q}$.

For all the FPP models under consideration in this chapter, the vertex set $\mathcal{V}$ is *asymptotically omnidirectional* in the sense that, for all finite K,

$$\{\mathbf{v}/|\mathbf{v}| : \mathbf{v} \in \mathcal{V} \text{ and } |\mathbf{v}| > K\}$$

is dense in S^{d-1}. Theorem 5.2 is an immediate consequence of Theorem 5.3 and the next theorem.

Theorem 5.4. *Suppose $\mathcal{T}$ is a tree whose vertex set $\mathcal{V} \subset \mathbf{R}^d$ is locally finite but asymptotically omnidirectional and such that every vertex has finite degree. Suppose further that for some $\mathbf{u} \in \mathcal{V}$, $\mathcal{T}$ is h-straight at $\mathbf{u}$, where $h(n) \to 0$ as $n \to \infty$. Then $\mathcal{T}$ satisfies the following properties: (i) every semi-infinite path in $\mathcal{T}$ starting from $\mathbf{u}$ has an asymptotic direction; (ii) for every $\hat{\mathbf{x}} \in S^{d-1}$, there is at least one semi-infinite path in $\mathcal{T}$ starting from $\mathbf{u}$ with asymptotic direction $\hat{\mathbf{x}}$; (iii) the set $V(\mathbf{u})$ of $\hat{\mathbf{x}}$'s such that there is more than one semi-infinite path starting from $\mathbf{u}$ with asymptotic direction $\hat{\mathbf{x}}$ is dense in S^{d-1}.*

To see (i), let $M = (\mathbf{u} = \mathbf{u}_0, \mathbf{u}_1, \dots)$ be a semi-infinite path in $\mathcal{T}$. Then h-straightness implies that for large m, the angle $\theta(\mathbf{u}_n - \mathbf{u}, \mathbf{u}_m - \mathbf{u}) \leq h(|\mathbf{u}_m - \mathbf{u}|)$ for $n \geq m$. Since $|\mathbf{u}_m| \to \infty$ as $m \to \infty$ (because $\mathcal{V}$ is locally finite), it follows that $\mathbf{u}_n/|\mathbf{u}_n|$ converges, proving (i). Fix any $\hat{\mathbf{x}}$. Since $\mathcal{V}$ is asymptotically omnidirectional, we may choose $\mathbf{u}_0, \mathbf{u}_1, \dots$ with $\mathbf{u}_n \to \infty$ and $\mathbf{u}_n/|\mathbf{u}_n| \to \hat{\mathbf{x}}$. Since each vertex has finite degree, it follows (as in Theorem 5.1) that starting from $\mathbf{v}_0 = \mathbf{u}$, one can inductively construct a semi-infinite path $\mathbf{v}_0, \mathbf{v}_1, \dots$ in $\mathcal{T}$ such that for each j, $\mathcal{T}_{out}(\mathbf{u}, \mathbf{v}_j)$ contains infinitely many of the $\mathbf{u}_j$. But (i) shows that $\mathbf{v}_j/|\mathbf{v}_j|$ tends to some $\hat{\mathbf{y}}$ and then h-straightness implies $\theta(\hat{\mathbf{x}}, \mathbf{v}_j - \mathbf{u}) \leq h(|\mathbf{v}_j - \mathbf{u}|)$ for large j. Letting $j \to \infty$ yields $\hat{\mathbf{x}} = \hat{\mathbf{y}}$, proving (ii).

Given any (large) finite K, one can consider those (finitely many) vertices $\mathbf{v}$ with $|\mathbf{v}| > K$ such that no other vertex $\mathbf{w}$ on the path from $\mathbf{u}$ to $\mathbf{v}$ has $|\mathbf{w}| > K$. Calling these vertices $\mathbf{v}_1, \dots, \mathbf{v}_{m(K)}$, one has that each $|\mathbf{v}_j| > K$, and the $\mathcal{T}^{out}(\mathbf{u}, \mathbf{v}_j)$'s are disjoint and their union includes all but finitely many vertices of $\mathcal{V}$ (from among those within distance K of $\mathbf{u}$). For a given K, let G_j denote the set of $\hat{\mathbf{x}}$'s such that some semi-infinite path from $\mathbf{u}$ passing through $\mathbf{v}_j$ has asymptotic direction $\hat{\mathbf{x}}$. Then by (ii), $\cup_j G_j = S^{d-1}$. On the other hand, by h-straightness, each G_j is a subset of the (small) spherical cap $\{\hat{\mathbf{x}} : \theta(\hat{\mathbf{x}}, \mathbf{v}_j) \leq h(|\mathbf{v}_j - \mathbf{u}|) \leq \varepsilon(K)\}$ where $\varepsilon(K) \to 0$ as $K \to \infty$ (since $|\mathbf{v}_j| > K$). Furthermore, by the same arguments that proved (ii), each G_j is a *closed* subset of S^{d-1}. It follows that $V(\mathbf{u})$ contains, for each K, $\cup_{j \leq m(K)} \partial G_j$, where ∂G_j denotes the usual boundary (G_j less its interior). Since $\varepsilon(K) \to 0$ as $K \to \infty$, we obtain (iii) by standard arguments.

5.2 Coalescence and Another Spanning Tree in 2 Dimensions

We know that for each $\mathbf{q} \in Q$ and each direction $\hat{\mathbf{x}}$, there is at least one $\hat{\mathbf{x}}$-geodesic starting at $\mathbf{q}$. It is natural to ask if, for deterministic $\hat{\mathbf{x}}$, this geodesic is unique. Additionally, given another $\hat{\mathbf{x}}$-geodesic starting from a different particle $\mathbf{q}' \in Q$, do these geodesics ever meet or even coalesce. (Geodesics $(\mathbf{q}_0, \mathbf{q}_1, \dots)$ and $(\mathbf{q}_0', \mathbf{q}_1', \dots)$ *coalesce* if $(\mathbf{q}_i, \mathbf{q}_{i+1}, \dots) = (\mathbf{q}_j', \mathbf{q}_{j+1}', \dots)$ for some i and j.) In dimension 2, notwithstanding (iii) of Theorem 5.2, we have the following theorem for Euclidean FPP (see [30]).

Theorem 5.5. *Suppose $d = 2$ and $\alpha \geq 2$. Then for any deterministic direction $\hat{\mathbf{x}}$ the following two statements are true almost surely: (i) for every $\mathbf{q} \in Q$ there is a unique $\hat{\mathbf{x}}$-geodesic; (ii) any two $\hat{\mathbf{x}}$-geodesics starting at different $\mathbf{q}, \mathbf{q}' \in Q$ coalesce.*

The $\alpha \geq 2$ condition (versus $\alpha > 1$), which ought to be unnecessary, is used to insure that geodesics that cross each other cross at a particle in Q, as stated in the following proposition.

Proposition 5.6. *Suppose $d = 2$ and $\alpha \geq 2$. For almost every configuration Q: if particles $\mathbf{q}_1$ and $\mathbf{q}_2$ appear consecutively on one geodesic and particles $\mathbf{q}_1'$ and $\mathbf{q}_2'$ appear consecutively on another, then either $\overline{\mathbf{q}_1 \mathbf{q}_2}$ and $\overline{\mathbf{q}_1' \mathbf{q}_2'}$ are disjoint, or they coincide, or their intersection consists of one point which is an endpoint of both line segments.*

If $\mathbf{q}_1$ and $\mathbf{q}_2$ appear consecutively on a geodesic, then $R_\alpha(\mathbf{q}_1, \mathbf{q}_2)$ is devoid of Poisson particles (see (3.12)). But, for $\alpha \geq 2$, $R_\alpha(\mathbf{q}_1, \mathbf{q}_2)$ contains the disk with diameter $\overline{\mathbf{q}_1 \mathbf{q}_2}$, which is therefore also devoid of particles. Similarly, the disk with diameter $\overline{\mathbf{q}_1' \mathbf{q}_2'}$ is devoid of Poisson particles. The proposition follows easily from the following geometric fact: If D and D' are diameters with unequal length of disks B and B' such that D and D' intersect at a point that is not an endpoint of either D or D', then the interior of B' contains an endpoint of D or the interior of B contains an endpoint of D'. (We use here that if any three of $\mathbf{q}_1$, $\mathbf{q}_2$, $\mathbf{q}_1'$ and $\mathbf{q}_2'$ are distinct, then almost surely $|\mathbf{q}_1 - \mathbf{q}_2| \neq |\mathbf{q}_1' - \mathbf{q}_2'|$.)

Statements (i) and (ii) in Theorem 5.5 sound like they are related; in fact, for Euclidean FPP, they are equivalent in all dimensions. That (ii) implies (i) is straightforward. Assume (ii) and suppose $(\mathbf{q}_0 = \mathbf{q}, \mathbf{q}_1, \dots)$ and $(\mathbf{q}_0' = \mathbf{q}, \mathbf{q}_1', \dots)$ are two distinct $\hat{\mathbf{x}}$-geodesics. Let i be maximal with $\mathbf{q}_j = \mathbf{q}_j'$ for all $j \leq i$. The case $\mathbf{q}_k = \mathbf{q}_j$ for some $k > i$ and $j > i$ violates the uniqueness of minimizing paths, so $(\mathbf{q}_{i+1}, \mathbf{q}_{i+2}, \dots)$ and $(\mathbf{q}_{i+1}', \mathbf{q}_{i+2}', \dots)$ must be disjoint $\hat{\mathbf{x}}$-geodesics. But this contradicts (ii).

That (i) implies (ii) is a little harder and requires a local change of configuration argument of the sort used in [49], [30] and, earlier, in [56]. In fact, in [49], a different version of this general type of argument is used to prove Theorem 5.5 for Standard FPP in 2 dimensions, but only for Lebesgue-a.e. $\hat{\mathbf{x}}$, where no specific directions (e.g., the coordinate directions) are known to satisfy the theorem. We will use this type of argument to show that if (ii) does not hold, then (i) does not hold. Suppose, then, that (ii) does not hold. Two $\hat{\mathbf{x}}$-geodesics starting from distinct $\mathbf{q}, \mathbf{q}' \in Q$ are either disjoint, meet and coincide for a while then forever separate, or they coalesce. (If they met, separated, then met again, path-uniqueness would be violated.) If there were positive probability of them meeting for a while then separating forever at some particle (call it $\mathbf{q}^*$), then (i) would not hold at $\mathbf{q} = \mathbf{q}^*$. Suppose, alternatively, that with positive probability there are $\hat{\mathbf{x}}$-geodesics that are disjoint. Then for some large L, with positive probability there are $\mathbf{q}, \mathbf{q}' \in Q \cap B(0, L)$ with disjoint $\hat{\mathbf{x}}$-geodesics $M_{\mathbf{q}} \equiv (\mathbf{q}_0 = \mathbf{q}, \mathbf{q}_1, \dots)$ and $M_{\mathbf{q}'} \equiv (\mathbf{q}_0' = \mathbf{q}', \mathbf{q}_1', \dots)$. Now take a much larger L' and alter the configuration Q inside $B(0, L')$ as follows: (1) delete all particles in $Q \cap B(0, L')$ except for those on $M_{\mathbf{q}}$ and $M_{\mathbf{q}'}$; (2) add a large number of particles (nearly) evenly spaced (nearly) on the straight line segment $\overline{\mathbf{q}\,\mathbf{q}'}$. Denote these added particles going from $\mathbf{q}$ to $\mathbf{q}'$ by $\tilde{\mathbf{q}}_1, \dots, \tilde{\mathbf{q}}_n$. If this is done properly, in this new configuration both $M_{\mathbf{q}}$ and $(\mathbf{q}, \tilde{\mathbf{q}}_1, \dots, \tilde{\mathbf{q}}_n, \mathbf{q}_0', \mathbf{q}_1', \dots)$ will be $\hat{\mathbf{x}}$-geodesics starting from $\mathbf{q}$. The point is that this type of configuration, which produces two distinct $\hat{\mathbf{x}}$-geodesics starting at a common point, can occur naturally (without any configuration changes) with positive probability. Thus, if (ii) does not hold, with positive probability (hence, by ergodicity, with probability one) there will be a $\mathbf{q}$ with two distinct $\hat{\mathbf{x}}$-geodesics.

To see that (i) holds, we follow Howard and Newman in [30]. Let $\tilde{e} = (\mathbf{q}, \mathbf{q}')$. If one or more of the semi-infinite geodesics in $\mathcal{T}(\mathbf{q})$ begins with $\tilde{e}$, then we will define a particular one, denoted $r^+(\tilde{e})$; otherwise $r^+(\tilde{e})$ will be undefined. The geodesic $r^+(\tilde{e}) = (\mathbf{q}_1, \mathbf{q}_2, \mathbf{q}_3, \dots)$ (where $\mathbf{q}_1 = \mathbf{q}$ and $\mathbf{q}_2 = \mathbf{q}'$) is obtained by a counterclockwise search algorithm within $\mathcal{T}(\mathbf{q})$. That is, if the first k vertices of $r^+(\tilde{e})$ are $(\mathbf{q}_1, \dots, \mathbf{q}_k)$, $\mathbf{q}_{k+1}$ is the next vertex on the semi-infinite geodesic which, among all semi-infinite geodesics extending $(\mathbf{q}_1, \dots, \mathbf{q}_k)$, maximizes the angle (in $(-\pi, \pi)$) from $\mathbf{q}_k - \mathbf{q}_{k-1}$ to $\mathbf{q}_{k+1} - \mathbf{q}_k$.

If there are two distinct $\hat{\mathbf{x}}$-geodesics r_1 and r_2 starting from some particle $\mathbf{q}$, they must bifurcate at some particle $\tilde{\mathbf{q}}$, going respectively to $\mathbf{q}^*$ and $\mathbf{q}^{**}$ in their next steps. After $\tilde{\mathbf{q}}$, the polygonal paths of r_1 and r_2 never touch by route-uniqueness and Proposition 5.6. We assume, without loss of generality, that r_1 is asymptotically counterclockwise to r_2. Then $r^+((\tilde{\mathbf{q}}, \mathbf{q}^{**}))$, which is caught "between" r_1 and r_2, is an $\hat{\mathbf{x}}$-geodesic (see Figure 7). (Note that possibly $r^+((\tilde{\mathbf{q}}, \mathbf{q}^{**})) = r_2$.) We conclude that $U(\hat{\mathbf{x}})$, the event that the $\hat{\mathbf{x}}$-geodesic starting at $\mathbf{q}$ is unique, occurs unless the event $G(\hat{\mathbf{x}})$, that for some $\tilde{e}$, $r^+(\tilde{e})$ is defined and is an $\hat{\mathbf{x}}$-geodesic, occurs. Since there are only countably many such $\tilde{e}$, only countably many $\hat{\mathbf{x}}$'s have the property that some $r^+(\tilde{e})$ is defined and is an $\hat{\mathbf{x}}$-geodesic. Denoting the uniform measure on the $\hat{\mathbf{x}}$'s by $d\hat{\mathbf{x}}$, we have, by this fact and Fubini's Theorem, that

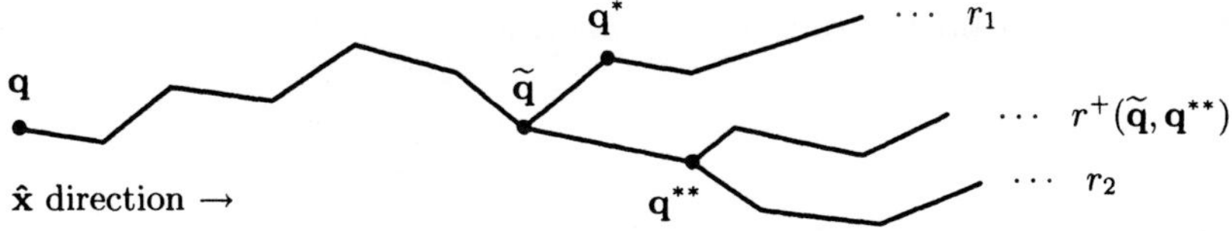

Fig. 7. If r_1 and r_2 are $\hat{\mathbf{x}}$-geodesics, then so also is $r^+((\widetilde{\mathbf{q}}, \mathbf{q}^{**}))$.

$$1 \geq \int P[U(\hat{\mathbf{x}})]\,d\hat{\mathbf{x}} \geq 1 - \int P[G(\hat{\mathbf{x}})]\,d\hat{\mathbf{x}}$$

$$= 1 - \int\int I_{G(\hat{\mathbf{x}})}\,d\hat{\mathbf{x}}\,dP = 1 - \int 0\,dP = 1.$$

This proves that $P[U(\hat{\mathbf{x}})]$ must equal 1 for Lebesgue-a.e. $\hat{\mathbf{x}}$. But by isotropy, $P[U(\hat{\mathbf{x}})]$ is independent of $\hat{\mathbf{x}}$ and so equals 1 for every $\hat{\mathbf{x}}$, as desired.

Letting $M_{\mathbf{q}}(\hat{\mathbf{x}})$ denote the unique $\hat{\mathbf{x}}$-geodesic, Theorem 5.5 yields the following (see [32] for the details of the argument).

Theorem 5.7. *Suppose $d = 2$, $2 \leq \alpha < \infty$, and $\hat{\mathbf{x}}$ is a deterministic unit vector (in S^1). Then the following are all valid almost surely. For any $\mathbf{q} \in Q$ and any $\mathbf{q}_1, \mathbf{q}_2, \cdots \in Q$ such that $\mathbf{q}_k/|\mathbf{q}_k| \to \hat{\mathbf{x}}$, the finite geodesic $M(\mathbf{q}, \mathbf{q}_k)$ converges as $k \to \infty$ to the unique $\hat{\mathbf{x}}$-geodesic starting from $\mathbf{q}$. Thus the spanning trees $T(\mathbf{q}_k) \to T^*(\hat{\mathbf{x}})$ as $k \to \infty$, where $T^*(\hat{\mathbf{x}})$ has vertex set Q and edge set $\cup_{\mathbf{q} \in Q}\{edges\ on\ M_{\mathbf{q}}(\hat{\mathbf{x}})\}$. $T^*(\hat{\mathbf{x}})$ is a spanning tree on Q with every vertex having finite degree and with a single infinite path from each $\mathbf{q}$.*

Monotonicity Revisited. Theorems 5.5 and 5.7 imply an additional monotonicity result for Euclidean FPP in dimension 2 for $\alpha \geq 2$, as observed by Kesten [42]. These theorems imply that

$$\lim_{n\to\infty} [T(-n\hat{\mathbf{x}}, \Delta n\hat{\mathbf{e}}_1) - T(-n\hat{\mathbf{x}}, \mathbf{0})]$$

exists almost surely. (Here, n and Δn are continuous variables.) Since

$$T(-n\hat{\mathbf{x}}, \Delta n\hat{\mathbf{e}}_1) - T(-n\hat{\mathbf{x}}, \mathbf{0}) \overset{\mathrm{d}}{=} T(\mathbf{0}, (n + \Delta n)\hat{\mathbf{e}}_1) - T(\mathbf{0}, n\hat{\mathbf{e}}_1),$$

and the family $T(\mathbf{0}, (n + \Delta n)\hat{\mathbf{e}}_1) - T(\mathbf{0}, n\hat{\mathbf{e}}_1)$ is uniformly integrable, we have that

$$\lim_{n\to\infty} ET(\mathbf{0}, (n + \Delta n)\hat{\mathbf{e}}_1) - ET(\mathbf{0}, n\hat{\mathbf{e}}_1) = \mu\Delta n.$$

5.3 Doubly-Infinite Geodesics

The existence or non-existence, as the case may be, of doubly-infinite geodesics is an open problem in FPP. For Euclidean FPP, we do have the next theorem.

Theorem 5.8. *In dimension 2 with $\alpha \geq 2$, for any deterministic directions $\hat{\mathbf{x}}$ and $\hat{\mathbf{y}}$, almost surely there are no $(\hat{\mathbf{x}}, \hat{\mathbf{y}})$-geodesics.*

Here, an $(\hat{\mathbf{x}}, \hat{\mathbf{y}})$-geodesic $(\mathbf{q}_i : i \in \mathbf{Z})$ is one where

$$\lim_{i \to \infty} \frac{\mathbf{q}_i}{|\mathbf{q}_i|} = \hat{\mathbf{x}} \quad \text{and} \quad \lim_{i \to -\infty} \frac{\mathbf{q}_i}{|\mathbf{q}_i|} = \hat{\mathbf{y}}.$$

For Standard FPP, Theorem 5.8 also holds – but (presently) only for Lebesgue-a.e. $\hat{\mathbf{x}}$ and $\hat{\mathbf{y}}$ (see [49]). Theorem 5.8 appears in [30]. The proof there is essentially that of Licea and Newman in [49]; we follow it practically verbatim.

By Theorem 5.5, we may assume that $\hat{\mathbf{x}} \neq \hat{\mathbf{y}}$. If there were two distinct $(\hat{\mathbf{x}}, \hat{\mathbf{y}})$-geodesics, then two applications of Theorem 5.5 would show that they meet at two particles $\mathbf{q}$ and $\mathbf{q}'$ while being distinct in between. This would violate the uniqueness of the (finite) geodesic between $\mathbf{q}$ and $\mathbf{q}'$. Hence there is at most one $(\hat{\mathbf{x}}, \hat{\mathbf{y}})$-geodesic. Let A be the event that there is exactly one $(\hat{\mathbf{x}}, \hat{\mathbf{y}})$-geodesic; we must show that $P[A] = 0$. For $L > 0$ and $\mathbf{z} \in \mathbf{R}^2$, let $A(\mathbf{z}, L)$ be the event that there is exactly one $(\hat{\mathbf{x}}, \hat{\mathbf{y}})$-geodesic and it passes through a particle $\mathbf{q} \in \mathbf{z} + [-L, L]^2$. Now choose $\hat{\mathbf{w}} \neq \hat{\mathbf{x}}$ or $\hat{\mathbf{y}}$. By translation invariance, $P[A(k\hat{\mathbf{w}}, L)] = P[A(\mathbf{0}, L)]$ and, by ergodicity,

$$\lim_{n \to \infty} \frac{1}{n} \sum_{k=0}^{n-1} I_{A(k\hat{\mathbf{w}}, L)} = P[A(\mathbf{0}, L)] \quad \text{a.s.} \tag{5.4}$$

By the choice of $\hat{\mathbf{w}}$, any $(\hat{\mathbf{x}}, \hat{\mathbf{y}})$-geodesics can touch particles in at most finitely many of the $k\hat{\mathbf{w}} + [-L, L]^2$ yielding that $\sum_k I_{A(k\hat{\mathbf{w}}, L)} < \infty$ almost surely and, in conjunction with (5.4), that $P[A(\mathbf{0}, L)] = 0$. But $A(\mathbf{0}, L) \uparrow A$ as $L \uparrow \infty$, so $P[A] = 0$.

Clearly a doubly infinite geodesic that took a "turn" from one direction to another direction would violate the straightness implied by (4.5). That is the essence of the proof, in [32], of the following.

Theorem 5.9. *In Euclidean FPP in any dimension with any $\alpha > 1$, almost surely, doubly infinite geodesics other than $(\hat{\mathbf{x}}, -\hat{\mathbf{x}})$-geodesics do not exist.*

We remark that Theorem 5.8 says that in dimension 2, for *deterministic* $\hat{\mathbf{x}}$, there almost surely are no $(\hat{\mathbf{x}}, -\hat{\mathbf{x}})$-geodesics. This does not preclude the possibility of $(\hat{\mathbf{x}}, -\hat{\mathbf{x}})$-geodesics for Q-dependent $\hat{\mathbf{x}}$.

In the context of lattice FPP on the upper half-plane in two dimensions, Wehr and Woo [75] have proved that, almost surely, there are no doubly infinite geodesics.

6 Summary of Some Open Problems

Here we summarize the open problems discussed above, together with a few more. For many of these questions, there is general consensus regarding the correct answer; the issue is finding a rigorous proof.

1. Can $\mu(F, d)$ be explicitly computed for any non-trivial distribution F? (Section 2.2.)

2. Can existing bounds for μ be improved? (Section 2.2.)

3. Does the route length N_n^a satisfy $N_n^a/n \to C \in (0, \infty)$ as $n \to \infty$? (Section 2.3.)

4. Under what conditions on F is $\mu(F \oplus x)$ differentiable at $x = 0$? (Section 2.3.)

5. Beyond the simple properties of convexity and lattice symmetry, what, qualitatively, can be said about the asymptotic shape $\widetilde{W}$? (Section 3.2.)

6. Can any particular direction be shown to be a direction of curvature for any non-trivial F? (Section 3.2.)

7. Is $\widetilde{W}$ uniformly curved for any F? (Section 3.2.)

8. Do all "reasonable" definitions of the fluctuation exponents χ and ξ yield the same numerical value? (Sections 3.5, 4.1.)

9. For Standard FPP, are these exponents independent of direction? (Sections 3.5, 4.1.)

10. Do χ and ξ satisfy the scaling relation $\chi = 2\xi - 1$ in all dimension? (Section 4.1.)

11. What are the values of χ and/or ξ for any FPP model? (Sections 3 and 4.)

12. Is $\chi \leq 1/2$ for the Poisson models of Vahidi-Asl and Wierman, and Serafini? (Sections 3.3 and 3.4.)

13. Is there a limit law for $T(\mathbf{0}, n\hat{\mathbf{e}}_1)$? That is, does $(T(\mathbf{0}, n\hat{\mathbf{e}}_1) - a(n))/b(n)$ converge weakly for some sequences $a(n)$ and $b(n)$?

14. Is $ET(\mathbf{0}, n\hat{\mathbf{e}}_1)$ generally monotonic for (possibly) large enough n? (Section 3.6.)

15. Is $T(\mathbf{0}, n\hat{\mathbf{e}}_1)$ stochastically increasing in n for (possibly) large enough n? (Section 3.6.)

16. In Standard FPP, does every semi-infinite geodesic have an asymptotic direction? (Section 5.1.)

17. Does every direction have a semi-infinite geodesic running off in that direction? (Section 5.1.)

18. For deterministic $\hat{\mathbf{x}}$, do $\hat{\mathbf{x}}$-geodesics starting at different locations coalesce in all dimensions? (Section 5.2.)

19. Where geodesics coalesce, how quickly to they coalesce? For example, if the $\hat{\mathbf{e}}_1$-geodesics starting at $\mathbf{0}$ and $n\hat{\mathbf{e}}_2$ coalesce at $\mathbf{x}$, what is the typical order of magnitude of $\mathbf{x} \cdot \hat{\mathbf{e}}_1$ as a function of n? (Section 5.2.)

20. Do (possibly configuration-dependent) doubly-infinite geodesics exist for any FPP model in any dimension? (Section 5.3.)

21. Can doubly-infinite geodesics be ruled out, at least in dimension 2, for any model? (Section 5.3.)

Acknowledgement. I thank Arthur Downing, Eric Neubacher and Louisa Moy of the Baruch College library for their assistance.

References

1. Alexander, K.S. (1993): A note on some rates of convergence in first-passage percolation. Ann. Appl. Probab. **3**, 81–90
2. Alexander, K.S. (1997): Approximation of subadditive functions and convergence rates in limiting-shape results. Ann. Probab. **25**, 30–55
3. Alm, S.E. (1998): A note on a problem by Welsh in first-passage percolation. Combin. Probab. Comput. **7**, 11–15
4. Alm, S.E., Wierman, J.C. (1999): Inequalities for means of restricted first-passage times in percolation theory. Combin. Probab. Comput. **8**, 307–315
5. Baik, J., Deift, P., Johansson, K. (1999): On the distribution of the length of the longest increasing subsequence of random permutations. J. Amer. Math. Soc. **12**, 1119–1178
6. Barlow, M.T., Pemantle, R., Perkins, E.A. (1997): Diffusion-limited aggregation on a tree. Probab. Theory Related Fields **107**, 1–60
7. Benjamini, I., Kalai, G., Schramm, O. (Preprint): First passage percolation has sublinear distance variance
8. Benjamini, I., Peres, Y. (1994): Tree-indexed random walks on groups and first passage percolation. Probab. Theory Related Fields **98**, 91–112
9. van den Berg, J. (1983): A counterexample to a conjecture of J.M. Hammersley and D.J.A. Welsh concerning first-passage percolation. Adv. in Appl. Probab. **15**, 465–467
10. van den Berg, J., Kesten, H. (1993): Inequalities for the time constant in first-passage percolation. Ann. Appl. Probab. 3, 56–80
11. Boivin, D. (1990): First passage percolation: the stationary case. Probab. Theory Related Fields **86**, 491–499
12. Boivin, D. (1998): Ergodic theorems for surfaces with minimal random weights. Ann. Inst. H. Poincaré Probab. Statist. **34**, 567–599
13. Chayes, L. (1991) On the critical behavior of the first passage time in $d \geq 3$. Helv. Phys. Acta **64**, 1055–1071
14. Chayes, J.T., Chayes, L., Durrett, R. (1986): Critical behavior of the two-dimensional first passage time. J. Statist. Phys. **45**, 933–951
15. Chayes, L., Winfield, C. (1993): The density of interfaces: a new first-passage problem. J. Appl. Probab. **30**, 851–862
16. Cox, J.T. (1980): The time constant of first-passage percolation on the square lattice. Adv. Appl. Probab. **12**, 864–879
17. Cox, J.T., Durrett, R. (1981): Some limit theorems for percolation processes with necessary and sufficient conditions. Ann. Probab. **9**, 583–603
18. Cox, J.T., Kesten, H. (1981): On the continuity of the time constant of first-passage percolation. J. Appl. Probab. **18**, 809–819
19. Dhar, D. (1988): First passage percolation in many dimensions. Phys. Lett. A **130**, 308–310
20. Durrett, R., Liggett, T.M. (1981): The shape of the limit set in Richardson's growth model. Ann. Probab. **9**, 186–193

21. Eden, M. (1961): A two-dimensional growth process. In: Proc. Fourth Berkeley Symp. Math. Statist. Probab. **4**, 223–239, Univ. California Press

22. Fontes, L., Newman, C.M. (1993): First passage percolation for random colorings of $\mathbf{Z}^d$. Ann. Appl. Probab. **3**, 746–762

23. Gravner, J., Griffeath, D. (1999): Reverse shapes in first-passage percolation and related growth models. In: Bramson, M., Durrett, R. (eds) Perplexing problems in probability, 121–142. Birkhäuser, Boston-Basel-Berlin

24. Häggström, O., Meester, R. (1995): Asymptotic shapes for stationary first passage percolation. Ann. Probab. **23**, 1511–1522

25. Häggström, O., Pemantle, R. (1998): First passage percolation and a model for competing spatial growth. J. Appl. Probab. **35**, 683–692

26. Hammersley, J. M., Welsh, D. J. A. (1965): First-passage percolation, subadditive processes, stochastic networks, and generalized renewal theory. In: Neyman, J., LeCam, L.M. (eds) Bernoulli–Bayes–Laplace Anniversary Volume, 61–110. Springer-Verlag, New York

27. Higuchi, Y., Zhang, Y. (2000): On the speed of convergence for two-dimensional first passage Ising percolation. Ann. Probab. **28**, 353–378

28. Howard, C.D. (2000): Lower bounds for point-to-point wandering exponents in Euclidean first-passage percolation. J. Appl. Probab. **37**, 1061–1073

29. Howard, C.D. (2001): Differentiability and monotonicity of expected passage time in Euclidean first-passage percolation. J. Appl. Probab. **38**, 815-827

30. Howard, C.D., Newman, C.M. (1997): Euclidean models of first-passage percolation. Probab. Theory Related Fields **108**, 153–170

31. Howard, C.D., Newman, C.M. (1999): From greedy lattice animals to Euclidean first-passage percolation. In: Bramson, M., Durrett, R. (eds) Perplexing problems in probability, 107–119. Birkhäuser, Boston-Basel-Berlin

32. Howard, C.D., Newman, C.M. (2001): Geodesics and spanning trees for Euclidean first-passage percolation. Ann. Probab. **29**, 577–623

33. Janson, S. (1981): An upper bound for the velocity of first-passage percolation. J. Appl. Probab. **18**, 256–262

34. Janson, S. (1999): One, two and three times $\log n/n$ for paths in a complete graph with random weights. Combin. Probab. Comput. **8**, 347–361

35. Johansson, K. (2000): Transversal fluctuations for increasing subsequences on the plane. Probab. Theory Related Fields **116**, 445–456

36. Joshi, V.M. (1997): First-passage percolation on the plane square lattice. Sankhyā Ser. A **39**, 206–209

37. Kesten, H. (1980): On the time constant and path length of first-passage percolation. Adv. in Appl. Probab. **12**, 848–863

38. Kesten, H. (1985): First-passage percolation and a higher-dimensional generalization. In: Particle systems, random media and large deviations (Brunswick, Maine, 1984), 235–251, Contemp. Math., 41, Amer. Math. Soc., Providence, RI

39. Kesten, H. (1987): Surfaces with minimal random weights and maximal flows: a higher-dimensional version of first-passage percolation. Illinois J. Math. **31**, 99–166

40. Kesten, H. (1986): Aspects of first passage percolation. In: École d'Été de Probabilités de Saint-Flour XIV. Lecture Notes in Math., **1180**, 125–264. Springer, Berlin.

41. Kesten, H. (1993): On the speed of convergence in first-passage percolation. Ann. Appl. Probab. **3**, 296–338

42. Kesten, H.: Private communication.
43. Kesten, H., Zhang, Y. (1993): The tortuosity of occupied crossings of a box in critical percolation. J. Statist. Phys. **70**, 599–611
44. Kesten, H., Zhang, Y. (1997): A central limit theorem for "critical" first-passage percolation in two dimensions. Probab. Theory Related Fields **107**, 137–160
45. Kingman, J.F.C. (1968): The ergodic theory of subadditive stochastic processes. J. Roy. Statist. Soc. Ser. B **30**, 499-510
46. Kingman, J.F.C. (1973): Subadditive ergodic theory. Ann. Probab. **1**, 883-909
47. Kingman, J.F.C. (1976): Subadditive Processes. In: École d'Été de Probabilités de Saint-Flour V. Lecture Notes in Math., **539**, 167–223, Springer, Berlin
48. Krug, J., Spohn, H. (1991): Kinetic roughening of growing surfaces. In: Godrèche, C. (ed.) Solids far from equilibrium. Cambridge University Press, Cambridge
49. Licea, C., Newman, C.M. (1996): Geodesics in two-dimensional first-passage percolation. Ann. Probab. **24**, 399–410
50. Licea, C., Newman, C.M., Piza, M.S.T. (1996): Superdiffusivity in first-passage percolation. Probab. Theory Related Fields **106**, 559–591
51. Liggett, T. (1985): An improved subadditive ergodic theorem. Ann. Probab. **13**, 1279–1285
52. Lyons, R., Pemantle, R. (1992): Random walk in a random environment and first-passage percolation on trees. Ann. Probab. **20**, 125–136
53. Newman, C.M. (1995): A surface view of first-passage percolation. In: Proceedings of the International Congress of Mathematicians, 1017–1023. Birkhäuser, Basel
54. Newman, C.M. (1997): Topics in disordered systems. Birkhäuser Verlag, Basel
55. Newman, C.M., Piza, M.S.T. (1995): Divergence of shape fluctuations in two dimensions. Ann. Probab. **23**, 977–1005
56. Newman, C.M., Schulman, L.S. (1981): Infinite clusters in percolation models. J. Statist. Phys. **26**, 613–628
57. Pemantle, R. (1995): Tree-indexed processes. Statist. Sci. **10**, 200–213
58. Pemantle, R., Peres, Y. (1994): Planar first-passage percolation times are not tight. In: Grimmett, G. (ed) Probability and phase transition, 261–264. Kluwer, Dordrecht
59. Reh, W. (1979): First-passage percolation under weak moment conditions. J. Appl. Probab. **16**, 750–763
60. Richardson, D. (1973): Random growth in a tesselation. Proc. Cambridge Philos. Soc. **74**, 515–528
61. Serafini, H.C. (1997): First-passage percolation in the Delaunay graph of a d-dimensional Poisson process. Ph.D. Thesis, New York University, New York
62. Seppäläinen, T. (1998): Exact limiting shape for a simplified model of first-passage percolation on the plane. Ann. Probab. **26**, 1232–1250
63. Sidoravicius, V., Vares, M.E., Surgailis, D. (1999): Poisson broken lines process and its application to Bernoulli first passage percolation. Acta Appl. Math. **58**, 311–325
64. Smythe, R.T. (1976): Remarks on Renewal Theory for Percolation Processes. J. Appl. Probab. **13**, 290–300
65. Smythe, R.T. (1980): Percolation models in two and three dimensions. In: Biological growth and spread. Lecture notes in Biomath., **38**, 504–511, Springer, Berlin - New York

66. Smythe, R.T., Wierman, J.C. (1977): First-passage percolation on the square lattice. I. Adv. in Appl. Probab. **9**, 38–54

67. Smythe, R.T., Wierman, J.C. (1978): First-passage percolation on the square lattice. Lecture Notes in Mathematics, 671. Springer, Berlin

68. Smythe, R.T., Wierman, J.C. (1978): First-passage percolation on the square lattice. III. Adv. in Appl. Probab. **10**, 155–171

69. Steele, M.J., Zhang, Y. (Preprint): Nondifferentiability of the time constants of first-passage percolation

70. Sznitman, Alain-Sol (1998): Brownian Motion, Obstacles and Random Media. Springer, Berlin

71. Vahidi-Asl, M.Q., Wierman, J.C. (1990): First-passage percolation on the Voronoi tessellation and Delaunay triangulation. In: Koroński, M., Jaworski, J., Ruciński, A. (eds) Random graphs '87, 341–359, Wiley, New York

72. Vahidi-Asl, M.Q., Wierman, J.C. (1992): A shape result for first-passage percolation on the Voronoi tessellation and Delaunay triangulation. In: Frieze, A., Luczak, T. (eds.) Random graphs '89, 247–262. Wiley, New York

73. Vahidi-Asl, M.Q., Wierman, J.C. (1993): Upper and lower bounds for the route length of first-passage percolation in Voronoi tessellations. Bull. Iranian Math. Soc. **19**, 15–28

74. Wehr, J., Aizenman, J. (1990): Fluctuations of extensive functions of quenched random couplings. J. Stastic. Phys. **60**, 287–306

75. Wehr, J., Woo, J. (1998): Absence of geodesics in first-passage percolation on a half-plane. Ann. Probab. **26**, 358–367

76. Wierman, J.C. (1977): First-passage percolation on the square lattice. II. Adv. in Appl. Probab. **9**, 283–295

77. Wierman, J.C. (1980): Weak moment conditions for time coordinates in first-passage percolation models. J. Appl. Probab. **17**, 968–978

78. Wierman, J.C., Reh, W. (1978): On conjectures in first passage percolation theory. Ann. Probab. **6**, 388–397

79. Wüthrich, M.V. (1998): Fluctuation results for Brownian motion in a Poissonian potential. Ann. Inst. H. Poincaré Probab. Statist. **34**, 279–308

80. Wüthrich, M.V. (1998): Superdiffusive behavior of two-dimensional Brownian motion in a Poissonian potential. Ann. Probab. **26**, 1000–1015

81. Wüthrich, M.V. (1998): Scaling identity for crossing Brownian motion in a Poissionian potential. Probab. Th. Rel. Fields **112**, 299–319

82. Zhang, Y. (1995): Supercritical behaviors in first-passage percolation. Stochastic Process. Appl. **59** 251–266

83. Zhang, Y. (1999): Double behavior of critical first-passage percolation. In: Bramson, M., Durrett, R. (eds) Perplexing problems in probability, 143–158. Birkhäuser, Boston-Basel-Berlin

84. Zhang, Y., Zhang, Y.C. (1984): A limit theorem for N_{0n}/n in first-passage percolation. Ann. Probab. **12**, 1068–1076

Relaxation Times of Markov Chains in Statistical Mechanics and Combinatorial Structures

Fabio Martinelli

Abstract. In Markov chain Monte Carlo theory a particular Markov chain is run for a very long time until its distribution is close enough to the equilibrium measure. In recent years, for models of statistical mechanics and of theoretical computer science, there has been a flourishing of new mathematical ideas and techniques to rigorously control the time it takes for the chain to equilibrate. This has provided a fruitful interaction between the two fields and the purpose of this paper is to provide a comprehensive review of the state of the art.

1 Introduction

In Markov Chain Monte Carlo (MCMC) simulations of lattice models of classical statistical mechanics, in order to approximately sample from the Gibbs measure μ of the model and to be able to compute some interesting thermodynamics quantities like the spontaneous magnetization, the free energy or the specific heat, one usually runs for a very long time T a suitable ergodic Markov chain on the configuration space Ω of the system. In order to correctly approach the Gibbs measure as $T \to \infty$, the Markov chain must be ergodic and have μ as its unique invariant measure. Typically, the latter requirement is automatically satisfied by requiring the chain to be *reversible* w.r.t. μ, i.e. the transition matrix $\{P(\sigma,\eta)\}_{\sigma,\eta \in \Omega}$ must satisfy

$$\mu(\sigma)P(\sigma,\eta) = \mu(\eta)P(\eta,\sigma), \quad \forall\, \sigma,\eta \in \Omega \tag{1.1}$$

Moreover, in order to be able to actually run the chain on a computer, the single moves of the chain must be simple enough. For, e.g. ± 1 spin models on a finite subset Λ of the cubic lattice $\mathbb{Z}^d$ with $\Omega = \{-1,1\}^\Lambda$, a move of a typical chain like the Metropolis or Heat Bath algorithm, consists in choosing a site $x \in \Lambda$ according to some simple random rule, e.g. uniformly, and to flip the value of the configuration at x with a probability that depends on the local configuration around x. A key instance in the above context is to be able to rigorously determine how large one should choose the running time T, depending on the size of Λ and on the thermodynamic parameters of the model, in order to sample from μ within a predefined error (see, e.g. [148]).

In the last decade, MCMC methods have also attracted the increasing attention of theoretical computer scientists because of their relevance in finding (efficient) approximate counting schemes for complex combinatorial structures [90, 146]. Suppose that $N : \Sigma^* \mapsto \mathbb{N}$ is a function mapping problem instances (encoded as words of some alphabet Σ) to natural numbers. For example Σ^* could be a collection of graphs and N the function that associates with any given graph G the number $N(G)$ of perfect matchings in G. Another example is the function N that, given $q \in \mathbb{N}$, associates with G the number of proper q-colorings of G.

It turns out that *efficient exact* counting schemes to compute $N(x)$, $x \in \Sigma^*$, exist only for a small class of interesting structures. The word *efficient* here means that the run time of the algorithm grows only polynomially with the size of x. However, even if for a given (N, Σ^*) there is no efficient exact counting scheme available, one can expect to find (efficient) *randomized approximate* schemes, namely a randomized algorithm that, given an error bound $\varepsilon > 0$ and an instance $x \in \Sigma^n$, generates a random variable Y such that

$$\mathbb{P}\big((1 - \varepsilon)N(x) \leq Y \leq (1 + \varepsilon)N(x) \big) \geq \frac{3}{4} \tag{1.2}$$

Here the number 3/4 is just a conventional choice for a number larger than 1/2 and efficient means *fully polynomial*, i.e. the running time $T(x, \varepsilon)$ to achieve the bound (1.2) grows at most polynomially in n and ε^{-1}. A randomized algorithm with this property is usually referred to as *FPRAS* (fully polynomial randomized approximation scheme).

It turns out, and this was one of the great achievements of the last decade, that in many cases of interest for theoretical computer science for which no efficient exact counting scheme is available, it is possible to prove the existence of FPRAS (see, e.g. [146]). Moreover, quite often, many of the algorithms one would like to prove to be FPRAS are Markov chains very similar to those considered in statistical mechanics; thus the increasing reciprocal interest between the two fields.

It is, however, important to realize that much of the rigorous analysis of Markov chains arising in MCMC for lattice models of statistical mechanics done in the last fifteen years has been motivated by reasons that go beyond computational problems. On one hand, many of these chains, particularly spin flip and spin exchange dynamics (see below for more details), have been put forward as Markovian models of *nonequilibrium* statistical mechanics in order to understand, starting from a *microscopic model*, nonequilibrium phenomena like nucleation, metastability, dynamical phase transitions and interface motion, or to derive *macroscopic evolution equations* in some appropriate scaling limit (see, e.g. [162] and [95]). On the other hand, they have been extensively studied from an analytic and probabilistic point of view as Markov processes in an infinite (or very large) dimensional space. This approach (see e.g [105], [74] or [153]) has provided some of the most important tools to rigorously analyze the mixing times of these chains and it has stimulated very interesting research in various directions like equilibrium statistical mechanics, infinite dimensional analysis and probability theory.

Let us now describe the main common goals and stress the main differences between the analytic–probabilistic and the algorithmic approaches to MCMC applied to statistical mechanics models.

As we already said, a key issue in MCMC is to provide tight bounds on the speed of relaxation of the Markov chain to its invariant measure. This question is obviously relevant for a rigorous approach to simulation or to approximate counting but it also plays a crucial role in many circumstances of nonequilibrium statistical mechanics, like for example in the approach to the hydrodynamic limit of nongradient systems [95], [162]), and it raises very interesting mathematical questions in probability theory and infinite (or very large) dimensional analysis. Therefore substantial efforts were made from both sides to solve this problem with, however, some important differences.

The first, although less relevant, difference is that theoretical computer scientists prefer to work with *discrete time* Markov chains, while rigorous analysis in mathematical statistical mechanics and probability theory has been mostly done for *continuous-time* Markov chains because of the possibility of extending the model to an infinite dimensional Markov process.

More importantly, theoretical computer scientists have mostly considered models related to statistical mechanics, like proper coloring, independent sets or domino tilings, for *very general* finite graphs G and, sometimes, with *hard-core* conditions on the configuration space, with the goal of relating the polynomial or faster than polynomial (in the size of G) growth of the mixing time to some *general feature* of the graph itself. Moreover, in most cases, the underlying Gibbs measure was just the uniform measure and people were not so concerned with the determination of the *exact* polynomial growth of the mixing time, the main issue being to be able to distinguish between polynomial and nonpolynomial growth. The latter was often referred to as *exponential* even in those cases in which it was exponential in $|G|^\alpha$ with $0 < \alpha < 1$.

The analytic and probabilistic efforts were instead mostly concentrated on both general and specific statistical mechanics models, like the Ising and Potts models or various types of spin glasses, without hard-core conditions on *regular* subgraphs of $\mathbb{Z}^d$ (e.g. large cubes), with the main goal of relating the speed of relaxation of the underlying Markov chain to the equilibrium properties of the model (absence/presence of a phase transition, decay of correlations, equilibrium large deviations etc.). In particular a great deal of research was devoted to proving a general statement of the form: rapid mixing of the Markov chain is equivalent to absence of phase transition (see theorem 5.1). Moreover, *optimal bounds* on mixing times in, e.g. finite cubes were emphasized, together with the physical mechanism leading to them, regardless of whether the corresponding growth in the size of the cube was polynomial or nonpolynomial. The above issue is of the greatest importance if one aims to distinguish between the speed of relaxation outside the phase coexistence region, exactly at a critical point and inside a pure phase. The test polynomial versus nonpolynomial mixing time is usually too rough.

A third important difference between the two approaches lies in the technical tools developed to establish the main results. It is probably not too unfair to say that MCMC applied to combinatorial structures have been mostly analyzed by coupling and path methods (see sections 2.3 and 2.5), particularly after the discovery of the path coupling technique [26]. That happened, I guess, for two different reasons. The first one is that coupling techniques are really designed to analyze that definition of the mixing time of a Markov chain that is behind the notion of FPRAS (see section 2). The second reason is that coupling methods and also path techniques are somehow more algorithmic in nature and require constructions that, even if very complicated and difficult, may look more natural for people working on combinatorial structures than those techniques more familiar for statistical mechanics. Most of the results obtained for MCMC for lattice models of statistical mechanics have instead been obtained by either analytic techniques (see, e.g. section 5 and section 9) or by combination of analytic and coupling methods.

The whole situation has been changing in the last few years and the two different approaches started to better appreciate the other side's, goals and methods (see, e.g. [94] and [149]). The aim of this paper is to present an, obviously unbalanced, review of the state of the art. A quick check of the existing literature shows that there exist already several different review papers that deal in one way or another with the material presented here. We have in mind in particular the course on finite Markov chains illustrated in [139], the basic book on MCMC for combinatorial structures [146], the courses on logarithmic Sobolev inequalities in [153] and [74], the course on Glauber dynamics for discrete spin systems of [121], the review paper [135] and, last but not least, the not yet published but web-available [4]. Given the situation, a substantial overlap with the existing literature was unavoidable. We hope, however, that, because of the choice of the material, including several recent results not contained in any of the above-quoted papers like for example the analysis of the most widely studied models with a conservation law, and the way it is presented, will help the reader to have a more clear perspective of this beautiful field of research and of its possible future developments.

The paper is divided into four main parts. A first part introduces the basic material to analyze the speed of convergence to equilibrium for finite, continuous-time, reversible Markov chains. The second part describes some statistical mechanics and combinatorial models, including a class of quantum spin systems, whose equilibrium Gibbs measure will serve as invariant measures of suitable Markov chains. In the third part we define and analyze the so-called *Glauber dynamics* for the models just described. Finally, in the last part we review the relaxation behavior of some *conservative chains*, i.e. Markov chains whose evolution satisfies a conservation law like the number of particles in an exclusion process, whose analysis requires new ideas and techniques compared to the Glauber case.

2 Mixing times for reversible, continuous-time Markov chains

In this section, following [139], we recall some basic facts (spectral gap, logarithmic Sobolev constant, mixing and coupling times, canonical paths) about reversible, continuous-time, finite Markov chains that allow us to control in a quantitative way their speed of relaxation toward the invariant measure.

Let X be a finite set and let $\mathcal{L}$ be the generator of an irreducible continuous-time Markov chain $(x_t)_{t\geq0}$ on X, reversible with respect to the positive probability measure π. That means that the matrix elements of $\mathcal{L}$ satisfy:

(i) $\sum_{y\in X}\mathcal{L}(x,y) = 0$ for any $x \in X$;

(ii) $\mathcal{L}(x,y) \geq 0$ for any $x \neq y \in X$;

(iii) $\pi(x)\mathcal{L}(x,y) = \pi(y)\mathcal{L}(y,x)$ for any $x,y \in X$ (detailed balance condition);

(iv)   for any pair $x \neq y \in X$ there exists $n \in \mathbb{N}$ such that $(\mathcal{L}^n)(x,y) > 0$;

and that π is the unique invariant measure for the chain, i.e. $\pi(P_t f) = \pi(f)$ for any f, where $P_t = e^{t\mathcal{L}}$ is the Markov semigroup associated with $\mathcal{L}$ and

$\pi(f) := \sum_{x \in X} \pi(x) f(x)$ denotes the mean of f according to π. Notice that

$$\mathcal{L}(x, y) = \frac{d}{dt} P_t(x, y)|_{t=0}$$

which justifies the name *jump rate from x to y* for the matrix element $\mathcal{L}(x, y)$, $x \neq y$. In the sequel we will refer shortly to the above process as the chain $(\mathcal{L}, \pi)$.

We denote by $\mathcal{E}(f, f)$ the associated Dirichlet form, i.e. the symmetric, closed quadratic form of $-\mathcal{L}$ on $\ell^2(X, \pi)$. As is well known, the Dirichlet form can be written as

$$\mathcal{E}(f, f) = \frac{1}{2} \sum_{x, y} \pi(x) \, \mathcal{L}(x, y) [f(x) - f(y)]^2$$

because of the reversibility condition (iii) above. Moreover, if $\|f\|_2$ denotes the $\ell^2(X, \pi)$ norm of a function f and $\mathrm{Var}_\pi(f) := \|f - \pi(f)\|_2^2$, then

$$\frac{d}{dt} \mathrm{Var}_\pi(f) = -2 \, \mathcal{E}(P_t f, P_t f) \tag{2.1}$$

The above two equalities are crucial in most approaches to quantitative estimates of the relaxation speed to equilibrium of the chain. The first equality shows that the Dirichlet form is a sum of positive terms and it allows us to estimate $\mathcal{E}(f, f)$ by means of geometric quantities and to compare it to different Dirichlet forms symmetric w.r.t. the measure π. The second equality suggests the use of the functional coercive inequalities like the Poincaré inequality or logarithmic Sobolev inequalities to obtain meaningful bounds on the long time behavior of the semigroup P_t. As is well known, the Perron–Frobenius theorem (see, e.g. [139]) implies that $\lim_{t \to \infty} \|P_t f - \pi(f)\|_\infty = 0$ and that the limit is attained exponentially fast. However, the standard proofs of the Perron–Frobenius theorem do not provide any clue (except a usually very lousy one) on how large the time t should be for the distribution of the chain at time t to be close, in some sense to be specified, to the invariant measure π. This is a crucial problem in several applications and particularly in the Markov chain approach to hard combinatorial problems or in statistical mechanics.

In order to attack the problem we first need to recall how one measures the distance between two measures μ and π on a finite set X. The first interesting notion is that of the total variation distance, here denoted by $\|\mu - \pi\|$, defined by

$$\|\mu - \pi\| := \frac{1}{2} \sum_{x \in X} |\mu(x) - \pi(x)|$$

Notice that $\|\mu - \pi\| = \sup\{\, |\mu(f) - \pi(f)| \, : \, |f| \leq 1 \,\}$. In many cases the total variation distance is a very natural distance and, as we will see later on, it is closely related to coupling techniques.

All other notions of distance between μ and π are expressed in terms of the relative density $h(x) := \frac{\mu(x)}{\pi(x)}$ of μ w.r.t. π and they appear naturally in the context of coercive inequalities for reversible, irreducible Markov chains.

The ℓ^p distances, $p \geq 1$, are defined by

$$\|h - 1\|_p := \left(\sum_{x \in X} \pi(x)|h(x) - 1|^p \right)^{\frac{1}{p}} , \qquad \|h - 1\|_\infty := \sup_x |h(x) - 1|$$

while the Kullback–Leibler separation or relative entropy is given by

$$\mathrm{Ent}_\pi(h) := \sum_x \pi(x)\, h(x) \log h(x)$$

The next result (see Lemma 2.4.1 in [139]) collects some inequalities between these different notions.

Lemma 2.1. *Let π and $\mu = h\pi$ be two probability measures on a finite set X. Set $\pi_* = \min_x \pi(x)$. Then:*

(a)

$$\|h - 1\|_1 = 2\|\mu - \pi\|$$

(b) For any $1 \leq p \leq q \leq \infty$,

$$\|h - 1\|_p \leq \|h - 1\|_q \leq \pi_*^{\frac{1}{q} - \frac{1}{p}} \|h - 1\|_p$$

(c)

$$\frac{1}{2}\|h - 1\|_1^2 \leq \mathrm{Ent}_\pi(h) \leq \frac{1}{2} \left(\|h - 1\|_1 + \|h - 1\|_2^2 \right)$$

2.1 Analytic methods

Here we describe the main analytic tools to get meaningful bounds on the rate of convergence to equilibrium for our continuous-time Markov chain. In what follows $h_t^x(y) := \frac{P_t(x,y)}{\pi(y)}$ will denote the relative density w.r.t. to the equilibrium measure π of the distribution at time t of the chain started at x.

The first quantity of interest is the *spectral gap*, denoted by λ in what follows. Since the generator $\mathcal{L}$ is a nonpositive self-adjoint operator on $\ell^2(X, \pi)$, its spectrum consists of discrete eigenvalues of finite multiplicity which can be arranged in decreasing order as $0, -\lambda_1, -\lambda_2, \ldots, \lambda_{n-1}$, $|X| = n$, with $\lambda_i > 0$, $i = 1 \ldots n - 1$.

Definition 2.2. *The spectral gap $\lambda = \lambda_1$ is the absolute value of the first nonzero eigenvalue λ_1 and it satisfies*

$$\lambda := \inf \left\{ \frac{\mathcal{E}(f,f)}{\mathrm{Var}_\pi(f)}; \ \mathrm{Var}_\pi(f) \neq 0 \right\}$$

Elementary eigenfunction decomposition together with (2.1) shows that

$$\mathrm{Var}_\pi(P_t f) \le e^{-2\lambda t}\,\mathrm{Var}_\pi(f)$$

which justifies the name *relaxation time* for the quantity λ^{-1}. Clearly λ^{-1} is the best constant c in the *Poincaré inequality*

$$\mathrm{Var}_\pi(f) \le c\,\mathcal{E}(f,f), \quad \forall f \in \ell^2(X,\pi) \tag{2.2}$$

Next, given $1 \le p \le \infty$, we define

$$T_p := \min\left\{ t > 0 : \sup_x \|h_t^x - 1\|_p \le \frac{1}{e} \right\} \tag{2.3}$$

Usually T_1 is called *the mixing time*, although other notions of "mixing time" are possible and in many instances more natural (see, e.g. [6] and [107]). The following has been shown in [139]:

Theorem 2.3. *Let $(\mathcal{L}, \pi)$ be a continuous-time, reversible Markov chain on a finite set X with spectral gap $\lambda > 0$. If π_* is as in lemma 2.1, it holds that*

$$\frac{1}{\lambda} \le T_p \le \frac{1}{2\lambda}\left(2 + \log\frac{1}{\pi^*} \right), \quad \forall\, 1 \le p \le 2$$

whereas

$$\frac{1}{\lambda} \le T_p \le \frac{1}{\lambda}\left(1 + \log\frac{1}{\pi^*} \right), \quad \forall\, 2 < p \le \infty$$

The key point in the above theorem is the presence of the term $\log\frac{1}{\pi^*}$ which can be very large (e.g. if $|X| \gg 1$) thus worsening the tightness of the above bounds. Consider the following example of a continuous-time random walk (birth and death process) on $X = \{0, 1, \dots n\}$ reversible w.r.t. to the measure $\pi(x) = \frac{1}{Z}e^{-x}$, $x \in X$, Z being a normalization constant. The walk jumps to the left with rate one and to the right with rate e^{-1} with a reflecting boundary condition at $x = 0, n$. The spectral gap λ of the chain turns out to be of $O(1)$ uniformly in n but $T_1 \approx n$ because it takes approximately $O(n)$ steps to hit 0 starting from $x = n$. In other words the r.h.s in theorem 2.3 gives the right order of magnitude for T_1. On the other hand, if π is the uniform measure on X and we consider the symmetric walk on X with rates $\mathcal{L}(x,y) = \frac{1}{2}$ if $y = x \pm 1$, then $\lambda = 1 - \cos\frac{\pi}{n+1}$ and $T_2 \le \frac{3}{4}(n+1)^2$ (see [139]). Thus in this case the lower bound in theorem 2.3 gives the right order of magnitudes while the upper bound is off by a factor $\log n$.

Before discussing another important quantity related to a new coercive inequality for the chain $(\mathcal{L}, \pi)$, we would like to point out the following. In the definition of the mixing times T_p we only considered the worst possible case over the choice of the initial condition x. There are cases, however, in which one would like to measure the speed of relaxation to equilibrium when the initial condition is distributed according to some specified known probability measure μ, e.g. uniformly on X. For this purpose we define $h_t^\mu := \sum_{x \in X} \mu(x) h_t^x$ and let

$$T_1^\mu := \min \left\{ t > 0 : \sup_{s \geq t} \|h_s^\mu - 1\|_1 \leq \frac{1}{e} \right\} \tag{2.4}$$

Clearly $\|h_t^\mu - 1\|_1 \leq \frac{1}{e}$ for any $t \geq T_1^\mu$. The point now is that for Markov chains with rare but very deep traps in an otherwise fairly smooth environment, it may happen that the mixing time T_1 is much larger than the mixing time starting from a particular μ. That is actually the case for certain models of spin glasses in statistical mechanics with μ equal to the uniform measure (see [125]).

The way to bound the mixing time T_1^μ is via *generalized Poincaré inequalities* [125]. The following holds (see [124]):

Proposition 2.4. *Let $p \in (0,1)$ and define*

$$\lambda(\mu, p) := \inf \left\{ \frac{\mathcal{E}(f,f) \|f\|_\infty^{\frac{2-2p}{p}}}{\mu(|f|)^{\frac{2}{p}}} \, ; \, \pi(f) = 0 \right\}$$

Then

$$\log T_1^\mu \leq \frac{2}{p} + \log \frac{1}{\lambda(\mu, p)}$$

We now turn to define the logarithmic Sobolev constant c_s of the pair $(\mathcal{L}, \pi)$, another key quantity to measure the speed of relaxation to equilibrium. The definition is similar to that of λ^{-1} as provided by the Poincaré inequality, but with the variance $\mathrm{Var}_\pi(f)$ replaced by

$$\mathcal{I}_\pi(f) := \mathrm{Ent}_\pi \left(\frac{f^2}{\|f\|_2^2} \right) \tag{2.5}$$

Notice that $\mathcal{I}_\pi(f) \geq 0$ and that $\mathcal{I}_\pi(f) = 0$ iff f is constant.

Definition 2.5. *The logarithmic Sobolev constant c_s of the chain $(\mathcal{L}, \pi)$ is the best constant c in the logarithmic Sobolev inequality*

$$\mathcal{I}_\pi(f) \leq c\,\mathcal{E}(f,f), \quad \forall f \tag{2.6}$$

Remark 2.6. The definition of the logarithmic Sobolev constant is not uniform over the literature. In several important references (see, e.g. [139] and [47]) the logarithmic Sobolev constant is defined to be the *inverse* of our constant c_s. In this way the spectral gap and the logarithmic Sobolev constant are treated on the same footing. On the other hand, as we will see next, our notation emphasizes the interpretation of c_s as a "natural" time scale for the relaxation process of the chain. It is, however, fair to say that our own original "imprinting" on logarithmic Sobolev constants played a major role in the choice of the definition.

As is well known, as long as X is finite the logarithmic Sobolev constant c_s is finite and satisfies

$$2\lambda^{-1} \le c_s \le \frac{\log\big[(1-\pi_*)/\pi_*\big]}{1-2\pi_*}\,\lambda^{-1} \tag{2.7}$$

(see lemma 2.2.2, theorem 2.2.3 and corollary 2.2.10 in [139]).

It is quite simple to prove that both bounds can be saturated. Consider for instance the upper bound and define $(\mathcal{L}',\pi)$ to be the simple chain in which $\mathcal{L}'(x,y) = \pi(y)$ for all $x \ne y$. Then the associated Dirichlet form $\mathcal{E}'(f,f)$ is nothing but the variance $\mathrm{Var}_\pi(f)$ so that the corresponding spectral gap is one. Moreover, it is not difficult to prove that c_s is instead equal to the r.h.s. of (2.7). The above trivial chain can actually be used to prove the general upper bound in the r.h.s. of 2.7. In fact, if $\mathcal{E}(f,f)$ is the Dirichlet form of any other chain $(\mathcal{L},\pi)$ with spectral gap λ, the Poincaré inequality for $(\mathcal{L},\pi)$ implies that

$$\mathcal{I}_\pi(f) \le \frac{\log\big[(1-\pi_*)/\pi_*\big]}{1-2\pi_*}\,\mathrm{Var}_\pi(f) \le \frac{\log\big[(1-\pi_*)/\pi_*\big]}{1-2\pi_*}\,\lambda^{-1}\,\mathcal{E}(f,f)$$

which, in turn, implies the r.h.s. of (2.7).

Remark 2.7. The above reasoning is the first and simplest example of a general technique known as *block dynamics*, that proved to be quite successful in treating large Markov chains arising in statistical mechanics models or in sampling problems from large combinatorial structures. Later on we will discuss more sophisticated examples of such a technique.

The main interest for the logarithmic Sobolev constant c_s comes from its tight relation with the hypercontractivity properties of the Markov semigroup P_t (see [73]) but see [13] for its connection with concentration bounds for the measure π and related topics.

Definition 2.8. *Given a strictly increasing function* $q : \mathbb{R}^+ \to [q(0),\infty]$, *we say that the Markov semigroup* P_t *is* hypercontractive *with contraction function* q *iff for any function* f *and any* $t \ge 0$

$$\|P_t f\|_{q(t)} \le \|f\|_{q(0)}$$

The following summarizes a number of results (see section 2.2 in [139]).

Theorem 2.9. *Assume that the chain* $(\mathcal{L},\pi)$ *has logarithmic Sobolev constant equal to* c_s. *Then:*

(i) P_t *is hypercontractive with contraction function* $q(t) = 1 + e^{\frac{4}{c_s}t}$.

(ii) $\mathrm{Ent}_\pi(P_t f) \le e^{-\frac{4t}{c_s}}\,\mathrm{Ent}_\pi(f)$ *for any non-negative* $f : X \to \mathbb{R}$.

(iii) $\|h_t^x - 1\|_2 \le e^{1-c}$ *for all* $c \ge 0$ *and* $t = \frac{c_s}{4}\big(\log_+\log(\frac{1}{\pi(x)})\big) + \frac{c}{\lambda}$ *where* $\log_+ t := \max\{\log t, 0\}$

(iv) Let T_p be as in theorem 2.3. Then

$$\frac{1}{\lambda} \leq T_p \leq \frac{c_s}{4}\left(4 + \log_+ \log \frac{1}{\pi_*}\right), \quad \forall\, 1 \leq p \leq 2$$

$$\frac{c_s}{2} \leq T_p \leq \frac{c_s}{2}\left(3 + \log_+ \log \frac{1}{\pi_*}\right), \quad \forall\, 2 < p \leq \infty \qquad (2.8)$$

The theorem shows that the mixing times T_p, $p \geq 2$, are more closely related to the logarithmic Sobolev constant than to the inverse of the spectral gap.

Remark 2.10. In our basic reference [139] it is erroneously stated that $T_p \geq c_s/2$ for $1 \leq p \leq 2$, whereas in the original paper [47] the proof is correctly given only for $p > 2$. It is not difficult to cook up examples in which the logarithmic Sobolev constant is much larger than the mixing time T_1.

2.2 Tensorization of the Poincaré and logarithmic Sobolev inequalities

Unlike classical Sobolev inequalities, (2.2) and (2.6) are dimension independent and remain both meaningful and valid in infinite dimensions. This fundamental feature is based on the following tensorization property of the variance and entropy (see, e.g. [13, 139] and references therein) which plays a key role in many application to different topics in probability theory and it represents the keystone for the extension of these ideas to infinite dimensional measures like Gibbs measures.

Theorem 2.11. *Let $(E_i, \mathcal{F}_i, \pi_i)$, $i = 1, \ldots, n$, be n probability spaces and let $(E^n, \mathcal{F}^n, \pi^n)$ be the associated product space, with $\pi^n := \otimes_i \pi_i$. Then*

$$\mathrm{Var}_{\pi^n}(f) \leq \sum_i \mu^n\left(\mathrm{Var}_{\pi_i}(f)\right)$$

$$\mathrm{Ent}_{\pi^n}(f) \leq \sum_i \mu^n\left(\mathrm{Ent}_{\pi_i}(f)\right) \qquad (2.9)$$

where the notation $\mathrm{Var}_{\pi_i}(f)$ and $\mathrm{Ent}_{\pi_i}(f)$ means that only the ith variable is being integrated.

Recently, the following nice and important generalization of the above result has been proved [40]. Let $(\Omega, \mathcal{F}, \pi)$ be a probability space and let $\mathcal{F}_1$, $\mathcal{F}_2$ be two sub-σ-algebras of $\mathcal{F}$. Define the conditional variance and the conditional entropy of f as

$$\mathrm{Var}_\pi(f \,|\, \mathcal{F}_i) := \pi(f^2 \,|\, \mathcal{F}_i) - \pi(f \,|\, \mathcal{F}_i)^2 \qquad\qquad f \in L^2(\pi)$$

$$\mathrm{Ent}_\pi(f \,|\, \mathcal{F}_i) := \pi[f \log f \,|\, \mathcal{F}_i] - \pi(f \,|\, \mathcal{F}_i) \log \pi(f \,|\, \mathcal{F}_i) \quad f \geq 0, f \log f \in L^1(\pi),$$

$$(2.10)$$

It follows from (2.9) that if $\mathcal{F}_1$ and $\mathcal{F}_2$ are independent we have

$$\mathrm{Var}_\pi(f) \leq \pi[\mathrm{Var}_\pi(f\,|\,\mathcal{F}_1) + \mathrm{Var}_\pi(f\,|\,\mathcal{F}_2)] \tag{2.11}$$

$$\mathrm{Ent}_\pi(f) \leq \pi[\mathrm{Ent}_\pi(f\,|\,\mathcal{F}_1) + \mathrm{Ent}_\pi(f\,|\,\mathcal{F}_2)], \tag{2.12}$$

It is natural to guess that inequalities (2.11), (2.12) are stable against appropriate "perturbations" of the hypothesis of independence of the σ-algebras $\mathcal{F}_1$, $\mathcal{F}_2$. The independence assumption can be stated by saying that $\pi(f\,|\,\mathcal{F}_2)$ is a.s. equal to $\pi(f)$ whenever f is measurable w.r.t. $\mathcal{F}_1$. Hence one may look for a "weak dependence" condition of the form

$$|\pi(f\,|\,\mathcal{F}_2) - \pi(f)| \quad \text{"is small in some sense"} \quad \forall f \in L^1(\Omega, \mathcal{F}_1, \pi)$$

In [40] the following result was established.

Theorem 2.12. *There exists* $\vartheta : [0,1) \mapsto \mathbb{R}_+$, *with* $\limsup_{\varepsilon \to 0}(\vartheta(\varepsilon)/\varepsilon) < \infty$, *such that the following holds. Assume that for some* $\varepsilon \in [0, \sqrt{2}-1)$, $p \in [1,\infty]$, *we have*

$$\|\pi(g\,|\,\mathcal{F}_1) - \pi(g)\|_p \leq \varepsilon \|g\|_p \quad \forall g \in L^p(\Omega, \mathcal{F}_2, \pi)$$

$$\|\pi(g\,|\,\mathcal{F}_2) - \pi(g)\|_p \leq \varepsilon \|g\|_p \quad \forall g \in L^p(\Omega, \mathcal{F}_1, \pi)$$

Then

$$\mathrm{Var}_\pi(f) \leq (1 + \vartheta(\varepsilon))\,\pi\left(\mathrm{Var}_\pi(f\,|\,\mathcal{F}_1) + \mathrm{Var}_\pi(f\,|\,\mathcal{F}_2)\right) \quad \forall f \in L^2(\pi)$$

Moreover if

$$\|\pi(g\,|\,\mathcal{F}_2) - \pi(g)\|_\infty \leq \varepsilon \|g\|_1 \quad \forall g \in L^1(\Omega, \mathcal{F}_1, \pi)$$

then, for all functions f *such that* $f^2 \left[\log f^2 \vee 0\right] \in L^1(\pi)$, *we have*

$$\mathrm{Ent}_\pi(f^2) \leq (1 + \vartheta(\varepsilon))\pi\left(\mathrm{Ent}_\pi(f^2\,|\,\mathcal{F}_1) + \mathrm{Ent}_\pi(f^2\,|\,\mathcal{F}_2)\right)$$

In the next three sections we will describe some basic techniques to estimate the spectral gap, the logarithmic Sobolev constant or, more generally the mixing time of reversible, irreducible, continuous-time finite Markov chains described in the previous section. We emphasize that the main interest of these techniques is actually represented by their application to concrete interesting examples (see, e.g. [145] and references therein). Moreover, as happens in many cases, these techniques often get combined in order to obtain the best results.

Before entering into some details let us observe that, without loss of generality, we can always assume that the infinitesimal generator $\mathcal{L}$ is such that $\mathcal{L} + \mathbb{1}$ is a stochastic matrix. If that is not the case we can divide $\mathcal{L}$ by $q = \max_x \sum_{y \neq x} \mathcal{L}(x,y)$ and obtain a new generator with the required property. Clearly, dividing by q simply amounts to a global rescaling of the time.

188   F. Martinelli

2.3 Geometric tools

Geometric ideas in the study of the relaxation process of irreducible Markov chains were introduced years ago in [92], [147] and [91] in the framework of a Markov chain approach to computational problems. Since then there has been a lot of activity in the area with very many interesting applications (see, e.g. [139], [145], [146] and references therein).

We first need to establish some handy notation. Given a chain $(\mathcal{L}, \pi)$ with state space X and $\mathcal{L} + \mathbb{I}$ a stochastic matrix, we define G, the directed graph associated with the chain, to have vertex set X and edge set $E := \{(x, y) \in X \times X : \mathcal{L}(x, y) > 0\}$. A generic edge in E will be denoted by e. Notice that, by reversibility and irreducibility, if $e = (x, y) \in E$ then also $e' = (y, x) \in E$. Given $e = (x, y) \in E$ and $f : X \to \mathbb{R}$ we let $df(e) := f(y) - f(x)$ and $Q(e) := \pi(x)\mathcal{L}(x, y)$. With these definitions the Dirichlet form becomes

$$\mathcal{E}(f, f) = \frac{1}{2} \sum_{e \in E} Q(e) \, df(e)^2$$

Finally, we define a path $\gamma := (x_0, x_1, \ldots, x_n)$ of length n as a sequence of $(n + 1)$ vertices in X such that $(x_{i-1}, x_i) \in E$ for any $i = 1, \ldots, n$. Given two vertices x, y we denote by $\Gamma(x, y)$ the set of all paths without self-repetition (each edge in the path appears exactly once) that join x to y. A specific choice of an element $\gamma(x, y) \in \Gamma(x, y)$ for each $x, y \in X \times X$ will be referred to as a choice of *canonical paths*. The first and simplest result relating canonical paths and speed of convergence is the following [139].

Proposition 2.13. *Given a chain $(\mathcal{L}, \pi)$ and a choice of canonical paths the spectral gap λ satisfies $\lambda \geq \frac{1}{A}$ where*

$$A := \max_{e \in E} \left\{ \frac{1}{Q(e)} \sum_{\substack{x, y \in X \\ \gamma(x, y) \ni e}} |\gamma(x, y)| \pi(x)\pi(y) \right\}$$

The above result can be refined in various ways. One could for example measure the length of the path using a weight function on the set of edges E that is not just constant (see theorem 3.2.3 and example 3.2.5 in [139]), or one could use more than one path for each pair $(x, y) \in X \times X$ (see theorem 3.2.5 in [139]), or, finally, combine the two techniques (see Theorem 3.2.9 in [139]). We also mention a recent nice extension to the logarithmic Sobolev constant c_s based on the link between the entropy of f and the Orlicz norm of f [13]. With the same notation

$$c_s \leq \max_{e \in E} \left\{ \frac{1}{Q(e)} \sum_{\substack{x, y \in X \\ \gamma(x, y) \ni e}} \pi(x)\pi(y) \log \frac{1}{\pi(y)} \right\}$$

Another way in which geometric ideas come into play in order to bound the spectral gap of a reversible chain $(\mathcal{L}, \pi)$ is via isoperimetric inequalities like the Cheeger inequality (which was actually introduced in a different setting in [44]). Earlier references go back to [11], [12], [145], [52], [102] and, as for all this section, we refer to section 3.3 of our basic reference [139].

Given a set $S \subset X$ we define the *capacity* of S as the quantity $C_S := \pi(S)$ and the *ergodic flow* out of S as the quantity

$$F_S := \sum_{\substack{e = (x, y) \in E \\ x \in S, y \in S^c}} Q(e)$$

Notice that $\Phi_S := \frac{F_S}{C_S} \leq 1$ because of the hypothesis that $\mathcal{L} + \mathbb{1}$ is a stochastic matrix. The ratio Φ_S can be interpreted as the conditional probability that the stationary process escape from S in one step, given that it starts in S. Finally, we define the *conductance* of the chain as the quantity

$$\Phi := \min_{\substack{S \subset X \\ C_S \leq 1/2}} \Phi_S \tag{2.13}$$

The conductance is also referred to as the *isoperimetric constant* of the chain $(\mathcal{L}, \pi)$ and it satisfies $\frac{\Phi}{2} \leq I' \leq \Phi$ where

$$I' = \inf_f \left\{ \frac{\sum_e Q(e) \, |df(e)|}{\sum_x \pi(x) \, |f(x) - \pi(f)|} \right\}$$

Notice that the above variational characterization of I' looks like an ℓ^1 version of the Poincaré inequality; this suggests the use of path techniques to bound the conductance (see, e.g. [113] for an application to the Ising model).

Proposition 2.14. *Given a chain $(\mathcal{L}, \pi)$ and a choice of canonical paths the conductance Φ satisfies*

$$\Phi^{-1} \leq \max_{e \in E} \left\{ \frac{1}{Q(e)} \sum_{\substack{x, y \in X \\ \gamma(x, y) \ni e}} \pi(x)\pi(y) \right\}$$

Next we relate the conductance to the spectral gap. Intuitively, if the conductance is very small, then the chain should relax very slowly because of the presence of a *bottleneck*. This can be made into a theorem (Cheeger inequality).

Theorem 2.15. *The spectral gap and the conductance of the chain $(\mathcal{L}, \pi)$ are related by*

$$\frac{\Phi^2}{2} \leq \lambda \leq \Phi$$

It follows from the Cheeger inequality that the mixing time T_1 is bounded from above by

$$c\,\Phi^2 \log \frac{1}{\pi_*}$$

for some universal finite constant c. The penalty $\log \frac{1}{\pi_*}$ comes from starting at a fixed point $x \in X$. In [106] the above bound was improved in a way that turned out to be important in some nontrivial examples like sampling from a convex body [106] and in the analysis of the mixing time of a simple random walk on the supercritical percolation cluster in $\mathbb{Z}^d$ [15].

Define the *conductance function*

$$\Phi(\lambda) := \min_{\substack{S \subset X \\ 0 < c_S \leq \lambda}} \Phi_S$$

Then the following holds.

Theorem 2.16.

$$T_1 \leq 500 \int_{\pi_*}^{3/4} \frac{d\lambda}{\lambda \Phi(\lambda)} \tag{2.14}$$

Remark 2.17. The above result has been recently strengthened in a significant way, by replacing in (2.14) the mixing time T_1 with the larger mixing time T_2, in [126]. It follows in particular that the logarithmic Sobolev constant can be bounded from above by twice the r.h.s. of (2.14).

2.4 Comparison methods

We briefly recall here the main ideas of a useful technique known as the *comparison technique* to bound from above mixing times of a given chain $(\mathcal{L}, \pi)$ (see [48], [139] and references therein). For application of these ideas to statistical mechanics models we refer to [1], [80], [116], [113], [121] and [104].

The main idea in its simplest form can be explained as follows. Given a chain $(\mathcal{L}, \pi)$ with state space X, we suppose that we are able to construct another chain $(\mathcal{L}', \pi')$ with the same state space and with relaxation properties (e.g. spectral gap or logarithmic Sobolev constant) well under control. It is quite clear that if we are able to bound from above the Dirichlet form $\mathcal{E}'(f, f)$ in terms of the Dirichlet form of the original chain $\mathcal{E}(f, f)$ and if the density of π w.r.t. π' is bounded, then we can bound the spectral gap and logarithmic Sobolev constant of the chain $(\mathcal{L}, \pi)$ in terms of those of $(\mathcal{L}', \pi')$. More precisely, if for some positive A, a

$$\mathcal{E}'(f, f) \leq A \mathcal{E}(f, f) \quad \forall f \quad \text{and} \quad a\pi(x) \leq \pi'(x) \quad \forall x \in X$$

then

$$\lambda \geq \frac{a\lambda'}{A}, \quad \text{and} \quad c_s \leq \frac{A\, C_s'}{a}$$

For a more general formulation see theorem 4.1.1 in [139].

The simplest example of a comparison chain $(\mathcal{L}', \pi')$ is the chain for which, with rate one and independently of the starting point x, a new state y is chosen with probability $\pi(y)$. For such a chain the Dirichlet form $\mathcal{E}'(f,f)$ is just the variance of f so that the spectral gap is one while $c'_s = \frac{\log[(1-\pi_*)/\pi_*]}{1-2\pi_*}$ (we are assuming here $\pi_* < \frac{1}{2}$). Thus in this case $a = 1$ and $A = \lambda^{-1}$. Although rather trivial, the above example, combined with proposition 2.13, suggests that path methods could be useful to compare different Dirichlet forms.

Proposition 2.18. *Let E be the edge set for the chain $(\mathcal{L}, \pi)$ with state space X. Let $(\mathcal{L}', \pi')$ be an auxiliary pair on X with edge set E'. For any $e = (x, y) \in E'$ choose one path $\gamma(x, y) \in \Gamma(x, y)$. Then*

$$\mathcal{E}'(f,f) \leq A\,\mathcal{E}(f,f)$$

where

$$A = \max_{e \in E} \left\{ \frac{1}{Q(e)} \sum_{\substack{(x,y)\,\in\,E' \\ \gamma(x,y)\,\ni\,e}} |\gamma(x,y)|\pi'(x)\mathcal{L}'(x,y) \right\}$$

A nice example of the comparison method is represented by the simple exclusion process in the interval $\Lambda := [1, 2, \ldots, L] \subset \mathbb{Z}$. The process can be described as follows. We have $n < L$ particles in Λ with at most one particle per site, and each particle jumps to the nearest empty site with rate one. The auxiliary chain, known as the Bernoulli–Laplace chain, is similar but now each particle jumps with rate one to any other empty site (not necessarily a nearest neighbor). The invariant measure is simply the uniform measure on $X = \left\{ \eta \in \{0,1\}^L : \sum_i \eta(i) = n \right\}$ and the two Dirichlet forms are given by

$$\mathcal{E}(f,f) = \frac{1}{2} \sum_{i=2}^{L} \pi\left([f(\eta^{i-1,i}) - f(\eta)]^2\right)$$

$$\mathcal{E}'(f,f) = \frac{1}{2L} \sum_{i=1}^{L} \sum_{j=1}^{L} \pi\left([f(\eta^{i,j}) - f(\eta)]^2\right)$$

where $\eta^{i,j}$ denotes the configuration in which the occupation numbers at the two sites i and j have been exchanged.

The spectral gap of the auxiliary chain is equal to $\frac{1}{2}$. It is not difficult to prove (see [168] for a proof in a more general context) that

$$\pi\left([f(\eta^{i,j}) - f(\eta)]^2\right) \leq C|i - j| \sum_{\ell=i}^{j-1} \pi\left([f(\eta^{\ell,\ell+1}) - f(\eta)]^2\right)$$

for a suitable constant C independent of n and L. Therefore

$$\mathcal{E}'(f,f) \leq C' L^2 \mathcal{E}(f,f)$$

so that the spectral gap of the simple exclusion is bounded from below by $\frac{1}{C'L^2}$.

2.5 Coupling methods and block dynamics

Last but not least we discuss the coupling technique approach to bound mixing times (see, e.g. [5]) that proved to be quite successful for Markov chain algorithms used in combinatorial problems [90], especially when combined with comparison methods [135].

Coupling methods work as follows. Given a chain $(\mathcal{L}, \pi)$ on X, a coupling is a new Markov chain on $X \times X$ such that:

(a)   each replica evolves as the original chain;
(b)   if the two copies agree at time $t \geq 0$ then they necessarily must agree for all future times $s \geq t$.

Given $x, y \in X$ and a coupling $(x_t, y_t)_{t \geq 0}$ such that $x_0 = x$ and $y_0 = y$, we define the stopping time

$$T_{xy} := \min\{t \geq 0\,;\, x_t = y_t\}$$

and the *coupling time* T_{coupling} as

$$T_{\text{coupling}} := \max_{x, y \in X} \mathbb{E}\,(T_{xy}) \tag{2.15}$$

As is well known (see [5]) the mixing time T_1 can be bounded from above by $C T_{\text{coupling}}$ with C a universal constant (in the discrete time setting $C = 12$) and therefore the whole point of the method is to design a coupling with the smallest possible coupling time. In order to bound the coupling time the usual approach consists in choosing a distance function $\Phi(x, y)$ on $X \times X$ with integers values in $\{0, 1, \dots, D\}$ such that $\Phi(x = y) = 0$ and prove that the expected rate of change of $\Phi(t) := \Phi(x_t, y_t)$ is nonpositive. More precisely, let $\Delta\Phi(t)$ be the change of $\Phi(t)$ after one step of the discrete time Markov chain underlying the evolution of the coupled process (x_t, y_t). Then (see, e.g. [109]):

Lemma 2.19. *If*

$$\mathbb{E}\,(\Delta\Phi(t)\,|\,(x_t, y_t)) \leq 0$$

and whenever $\Phi(t) > 0$

$$\mathbb{E}\,\big([\Delta\Phi(t)]^2\,|\,(x_t, y_t)\big) \geq V > 0$$

then the coupling time is bounded from above by $C\Phi(0)\frac{(2D - \Phi(0))}{V}$.

A further refinement of the coupling technique is represented by the *path coupling* method introduced in [26] which is behind several recent results on mixing times for Markov chain algorithms in combinatorial problems. The path coupling method goes as follows.

Let $S \subset X \times X$ and, for any given pair $x, y \in X$, let $\Gamma_S(x, y) \subset \Gamma(x, y)$ be the set of paths $\gamma(x, y)$ between x and y such that each edge in γ belongs to S.

Theorem 2.20 ([26]). *Let Φ be an integer metric on $X \times X$ with values in $\{0, 1, , \ldots, D\}$ such that, for all $x, y \in X$ there exists a path $\gamma = \{x = x_0, x_1, \ldots, x_k = y\} \in \Gamma_S(x, y)$ with*

$$\Phi(x, y) = \sum_i \Phi(x_i, x_{i+1})$$

Assume that there exists $\beta < 1$ and a coupling (x_t, y_t) of the Markov chain such that, for all $(x, y) \in S$

$$\mathbb{E}\left(\Delta\Phi(t) \,|\, (x_t, y_t)\right) \leq -(1 - \beta)\Phi(t)$$

where $\Delta\Phi(t)$ is as in lemma 2.19. Then the coupling time is bounded by $T_1 \leq \frac{C \log(D)}{1 - \beta}$ for some universal constant C.

Remark 2.21. In the original discrete time setting in [26] the constant C can be taken equal to one. Here it appears only because we choose to work in the continuous-time case.

It turns out that in several applications arising in combinatorial problems or in statistical mechanics, one is not able to find directly a good coupling for some natural chain associated with the problem. That is so because the most natural chains are typically built by means of very elementary transitions which only move one basic dynamical variable at a time. Rather, in order to make coupling analysis more feasible, one needs to add certain transitions to the original chain that, in one step, are able to change not just one but a whole collection (block) of dynamical variables. For this reason the corresponding chain is usually referred to as *block dynamics*, and the original chain as *single site dynamics*. If cleverly chosen, the block moves enhance the speed of relaxation to equilibrium and in many examples block dynamics is more appropriate for a coupling analysis because of the "coarse grained" structure of its transitions. Finally, in most applications, the block moves can be viewed as special sequences of suitable moves of the original (more basic) chain and one can therefore try to compare the two chains by means of paths methods. In conclusion, a way to bound the mixing time for the original "single site dynamics" is:

(i)   first construct block dynamics on the same state space X, reversible w.r.t. the same measure π, and whose mixing time can be bounded using coupling techniques;

(ii)   compare the spectral gap and logarithmic Sobolev constant of the original chain to those of the block dynamics via path methods.

We refer to [135] for a nice review of this approach and its concrete applications to lozenge tilings and random triangulations of a convex polygon.

In many applications in statistical mechanics, particularly in the analysis of Glauber dynamics in the one-phase region, the above strategy provides bounds that are worse than the bounds obtained via more analytic techniques. The technique has, however, proved invaluable to analyze in a very detailed way the behavior of Glauber dynamics in the phase coexistence region (see, e.g. [121]).

3 Statistical mechanics models in $\mathbb{Z}^d$

In this section we first recall some basic notions and results for classical lattice spin systems and then we discuss some concrete examples for which we will later construct a reversible Markov process whose relaxation time will be studied by means of some of the techniques discussed in the previous section. By no means can our necessarily short presentation be considered exhaustive and we refer to classical references on this subject for more details (see e.g [69], [136], [144], [160]).

3.1 Notation

(i) *The lattice.* We consider the d-dimensional lattice $\mathbb{Z}^d$ with sites $x = \{x_1, \ldots, x_d\}$ and norm $\|x\|^2 = \sum_i |x_i|^2$. The associated distance function is denoted by $d(\cdot, \cdot)$. By Q_l we denote the cube of all $x = (x_1, \ldots, x_d) \in \mathbb{Z}^d$ such that $x_i \in \{0, \ldots, l-1\}$. If $x \in \mathbb{Z}^d$, $Q_l(x)$ stands for $Q_l + x$. We also let B_l be the ball of radius l centered at the origin, i.e. $B_l = Q_{2l+1}((-l, \ldots, -l))$.

If Λ is a finite subset of $\mathbb{Z}^d$ we write $\Lambda \subset\subset \mathbb{Z}^d$. The cardinality of Λ is denoted by $|\Lambda|$. $\mathbb{F}$ is the set of all nonempty finite subsets of $\mathbb{Z}^d$.

We finally define the exterior n-boundary as $\partial_n^+ \Lambda = \{x \in \Lambda^c : d(x, \Lambda) \leq n\}$. Given $r \in \mathbb{Z}_+$, we say that a subset V of $\mathbb{Z}^d$ is r-connected if for any two sites, y in V there exists $\{x^1, \ldots, x^n\} \subset V$ such that $x^1 = y$, $x^n = z$ and $|x^{i+1} - x^i| \leq r$ for $i = 2, \ldots, n$.

(ii) *Regular sets.* A finite subset Λ of $\mathbb{Z}^d$ is said to be l-regular, $l \in \mathbb{Z}_+$, if there exists $x \in \mathbb{Z}^d$ such that Λ is the union of a finite number of cubes $Q_l(x^i + x)$ where $x^i \in l\mathbb{Z}^d$. We denote the class of all such sets by $\mathbb{F}_l$. Notice that any set is 1-regular i.e. $\mathbb{F}_{l=1} = \mathbb{F}$.

(iii) *The configuration space.* Our *configuration space* is $\Omega = S^{\mathbb{Z}^d}$, where S is a finite set, typically $S = \{-1, +1\}$, or $\Omega_V = S^V$ for some $V \subset \mathbb{Z}^d$. The single spin space S is endowed with its natural topology and Ω with the corresponding product topology. Given $\sigma \in \Omega$ and $\Lambda \subset \mathbb{Z}^d$ we denote by σ_Λ the natural projection over Ω_Λ. If U, V are disjoint, $\sigma_U \eta_V$ is the configuration on $U \cup V$ which is equal to σ on U and η on V.

(iv) *Local functions.* If f is a measurable function on Ω, the support of f, denoted by Λ_f, is the smallest subset of $\mathbb{Z}^d$ such that $f(\sigma)$ depends only on σ_{Λ_f}. f is called *local* if Λ_f is finite. $\mathcal{F}_\Lambda$ stands for the σ-algebra generated by the set of projections $\{\pi_x\}$, $x \in \Lambda$, from Ω to S, where $\pi_x : \sigma \mapsto \sigma(x | \mathcal{F}_i)$. When $\Lambda = \mathbb{Z}^d$ we set $\mathcal{F} \equiv \mathcal{F}_{\mathbb{Z}^d}$ and $\mathcal{F}$ coincides with the Borel σ-algebra on Ω with respect to the topology introduced above. By $\|f\|_\infty$ we mean the supremum norm of f.

Remark 3.1. We have deliberately avoided here the more natural and general choice of the single spin space S as a compact metric space like $S = S^1$ because most of the results on the speed of relaxation for models of statistical mechanics have been established for discrete spin models. The important case of $S = \mathbb{R}$ will be discussed separately later on.

3.2 Grand canonical Gibbs measures

The translation group $\mathbb{Z}^d$ acts on the infinite volume configuration space Ω by

$$(T_x\sigma)(y) := \sigma(y - x) \qquad \text{for all } x \in \mathbb{Z}^d$$

Definition 3.2. *A finite-range, translation-invariant potential $\{\Phi_\Lambda\}_{\Lambda \in \mathbb{F}}$ is a collection of real, local continuous functions on Ω with the following properties:*

(1) $\Phi_\Lambda(\sigma) = \Phi_{\Lambda+x}(T_x\sigma)$ for all $\Lambda \in \mathbb{F}$ and all $x \in \mathbb{Z}^d$.
(2) For each Λ the support of Φ_Λ coincides with Λ.
(3) There exists $r > 0$, called the range of the interaction, such that $\Phi_\Lambda = 0$ if $\operatorname{diam}\Lambda > r$.
(4) $\|\Phi\| \equiv \sum_{\Lambda \ni 0} \|\Phi_\Lambda\|_\infty < \infty$.

Remark 3.3. Notice that we do not allow here interactions that are infinite for certain configurations (hard-core interactions). Nevertheless we will discuss later on a particular model known as "the hard-core model" for which certain configurations have infinite energy and become therefore forbidden. Systems with random interactions, like spin glasses or diluted magnetic models, are also not covered and will be discussed separately.

Definition 3.4. *Given a potential or interaction Φ with the above four properties and $V \in \mathbb{F}$, we define the Hamiltonian $H^\Phi_{V,free}$ by*

$$H^\Phi_{V,free} = \sum_{\Lambda:\, \Lambda \subset V} \Phi_\Lambda$$

Free boundary conditions are, however, not sufficient; for many purposes one needs Hamiltonians in which the interior of V is allowed to interact with the exterior of V. To this end we define

$$W^\Phi_{V,V^c} = \sum_{\substack{\Lambda:\, \Lambda \cap V \neq \emptyset \\ \Lambda \cap V^c \neq \emptyset}} \Phi_\Lambda$$

Note that, because of the finite-range condition on the interaction, the above sum is always finite.

196   F. Martinelli

Definition 3.5. *Given a potential or interaction Φ with the above four properties and $V \in \mathbb{F}$, we define the Hamiltonian H_V^Φ with general boundary condition by*

$$H_V^\Phi = H_{V,free}^\Phi + W_{V,V^c}^\Phi$$

It is convenient to think of the configuration outside V as fixed and the configuration inside V as variable. With this in mind, for $\sigma, \tau \in \Omega$ we let $H_V^{\Phi,\tau}(\sigma) = H_V^\Phi(\sigma_V \tau_{V^c})$ and we call τ the *boundary condition*.

Definition 3.6. *For each $V \in \mathbb{F}$, $\tau \in \Omega$ the (finite volume) conditional (grand canonical) Gibbs measures on $(\Omega, \mathcal{F})$, are given by*

$$d\mu_V^{\Phi,\tau}(\sigma) = \begin{cases} \left(Z_V^{\Phi,\tau}\right)^{-1} \exp[\,-H_V^{\Phi,\tau}(\sigma)\,] \prod_{x \in \Lambda} d\mu_0\big(\sigma(x)\big) & \text{if } \sigma(x) = \tau(x) \\ & \text{for all } x \in V^c \\ 0 & \text{otherwise} \end{cases}$$

where $Z_V^{\Phi,\tau}$ is the proper normalization factor a called partition function and $\mu_0(\cdot)$ is some a priori probability measure on S.

Notice that in (3.6) we have absorbed in the interaction Φ the usual inverse temperature factor β in front of the Hamiltonian. In most notation we will drop the superscript Φ if that does not generate confusion.

Given a measurable bounded function f on Ω, $\mu_V(f)$ denotes the *function* $\sigma \mapsto \mu_V^\sigma(f)$ where $\mu_V^\sigma(f)$ is just the average of f w.r.t. μ_V^σ. Analogously, if $X \in \mathcal{F}$, $\mu_V^\tau(X) \equiv \mu_V^\tau(\mathbb{1}_X)$, where $\mathbb{1}_X$ is the characteristic function on X. $\mu_V^\tau(f,g)$ stands for the covariance or *truncated correlation* (with respect to μ_V^τ) of f and g. The set of measures (3.6) satisfies the DLR compatibility conditions

$$\mu_\Lambda^\tau(\mu_V(X)) = \mu_\Lambda^\tau(X) \qquad \forall X \in \mathcal{F} \qquad \forall V \subset \Lambda \subset\subset \mathbb{Z}^d \qquad (3.1)$$

Definition 3.7. *A probability measure μ on $(\Omega, \mathcal{F})$ is called a Gibbs measure for Φ if*

$$\mu(\mu_V(X)) = \mu(X) \qquad \forall X \in \mathcal{F} \qquad \forall V \in \mathbb{F}$$

Remark 3.8. In the above definition we could have replaced the σ-algebra $\mathcal{F}$ with $\mathcal{F}_V$ (see section 2.3.2 in [160]).

The set of all Gibbs measures relative to a fixed given potential Φ will be denoted by $\mathcal{G}$. It can be proved that $\mathcal{G}$ is a nonempty, convex compact set. We will say that the discrete spin system described by the potential Φ has multiple phases if $\mathcal{G}$ contains more than one element. The reader is referred to [69] and [160] for a much more advanced discussion of Gibbs measures.

3.3 Mixing conditions and absence of long-range order

As a next step we define two similar, but at the same time deeply different, notions of *weak dependence of the boundary conditions* for finite volume Gibbs measures (see [116]). These notions will be refered to in the sequel as *weak* and *strong mixing* (not to be confused with the classical notion of strong-mixing for random fields), respectively. They both imply that there exists a unique infinite-volume Gibbs state with exponentially decaying covariances. Actually the validity of our strong mixing condition on, e.g. all squares implies much more, namely analyticity properties of the Gibbs measure, the existence of a convergent cluster expansion (see [129] and [130]) and good behavior under the renormalization–group transformation known as the "decimation transformation" (see [118] and [115]). Moreover, and this is our main motivation, both notions play a key role in the discussion of the relaxation time of a Glauber dynamics for discrete lattice spin systems.

Roughly speaking, the *weak mixing* condition implies that if in a finite volume V we consider the Gibbs state with boundary condition τ, then a local (e.g. in a single site $y \in V^c$) modification of the boundary condition τ has an influence on the corresponding Gibbs measure which decays exponentially fast inside V with the *distance from the boundary* $\partial^+ V$.

The *strong mixing* condition, instead, implies, in the same setting as above, that the influence of the perturbation decays in V exponentially fast with the distance from the *support of the perturbation* (e.g. the site y).

This distinction is very important since, even if we are in the one-phase region with a unique infinite-volume Gibbs state with exponentially decaying covariances, it may happen that, if we consider the same Gibbs state in a finite volume V, a local perturbation of the boundary condition radically modifies the Gibbs measure on the *whole region* close to the boundary while leaving it essentially unchanged in the bulk and this "long-range order effect" at the boundary persists even when V becomes arbitrarily large. We will refer to this phenomenon as a "boundary phase transition". It is clear that if a boundary phase transition takes place, then our Gibbs measure may satisfy a weak mixing condition but not a strong one.

A boundary phase transition is apparently not such an exotic phenomenon since it is expected to take place in the three-dimensional ferromagnetic Ising model at low temperatures and small enough magnetic field (depending on the temperature) [42]. On the contrary, for finite-range two-dimensional systems and for regular volumes (e.g. squares) we do not expect any boundary phase transition since the boundary is one-dimensional and, unless the interaction is itself long range, no long-range order is possible. Thus in two dimensions weak mixing should be equivalent to strong mixing. That is precisely the content of theorem 3.12 below.

We conclude by pointing out that it may happen, also for very natural model like the Ising model at low temperature and positive external field, that strong mixing holds for "regular" volumes, like all multiples of a given large enough cube, but fails for other sets (see [116]). This fact led to a revision of the theory of "completely analytical Gibbsian random fields" (see [56], [57]) and it plays an important role in the discussion of pathologies of renormalization group transformations in statistical mechanics (see [160]).

Let us now define our two conditions. Given $\Delta \subset V \subset\subset \mathbb{Z}^d$ and a Gibbs measure μ_V^τ on Ω_V, we denote by $\mu_{V,\Delta}^\tau$ the projection of the measure μ_V^τ on Ω_Δ, i.e.

$$\mu_{V,\Delta}^\tau(\sigma) = \sum_{\eta;\,\eta_\Delta=\sigma_\Delta} \mu_V^\tau(\eta)$$

We are now in a position to define strong mixing and weak mixing.

Definition 3.9. *We say that the Gibbs measures μ_V satisfy the weak mixing condition in V with constants C and m if for every subset $\Delta \subset V$*

$$\sup_{\tau,\tau'} \|\mu_{V,\Delta}^\tau - \mu_{V,\Delta}^{\tau'}\| \leq C \sum_{x\in\Delta,\,y\in\partial_r^+ V} e^{-md(x,y)} \tag{3.2}$$

We denote this condition by $WM(V,C,m)$.

Definition 3.10. *We say that the Gibbs measures μ_V satisfy the strong mixing condition in V with constants C and m if for every subset $\Delta \subset V$ and every site $y \in V^c$*

$$\sup_{\tau} \|\mu_{V,\Delta}^\tau - \mu_{V,\Delta}^{\tau^y}\| \leq Ce^{-md(\Delta,y)} \tag{3.3}$$

We denote this condition by $SM(V,C,m)$.

Remark 3.11. It is clear that either one of the above properties becomes interesting when it holds with the *same* constants C and m for an infinite class of finite subsets of $\mathbb{Z}^d$, e.g. all cubes. It is also worth mentioning that in $d=1$ for any translation invariant, finite-range interaction both conditions are satisfied for some choice of the constants C,m.

It is a relatively easy task to show that strong mixing is more stringent than weak mixing in the sense that, for example, strong mixing for all cubes implies weak mixing for all cubes. The converse of the above result, namely weak mixing implies strong mixing, is in general expected to be false in dimensions greater than two. In two dimensions we have instead the following (see [119]):

Theorem 3.12. *In two dimensions, $WM(V,C,m)$ for every $V \subset\subset \mathbb{Z}^d$ implies $SM(Q_L,C',m')$ for every square Q_L, for suitable constants C' and m'.*

Remark 3.13. It is very important to notice that it is known, by means of explicit examples, that the above result becomes false if we try to replace in the above theorem *for all squares* with *for all finite subsets of $\mathbb{Z}^2$* (see [116]). We refer the reader to [7] and [8] for further results on weak mixing in two dimensions.

It is not difficult to realize that fast decay of the influence of boundary conditions is intimately related to rapid decay of covariances between faraway observables. To make this connection precise we first need two more definitions. Let

$$h_x^s(\sigma) = \exp\left(-\sum_{\Lambda \ni x}[\Phi_\Lambda(\sigma^{x,s}) - \Phi_\Lambda(\sigma)]\right) \tag{3.4}$$

where $\sigma^{x,s}$ denotes the configuration obtained from σ by replacing the spin $\sigma(x)$ with an admissible value s.

Definition 3.14. *Given $V \in \mathcal{F}$, $\ell, \alpha > 0$, we say that condition $SMT(V, \ell, \alpha)$ holds if for all local functions f and g on Ω such that $d(\Lambda_f, \Lambda_g) \geq \ell$ we have*

$$\sup_{\tau \in \Omega} |\mu_V^\tau(f, g)| \leq |\Lambda_f| |\Lambda_g| \|f\|_\infty \|g\|_\infty \exp[-\alpha\, d(\Lambda_f, \Lambda_g)]$$

Then we have [121]:

Theorem 3.15. *The following are equivalent.*

(i)   *There exist C, m and L_0 such that $SM(\Lambda, C, m)$ holds for all Λ multiples of Q_{L_0}.*

(ii) *There exist ℓ, m and L_0 such that $SMT(\Lambda, \ell, m)$ holds for all Λ multiples of Q_{L_0}.*

(iii) *There exist ℓ, m and L_0 such that*

$$\sup_{\tau \in \Omega}\sup_{s,s' \in S} |\mu_\Lambda^\tau(h_x^s, h_y^{s'})| \leq \exp(-m|x - y|)$$

holds for all Λ multiples of Q_{L_0}.

Remark 3.16. It is not difficult to check that any of the above three conditions implies the slightly more precise and better looking bound

$$|\mu_V^\tau(f, g)| \leq \mathrm{Var}_{\mu_V^\tau}(f)^{1/2}\, \mathrm{Var}_{\mu_V^\tau}(g)^{1/2} \sum_{x \in \partial_r^+ \Lambda_f} \sum_{y \in \partial_r^+ \Lambda_g} e^{-m|x-y|}, \quad \forall \tau \in \Omega$$

provided that V is a multiple of Q_{L_0} and $d(\Lambda_f, \Lambda_g) \geq \ell$.

Remark 3.17. The alert reader may wonder how, in concrete cases, one can compare finite-volume Gibbs measures and prove either one of the above forms of weak dependence on the boundary conditions. Of key importance are, in this respect, *finite-volume* conditions, i.e. conditions that, if satisfied for all the Gibbs measures in volumes with "size" not larger than some fixed constant, imply either strong mixing or weak mixing for some *infinite* collection of regions Λ_i whose union is the whole lattice $\mathbb{Z}^d$. Because of the discrete nature of the single spin space these conditions can be, at least in principle, be proved or disproved in a finite number of steps on a computer.

The first and most famous finite-volume condition is certainly the Dobrushin uniqueness condition [54] which can be formulated as follows. Define the Dobrushin matrix C_{xy} by

$$C_{xy} := \sup_{\substack{\tau,\tau'\,:\,\tau(z)=\tau'(z) \\ \forall z \neq y}} \|\mu^\tau_{\{x\}} - \mu^{\tau'}_{\{x\}}\|$$

Then, if $\sup_x \sum_y C_{xy} < 1$ strong mixing $SM(\Lambda,C,m)$ holds, for some C,m and *any* $\Lambda \subset \mathcal{F}$. Notice that in the above sum only sites y with $d(x,y) \leq r$, r being the range of the interaction, contribute. Unfortunately, as soon as we change the interaction Φ (e.g. by lowering the temperature) and we come close to a phase transition, there is no hope of satisfying the Dobrushin uniqueness condition because the characteristic length over which the influence of boundary conditions becomes negligible may get very large. Dobrushin and Shlosman in two famous papers [55],[57] generalized the Dobrushin uniqueness condition to blocks larger than just a single site but still finite in order to overcome the above problem and to be able to get, in principle, arbitrarily close to the phase transition point. We refer the reader to [116],[129],[130] and [172] for more detailed reviews of finite-size conditions, to [159] for an approach to the uniqueness problem via disagreement percolation and to [166] for an interesting "Markov chain" approach to the Dobrushin and Dobrushin–Shlosman uniqueness conditions and for their generalization to graphs other than $\mathbb{Z}^d$. We conclude with one form of finite-size condition whose proof is nothing but an appropriate use of the Markov property for Gibbs measures (see [114]V).

For any integer l denote by $\mathcal{F}^0_l$ the collection of those cubes $Q_l(x^i)$, with $x^i \in l\mathbb{Z}^d$ and $|x^i| \leq l$. Set

$$\varphi(l) =: \sup_{x \in \partial^+_r Q_l} \sup_{\tau \in \Omega} \sup_{s,t \in S} \sup_{\substack{V \in \mathbb{F}^0_l\,: \\ x \in \partial^+_r V \\ Q_l \subset V}} \sum_{z \in \partial^+_r B_{2l-1}} |\mu^\tau_V(h^s_x, h^t_z)|$$

$$\psi(l) =: \sup_{x \in \partial^+_r Q_l} \sup_{\tau \in \Omega} \sup_{s,t \in S} \sup_{\substack{V \subset B_{2l-1} \\ \partial^+_r V \ni x}} \sum_{z \in \partial^+_r B_{2l-1}} |\mu^\tau_V(h^s_x, h^t_z)|$$

Notice that in the definition of $\varphi(l)$ and $\psi(l)$ appear only covariances w.r.t. Gibbs measures on subsets of B_{2l-1}. However, and this is also the main difference between the two quantities, in the definition of $\varphi(l)$ covariances are computed in l-regular volumes while in the definition of $\psi(l)$ subsets of B_{2l-1} of arbitrary shape appear. This fact implies, in particular, that $\psi(l) \geq \varphi(l)$. With this notation the following hold.

Theorem 3.18. *There exists $\delta > 0$ such that*

(1) if $\ell^{d-1}\varphi(\ell) \leq \delta e^{\|\Phi\|}$ then strong mixing $SM(\Lambda, C, m)$ holds for some C, m and all regular sets in $\mathbb{F}_\ell$.

(2) if $\ell^{d-1}\psi(\ell) \leq \delta e^{\|\Phi\|}$ then strong mixing $SM(\Lambda, C, m)$ holds for some C, m and all sets in $\mathbb{F}$.

3.4 Canonical Gibbs measures for lattice gases

Here we consider the so-called lattice gas models, $S = \{0, 1\}$, with the convention that $\sigma(x) = 1$ means that at site x there is a particle, while $\sigma(x) = 0$ means that the site x is empty. Given $\Lambda \in \mathcal{F}$, let $N_\Lambda(\sigma) := \sum_x \sigma(x)$ denote the number of particles in Λ for the configuration σ. Then the *canonical Gibbs measure* with $N \in \{0, 1, \ldots, |\Lambda|\}$ particles and boundary condition τ associated with an interaction Φ is defined as

$$\nu_{\Lambda,N}^{\Phi,\tau} = \mu_\Lambda^{\Phi,\tau}(\cdot \mid N_\Lambda = N) \tag{3.5}$$

In what follows we will suppress the superscript Φ in our notation. Given $\lambda \in \mathbb{R}$ we will denote by $\mu_\Lambda^{\lambda,\tau}$ the grand canonical Gibbs measure in Λ with boundary condition τ corresponding to the interaction Φ^λ defined by

$$\Phi_V^\lambda = \begin{cases} \Phi_V & \text{if } |V| \geq 2 \\ \Phi_{\{x\}} + \lambda\sigma(x) & \text{if } V = \{x\} \end{cases}$$

The parameter λ is usually called the chemical potential. Notice that $\nu_{\Lambda,N}^{\Phi,\tau} = \nu_{\Lambda,N}^{\Phi^\lambda,\tau}$ for all λ. Then we have the following sharp equivalence between the canonical measure $\nu_{\Lambda,N}^\tau$ and its grand canonical counterpart $\mu_\Lambda^{\lambda,\tau}$ [29].

Theorem 3.19. *Assume condition $SM(\Lambda, C, m)$ for all Λ multiples of Q_{L_0}. Then, for any ε small enough there exists a constant C' such that for all Λ multiples of Q_{L_0}, for all local function f with $|\Delta_f| \leq |\Lambda|^{1-\varepsilon}$ and for all $N \in \{0, 1, \ldots, |\Lambda|\}$*

$$|\nu_{\Lambda,N}^\tau(f) - \mu_\Lambda^{\tau,\lambda}(f)| \leq C'\|f\|_\infty \frac{|\Delta_f|}{|\Lambda|}$$

provided that $\lambda := \lambda(\Lambda, \tau, N)$ is such that

$$\mu_\Lambda^{\lambda,\tau}(N_\Lambda) = N$$

3.5 The ferromagnetic Ising and Potts models

In the standard ferromagnetic Ising model $S = \{-1, +1\}$ and the interaction Φ is given by

$$\Phi_\Lambda(\sigma) = \begin{cases} -\beta\sigma(x)\sigma(y) & \text{if } \Lambda = \{x, y\} \text{ with } \|x - y\| = 1 \\ -\beta h\sigma(x) & \text{if } \Lambda = \{x\} \\ 0 & \text{otherwise} \end{cases} \tag{3.6}$$

where $\beta \geq 0$ and $h \in \mathbb{R}$ are two thermodynamic parameters representing the inverse temperature and the external magnetic field respectively. The Gibbs measure associated with the spin system with boundary conditions τ is denoted for convenience by $\mu_\Lambda^{\beta,h,\tau}$. If the boundary conditions are uniformly equal to $+1$ (resp. -1), the Gibbs measure will be denoted by $\mu_\Lambda^{\beta,h,+}$ (resp. $\mu_\Lambda^{\beta,h,-}$). The phenomenon of multiplicity of phases occurs for $d \geq 2$ at low temperature, i.e. large β, and zero external field h and it is characterized by the appearance of a spontaneous magnetization in the thermodynamic limit. There is a critical value β_c such that

$$\forall \beta > \beta_c, \qquad \lim_{\Lambda \to \mathbb{Z}^d} \mu_\Lambda^{\beta,0,+}(\sigma(0)) = - \lim_{\Lambda \to \mathbb{Z}^d} \mu_\Lambda^{\beta,0,-}(\sigma(0)) = m^*(\beta) > 0 \tag{3.7}$$

Furthermore, in the thermodynamic limit the measures $\mu_\Lambda^{\beta,0,+}$ and $\mu_\Lambda^{\beta,0,+}$ converge (weakly) to two distinct extremal Gibbs measures $\mu^{\beta,+}$ and $\mu^{\beta,-}$ which are measures on the space $\{\pm 1\}^{\mathbb{Z}^d}$. We refer the reader to, e.g. [105] or [144].

The ferromagnetic q-state Potts model [133], $q \in \mathbb{N}$, is characterized by $S = \{1, 2, \ldots, q\}$ and

$$\Phi_\Lambda(\sigma) = \begin{cases} -\beta\delta_{\sigma(x),\sigma(y)} & \text{if } \Lambda = \{x, y\} \text{ with } \|x - y\| = 1 \\ 0 & \text{otherwise} \end{cases} \tag{3.8}$$

where $\delta_{u,v}$ is the Kronecker delta. The case $q = 2$ coincides with the Ising model without external field and an amended value of β since $\sigma(x)\sigma(y) = 2\delta_{\sigma(x),\sigma(y)} - 1$ for $\sigma(x), \sigma(y) \in \{-1, +1\}$. Ferromagnetic Potts models have been extensively studied over the last two decades and much is known about their phase diagrams mostly via a graphical representation (see below). In particular it has been proved [99] that there exists an order/disorder phase transition marking the coexistence of q low-energy ordered states and a high-entropy disordered state.

3.6 FK representation of Potts models

It was realized long ago [65] that ferromagnetic Potts models can be formulated as "random-cluster models". Such a representation provides a unified way of studying percolation and ferromagnetic models and proved to be a key step in obtaining fundamental new results in this field (see, e.g. [72] and references therein). Moreover, it was instrumental in the construction of the Swendsen–Wang algorithm [165] for generating random configurations out of the Potts–Gibbs measure.

Perhaps the neatest way to define the random-cluster model is that described in [61]. Let $G = (V, E)$ be a finite graph and define for q a positive integer

$$\Sigma_V := \{1, 2, \ldots, q\}^V$$
$$\Omega_E := \{0, 1\}^E$$

Next we define a probability measure on $\Omega_G := \Sigma_V \times \Omega_E$ by

$$\mu(\sigma, \omega) \propto \prod_{e \in E} \left\{ (1 - p)\delta_{\omega(e),0} + p\delta_{\omega(e),1}\delta_e(\sigma) \right\}$$

where $0 \le p \le 1$ and $\delta_e(\sigma) := \delta_{\sigma(x),\sigma(y)}$ if $e = (x, y) \in E$. It is not difficult to check that:

(i) The *marginal on* Σ_V is given by

$$\mu(\sigma) \propto \exp\left(\beta \sum_{e \in E} \delta_e(\sigma) \right)$$

provided that $p = 1 - e^{-\beta}$. In other words the marginal on Σ_V coincides with the Gibbs measure of the ferromagnetic Potts model with *free* boundary condition on G

(ii) The *marginal on* Ω_E, known as the *random cluster measure* $\mu_{p,q}$, is given by

$$\mu_{p,q}(\omega) \propto \left\{ \prod_{e \in E} p^{\omega(e)}(1 - p)^{1 - \omega(e)} \right\} q^{k(\omega)}$$

where $k(\omega)$ is the number of connected components (clusters) of the new graph with vertex set V and edge set $\{e \in E : \omega(e) = 1\}$.

(iii) The *conditional measure* on Σ_V given the $\omega(e)$'s is obtained by assigning (uniformly) a common value $j \in \{1, \ldots, q\}$ to all spins in a given cluster, independently for each cluster. On the other hand, given σ, the conditional measure on Ω_E is obtained by setting $\omega(e) = 0$ if $\delta_{\sigma(e)} = 0$ and $\omega(e) = 1$ with probability p if $\delta_{\sigma(e)} = 1$, independently for each edge e.

Remark 3.20. Notice that the random cluster measure makes sense for any positive real q and not just $q \in \mathbb{N}$. In particular for $q = 1$ it coincides with the usual independent bond percolation on G with parameter p. The main interest of the random cluster representation is that it brings percolation and stochastic geometry type of questions into the study of phase transitions in a very natural way.

We conclude this part by briefly discussing boundary condition in the framework of the random cluster measure. We do this when the graph G is a finite box Λ of the cubic lattice $\mathbb{Z}^d$ with its natural graph structure. The set of edges of $\mathbb{Z}^d$ is denoted by E, while the edges of G are denoted by E_Λ. Given $\{\omega(e)\}_{e \in E_\Lambda}$ and $e \in E$, we set $\eta(\omega)(e) = \omega(e)$ if $e \in E_\Lambda$ and $\eta(\omega)(e) = 1$ otherwise. If $k(\omega, \Lambda)$ denotes the number of clusters (connected components) for the *infinite* edge configuration $\eta(\omega)$ we define the random cluster measure on E_Λ with *wired boundary conditions* by

$$\mu_{p,q}^w(\omega) \propto \left\{ \prod_{e \in E_\Lambda} p^{\omega(e)} (1 - p)^{1 - \omega(e)} \right\} q^{k(\omega, \Lambda)}$$

Remark 3.21. It is possible to introduce inside the FK representation external fields like the magnetic field for the Ising model and more general boundary conditions than just the free or wired b.c. Moreover the FK representation can be extended to *nonferromagnetic interactions* like, e.g. spin glasses [128].

3.7 Antiferromagnetic models on an arbitrary graph: Potts and hard-core models

Unlike ferromagnetic models like the Ising or Potts models for which nearest neighbors spins prefer to be in the same state and therefore, roughly speaking, only one kind of ordered phase is possible, in antiferromagnetic systems nearest neighbors spins prefer to be in a different state and the structure of the phase diagram is much more subtle and may depend in a very delicate way on the value of some parameters, e.g. the number of states q in the antiferromagnetic Potts model) and on graph structure of the underlying lattice (e.g. its degree or whether the graph is bipartite) on which they are defined. For this reason there has been a great deal of research on antiferromagnetic models on an arbitrary (finite) graph $G = (V, E)$ beyond the more standard cubic lattice $\mathbb{Z}^d$. Two models are of great interest for both people working in statistical physics and in theoretical computer science: the antiferromagnetic Potts model (related to the q-coloring of the vertices of the graph) and the hard-core model ("independent-sets model" in the graph-theory language).

We begin by defining the *"soft-core" gas model*. Given a general finite graph $G = (V, E)$, consider the lattice gas model ($S = \{0, 1\}$) with interaction Φ

$$\Phi_\Lambda(\sigma) = \begin{cases} \beta \sigma(x) \sigma(y) & \text{if } \Lambda = \{x, y\} \in E \\ -\log(\lambda) \sigma(x) & \text{if } x \in V \\ 0 & \text{otherwise} \end{cases}$$

where $\beta > 0$ is the inverse temperature and $\lambda > 0$ is called the *fugacity*. It is not difficult to prove that the associated grand canonical Gibbs measure with free boundary condition $\mu_V^{\Phi, free}$ converges, as $\beta \to \infty$, to the so-called "hard-core model" Gibbs measure on G given by

$$\mu(\sigma) := \frac{1}{Z(\lambda)} \lambda^{|\sigma|}; \qquad Z(\lambda) = \sum_{\substack{\sigma \subset V \\ \sigma \text{ independent}}} \lambda^{|I|} \tag{3.9}$$

for every independent (i.e. containing no adjacent vertices) set $\sigma \subset V$. When G is countably infinite the hard-core Gibbs measure can be defined by the standard Dobrushin–Lanford–Ruelle prescription [69]. If Δ denotes the maximum degree of the graph G it has been proved that for $0 < \lambda < \frac{1}{\Delta - 1}$ the Dobrushin uniqueness theorem applies and the unique infinite-volume Gibbs measure has covariances that decay exponentially fast. The bound was then improved to $\lambda < \frac{2}{\Delta - 2}$ for general nonamenable graphs in [59] (see also [164] and [166]). Finally, in [149](conjecture 3.10) the following conjecture was put forward:

Conjecture. *For any countable infinite graph G of maximum degree Δ, the hard-core lattice gas on G has a unique Gibbs measure whenever*

$$0 < \lambda < (1 - \varepsilon) \frac{(\Delta - 1)^{\Delta - 1}}{(\Delta - 2)^\Delta}$$

for some $\varepsilon > 0$.

Remark 3.22. The main motivation behind the conjecture is the fact that $\frac{(\Delta - 1)^{\Delta - 1}}{(\Delta - 2)^\Delta}$ is the critical value λ_c for the complete rooted tree with branching factor $\Delta - 1$ [149].

On the other hand, in [53] it was proved that the hard-core gas on $\mathbb{Z}^d$, $d \geq 2$, has multiple phases for large enough values of λ. The λ required in [53] grows with the dimension d, whereas one would expect a phase transition to occur sooner (i.e., at lower fugacity) as dimension increased, since the boundary is closer to the origin for a fixed volume, and many efforts in the statistical mechanics and in the discrete mathematics community were devoted to obtaining better bounds. Recently, a rather spectacular upper bound on the critical fugacity has been announced in [67].

Let $\lambda(d) = \sup\{\lambda : \text{there is only one Gibbs measure } \lambda\}$. Then:

Theorem 3.23. $\lambda(d) \leq O(d^{-1/4} \log^{3/4} d)$.

Let us turn to antiferromagnetic Potts models. Let $G = (V, E)$ be a general finite graph and q be a positive integer. The q-state *antiferromagnetic* Potts model on G at inverse temperature β and free boundary conditions is described by the Gibbs measure

$$\mu^\beta(\sigma) := \frac{1}{Z} e^{-\beta \sum_{(x,y) \in E} \delta_{\sigma(x), \sigma(y)}}, \qquad \sigma \in \{1, \ldots, q\}^V$$

A *proper q-coloring* of G is a map $\sigma : V \to \{1, 2, \ldots, q\}$ such that $\sigma(x) \neq \sigma(y)$ for any pair of adjacent vertices x, y. If μ denotes the *uniform* measure on the set of proper q-colorings (if nonempty) then it is clear that $\mu = \lim_{\beta \to \infty} \mu^\beta$ and the resulting model will be referred to as the *q-coloring model on G*.

When G is countably infinite, Gibbs measures for the q-coloring model can be defined via the Dobrushin–Lanford–Ruelle prescription. There are a number of equilibrium results proving that, for q large enough compared to the maximum degree Δ of G, the Gibbs measure for the q-coloring model is unique with exponentially decaying covariances. More precisely we have (see [98], [137]):

Theorem 3.24. *If $q > 2\Delta$ then the Dobrushin uniqueness theorem applies and there exists a unique infinite-volume Gibbs state with exponentially decaying covariances.*

The same result applies to the positive temperature antiferromagnetic Potts model [138]. At least for amenable graphs the bound $q > 2\Delta$ was improved to $q > \frac{11}{6}\Delta$ in [163].

3.8 Model with random interactions

Let us start by first describing random Ising-like systems with single spin space $S = \{-1, +1\}$ on the lattice $\mathbb{Z}^d$. We consider an abstract probability space $(\Theta, \mathcal{B}, \mathbb{P})$ and a set of real-valued random variables $J = \{J_A\}$ with $A \in \mathbb{F}$, with the properties:

(a) J_A and J_B are independent if $A \neq B$.
(b) J_A and J_{A+x} are identically distributed for all $A \in \mathbb{F}$ and all $x \in \mathbb{Z}^d$.
(c)   There exists $r > 0$ such that with $\mathbb{P}$-probability 1, $J_A = 0$ if $\operatorname{diam} A > r$.
(d)   Let $\|J\|_x \equiv \sum_{A \ni x} |J_A|$. Then $\|J\| = \sup\{\|J\|_x : x \in \mathbb{Z}^d\} < \infty$.

The simplest example is the so-called *dilute ferromagnetic Ising model*. In this case the random couplings $\{J_A\}$ are different from zero only if $A = \{x, y\}$ with $\|x - y\| = 1$ and in that case they take (independently for each bond $e = (x, y)$) only two values, $J_{xy} = 0$ and $J_{xy} = \beta > 0$ with probability $1 - p$ and p respectively. In a more pictorial form one starts from the standard Ising model and removes, independently for each bond $e = (x, y)$, the coupling J_{xy} with probability $1 - p$.

Since the $\{J_{xy}\}$ are uniformly bounded, at sufficiently high temperatures (i.e. sufficiently small values of β) Dobrushin's uniqueness theory applies and detailed information about the unique Gibbs measure are available using the concept of complete analyticity. This regime is usually referred to as the *paramagnetic* phase and, at least for the two-dimensional dilute Ising model, it is known to cover the whole interval $\beta < \beta_c$ where β_c is the critical value for the "pure" Ising system.

There is then a range of temperatures, below the paramagnetic phase, where, even if the Gibbs state is unique, certain characteristics of the paramagnetic phase like the analyticity of the free energy as a function of the external field disappear. This is the so-called *Griffiths regime* [71] (see also [66] for additional discussion on this and many other related topics). This "anomalous behavior" is caused by the presence of arbitrarily large clusters of bonds associated with "strong" couplings J_{xy}, which can produce a long-range order inside the cluster. Even above the percolation threshold, i.e. when one of such clusters is infinite with probability one, there may be a Griffiths phase for values of $\beta \in (\beta_c, \beta_c(p))$, where β_c is the critical value for the Ising model on $\mathbb{Z}^d$ and $\beta_c(p)$ the critical value of the dilute model above which there is the phenomenon of spontaneous magnetization (see [66]). What happens is that for almost all realizations of the disorder J and for all sites x there is a finite (random) length scale $\ell(J, x)$, such that correlations between $\sigma(x)$ and $\sigma(y)$ start decaying exponentially at distances greater than $\ell(J, x)$. We will see later that the presence of large clusters of strongly interacting spins has major effects on the relaxation time for, e.g. the Glauber dynamics.

Another popular choice for the random couplings $\{J_A\}_{A \in \mathcal{F}}$ is to take only nearest neighbor couplings that form a collection of i.i.d. random variables with the fair Bernoulli distribution on $\{-\beta, +\beta\}$, $\beta > 0$ (the short-range spin glass at inverse temperature β).

Finally, the class of models that continues to attract, since many years, the attention of both physicists and mathematicians are mean-field spin glass models. The setting is as follows. Let $\Omega_N := \{-1, +1\}^N$ and let $\Theta_N := \mathbb{R}^{\Omega_N}$. An environment is an element $\{\vartheta(\sigma)\}_{\sigma \in \Omega_N} \in \Theta_N$ to be interpreted as the (random) Hamiltonian of the system. With each environment ϑ we associated the Gibbs measure

$$\mu^\vartheta(\sigma) = \frac{e^{-\beta\vartheta(\sigma)}}{Z^{\vartheta,\beta}}$$

where $\beta \geq 0$ is the inverse temperature.

Two models in particular became very popular, the Random Energy Model (REM) [46] and the Sherrington–Kirkpatrick model (SK) [142].

The REM is specified by choosing $\vartheta(\sigma) := \sqrt{N}\hat{\vartheta}(\sigma)$, where $\{\hat{\vartheta}(\sigma)\}_{\sigma \in \Omega_N}$ form a collection of i.i.d. $N(0, 1)$ random variables.

In the *SK model*, instead $\vartheta(\sigma) := N^{-\frac{1}{2}} \sum_{i<j} J_{ij}\sigma(i)\sigma(j)$ where $\{J_{ij}\}_{1 \leq i < j \in \mathbb{N}}$ form a collection of i.i.d. $N(0, 1)$ random variables.

3.9 Unbounded spin systems

We conclude this short survey of models of statistical mechanics by mentioning one basic model of noncompact spins for which very interesting results were obtained on the relaxation time for an associated Markovian dynamics following the results for compact spins [174], [172], [171], [170], [19], [103].

The setting is as follows. Let $U : \mathbb{R} \to \mathbb{R}$ be such that $U(x) = v(x) + w(x)$ with $v''(x) > c > 0$ and $\|w\|_\infty + \|w'\|_\infty < \infty$. Let $J(x, y)$ be a smooth, symmetric function on $\mathbb{R}^2$, e.g. $J(x,y) = Jxy$ where $J \in \mathbb{R}$ or $J(x,y) = V(x-y)$ where V is an even function. Let also $J := \|\partial_{xx}J\|_\infty + \|\partial_{xy}J\|_\infty$.

Given $\Lambda \subset \mathbb{Z}^d$, let ϱ_Λ be the product measure of $d\varrho(\varphi) := \frac{1}{Z}e^{-U(\varphi)}$, $\varphi \in \mathbb{R}$, and define, for a given $\tau \in \mathbb{R}^{\mathbb{Z}^d}$

$$d\mu_\Lambda^\tau(\varphi) := \frac{1}{Z_\Lambda^\tau}e^{-H_\Lambda^\tau(\varphi)}d\varrho_\Lambda(\varphi) \tag{3.10}$$

where

$$H_\Lambda^\tau(\varphi) := \sum_{\substack{(x,y) \cap \Lambda \neq \emptyset \\ \|x-y\| = 1}} J(\varphi_x, \varphi_y), \qquad \varphi = \{\varphi_x\}_{x \in \mathbb{Z}^d}$$

and $\varphi_x = \tau_x$ if $x \notin \Lambda$.

Using the above notation it can be proved (see, e.g. [103] and references therein) that if

$$(c - 2dJ)e^{-4\|w\|_\infty} - 2d(1+e)J \geq \vartheta > 0$$

then the covariance w.r.t. μ_Λ^τ between $f(\varphi_x)$ and $g(\varphi_y)$ decays exponentially fast uniformly in Λ and τ. More precisely:

$$|\mu_\Lambda^\tau(f,g)| \leq \vartheta^{-1}e^{-\|x-y\|}\left(\mu_\Lambda^\tau\big((f')^2\big)\right)^{\frac{1}{2}}\left(\mu_\Lambda^\tau\big((g')^2\big)\right)^{\frac{1}{2}}$$

This result can be formulated also for more general functions and it is the exact analogue of the condition $SMT(\Lambda, l, m)$ for all $\Lambda \in \mathcal{F}$ discussed in the context of discrete spin models.

3.10 Ground states of certain quantum Heisenberg models as classical Gibbs measures

Some years ago it was discovered by Alcaraz [3] that a class of asymmetric reversible simple exclusion processes on $\mathbb{Z}^d$, related to models of diffusion-limited chemical reactions, are unitarily equivalent to certain anisotropic quantum Heisenberg Hamiltonians, known as XXZ models, that have received in recent years increasing attention in connection with the analysis of quantum domain walls (see [2], [127], [151] and references therein). Such an equivalence implies that the spectrum of (minus) the Markov generator of the process coincides with the spectrum of the quantum Hamiltonian. In particular the energy gap above the quantum ground state, a key quantity in the theory of quantum spin systems, becomes identical to the spectral gap of the process and a variety of probabilistic techniques come into play in order to obtain meaningful estimates. Such an observation was exploited recently in [35] and [34] to prove sharp bounds on the energy of low-lying excitations above the ground state. The setup is the following. Given $q \in (0,1)$ and two natural numbers L, H let

$$\Lambda = \{(i,h) \in \mathbb{Z}^2 : i = 1,\dots,L \text{ and } h = 1,\dots,H\} \tag{3.11}$$

and define the product probability measure μ^λ on $\Omega := \{0,1\}^\Lambda$ by

$$\mu^\lambda(\alpha) = \prod_{i=1}^{L} \prod_{h=1}^{H} \frac{q^{(2h-\lambda)\alpha_{(i,h)}}}{1+q^{2h}} \tag{3.12}$$

where $\lambda \in \mathbb{R}$ can be interpreted as a chemical potential. According to μ^λ, particles (sites with $\alpha = 1$) accumulate inside the region $h \leq \lambda$, i.e. the measure μ^λ describes a sharp profile around height λ if we interpret h as a vertical coordinate. Thus μ^λ can be looked upon as the Gibbs measure of a noninteracting lattice gas with linear chemical potential. We then define the associated canonical measure

$$\nu = \nu_N = \mu\left(\,\cdot\,\Big|\, \sum_{i=1}^{L} n_i = N\right) \tag{3.13}$$

As we will see later the measure ν is the reversible measure of a very natural simple exclusion process on Λ. In order to make a link with quantum Heisenberg models we introduce the horizontal sums of the basic variables $\alpha_{i,h}$ given by

$$\omega_h = \sum_{i=1}^{L} \alpha_{(i,h)}\,, \qquad h = 1,\ldots,H$$

and we denote by $\hat{\nu}$ the marginal of ν on $\omega = \{\omega_h\}$. The weight $\hat{\nu}(\omega)$ of a single $\omega \in \hat{\Omega}$ compatible with the global constraint $\sum_h \omega_h = N$ is easily computed to be

$$\hat{\nu}(\omega) = \frac{1}{Z} \prod_{h=1}^{H} \binom{L}{\omega_h} q^{2h\omega_h} \qquad Z = \sum_{\substack{\omega \in \hat{\Omega}: \\ \sum_h \omega_h = N}} \prod_{h=1}^{H} \binom{L}{\omega_h} q^{2h\omega_h} \tag{3.14}$$

The connection with quantum spin models goes as follows.

Given $S \in \frac{1}{2}\mathbb{N}$, $H \in \mathbb{N}$, consider the Hilbert space $\mathfrak{H} = \otimes_{h=1}^{H} \mathbb{C}^{2S+1}$. The spin-S XXZ chain on $[1,H] \cap \mathbb{Z}$ with *kink* boundary conditions is defined by the operator

$$\mathcal{H}^{(S)} = \sum_{h=1}^{H-1} \mathcal{H}^{(S)}_{h,h+1} \tag{3.15}$$

$$\mathcal{H}^{(S)}_{h,h+1} = S^2 - \Delta^{-1}\left(S_h^1 S_{h+1}^1 + S_h^2 S_{h+1}^2\right) - S_h^3 S_{h+1}^3 + S\sqrt{1-\Delta^{-2}}\left(S_{h+1}^3 - S_h^3\right)$$

Here S_h^i, $i = 1,2,3$, are the spin-S operators (the $2S+1$-dimensional irreducible representation of $SU(2)$) at every h, and the constant S^2 has been added in order to have zero ground state energy. The parameter $\Delta \in (1,\infty)$ measures the anisotropy along the third axis. The kink boundary condition is obtained through the telescopic sum $S_H^3 - S_1^3 = \sum_{h=1}^{H-1}\left(S_{h+1}^3 - S_h^3\right)$ and the pre-factor $S\sqrt{1 - \Delta^{-2}}$ is chosen in order to obtain nontrivial ground states describing quantum domain walls (see [2], [151] and references therein).

We choose the basis of $\mathfrak{H}$ labeled by the $2S+1$ states of the third component of the spin at each site and we write it in terms of configurations

$$m = (m_1, \ldots, m_H) \in \{-S, -S+1, \ldots, S-1, S\}^H =: \mathcal{Q}_S$$

so that $|m\rangle = \otimes_{h=1}^{H} |m_h\rangle$ stands for the generic basis vector in $\mathfrak{H}$. With these notations, and introducing the stair-operators $S^{\pm} = S^1 \pm iS^2$, the action of S^i, $i = 1, 2, 3$, is given by

$$S_h^3 |m_h\rangle = m_h |m_h\rangle\,, \quad S_h^{\pm} |m_h\rangle = c_{\pm}(S, m_h) |m_h \pm 1\rangle \qquad (3.16)$$

$$c_{\pm}(S, m_h) := \sqrt{(S \mp m_h)(S \pm m_h + 1)}$$

The action of $\mathcal{H}^{(S)}$ is explained by rewriting the pair-interaction terms as

$$\mathcal{H}^{(S)}_{h,h+1} = S^2 - (2\Delta)^{-1}\left(S_h^+ S_{h+1}^- + S_h^- S_{h+1}^+\right) - S_h^3 S_{h+1}^3 + S\sqrt{1-\Delta^{-2}}\left(S_{h+1}^3 - S_h^3\right) \tag{3.17}$$

The Hamiltonian $\mathcal{H}^{(S)}$ commutes with the total third component of the spin

$$S_{\text{tot}}^3 = \sum_{h=1}^{H} S_h^3$$

We shall divide the space $\mathfrak{H}$ into sectors $\mathfrak{H}_n$, $n \in \{-SH, -SH + 1, \ldots, SH - 1, SH\}$, given by the eigenspaces corresponding to the eigenvalue n of S_{tot}^3. It is known [2] that for each n there is a unique (up to multiplicative constants) vector $\psi_n \in \mathfrak{H}_n$ such that $\mathcal{H}^{(S)}\psi_n = 0$, which is given by

$$\psi_n = \sum_{\substack{m \in \mathcal{Q}_S: \\ \sum_h m_h = n}} \psi(m) |m\rangle$$

$$\psi(m) = \prod_h q^{hm_h} \sqrt{\binom{2S}{S + m_h}} \tag{3.18}$$

Here $q \in (0, 1)$ is the anisotropy parameter linked to Δ by the equation

$$\Delta = \frac{1}{2}(q + q^{-1}) \tag{3.19}$$

Setting $L = 2S$ and $N = SH + n$ we see that the measure $\hat{\nu}$ in (3.14) can be written using (3.18) with $m = \omega - S$:

$$\hat{\nu}(\omega) = \frac{1}{\tilde{Z}}\left[\psi(\omega - S)\right]^2 \qquad \tilde{Z} = \sum_{\substack{\omega \in \hat{\Omega}: \\ \sum_h \omega_h = SH+n}} \left[\psi(\omega - S)\right]^2 \tag{3.20}$$

In other words the square of the ground state wave function is nothing else but a canonical lattice gas Gibbs measure. We will see later that this identification allows us to transform the quantum Hamiltonian into the generator of a particular kind of simple exclusion process on Λ.

4 Glauber dynamics in $\mathbb{Z}^d$

In this section, given a finite space S and a finite-range, translation invariant interaction Φ on $\Omega := S^{\mathbb{Z}^d}$, we first define a special class of continuous-time Markov chains, reversible w.r.t. the Gibbs measure μ associated with Φ, known as *Glauber dynamics*. We then analyze the role of the mixing properties of the Gibbs measure in the estimate of the mixing times of such chains. Finally, we analyze several specific models. In what follows, for notation convenience, the interaction Φ will not appear in our notation whenever no confusion is possible.

4.1 The dynamics in a finite volume

The continuous-time Markov chain we want to study is determined by the Markov generators $\mathcal{L}_\Lambda^\tau$, $\Lambda \subset\subset \mathbb{Z}^d$ and $\tau \in \Omega$, defined by

$$(\mathcal{L}_\Lambda^\tau f)(\sigma) = \sum_{x \in \Lambda} \sum_{s \in S} c(x, s, \sigma)[f(\sigma^{x,s}) - f(\sigma)] \tag{4.1}$$

where $\sigma \in \Omega$ is such that $\sigma(y) = \tau(y)$ for any $y \in \Lambda^c$ and we recall that $\sigma^{x,s}$ denotes the configuration obtained from σ by replacing the spin $\sigma(x)$ with an admissible value s. The nonnegative real quantities $c(x, s, \sigma)$ are called the *jump rates* for the process and are assumed to satisfy the following general conditions.

(i) *Finite range.* If $\sigma(y) = \sigma'(y)$ for all y such that $d(x, y) \leq r$, then $c(x, s, \sigma) = c(x, s, \sigma')$.

(ii) *Detailed balance.* For all $\sigma \in \Omega$, $x \in \mathbb{Z}^d$ and $s \in S$

$$\exp\left[-H_{\{x\}}^\tau(\sigma)\right] c(x, s, \sigma) = \exp\left[-H_{\{x\}}^\tau(\sigma^{x,s})\right] c(x, \sigma(x), \sigma^{x,s}) \tag{4.2}$$

(iii) *Positivity and boundedness.* There exist positive real numbers c_m and c_M such that

$$0 < c_m \leq \inf_{x,s,\sigma} c(x, s, \sigma) \quad \text{and} \quad \sup_{x,s,\sigma} c(x, s, \sigma) \leq c_M \tag{4.3}$$

(iv) *Translation invariance.* If, for some $k \in \mathbb{Z}^d$, $\sigma'(y) = \sigma(y + k)$ for all $y \in \mathbb{Z}^d$ then $c(x, s, \sigma') = c(x + k, s, \sigma)$ for all $x \in \mathbb{Z}^d$ and all $s \in S$.

The chain defined by the generator $\mathcal{L}_\Lambda^\tau$ has a unique invariant measure, the (finite volume) Gibbs measure μ_Λ^τ which is moreover *reversible* for the process. Usually, the variance and entropy of a function f w.r.t. μ_Λ^t will be denoted by $\mathrm{Var}_\Lambda^\tau(f)$ and $\mathrm{Ent}_\Lambda^\tau(f)$ respectively. Moreover, according to our general notation, the chain will be denoted by $(\mathcal{L}_\Lambda^\tau, \mu_\Lambda^\tau)$ and its spectral gap and logarithmic Sobolev constant by $\mathrm{gap}(\mathcal{L}_\Lambda^\tau)$ and $c_s(\mathcal{L}_\Lambda^\tau)$ respectively.

Remark 4.1. Notice that, in general, the generator $\mathcal{L}_\Lambda^\tau$ will not be of the form $K - \mathbb{1}$ for some stochastic kernel K, since the quantity $q = \sup_\sigma \sum_{x,s} c(x, s, \sigma)$ will be of order $|\Lambda|$. In other words for each unit time interval each spin has a positive chance to change value. In the evaluation of the mixing times, particularly when comparing mixing times of our continuous-time Glauber dynamics to those of discrete time dynamics (defined in the obvious way), one should always remember that there is an overall conversion factor of the order of the cardinality of Λ.

A natural and popular choice of jump rates satisfying the above conditions goes under the name *heat bath dynamics* or *Gibbs sampler* and it is as follows:

$$c(x, s, \sigma) := \mu_{\{x\}}^\sigma(s)$$

The heat bath chain can be interpreted as follows. Each site $x \in \Lambda$ waits an exponential time of mean one and then the existing configuration σ is replaced by the new configuration $\sigma^{x,s}$ in which the new value of the spin at x is chosen according to the equilibrium measure at x given σ outside x. Notice that with probability one at each time t at most one spin changes its value.

A natural generalization of the heat bath chain is a process in which more than one spin can change value at the same time. For this purpose, let $\mathcal{D} = \{\Lambda_1, \ldots, \Lambda_n\}$ be an arbitrary collection of finite sets $\Lambda_i \in \mathbb{F}$ such that $\Lambda = \cup_i \Lambda_i$. Then we will denote by the term *block dynamics* with blocks $\{\Lambda_1, \ldots, \Lambda_n\}$ the continuous-time Markov chain on Ω_Λ in which each block waits an exponential time of mean one and then the configuration inside the block is replaced by a new configuration distributed according to the Gibbs measure of the block given the previous configuration outside the block. More precisely, the generator of the Markov process corresponding to $\mathcal{D}$ is defined as

$$\mathcal{L}_{\text{blocks}}^\tau f = \sum_{i=1}^n (\mu_{\Lambda_i}^\tau(f) - f) \tag{4.4}$$

From the DLR condition it easily follows that the block dynamics is reversible w.r.t. the Gibbs measure μ_Λ^τ. Moreover, the Dirichlet form associated with $\mathcal{L}_{\text{blocks}}^\tau$ is easily computed to be equal to

$$\mathcal{E}_{\text{blocks}}^\tau(f, f) = wv2\mu_\Lambda^\tau \left(\sum_i \text{Var}(f \mid \mathcal{F}_i) \right)$$

where $\mathcal{F}_i$ is the σ-algebra generated by $\{\sigma(x)\}_{x \in \Lambda \setminus \Lambda_i}$. One can then compare the Dirichlet form of the block dynamics to that of the Glauber dynamics $\mathcal{E}^\tau(f, f)$ to get

$$\mathcal{E}_{\text{blocks}}^\tau(f, f) \le \max_{i,\tau} \text{gap}(\mathcal{L}_{\Lambda_i}^\tau)^{-1} \mu_\Lambda^\tau \left(\sum_{\substack{x \in \Lambda \\ s \in S}} N_x \, c(x, s, \sigma) \, [f(\sigma^{x,s}) - f(\sigma)]^2 \right) \tag{4.5}$$

where $N_x := \#\{i : \Lambda_i \ni x\}$. In particular

$$\mathrm{gap}(\mathcal{L}_\Lambda^\tau) \geq \mathrm{gap}(\mathcal{L}_{\mathrm{blocks}}^\tau) \inf_{i,\tau \in \Omega} \mathrm{gap}(\mathcal{L}_{\Lambda_i}^\tau) \left(\sup_{x \in \Lambda} N_x\right)^{-1} \qquad (4.6)$$

$$c_s(\mathcal{L}_\Lambda^\tau) \leq c_s(\mathcal{L}_{\mathrm{blocks}}^\tau) \max_{i,\tau \in \Omega} \mathrm{gap}(\mathcal{L}_{\Lambda_i}^\tau)^{-1} \sup_{x \in \Lambda} N_x \qquad (4.7)$$

Remark 4.2. The above result can be understood as follows. The relaxation time (in what follows identified with either the inverse of the spectral gap or with the logarithmic Sobolev constant) of the single-site Glauber dynamics $(\mathcal{L}_\Lambda^\tau, \mu_\Lambda^\tau)$ is not larger than a factor that measures the maximum number of blocks that contribute to the updating of a single site multiplied by the largest among the relaxation times of the same dynamics restricted to each of the blocks of some block-dynamics for the same Gibbs measure multiplied by the relaxation time of the block-dynamics itself. It is important to observe that in general there is no result connecting the speed of exponential relaxation to equilibrium of the single site dynamics to that of a block dynamics. For example, by assuming strong mixing, it is possible to prove by coupling arguments that the block dynamics with cubic blocks of side 2ℓ and centers on the rescaled lattice $\ell \mathbb{Z}^d$ is uniformly exponentially ergodic [117]. However, there is yet no direct proof that this result alone implies uniform ergodicity of the single site dynamics, at least for general systems. Quite recently we learned [131] of very interesting progress in this direction for *attractive* dynamics (see below for a definition) and we believe that this is an interesting direction to explore.

4.2 The dynamics in an infinite volume

Let μ be a Gibbs measure for the interaction Φ. Since the transition rates are bounded and of finite range, the infinite-volume generator $\mathcal{L}$ obtained by choosing $\Lambda = \mathbb{Z}^d$ in (4.1) is well defined on the set of functions f such that

$$|||f||| := \sum_{x \in \mathbb{Z}^d} \sup_{s,\sigma} |f(\sigma^{x,s}) - f(\sigma)| < \infty$$

We can then take the closure of $\mathcal{L}$ in $C(\Omega)$, the metric space of all continuous functions on Ω with the sup-distance, and get a Markov generator (see, for instance Theorem 3.9 in Chapter I in [105]) or take the closure in $L^2(\Omega, d\mu)$ and get a self-adjoint Markov generator in $L^2(\Omega, \mu)$ (see Theorem 4.1 in Chapter IV of [105]) that will be denoted by $\mathcal{L}$. In the latter case, since the generator is self-adjoint on $L^2(\Omega, d\mu)$ the associated Markov process is reversible w.r.t. the Gibbs measure μ. We conclude with a general result relating the set of invariant measures of the infinite-volume Glauber dynamics with the set of Gibbs measures for the given interaction Φ (see [105]).

Theorem 4.3. *Assume (i)...(iv) on the jump rates. Then:*

(a) If $d = 1, 2$ the set of invariant measures for the above Markov process coincides with the set of Gibbs measures $\mathcal{G}$.

(b) If $d \geq 3$ then:

  (i)   any invariant measure which is also translation invariant is a Gibbs measure;

  (ii) the set of Gibbs measures coincides with the set of reversible invariant measures;

  (iii) if the process is attractive (see below) then the process is ergodic if and only if there is a unique Gibbs measure.

4.3 Graphical construction

We briefly describe here a very convenient way introduced in [140] to realize simultaneously on the same probability space all Markov processes whose generator is $\mathcal{L}_\Lambda^\tau$, as the initial configuration and the boundary condition τ vary in Ω. As a byproduct of the construction we will get, in a rather simple way, a key result which shows that "information" propagates through the system at most with finite speed.

Let $|S|$ be the cardinality of the single spin space S. With each site $x \in \mathbb{Z}^d$ we associate $|S|$ independent Poisson processes, each one with rate c_M, and we assume independence as x varies in $\mathbb{Z}^d$. We denote by $\{t_{x,n}^s\}_{n=1,2...}$ the successive arrivals after time $t = 0$ of the process indexed by $s \in S$. We say that at time t there has been an s-mark at x if $t_{x,n}^s = t$ for some n. Notice that, with probability one, all the arrival times are different. Next we associate with each arrival time $t_{x,n}^s$ a random variable $U_{x,n}^s$ uniformly distributed in the interval $[0, 1]$. We assume that these random variables are mutually independent and independent from the Poisson processes. This completes the construction of the probability space. The corresponding probability measure and expectation are denoted by $\mathbb{P}$ and $\mathbb{E}$ respectively.

Given now $\Lambda \subset\subset \mathbb{Z}^d$, a boundary condition $\tau \in \Omega$ and an initial condition $\eta \in \Omega$ that agrees with τ outside Λ, we construct a Markov process $\{\sigma_t^{\Lambda,\tau,\eta}\}_{t \geq 0}$ on the above probability space according to the following updating rules. Let us suppose that $t = t_{x,n}^s$ for some $x \in \Lambda$, $n \in \mathbb{Z}_+$ and $s \in S$, and assume that the configuration immediately before t was σ. Then:

(1)   The spins $\sigma(y)$ with $y \neq x$ do not change.

(2)   If $\sigma(x) = s$ then $\sigma(x)$ does not change.

(3)   If $\sigma(x) \neq s$ then $\sigma(x)$ changes to s if and only if $c(x, s, \sigma) \geq U_{x,n}^s c_M$.

One can easily check that the above continuous Markov chain on Ω_Λ has indeed the correct jump rates $c(x, s, \sigma)$ so that the above construction represents a global coupling among all processes generated by $\mathcal{L}_\Lambda^\tau$ as the boundary condition τ and the initial condition η vary. Using the graphical construction one can investigate how the process $\sigma_t^{\Lambda,\tau,\eta}(x)$ at site x is affected by a far away change either in the boundary condition τ or in the initial configuration η. The result is the following (see, e.g. Lemma 3.2 in [121]).

Lemma 4.4. *Let $P_t^{\Lambda,\tau}$ be the Markov semigroup associated with $\mathcal{L}_\Lambda^\tau$ on Ω_Λ. There exists a constant $k = k(d, r, c_M)$ such that for all local functions f and all $t \geq 0$ the following holds.*

(1) For all pairs $\Lambda_1 \subset\subset \mathbb{Z}^d$ and $\Lambda_2 \subset\subset \mathbb{Z}^d$, with $d(\Lambda_i^c, \Lambda_f) \geq kt$, $i = 1, 2$,

$$\sup_{\tau_1,\tau_2\in\Omega} \|P_t^{\Lambda_1,\tau_1} f - P_t^{\Lambda_2,\tau_2}(t)f\|_\infty \leq \|\|f\|\| e^{-t}$$

(2) For all $\Lambda \subset\subset \mathbb{Z}^d$ with $d(\Lambda^c, \Lambda_f) \geq kt$ and all $\eta_1, \eta_2 \in \Omega_\Lambda$, with $\eta_1(x) = \eta_2(x)$ for all x such that $d(x, \Lambda_f) \leq kt$,

$$\sup_{\tau\in\Omega} | P_t^{\Lambda,\tau}(t)f(\sigma_1) - P_t^{\Lambda,\tau}(t)f(\sigma_2) | \leq \|\|f\|\| e^{-t}$$

4.4 Uniform ergodicity and logarithmic Sobolev constant

Most of the research on Glauber dynamics in the cubic lattice $\mathbb{Z}^d$ in the late eighties and in the first half of the nineties (see [83], [82], [81], [80], [1], [85], [84], [79], [152], [156], [155], [154] [116],[117], [172], [74], [121]) was directed to proving the equivalence between uniqueness of the Gibbs measure and rapid mixing (in a sense to be made precise) of the Glauber dynamics. For other graphs like, e.g. trees, the main issue may change [94] and it has been much less investigated. In $\mathbb{Z}^d$, when the infinite-volume Gibbs measure is unique, one expects that equilibrium is reached by the Glauber dynamics via a homogeneous process: far apart regions equilibrate in finite time without exchanging almost any information, very much like an infinite collection of non-interacting continuous-time ergodic Markov chains. The best results confirming this intuition are known only for the rather special, though important, class of *attractive* dynamics. These are defined as follows.

Let the single spin space S be of the form $S = \{1, 2, \ldots, N\}$ and let us introduce a partial order on the configuration space Ω by saying that $\sigma \leq \eta$ iff $\sigma(x) \leq \eta(x) \; \forall x \in \mathbb{Z}^d$. A function $f : \Omega \mapsto \mathbb{R}$ is called *monotone increasing (decreasing)* if $\sigma \leq \sigma'$ implies $f(\sigma) \leq f(\sigma')$ $(f(\sigma) \geq f(\sigma'))$.

Definition 4.5. *We say that the jump rates $\{c(x, s, \sigma)\}$, $x \in \mathbb{Z}^d$, $s \in S, \sigma \in \Omega$ define an attractive dynamics iff $\sigma(x) \geq \eta(x)$ for all x implies that*

(i) $\sum_{b\leq a} c(x, b, \sigma) \leq \sum_{b\leq a} c(x, b, \eta)$ for any $a \leq \eta(x)$.
(ii) $\sum_{b\geq a} c(x, b, \sigma) \geq \sum_{b\leq a} c(x, b, \eta)$ for any $a \geq \sigma(x)$.

It is easy to show (see [105]) that attractivity is equivalent to the condition that the Markov semigroup P_t leaves invariant the set of increasing (decreasing) functions on Ω.

Assuming attractivity, it is not difficult to check that condition $WM(\Lambda, C, m)$ for all $\Lambda \subset\subset \mathbb{Z}^d$ is equivalent to the following:

There exist positive constants C and m such that, for any integer L

$$\mu_{B_L}^{+}(\sigma(0)) - \mu_{B_L}^{-}(\sigma(0)) \leq Ce^{-mL}$$

where $+(-)$ denotes the constant configurations identically equal to the maximum (minimum) value of the spin in S.

In [116] the following result was proved.

Theorem 4.6. *In the attractive case the following are equivalent:*

(i) *$WM(\Lambda, C, m)$ for all $\Lambda \subset \mathbb{Z}^d$.*

(ii) *There exists a positive constant m and for any local function f there exists a constant C_f such that:*

$$\|P_t(f) - \mu(f)\|_{\infty} \leq C_f \, e^{-mt}$$

Remark 4.7. The two key ingredients of the proof of theorem 4.6 are attractivity using coupling as, e.g. provided by the graphical construction and a basic old result by Holley [80] that says that if $\varrho(t) := \mathbb{P}\left(\sigma_t^{+}(0) \neq \sigma_t^{-}(0)\right)$ decays faster than t^{-d} then it must necessarily decay exponentially fast. Here σ_t^{+} denotes the infinite-volume process started from the configuration $+$ and similarly for σ_t^{-}.

Remark 4.8. As we have already discussed, for certain ferromagnetic models like the Ising model, condition (i) of the theorem holds throughout the whole one-phase region. Therefore, for these models the infinite-volume dynamics is uniformly exponentially ergodic inside the one-phase region. However, apart from certain special boundary conditions [141] or two-dimensional models, there is no result saying that under weak mixing the *finite-volume* Glauber dynamics is exponentially ergodic with constants independent of the volume. Actually there are very good reasons [41] to believe that a result of this kind in the above generality cannot be true.

For nonattractive systems that do not satisfy the single site Dobrushin condition the only way to control the mixing time in a finite volume uniformly in the boundary conditions is via the logarithmic Sobolev constant as we will show next (see, e.g. [74] or [121]).

Theorem 4.9. *Assume that $\sup_{\Lambda \subset \subset \mathbb{Z}^d} \sup_{\tau} c_s(\mathcal{L}_{\Lambda}^{\tau}) < \infty$. Then:*

(i) *There exists $m > 0$ and $C > 0$ such that for any local function f*

$$\sup_{\Lambda, \tau} \|P_t^{\tau, \Lambda} f - \mu_{\Lambda}^{\tau}(f)\|_{\infty} \leq C \, \|\|f\|\| \, e^{-mt}$$

(ii) *There exist positive constants m', C such that strong mixing $SM(\Lambda, C, m')$ holds for any $\Lambda \subset \mathbb{Z}^d$.*

(iii) *The infinite-volume dynamics is uniformly exponentially ergodic and there exists a unique infinite-volume Gibbs measure with exponentially decaying covariances.*

Remark 4.10. (i)   In the above theorem one could replace the supremum over all finite subsets Λ with the supremum over all cubes in $\mathbb{Z}^d$ and obtain similar results but, only for cubes.

(ii)   It follows from the theorem that the mixing time T_1 associated with $(\mathcal{L}_\Lambda^\tau, \mu_\Lambda^\tau)$ grows only logarithmically in the cardinality of Λ because of the definition of T_1 and the fact that if $f = \mathbb{I}_A$, $A \subset \Omega_\Lambda$, $|||f||| \leq |\Lambda|$.

(iii)   The main idea of proof of the first statement in the theorem is strongly related to the approximation lemma 4.4 and to the hypercontractivity of $P_t^{\Lambda,\tau}$. A sketchy proof goes as follows. Pick a local function f with, e.g. $0 \in \Lambda_f$, fix $t > 0$, k as in lemma 4.4 and choose $\Lambda_t := \Lambda \cap B_{kt}$ in such a way that

$$\|P_t^{\Lambda,\tau} f - P_t^{\Lambda_t,\tau} f\|_\infty \leq |||f||| e^{-t}$$

Then write $t = s + (t-s)$ and use $\|g\|_\infty \leq \left(\min_\sigma \mu_\Lambda^\tau(\sigma)\right)^{-\frac{1}{q}} \|g\|_q$ together with $\min_\sigma \mu_\Lambda^\tau(\sigma) \geq e^{-c(\Phi)|\Lambda|}$ for any g, Λ, τ and q to get

$$\|P_t^{\Lambda_t,\tau} f\|_\infty \leq e^{c(\Phi)\frac{|\Lambda|}{q}} \|P_{t-s}^{\Lambda_t,\tau} P_s^{\Lambda_t,\tau} f\|_q$$
$$\leq e^{c(\Phi)\frac{|\Lambda|}{q}} \|P_s^{\Lambda_t,\tau} f\|_2$$
$$\leq e^{c(\Phi)\frac{|\Lambda|}{q} - \mathrm{gap}(\mathcal{L}_{\Lambda_t}^\tau)t} \|f\|_2$$

provided that $q \leq 1 + e^{\frac{4(t-s)}{c_s(\mathcal{L}_{\Lambda_t}^\tau)}}$. Finally, observe that $|\Lambda_t| \approx t^d$ so that the choice $t - s = A \log t$ for a large enough constant A produces a $q = q(t)$ which is large enough to kill the term $|\Lambda_t|$ in the exponential factor in front of $\|f\|_2$.

(iv)   The last two implications of the theorem are also very closely related to the approximation lemma 4.4. We can in fact write

$$\mu_\Lambda^\tau(f,g) = \mu_\Lambda^\tau\big(P_t^{\Lambda,\tau}(fg)\big) - \mu_\Lambda^\tau(f)\mu_\Lambda^\tau(g)$$

Using lemma 4.4 we have

$$\|P_t^{\Lambda,\tau}(fg) - P_t^{\Lambda,\tau} f \, P_t^{\Lambda,\tau} g\|_\infty \leq e^{-a\,t} |||f||| \, |||g|||$$

as long as $d(\Lambda_f, \Lambda_g) \geq Bt$. Therefore

$$|\mu_\Lambda^\tau(f,g)| \leq |\mu_\Lambda^\tau\big(P_t^{\Lambda,\tau} f, \, P_t^{\Lambda,\tau} g\big)| + e^{-a\,t} |||f||| \, |||g|||$$
$$\leq \|P_t^{\Lambda,\tau} f - \mu_\Lambda^\tau(f)\|_2 \|P_t^{\Lambda,\tau} f - \mu_\Lambda^\tau(f)\|_2 + e^{-a\,t} |||f||| \, |||g|||$$
$$\leq e^{-2\,\mathrm{gap}(\mathcal{L}_\Lambda^\tau)t} \|f\|_2 \|g\|_2 + e^{-a\,t} |||f||| \, |||g|||$$
$$\leq C_f C_g e^{-md(\Lambda_f,\Lambda_g)}$$

for a suitable positive constant m independent of Λ and τ. Here we have used the bound $\mathrm{gap}(\mathcal{L}_\Lambda^\tau) \geq 2c_s(\mathcal{L}_\Lambda^\tau)^{-1}$ (see section 2.1) together with the hypothesis $\sup_{\Lambda \subset\subset \mathbb{Z}^d} \sup_\tau c_s(\mathcal{L}_\Lambda^\tau) < \infty$. Notice that in the above reasoning we only need $\inf_{\Lambda \subset\subset \mathbb{Z}^d} \mathrm{gap}(\mathcal{L}_\Lambda^\tau) > 0$. Quite recently an interesting paper has appeared [58] in which is provided a "combinatorial" (i.e. without the functional analysis involved above) proof of the statement that rapid mixing of the Glauber dynamics implies strong (spatial) mixing for the Gibbs measure together with the converse for attractive systems.

We conclude this section by recalling a nice, infinite-volume result [152] that shows that, at least for the cubic lattice $\mathbb{Z}^d$, as soon as one Gibbs measure μ satisfies a logarithmic Sobolev inequality then necessarily $\mathcal{G} = \{\mu\}$ and the infinite-volume Glauber dynamics is exponentially ergodic.

Theorem 4.11. *Let $\mu \in \mathcal{G}$ and assume that there exists a finite constant $c_s(\mu)$ such that, for all local functions f,*

$$\mathrm{Ent}_\mu(f^2) \le c_s(\mu)\mathcal{E}_\mu(f, f)$$

Then necessarily $\mathcal{G} = \{\mu\}$ and the infinite-volume Glauber dynamics is exponentially ergodic. The same conclusion holds provided that μ satisfies a local logarithmic Sobolev inequality of the form

$$\mathrm{Ent}_\mu(f^2) \le \sum_{x \in \mathbb{Z}^d} \beta_x \mu\big([\mu_x^{\cdot}(f) - f]^2\big)$$

provided that the growth of the local coefficients β_x is moderate (see theorem 1.4 in [152]).

5 Mixing property versus logarithmic Sobolev constant in $\mathbb{Z}^d$

One of the great achievements of the research on Glauber dynamics in $\mathbb{Z}^d$ outside the phase coexistence region is the result that says that a bound on the logarithmic Sobolev constant $c_s(\mathcal{L}_\Lambda^\tau)$ *or on the inverse spectral gap* $\mathrm{gap}(\mathcal{L}_\Lambda^\tau)^{-1}$ *uniform in the volume and in the boundary conditions* is equivalent to strong mixing. We refer the reader to [152], [156], [153], [117], [121], [40], [74], [108], [172].

Theorem 5.1. *The following are equivalent:*

(i) **Uniform logarithmic Sobolev constant.** *There exists a finite constant c_s such that $\sup_{\Lambda \subset\subset \mathbb{Z}^d} \sup_\tau c_s(\mathcal{L}_\Lambda^\tau) \le c_s$.*

(ii) **Uniform spectral gap.** *There exists a positive constant g such that $\inf_{\Lambda \subset\subset \mathbb{Z}^d} \inf_\tau \mathrm{gap}(\mathcal{L}_\Lambda^\tau) \ge g$.*

(iii) **Strong mixing condition.** *There exist positive constants C, m such that condition $SM(\Lambda, C, m)$ holds for all $\Lambda \subset\subset \mathbb{Z}^d$.*

Remark 5.2. Exactly as for theorem 4.9, there is a version of the above result only for ℓ regular volumes (see, e.g. [121]).

The fact that *(i)* implies *(ii)* is obvious. The implication *(ii)* $\Rightarrow$ *(iii)* was already discussed in the remark after theorem 4.9 and thus the really nontrivial implication to be analyzed is *strong mixing* $\Rightarrow$ *uniform logarithmic Sobolev constant.*

There have been in recent years several different approaches to the proof of such a result that we will try to briefly review. The common denominator of all these proofs is the fact that strong mixing implies that the Gibbs measure is more or less close to a product measure for which the tensorization of the logarithmic Sobolev inequality applies. However, the way the mixing condition enters in the various proofs is quite different and the degree of technicality can vary considerably.

5.1 The auxiliary chain and sweeping out relations method

The main idea of the first proof of a logarithmic Sobolev inequality for Glauber dynamics $(\mathcal{L}_\Lambda^\tau, \mu_\Lambda^\tau)$ (see, e.g [74] for a nice review) was to construct an auxiliary chain on the configuration space with a transition kernel Π satisfying the following conditions. For notation convenience we will omit in the sequel the volume Λ and the boundary conditions τ.

(a)   For any function f

$$\mu(\Pi f) = \mu(f)$$

(b)   There exists a positive finite constant $\bar{c}$ such that for any function f

$$\text{Ent}_\mu\left(\Pi(f^2 \log f^2)\right) - \Pi(f^2)\log(\Pi(f^2)) \leq 2\bar{c}\,\mathcal{E}(f, f)$$

(c)   There exists $\lambda \in (0, 1)$ such that

$$\mathcal{E}(\sqrt{\Pi(f^2)}, \sqrt{\Pi(f^2)}) \leq \lambda\mathcal{E}(f, f)$$

(d)   Given f, let $f_0 := f$ and let $f_n := \Pi f_{n-1}$, $n \in \mathbb{N}$. Then

$$\lim_{n \to \infty} f_n = \mu(f)$$

It is not difficult to show (see, e.g. [74]) that the above conditions imply that the logarithmic Sobolev constant of the heat bath dynamics on Λ is bounded from above by $\frac{\bar{c}}{1-\lambda}$. The problem is therefore to construct, using strong mixing, an auxiliary chain satisfying conditions (a), ..., (d).

Following [74] we fix an integer L and define, for any vector $\mathbf{k} \in \mathbb{Z}^d$, $X_{\mathbf{k}} := X_0 + \mathbf{k}$, where $X_0 := Q_{L+r}$, r being the range of the interaction. Let $\mathbf{v}_s \in \{0, 1\}^d$, $s = 0, \ldots, 2^d - 1$, and let $\mathcal{T}_s := \{\mathbf{k} \in \mathbb{Z}^d; \mathbf{k} = (L + 2r)\mathbf{v}_s + (2(L + 2r)\mathbb{Z})^d\}$. Set $\Gamma_s := \bigcup\{X_{\mathbf{k}} : \mathbf{k} \in \mathcal{T}_s\}$ and notice that Γ_s is the union of disjoint cubes of shape X_0 at distance one from the other equal to $2r$.

Moreover $\bigcup_s \Gamma_s = \mathbb{Z}^d$. The important fact is that for any $x \in \mathbb{Z}^d$ there exists $s \in \{0, \ldots, 2^d - 1\}$ and $X_\mathbf{k} \in \Gamma_s$ such that $x \in X_\mathbf{k}$ and $d(x, X_\mathbf{k}^c) \geq L/2$. Finally, we define (for simplicity we treat directly the case $\Lambda = \mathbb{Z}^d$)

$$\Pi f(\sigma) := \mathbb{E}_{2^d - 1}\big(\ldots \mathbb{E}_1\big(\mathbb{E}_0(f)\big)\big)$$

where $\mathbb{E}_s := \mu(\cdot \,|\, \mathcal{F}_s)$ and $\mathcal{F}_s$ is the σ-algebra generated by the variables $\{\sigma(x)\}_{x \notin \Gamma_s}$. Notice that, by construction, each measure $\mathbb{E}_s$ is a product measure over the cubes of Γ_s.

The key point at this stage is first to derive conditions $(\mathbf{a}), \ldots, (\mathbf{d})$ for the chain Π from the so-called *sweeping out relations* (see section 5.4.2 [74]) defined below and then to prove that the latter are implied by the strong mixing condition.

In order to define the sweeping out relations we need first an additional useful notation for the discrete gradient. We define

$$\nabla_x f(\sigma) := \int_S d\mu_0(s) f(\sigma^{x,s}) - f(\sigma)$$

where, as usual, $\sigma^{x,s}$ denotes the configuration obtained from σ by replacing the spin $\sigma(x)$ with an admissible value $s \in S$.

Definition 5.3. *We shall say that sweeping out relations are satisfied for a finite subset X_0 of $\mathbb{Z}^d$, if for any set Λ of the form $\Lambda = x + X_0$ for some $x \in \mathbb{Z}^d$, there exist nonnegative constants $\{\alpha_{zz'}\}_{z, z' \in \mathbb{Z}^d}$ satisfying*

$$\alpha_{zz'} \leq De^{-\varepsilon d(z, z')}$$

for some fixed constants D, ε independent of Λ, z, z', such that for any $y \in \mathbb{Z}^d$ with $d(y, \Lambda) \leq r$

$$|\nabla_y(\mu_\Lambda^\tau(f)^{\frac{1}{2}})| \leq \alpha_{yy}\big(\mu_\Lambda^\tau(|\nabla_y f^{\frac{1}{2}}|^2)\big)^{\frac{1}{2}} + \sum_{z \in \Lambda \cup \{y\}} \alpha_{yz}\big(\mu_{\{y\}}(\mu_\Lambda(|\nabla_z f^{\frac{1}{2}}|^2))\big)$$

5.2 The renormalization group approach

Here we describe a second approach to the implication *strong mixing $\Rightarrow$ uniform logarithmic Sobolev constant* that was developed in [117]. In this approach the proof is clearly divided into two distinct parts:

(i)   In the first part one proves that any Gibbs measure ν on a set Λ which is the (finite or infinite) union of certain "blocks" $\Lambda_1 \ldots \Lambda_j \ldots$ (e.g. cubes of side l or single sites of the lattice $\mathbb{Z}^d$) has a logarithmic Sobolev constant (w.r.t. the associated heat bath dynamics) not larger than a suitable constant which depends *only* on the maximum size of the blocks, provided that the interaction among the blocks, not necessarily of *finite range*, is very weak in a suitable sense. A simple example of such a situation is represented by a Gibbs state at high temperature, but the result is more general since it is not assumed that the interaction *inside* each block is weak.

(ii) It is in the second part that renormalization group ideas come into play. One uses a particular form of renormalization group transformation known as decimation (i.e. integration over a certain subset of the variables $\sigma(x)$), to show that, under the strong mixing hypothesis, the Gibbs state μ_Λ^τ after a finite (less than 2^d) number of decimations becomes a new Gibbs measure exactly of the type discussed in part (i). It is then a relatively easy task to derive the boundedness of the logarithmic Sobolev constant of μ_Λ^τ.

As is well known from the papers [129], [130], the strong mixing condition implies that if the decimation is done over blocks of a sufficiently large size, then it is possible to control, e.g. by a converging cluster expansion, the effective potential of the renormalized measure and to show that it satisfies the weak coupling condition needed in part (i). This is, however, more than what it is actually needed, since the hypotheses of part (i) are fulfilled by the renormalized measure as soon as the covariances of the *original* Gibbs measure μ_Λ^τ decay exponentially fast.

Remark 5.4. One important drawback of the above approach is the fact that in order to implement the first step one needs a priori a lower bound on the spectral gap of $\mathcal{L}_\Lambda^\tau$ uniform in Λ and in the boundary condition τ. Such an inconvenience was present also in the original version of the Zegarlinski approach but not in the later version given in [74]. Although strong mixing easily implies a lower bound on the spectral gap by, e.g. block dynamics and coupling methods or quasi-tensorization of the variance (see below), one would like to be able to establish a logarithmic Sobolev inequality without appealing to the weaker Poincaré inequality.

The decimation procedure used in [117] can easily be described in two dimensions as follows.

For any odd integer ℓ consider the renormalized lattice $\mathbb{Z}^2(\ell) := \ell\mathbb{Z}^2$ and collect together the blocks $Q_\ell(x)$, $x \in \mathbb{Z}^2(\ell)$, into four different families, denoted in the sequel by $\Gamma_1, \Gamma_2, \ldots, \Gamma_4$ according to whether the coordinates of their centers x are *(even, odd)*, *(even, even)*, *(odd, even)* or *(odd, odd)*.

Let finally $\Lambda(\ell)$ be a finite subset of $\mathbb{Z}^2(\ell)$, let $\Lambda = \bigcup_{x \in \Lambda(\ell)} Q_\ell(x)$ and let μ_Λ^τ be the Gibbs state in Λ with boundary condition τ. Out of μ_Λ^τ one constructs new Gibbs measures, denoted for simplicity by $\{\mu_i\}_{i=1}^4$, on the spin configurations (which agree with τ outside Λ) in Γ_i, $i = 1, \ldots, 4$, as follows.

The new measure μ_4 is simply obtained from the Gibbs measure μ_Λ^τ by conditioning on the spins in $\Gamma_1 \cup \Gamma_2 \cup \Gamma_3$. To construct μ_3 one first integrates out in μ_Λ^τ the spins in Γ_4 and then one conditions w.r.t. the spins in $\Gamma_1 \cup \Gamma_2$.

Similarly, to construct μ_2 one first integrates out in μ_Λ^τ the spins in $\Gamma_4 \cup \Gamma_3$ and then one conditions w.r.t. the spins in Γ_1. μ_1 is simply the marginal of μ_Λ^τ on the spins in Γ_1.

5.3 The martingale method

Here we present the method known as *the martingale approach* developed in [108]. It shares a common aspect with the *recursive method* to be described next in that it goes by induction from small spatial scales to larger ones. Martingale ideas then come into play in the induction step. For simplicity we sketch the main steps only for the Poincaré inequality in one dimension but we stress that dimensionality plays no role and that the logarithmic Sobolev inequality can be analyzed along the same lines.

Define $\gamma(L) := \sup_\Lambda^* \sup_\tau \mathrm{gap}(\mathcal{L}_\Lambda^\tau)^{-1}$ where $\sup_\Lambda^*$ is the supremum over all intervals of length at most L. The main idea is to prove that, for any L large enough, $\gamma(L)$ satisfies the following recursive inequality:

$$\gamma(2L) \leq \frac{3}{4}\gamma(L) + c \tag{5.1}$$

where c is a suitable constant. The above recursion easily implies that $\gamma(L)$ is uniformly bounded in L.

In order to prove (5.1) let us prove an upper bound on the inverse spectral gap for an interval with $2L + 1$ sites, in terms of $\gamma(L)$ and $\gamma(\log L)$. In what follows the letter c will denote a constant depending only on the norm of the interaction and on the mixing constants C, m that will vary from estimate to estimate.

Let $\Lambda = \Lambda_1 \cup \Lambda_2$ with $\Lambda_1 = \{-L+1, -L, \ldots 1\}$ and $\Lambda_2 = \{0, \ldots, L\}$, $L \in \mathbb{N}$. Let also $\mathcal{F}_j$, $j = 0, 1, \ldots, L$, be the σ-algebra generated by the spins $\sigma(x)$, $x \in \{j, j+1, \ldots L\}$, and define $f_j := \mu_\Lambda^\tau(f \,|\, \mathcal{F}_j)$. We start with the simple identity

$$\mathrm{Var}_\Lambda^\tau(f) = \mu_\Lambda^\tau \left(\mathrm{Var}_\Lambda^\tau(f \,|\, \mathcal{F}_0) + \sum_{j=0}^{L} \mathrm{Var}_\Lambda^\tau(f_j \,|\, \mathcal{F}_{j+1}) \right) \tag{5.2}$$

where, for an arbitrary σ-algebra $\mathcal{F}$,

$$\mathrm{Var}_\Lambda^\tau(f \,|\, \mathcal{F}) := \mu_\Lambda^\tau(f^2 \,|\, \mathcal{F}) - \mu_\Lambda^\tau(f \,|\, \mathcal{F})^2$$

denotes the usual conditional variance. We shall now estimate all the terms on the r.h.s. of (5.2).

Step 1. The first term on the r.h.s. of (5.2) is easily handled through the induction assumption. Thanks in fact to the Poincaré inequality applied for the Glauber in Λ_1, it is bounded from above by

$$\gamma(L) \sum_{x \in \Lambda_1} \sum_{s \in S} \mu_\Lambda^\tau \big(c(x, \sigma, s)[f(\sigma^{x,s}) - f(\sigma)]^2 \big)$$

which is the part of the global Dirichlet form associated with the interval Λ_1.

Step 2. The marginal of $\mu_\Lambda^\tau(\cdot \,|\, \mathcal{F}_{j+1})$ on the variable $\sigma(j)$ has a bounded density w.r.t. the uniform measure on the spin space S and therefore it satisfies a uniform (in L, τ) Poincaré inequality:

$$\mathrm{Var}(f_j \,|\, \mathcal{F}_{j+1}) \le c \sum_{s \in S} \mu_\Lambda^\tau \big(c(j, \sigma, s) \, [f_j(\sigma^{j,s}) - f_j(\sigma)]^2 \,|\, \mathcal{F}_{j+1} \big) \qquad (5.3)$$

Step 3. Remember that f_j is the conditional expectation of the function f given the spins $\sigma(j), \ldots \sigma(L)$. Therefore the discrete "gradient" operator appearing in the r.h.s of (5.3) will act on f but also on the conditional measure $\mu_\Lambda^\tau(\cdot \,|\, \mathcal{F}_{j+1})$. The action of the gradient on the conditional measure will produce a covariance term that has to be controlled by using the mixing assumption on the Gibbs measure.

More precisely, define $h_j^{(s)} := d\mu_\Lambda^\tau(\cdot \,|\, \sigma^{(j,s)}(j), \ldots, \sigma(L))/d\mu_\Lambda^\tau(\cdot \,|\, \sigma(j), \ldots, \sigma(L))$. Then it is easy to check that the r.h.s. of (5.3) is bounded from above by

$$c \sum_{s \in S} \mu_\Lambda^\tau \big(c(j, \sigma, s) \, [f(\sigma^{j,s}) - f(\sigma)]^2 \,|\, \mathcal{F}_{j+1} \big) + c\mu_\Lambda^\tau \left(\mu_\Lambda^\tau \left(f, h_j^{(s)} \,|\, \mathcal{F}_j \right)^2 \,|\, \mathcal{F}_{j+1} \right)$$

$$(5.4)$$

Notice that the first term in the above expression, upon averaging w.r.t. μ_Λ^τ and summing over j, will produce the piece of the global Dirichlet form associated with the interval Λ_2, multiplied by a fixed constant c, that did not appear in Step 1.

Step 4. It remains to bound the covariance part in (5.4). Here again martingale ideas come into play and it is here that the mixing condition plays a crucial role.

Fix $\ell, \alpha \in \mathbb{N}$ and let $\mathcal{F}_{j,\alpha} := \mathcal{F}^{(j,\alpha)} \cap \mathcal{F}_j$ where $\mathcal{F}^{(j,\alpha)}$ is the σ-algebra generated by the variables $\sigma(x)$ with $d(x, j) \ge \ell^\alpha$. Define $h_{j,\alpha}^{(s)}$ by

$$h_{j,\alpha}^{(s)} := \mu_\Lambda^\tau \left(h_j^{(s)} \,|\, \mathcal{F}_{j,\alpha} \right), \qquad h_{j,0}^{(s)} := h_j^{(s)}$$

and write

$$\mu_\Lambda^\tau \left(f, h_j^{(s)} \,|\, \mathcal{F}_j \right) = \sum_{\alpha=0}^\infty \mu_\Lambda^\tau \left(f, h_{j,\alpha}^{(s)} - h_{j,\alpha+1}^{(s)} \,|\, \mathcal{F}_j \right)$$

The partial averages $h_{j,\alpha}^{(s)}$ enjoy two key properties:

(i)   They form a martingale w.r.t. to the filtration $\{\mathcal{F}_{j,\alpha}\}_{\alpha=0}^{\infty}$ i.e.

$$\mu_\Lambda^\tau\left(h_{j,\alpha}^{(s)} - h_{j,\alpha+1}^{(s)} \mid \mathcal{F}_{j,\alpha+1}\right) = 0\,.$$

(ii) $\sup_{j,\Lambda,\tau}\sup_{\sigma,\eta}|h_{j,\alpha}^{(s)}(\sigma) - h_{j,\alpha}^{(s)}(\eta)| \le \text{const.}e^{-m\ell^\alpha}$ because of the mixing hypothesis.

These two properties plus a little bit of extra work (see section II of [108]), allow one to conclude that the covariance term contribution to the r.h.s of (5.2)

$$\sum_j \mu_\Lambda^\tau\left(\mu_\Lambda^\tau(f, h_j^{(s)} \mid \mathcal{F}_j)^2\right)$$

can be bounded from above by the total variance $\mathrm{Var}_\Lambda^\tau(f)$ multiplied by a small (with L) correction plus a suitable part of the global Dirichlet form of f of the form

$$\left(\frac{1}{100}\gamma(\log L) + c\right)\mathcal{E}_\Lambda^\tau(f,f)$$

Putting all together, we obtain that the r.h.s. of (5.2) is bounded from above by

$$\gamma(L)\sum_{x\in\Lambda_1}\sum_{s\in S}\mu_\Lambda^\tau\big(c(x,\sigma,s)[f(\sigma^{x,s}) - f(\sigma)]^2\big) + \varepsilon(L)\,\mathrm{Var}_\Lambda^\tau(f)$$

$$+ \left(\frac{1}{100}\gamma(\log L) + c\right)\mathcal{E}_\Lambda^\tau(f,f)$$

where $\lim_{L\to\infty}\varepsilon(L) = 0$.

Since the role of Λ_1 and Λ_2 can be clearly interchanged, we also obtain a similar bound with Λ_1 replaced by Λ_2. Finally, by taking the arithmetic mean between the two bounds, we obtain from (5.2) the estimate

$$\mathrm{Var}_\Lambda^\tau(f) \le \varepsilon(L)\,\mathrm{Var}_\Lambda^\tau(f) + \left(\frac{1}{2}\gamma(L) + \frac{1}{100}\gamma(\log L) + c\right)\mathcal{E}_\Lambda^\tau(f,f)$$

which implies (for any L large enough) that

$$\sup_\tau \mathrm{gap}(\mathcal{L}_\Lambda^\tau)^{-1} \le (1 + \varepsilon(L))\left(\frac{1}{2}\gamma(L) + \frac{1}{100}\gamma(\log L) + c\right) \le \frac{3}{4}\gamma(L) + c$$

which is an inequality like (5.1).

5.4 The recursive analysis

Here we describe a last approach to the proof of *strong mixing $\Rightarrow$ uniform logarithmic Sobolev constant*, based on the quasi-tensorization of the entropy discussed in theorem 2.12. The method was introduced in [121] and extended and considerably simplified to its present form in [40]. The beauty of this proof is that it avoids completely the technicalities of the other methods and it relies only on some basic property of the entropy.

Recall the definition of $\mu^\tau_{\Lambda,\Delta}$ as the restriction of μ^τ_Λ to $\mathcal{F}_\Delta$ and let $\varrho^t_{\Lambda,\Delta}(\sigma)$ be the Radon–Nikodym density of $\mu^t_{\Lambda,\Delta}$ w.r.t. $\otimes_{x\in\Delta}\mu_0(\sigma(x))$. It is not difficult to check that strong mixing implies that there exist $K > 0$, $m > 0$ such that for all $\Lambda \in \mathbb{F}$, $x \in \partial^+_r \Lambda$, $\Delta \subset \Lambda$, and for all $\sigma,\omega \in \Omega$ with $\sigma(y) = \omega(y)$ if $y \neq x$

$$\left\|\frac{\varrho^\omega_{\Lambda,\Delta}}{\varrho^\sigma_{\Lambda,\Delta}} - 1\right\|_u \leq K e^{-md(x,\Delta)} \tag{5.5}$$

As a corollary we get the following lemma. In what follows, for notation convenience, $c_s(\Lambda) := \sup_\tau c_s(\mathcal{L}^\tau_\Lambda)$.

Lemma 5.5. *Let $\Lambda \in \mathbb{F}$, and let Λ_1, Λ_2 be two subsets of Λ, such that $\Lambda = \Lambda_1 \cup \Lambda_2$. Let $l := d(\Lambda\backslash\Lambda_1, \Lambda\backslash\Lambda_2)$. Assume that*

$$|(\partial^+_r \Lambda_2) \cap \Lambda|\, K\, e^{-ml} \leq 1 \tag{5.6}$$

Then there exists $l_0 = l_0(K,m)$ such that for all $l \geq l_0$, for all $\tau \in \Omega$

$$\mathrm{Ent}_{\mu^\tau_\Lambda}(f^2) \leq \left(1 + K'e^{-ml}\right)[c_s(\Lambda_1) \vee c_s(\Lambda_2)]\left[\mathcal{E}^\tau_\Lambda(f,f) + \mu^\tau_\Lambda(\mathcal{E}_{\Lambda_1\cap\Lambda_2}(f,f))\right] \tag{5.7}$$

for a suitable K' independent of ℓ.

The above lemma suggests an iterative procedure to estimate the logarithmic Sobolev constant $c_s(\Lambda)$, which consists in dividing Λ roughly into two "halves" Λ_1, Λ_2, in such a way that $\Lambda = \Lambda_1 \cup \Lambda_2$ and Λ_1 and Λ_2 have an intersection "thick" enough so that (5.6) holds. Then, by (5.7) we "almost" have $c(\Lambda) \leq (1 + K'e^{-ml})(c_s(\Lambda_1) \vee c_s(\Lambda_2))$. The "almost" comes of course from the extra term $\mu^t_\Lambda(\mathcal{E}_{\Lambda_1\cap\Lambda_2}(f,f))$. A trivial upper bound for this term is $\mathcal{E}^\tau_\Lambda(f,f)$, but this is fatal to the argument since it yields $c(\Lambda) \leq 2(1 + K'e^{-ml})(c_s(\Lambda_1) \vee c_s(\Lambda_2))$. However, it was observed in [121] that one can write many, say r, different replicas of inequality (5.7), each corresponding to a different choice of Λ_1, Λ_2, and such that the sets $\Lambda_1 \cap \Lambda_2$ are disjoint for different replicas. At this point we can average over the number of replicas the inequalities obtained to get

$$c_s(\Lambda) \leq (1 + K'e^{-ml})(1 + 1/r)(c_(\Lambda_1) \vee c_s(\Lambda_2)) \tag{5.8}$$

Thus, if r is a function of the size of Λ which goes to zero fast enough as $\Lambda \to \mathbb{Z}^d$, a chance to obtain a convergent iteration from (5.8) becomes apparent.

The actual proof requires a simple geometric construction which was already used in [16] for obtaining a uniform lower bound for the spectral gap of a continuous gas.

Let $l_k := (3/2)^{k/d}$, and let $\mathbb{F}_k$ be the set of all $A \in \mathbb{F}$ which, modulo translations and permutations of the coordinates, are contained in

$$\big([0, l_{k+1}] \times [0, l_{k+2}] \times \cdots \times [0, l_{k+d}]\big) \cap \mathbb{Z}^d$$

Let also $G_k := \sup_{V \in \mathbb{F}_k} c(V)$. The idea behind this construction is that each volume in $\mathbb{F}_k \backslash \mathbb{F}_{k-1}$ can be obtained as a "slightly overlapping union" of two volumes in $\mathbb{F}_{k-1}$. More precisely we have:

Proposition 5.6. *For all $k \in \mathbb{Z}_+$, for all $\Lambda \in \mathbb{F}_k \backslash \mathbb{F}_{k-1}$ there exists a finite sequence $\{\Lambda_1^{(i)}, \Lambda_2^{(i)}\}_{i=1}^{s_k}$, where $s_k := \lfloor l_k^{1/3} \rfloor$, such that, letting $\delta_k := \frac{1}{8}\sqrt{l_k} - 2$,*

(i) $\Lambda = \Lambda_1^{(i)} \cup \Lambda_2^{(i)}$ and $\Lambda_1^{(i)}, \Lambda_2^{(i)} \in \mathbb{F}_{k-1}$, for all $i = 1, \ldots, s_k$;

(ii) $d(\Lambda \backslash \Lambda_1^{(i)}, \Lambda \backslash \Lambda_2^{(i)}) \geq \delta_k$; for all $i = 1, \ldots, s_k$;

(iii) $\Lambda_1^{(i)} \cap \Lambda_2^{(i)} \cap \Lambda_1^{(j)} \cap \Lambda_2^{(j)} = \emptyset$, if $i \neq j$.

The argument sketched above together with proposition 5.6 and the observation that $\sum_{i=1}^{s_k} \mu_\Lambda^\tau \big(\mathcal{E}_{\Lambda_1^{(i)} \cap \Lambda_2^{(i)}}(f, f) \big) \leq \mathcal{E}_\Lambda^\tau(f, f)$ allows us to conclude that

$$G_k \leq G_{k-1} \left(1 + K' e^{-m\delta_k}\right) \left[1 + \frac{1}{s_k}\right] \qquad \forall k \geq k_0$$

which implies $G_k \leq M G_{k_0}$ for all $k \geq k_0$, where

$$M := \prod_{k=k_0}^{\infty} \left\{ \left(1 + K' e^{-m\delta_k}\right) \left[1 + \frac{1}{s_k}\right] \right\} < \infty$$

Remark 5.7. Recently the recursive scheme has been successfully applied to establish spectral gap bounds for the Glauber dynamics of a *continuous gas* in $\mathbb{R}^d$ [16].

5.5 Rapid mixing for unbounded spin systems

We conclude this first part dedicated to rapid mixing of Glauber dynamics for lattice models by very briefly discussing the difficult subject of unbounded spin systems, e.g. like those introduced in section 3.9 (but see also the solid-on-solid (SOS) interface model discussed in [41]). We refer the reader to [174],[173], [172], [171], [170], [103], [19]. Strictly speaking, the usual dynamical models for continuous spin systems do not fit in the framework of continuous-time Markov chains since they are characterized by a symmetric generator which is a second order elliptic differential operator whose associated Dirichlet form is given by

$$\mathcal{E}_\Lambda^\tau(f, f) := \frac{1}{2} \sum_{x \in \Lambda} \mu_\Lambda^\tau(|\nabla_x f|^2) \tag{5.9}$$

on $\mathcal{C}_\Lambda := \{f \in C^\infty(\mathbb{R}^\Lambda) ; \ \sum_{x \in \Lambda} |\nabla_x f| < \infty\}$. Here μ_Λ^τ is the finite-volume Gibbs measure defined in section 3.9.

Despite the different nature of the dynamics, most of the relaxation properties of the diffusion process associated with (5.9), particularly the notion of exponential decay to equilibrium in the uniform norm, can be analyzed by means of the same analytical quantities like the spectral gap and logarithmic Sobolev constant, exactly as in the discrete case. In particular in [172] theorem 5.1 was extended to this new situation.

6 Torpid mixing in the phase coexistence region

In this section we discuss the important topic of the speed of relaxation to equilibrium for a Glauber dynamics when the thermodynamic parameters of the underlying interaction are such that the set of infinite-volume Gibbs measures consists of more than one element.

As we will see in some detail, at least in the case of the Ising model in $\mathbb{Z}^d$, the presence of multiple phases drastically modifies the behavior of the dynamics and new physical features slow down the relaxation; among those, the nucleation and the interface motions, until now only partially understood. Metastability is characteristic of these slow phenomena since the system is trapped for a very long period of time in a local equilibrium. In this case, the relaxation mechanism is so slow that the time of nucleation can be expressed in terms of equilibrium quantities. Later on we will review the same phenomenon for other models and other kinds of dynamics, including conservative ones.

6.1 Torpid mixing for the Ising model in $\Lambda \subset \mathbb{Z}^d$ with free boundary conditions

In order to be concrete (but also because it is the only model for which some detailed results are available), let us consider the usual Ising model in d dimensions $d \geq 2$ without external field h and inverse temperature β larger than the critical value β_c (see section 3.5). Then any associated infinite-volume Glauber dynamics is not ergodic and it is rather natural to ask how this absence of ergodicity is reflected if we look at the dynamics in a finite, but large cube Λ of side L, where ergodicity is never broken.

As a preliminary remark it is important to observe that the finite-norm condition on the interaction Φ (see (4) in definition 3.2) implies that

$$\text{gap}(\mathcal{L}_\Lambda^\tau) \geq e^{-cL^{d-1}}$$

for a suitable constant $c = c(\|\Phi\|)$. The proof is rather simple and one can either use a rough recursive argument or the canonical paths method discussed in section 2.3 (see [121]). The interesting question is whether the above rough bound can be saturated and, if yes, whether it is possible to find the precise value of the constant c in front of the surface term.

A first partial answer was provided in [158] many years ago for very low temperatures. In [158] it was proved that, if the boundary conditions are free, then the inverse spectral gap, $\text{gap}(\mathcal{L}_\Lambda^{free})$, diverges as $L \to \infty$, at least as an exponential of the *surface* L^{d-1}.

The reason for such a result is the presence of a rather tight "bottleneck" in the phase space. When in fact the boundary conditions are either free or periodic, the energy landscape determined by the energy function has only two absolute minima corresponding to the two configurations identically equal to either $+1$ or to -1. Thus the dynamics started, e.g. from all minuses, in order to relax to equilibrium, has to reach the neighborhood of the opposite minimum by necessarily crossing the set of configurations of zero magnetization (if the cardinality of Λ is even). Since the Gibbs measure gives to the latter a very small weight, of the order of a negative exponential of the *surface* of Λ, a bottleneck is present and the result follows by rather simple arguments. More precisely one takes the test function $f(\sigma) := \mathbb{1}_{\sum_{x \in \Lambda} \sigma(x) > 0}$ and proves that

$$\text{Var}_\Lambda^{free}(f) \approx \frac{1}{4} \qquad \text{as } \Lambda \to \mathbb{Z}^d$$

while

$$\mathcal{E}_\Lambda^{free}(f, f) \le k|\Lambda|\mu_\Lambda^{free}\left(\left|\sum_{x \in \Lambda} \sigma(x)\right| \le 1\right) \le e^{-cL^{d-1}}$$

for some constant $c(\beta)$. The result then follows from the variational characterization of the spectral gap.

The same reasoning also suggests that if the double well structure of the Gibbs measure is completely removed by the boundary conditions, e.g. by fixing equal to $+1$ all spins outside Λ, or if we measure the relaxation to equilibrium of a function f which is *even* w.r.t a global spin flip $\sigma \to -\sigma$, then the relaxation time should be much shorter than in the previous case since there are no bottlenecks to cross. We will come back to this interesting subject in a little while.

In a series of papers (see [113], [37], [112], [10], [78], [9]) the above and other related questions for the two-dimensional Ising model at inverse temperature β above β_c and without external field have been considered.

With free boundary conditions the bottleneck picture has been made much more precise and the result is ([113], [37])

$$\lim_{L \to \infty} -\frac{1}{\beta L} \log(\text{gap}) = \tau_\beta$$

where τ_β denotes the surface tension in the direction of, e.g. the horizontal axis. In this case, the picture of the relaxation behavior to the Gibbs equilibrium measure that comes out is the following. The system first relaxes rather rapidly to one of the two phases [112] and then it creates, via a large fluctuation, a thin layer of the opposite phase along one of the sides of Λ. Such a process already requires a time of the order of $\exp(\beta\tau_\beta L)$. After that, the opposite phase invades the whole system by moving, on a much shorter time scale, the interface to the side opposite to the initial one and equilibrium is finally reached. The time required for this final process can be computed to be of the order of at least L^3 in the SOS approximation (see [132]).

Once this picture is established it is not too difficult to show that, under a suitable stretching of the time by a factor $a(L) \approx \exp(\beta\tau_\beta L)$, the magnetization in the square Λ behaves in time as a continuous Markov chain with state space $\{-m^*(\beta), +m^*(\beta)\}$ and unitary jump rates, where $m^*(\beta)$ is the spontaneous magnetization ([113] and [112]).

It is interesting to observe that in the proof of the above picture for free boundary conditions the techniques of switching from single site dynamics to block dynamics plays a major role. Contrary to what happens when $\beta < \beta_c$, below the critical temperature ($\beta > \beta_c$) the variables $\{\sigma(x)\}_{x\in\Lambda}$ are no longer almost independent and they become strongly correlated on a *macroscopic scale* (the side of Λ). The slowest mode of the dynamics is connected with the physical process of creating a germ (droplet) of one phase inside the phase of the opposite sign. Moreover, in order not to die out following the dynamics of the majority of the spins, the droplet of the opposite phase must reach a macroscopic size. It is clear that to describe such a process using a microscopic dynamics is a very difficult task. Much simpler is the same task with a block dynamics with macroscopic blocks, since in that case dynamical large deviations become strongly related to equilibrium fluctuations and for the latter several powerful techniques (Peierls contours, Pirogov–Sinai theory, FK representation, cluster expansion, etc.) have been developed. Macroscopic blocks have, however, the disadvantage of contributing with possibly a very small factor to the formula relating the single site spectral gap to the spectral gap of block dynamics (see (4.6)). One has therefore to compromise and the results are blocks with mesoscopic volume, i.e. very large on a microscopic scale but very small on a macroscopic scale. The shape of the blocks is also crucial in order to get the best results and it is very tightly linked with the physical process driving the system to equilibrium (see [113], [121], [112]).

6.2 Interface driven mixing inside one phase

Let us examine now what happens when the boundary around the region Λ breaks the double well structure of the typical configurations by, e.g. fixing all the spins outside Λ to be $+1$. In this case it turns out that relaxation to equilibrium is driven by the slow motion of the interfaces.

When a droplet of the negative phase is surrounded by the positive phase, it tends to shrink according to its curvature under the action of the nonconservative dynamics on the spins close to the interface and the heuristics suggests that it should disappear in a time proportional to the square of its radius.

This subtle phenomenon has been studied rigorously only in rare instances (see [150] in the case of Ising model at zero temperature and [43]). Notice also that the motion by mean curvature plays a key role in the coarsening phenomenon, as has been shown recently in [64]. For positive temperatures, a mathematical derivation of similar results seems to be more challenging.

Quite recently [21] it was proved that for any dimension $d \geq 2$, zero external field h and below the critical temperature, the logarithmic-Sobolev constant for a domain of linear size L with positive boundary conditions diverges at least like L^2 (up to some logarithmic corrections). This can be considered as a first characterization of the slow down of the dynamics and is in agreement with the heuristics predicted by the motion by mean curvature. In the same setting but with $d = 2$, the same paper shows that the inverse of the spectral gap grows at least like L (up to logarithmic corrections). Although an almost exact computation using Hardy inequalities for a toy model mimicking mean curvature motion plus noise seems to confirm the above polynomial asymptotic [21], the mechanism behind the different scaling of the spectral gap and logarithmic Sobolev constant is not fully understood.

The proof given in [21] boils down to bounding the variational formula for the Poincaré and the log-Sobolev inequalities by choosing an appropriate test function. This reduces the problem to a computation under the equilibrium Gibbs measure and the main difficulty is to recover polynomial bounds by using only the exponential estimates provided by the equilibrium theory of phase segregation (see [20] and references therein). This is achieved by the choice of a subtle test function which was suggested some years ago by H.T. Yau.

It is important to stress that no matching upper bounds have been derived yet; the best existing bounds (see [113], [37], [121], [77] and [157]) are of the form

$$\mathrm{gap}(\mathcal{L}_\Lambda^+) \geq \exp\left(-\beta c \sqrt{L \log L}\right), \qquad \text{for } d = 2$$
$$\mathrm{gap}(\mathcal{L}_\Lambda^+) \geq \exp\left(-\beta c L^{d-2}(\log L)^2\right), \qquad \text{for } d \geq 3$$

where $c > 0$ is a suitable constant and $\beta \gg \beta_c$.

There is an interesting consequence for the infinite-volume dynamics inside one of the two pure phases of the upper bound on the spectral gap proved in [21] for $d = 2$.

Let us consider an arbitrary coupling (e.g. that provided by the graphical construction) of the Glauber dynamics in the infinite volume $\mathbb{Z}^2$. The two processes at time t are denoted by $(\sigma_t^\eta, \tilde{\sigma}_t^\omega)$, where (η, ω) are the initial spin configurations. The joint expectation of the process is denoted by $\hat{\mathbb{E}}$. The initial conditions will in general be chosen w.r.t. the product measure $d\hat{\mu}^{\beta,+}(\eta, \omega) = d\mu^{\beta,+}(\eta)d\mu^{\beta,+}(\omega)$, where $\mu^{\beta,+}$ is the Gibbs measure of the positive pure phase. In [21] the following theorem was proved.

Theorem 6.1. *There exist positive constants C_1, C_2 and κ independent of the choice of the coupling such that*

$$\forall t > 0, \quad \int d\hat{\mu}^{\beta,+}(\eta, \omega)\, \hat{\mathbb{E}}\left(\sigma_t^\eta(0) \neq \tilde{\sigma}_t^\omega(0)\right) \geq C_1 \exp\left(-C_2 \sqrt{t}\,(\log t)^\kappa\right) \quad (6.1)$$

Remark 6.2. Although one believes that the quantity considered in the theorem is a good measure of the time autocorrelation in the positive phase of the spin at the origin, predicted in [62] to decay as $e^{-\sqrt{t}}$ in $d = 2$, the latter is unfortunately only bounded from *above* by the LHS of (6.1). A related result at $\beta = +\infty$ was proved recently in [64] for the zero temperature dynamics (see theorem 1.2 there).

6.3 Torpid mixing for Potts model in $\mathbb{Z}^d$

In a recent paper [24] the problem of estimating the mixing time T_1 of Glauber dynamics inside the phase coexistence region for models other than the Ising model has been considered, particularly for the q-state Potts model (see section 3.5 for a definition).

In [24] the authors assume that the system is on a torus $\Lambda \subset \mathbb{Z}^d$ of side L (periodic boundary conditions) and prove the following.

Theorem 6.3. *For $d \geq 2$ and large enough q there exists $\beta_c = \beta_c(d, q)$ and $k > 0$ such that if $\beta > \beta_c$*

$$T_1 \geq e^{kL^{d-1}/\log L}$$

One of the interesting aspects of this paper, besides the main results, is that powerful techniques of statistical physics, like Pirogov–Sinai theory, are adapted and applied to control combinatorial quantities like the number of cut-sets in the graphical expansion of the model.

7 Glauber dynamics for certain random systems in $\mathbb{Z}^d$

In this section we discuss some aspects of the relaxational properties of the Glauber dynamics when we remove the assumption of *translation invariance* of the interaction and consider in particular short-range *random interactions*. The static and dynamic settings are those illustrated in section 3.8 and section 4.1, respectively.

We begin by discussing disordered systems in the so-called Griffiths region by analyzing the rather special but highly representative case of the dilute Ising model. We refer to [76], [75], [38] for a more general class of systems and the related physical literature.

7.1 Combination of torpid and rapid mixing: the dilute Ising model

The (bond) dilute Ising ferromagnet is obtained from the standard Ising model by removing, independently for each bond $[x, y]$, the coupling β with probability $1 - p$, $p \in (0, 1)$. It turns out (see, e.g. [66]) that for p above the percolation threshold there exists a critical value $\beta_c(p)$ such that for $\beta > \beta_c(p)$ the infinite-volume spontaneous magnetization is nonzero.

When $p \in [0,1]$ and $\beta \ll \beta_c(1)$ (in two dimensions even $\beta < \beta_c(1)$) strong mixing $SM(\Lambda, C, m)$ applies for either all volumes or all large enough regular volumes with constants C, m uniform in the disorder configuration, and the associated Glauber dynamics is rapid mixing in the sense discussed in the previous sections [39].

The more interesting phase region we want to analyze is instead the region of Griffiths singularities (see, e.g. [66]), namely $\beta \in (\beta_c(1), \beta_c(p))$. Here, as explained in section 3.8, the Gibbs state is still unique, but, e.g. the covariance between $\sigma(x)$ and $\sigma(y)$ starts to decay exponentially only for $d(x,y) \geq \ell(\vartheta, x)$, where $\ell(\vartheta, x)$ is a random length which is finite for almost all disorders ϑ. As we will see, in the Griffiths region the mixing time of the Glauber dynamics in a box Λ is the combination of the rapid relaxation of part of Λ plus the torpid relaxation of rare *bad* clusters inside Λ.

For this purpose it is very instructive to examine the simpler case of $p < p_c$, where p_c is the critical value of the independent bond percolation in $\mathbb{Z}^d$. A suitable coarse graining analysis (see [39] and [121]) shows that many (but not all) of the features of the case $p < p_c$ remain true also for $p \geq p_c$ and $\beta \in (\beta_c(1), \beta_c(p))$.

Let us first observe that, with probability one, the infinite-volume Glauber dynamics is a product dynamics for each of the percolation clusters $\{W_i\}$. Thus, if we consider a local function f that for simplicity we can take as the spin at the origin, we get that

$$\|P_t f\|_2 \leq e^{-\lambda_0 t}$$

where, for any $x \in \mathbb{Z}^d$,

$$\lambda_x = \mathrm{gap}(\mathcal{L}^\tau_{W_x})$$

if W_x is the cluster containing x. Since the clusters W_x are finite with probability one, we can immediately conclude that $\|P_t^{\Lambda, \tau} f\|_2$ converges exponentially fast to its equilibrium value but with an exponential rate, λ_0 in our case, that *depends* on the chosen local function through its support. It is important to outline here two important features of the dynamics in the present case:

(i) In $d = 2$ for any $\beta > \beta_c(1)$ or in $d \geq 3$ and any β large enough, $\inf_x \lambda_x = 0$ with probability one. By ergodicity we have in fact that, with probability one, for any $L \geq 1$ we can find $x(L)$ such that $Q_L(x(L)) = W_i$ for some i. Thanks to the results of the previous section the spectral gap of the cluster W_i is thus exponentially small in L^{d-1}. In particular the spectral gap of the infinite-volume dynamics is zero. We can say that such nonuniformity of the rates λ_x is a first signal of the Griffiths phase.

(ii) The fact that local functions relax exponentially fast, although with a nonuniform rate, is a specific feature of the dilute model and it does not extend to more general systems in which the interaction between clusters of strongly interacting spins is weak but nonzero (see [39]).

Although the analysis of the relaxation to equilibrium for a fixed realization of the disorder is certainly interesting, much more relevant from the physical point of view is the same analysis when one takes the average over the disorder. It is here that the differences between the dynamics in the paramagnetic phase and in the Griffiths phase appear more pronounced. Let us denote by $\mathbb{E}(\cdot)$ the average w.r.t the disorder. Then, using the exponential decay of the cluster size distribution together with attractivity and the bounds on the spectral gap for the pure Ising model below the critical temperature in a box Λ with free boundary conditions, is not difficult to prove [39] that (f is, as above, the spin at the origin), for any $\beta > \beta_c$ in $d = 2$ or for any β large enough in higher dimensions, there exist two positive constants C_1, C_2 such that, for any large enough time t

$$e^{-C_1 \log(t)^{\frac{d}{d-1}}} \leq \mathbb{E}\,\|P_t f\|_2 \leq e^{-C_2 \log(t)^{\frac{d}{d-1}}}$$

We would like to conclude this part with a short discussion of the almost sure scaling law of $\mathrm{gap}(\mathcal{L}_{Q_L}^\tau)$ as $L \to \infty$. One of the main results of [39] is that, almost surely, the inverse spectral gap in the box Q_L with boundary conditions τ grows, as $L \to \infty$, roughly as $e^{-k \log(L)^{\frac{d-1}{d}}}$ and similarly for the logarithmic Sobolev constant. Above the percolation threshold similar results hold true but with an extra factor in the exponential of the form $(\log \log L)^{d-1}$.

7.2 Relaxation to equilibrium for spin glasses

There are very few mathematical results describing the dynamics of mean field models of spin glasses (see for instance [25], [63] and [125]). Here we will briefly describe some results on the spectral gap of Glauber dynamics for two popular mean field models of spin glasses with N variables, namely the REM (*random energy model*) and the SK (*Sherrington–Kirkpatrick*) models (see section 3.8). The notation will be that of section 3.8 and we will denote by $\mathcal{L}_N$ the Glauber generator. One key feature of these models is the fact that, because of fluctuations of the environment, a large system creates deep traps, namely configurations which, when taken as the starting point of the dynamics, require a very large time to relax to the invariant measure. Although deep traps are not numerous and are randomly located in the configuration space, they tend to dominate the asymptotic scaling in N of, e.g. the spectral gap. For example, it is possible to deduce from [63] that for any $\beta > 0$ the scaling law of the spectral gap for the REM obeys

$$\lim_{N \to \infty} \frac{1}{N} \log(\mathrm{gap}(\mathcal{L}_N)) = -\beta \sqrt{2 \log 2}$$

in spite of a static transition (see, e.g [46]) at $\beta = \sqrt{2 \log 2}$. The main point of [125] is that, in order to feel the difference between high and low temperature, one needs to measure the relaxation time with the time T_1^ν given in (2.4), with ν the uniform measure on the configuration space because in this way the worst traps have very little weight in the computation of T_1^ν. In particular theorem 1.1 of [125] when applied to the REM proves that

$$\limsup_{N\to\infty} \frac{1}{N} \log(T_1^\nu) \le 2\beta^2 \qquad\qquad \text{if } \beta \le \sqrt{2\log 2}$$

$$\limsup_{N\to\infty} \frac{1}{N} \log(T_1^\nu) \le 2\beta\sqrt{2\log 2} \qquad\qquad \text{if } \beta \ge \sqrt{2\log 2}$$

Although the factor 2 in the r.h.s. is probably wrong, the two asymptotics above are able to distinguish between a high- and low-temperature phase. Similar results hold also for the SK model.

8 Glauber dynamics for more general structures

In this section we conclude the part on "nonconservative" dynamics by reviewing results and ideas related to the topics discussed so far, in the context of models that, either because the underlying graph is different from the ordinary cubic lattice or because the interaction is infinite for certain configurations or because the structure of the Markov chain is drastically different from that of the single site Glauber dynamics, do not fit the previous general assumptions. We have in mind here, e.g. the Ising model on trees, the hard-core model (independent sets) and the Swendsen–Wang dynamics for Potts models.

As a first general result we can quote the following. Let $G = (V, E)$ be a finite graph with maximal degree Δ, let for simplicity $S = \{0, 1\}$ and let μ be the Gibbs measure on $\Omega := S^V$ associated with a bounded, finite-range interaction Φ on G, i.e.

$$\mu(\sigma) \propto \exp(-H(\sigma))$$

where $H(\sigma) = \sum_{\Lambda \subset G} \Phi_\Lambda(\sigma)$ and $\Phi_\Lambda \equiv 0$ if the diameter of the subgraph Λ is larger than some fixed positive number r. Let $\|\Phi\| := \sup_{x \in V} \sum_{\Lambda \ni x} \|\Phi_\Lambda\|_\infty$ and assume condition $SMT(\Lambda, 1, m)$, i.e.

$$\sup_{\tau \in \Omega} \sup_{\Lambda \subset V} \sup_{s,s' \in S} |\mu_\Lambda^\tau(h_x^s, h_y^{s'})| \le e^{-md(x,y)}, \qquad \forall x, y \in V$$

where $d(x, y)$ is the distance between x and y along the graph and h_x^s has been defined in (3.4). Then the following holds (see for instance theorem 2.1 in [117] for a somewhat more general formulation).

Theorem 8.1. *Let $\mathcal{L}$ be the generator of a Glauber dynamics reversible w.r.t. μ, with rates satisfying the general conditions (i),(ii) and (iii) of section 4.1. Then there exists $m_0 > 0$ and for all $m > m_0$ a constant $\kappa = \kappa(\|\Phi\|, m, c_{min})$ such that the logarithmic Sobolev constant $c_s(\mathcal{L})$ satisfies $c_s(\mathcal{L}) \le \kappa$.*

Remark 8.2. The above is nothing but the "high temperature" or "very weak coupling" analogue for general graphs of the more sophisticated theorem 5.1 valid in $\mathbb{Z}^d$.

We will now discuss more specific models in order to have more detailed results.

8.1 Glauber dynamics on trees and hyperbolic graphs

The subjects of Glauber dynamics for Ising and Potts models and proper coloring on general graphs of bounded degree have been considered in a systematic way in [94] for a discrete time Gibbs sampler (heat bath dynamics) with the goal of relating the geometry of the graph with n vertices to the growth in n of the mixing time T_1. The setting is the following. Let $G = (V, E)$ be a finite graph with maximal degree Δ. When G is the b-ary tree of height r it will be denoted by T_r^b.

The Ising model on G with free boundary conditions, inverse temperature β and zero external field h is defined as in (3.6) but with the edges of $\mathbb{Z}^d$ replaced by the edges of G, and similarly for the Potts model. When $G = T_r^b$ the following construction of the Ising Gibbs measure with free boundary conditions is useful. Let $\varepsilon := \left(1 + e^{2\beta}\right)^{-1}$. Pick a random $\pm$ spin at the root of the tree uniformly. Scan the tree top-down, assigning to the vertex v a spin equal to the parent spin with probability $1 - \varepsilon$ and opposite with probability ε.

The equilibrium measure of proper coloring of G with q colors was defined in section 3.7 while its Glauber dynamics is as follows. With rate n, n being the number of vertices of G, a vertex v is chosen and the color $\sigma(v)$ in v is replaced with a new color chosen uniformly among all colors which are not assigned to the neighbors of v.

In order to discuss the results we need an extra definition.

Definition 8.3. *The* exposure $\mathcal{E}(G)$ *of the graph G is the smallest integer such that there exists a labeling $v_1, \ldots, v_n$ of the vertices of G with the property that for all $1 \leq k \leq n$, the number of edges connecting $\{v_1, \ldots, v_k\}$ to $\{v_{k+1}, \ldots, v_n\}$ is at most $\mathcal{E}(G)$.*

With this notation, two of the main results (translated in the continuous-time setting) of [94] are as follows.

Proposition 8.4. *(a) In the Ising case,* $\mathrm{gap}^{-1} \leq n e^{\beta(4\mathcal{E}(G)+2\Delta)}$.
(b) For proper coloring with $q \geq \Delta + 2$, $\mathrm{gap}^{-1} \leq (\Delta + 1)n\, q^{2\mathcal{E}(G)+1}$.

Theorem 8.5. *Consider the Ising model on the b-ary tree T_r^b. Then:*

(i) Low temperature. *If $1 - 2\varepsilon \geq \frac{1}{\sqrt{b}}$ then $\mathrm{gap}^{-1} \leq C\, n^{\log(b(1-2\varepsilon)^2)}$. Moreover, the relaxation time is polynomial in n at all temperatures with a diverging degree as $\beta \to \infty$.*

(ii) Intermediate and high temperature. *If $1 - 2\varepsilon < \frac{1}{\sqrt{b}}$ then gap^{-1} is bounded from above uniformly in r.*

Remark 8.6. The surprising outcome of these results, compared to their analogues for the cubic lattice, is that the relaxation time is bounded uniformly in the depth of the tree even for intermediate temperature when there are infinitely many Gibbs measures in the limit $r \to \infty$. In other words the usual appealing picture "uniqueness of the Gibbs measure $\Rightarrow$ rapid mixing" together with "nonuniqueness (phase coexistence) $\Rightarrow$ torpid mixing" does not apply to Glauber dynamics on trees.

Remark 8.7. The proof of the above proposition is based on canonical paths techniques and follows quite closely its cubic lattice analogue [121]. The proof of the theorem is based on a recursive scheme at low temperature and on block dynamics plus path coupling at intermediate and high temperatures. Quite interestingly, the proof of the boundedness of the relaxation time at high or intermediate temperatures is based on a key correlation inequality (see the proof of lemma 4.1 in [94]), valid for the Ising model on trees, that it has been conjectured to hold for any graph of bounded degree. The proof of such an inequality for $\mathbb{Z}^d$ would immediately imply, using the results of [116] and [141], rapid mixing and $SM(C, m)$ below β_c for the d-dimensional Ising model with zero external field.

The third interesting result of [94] has the same flavor of theorem 4.9, namely bounded relaxation time implies exponential decay of covariances in the Gibbs distribution, and it applies to any arbitrary graph G of bounded degree.

8.2 Glauber dynamics for the hard-core model

Here we analyze, following [164], the *hard model* or *independent set* with fugacity λ (see (3.9)) on a finite graph $G = (V, E)$ with maximum degree Δ. The associated Glauber dynamics is defined a priori on the set of all subsets of V, not just independent sets, but it will have the property that, if $\sigma_{t=0} \subset V$ is an independent set then σ_t is also an independent set for $t \geq 0$. As usual we present the continuous-time version of the chain described in [164].

Given $\sigma \subset V$ and a vertex $v \in V$ let

$$c_v^+(\sigma) = \begin{cases} \frac{\lambda}{1+\lambda} & \text{if no neighbors of } v \text{ are in } \sigma \\ 0 & \text{otherwise} \end{cases} \quad \text{and} \quad c_v^-(\sigma) = \frac{1}{1+\lambda}$$

The generator $\mathcal{L}_G$ takes then the following form:

$$\mathcal{L}_G f(\sigma) = \sum_{v \in V} c_v^+(\sigma)[f(\sigma \cup \{v\}) - f(\sigma)] + \sum_{v \in V} c_v^-(\sigma)[f(\sigma \setminus \{v\}) - f(\sigma)]$$

Notice that eventually the chain will enter the set of all the independent subsets of V and it will then stay there forever. The main result of [164] is the following.

Theorem 8.8. *Let $n := |V|$ and assume $\lambda < \frac{2}{\Delta-2}$. Then the mixing time T_1 is bounded by $C \log n$ for some constant C independent of n.*

Remark 8.9. In the discrete time setting the above result is usually quoted as "the mixing time is $O(n \log n)$". The proof given in [164], based on path coupling argument, does not prove directly that in the limiting case $\lambda = \frac{2}{\Delta-2}$ the mixing time is polynomial in n. That result is proved by some variant of the methods (see [163] and [59]). A weaker version of the above theorem limited to triangle free graphs was proved [110]. A key step to remove this restriction was the enlargement of the state space of the chain to all subsets of V. For $\Delta \geq 6$ and $\lambda = 1$ it has been proved in [60] that the mixing time is exponentially large in n.

8.3 Cluster algorithms: the Swendsen–Wang dynamics for Potts models

For ferromagnetic Potts models an alternative chain to the usual Glauber chain is represented by the famous Swendsen–Wang dynamics [165], [61] which is intimately related to the FK representation of the Gibbs measure discussed in section 3.6.

The Swendsen–Wang chain is a discrete time, highly nonlocal *cluster algorithm*, that turned out to be extremely successful in simulations because of its apparently very good mixing properties in many instances of practical interest, where instead Glauber dynamics mixes slowly.

The setting is that described in section 3.6 for the ferromagnetic Potts model with q colors at inverse temperatures β on a finite graph $G = (V, E)$ and a single step of the algorithm can be described as follows.

Let the current configuration of colors on V be σ. Then the new configuration σ' is obtained via the following updating rules:

(1) Let $B(\sigma) \subset E$ be the set of edges joining vertices with the same color. Delete each edge in $B(\sigma)$ independently with probability $e^{-\beta}$. Call $A(\sigma) \subset B(\sigma)$ the set of surviving edges.

(2) The new graph (V, A) consists of connected components (isolated vertices included) called "clusters". For each cluster and independently among all clusters a color is drawn at random uniformly from the q available colors and all the spins in the cluster are assigned that color.

It is not difficult to check that the above chain is ergodic and reversible w.r.t. the Gibbs measure of the Potts model on G with *free boundary conditions*. Other boundary conditions or external magnetic fields can also be accomplished.

At least when β is sufficiently small, $q = 2$ (Ising model) and G a finite box in $\mathbb{Z}^d$, the Swendsen–Wang dynamics is known be uniformly exponentially ergodic (mixing time $T_1 = O(\log n)$ if $|V| = n$) by coupling methods combined with multiscale analysis ([123], [122]).

Similar results were later obtained for more general graphs in [45] and [86]. For $q = 2$, large enough β and plus boundary conditions (any cluster attached to the boundary is part on just one cluster that has always the same color "+") at the boundary of a large box in $\mathbb{Z}^d$, it was proved in [120] that the speed of relaxation is exponentially fast in t^α, $\alpha = \frac{\log 2}{\log 3}$, after a time of the order of $\exp((\log n)^{1/2})$.

Quite interestingly, there have been recently a number of papers proving, contrary to a widespread heuristic, *slow mixing* in various cases of theoretical and practical interest. For the complete graph ("Curie–Weiss model"), $q \geq 3$ and $\beta = \beta_c(q)$, it was proved in [70] that there is a bottleneck in the phase space due to the occurrence of a first order phase transition and consequently the mixing time is exponentially large in n. Such a result was shown to persist for the random graph $G_{n,p}$ if $p = O(n^{-1/3})$ [45]. Finally, it was proved in [24] that the mixing time on the d-dimensional torus of side L satisfies

$$T_1 \geq e^{cL/(\log L)^2}$$

provided that $d \geq 2$, $\beta = \beta_c(q, d)$ and q is large enough.

9 Mixing time for conservative dynamics

In this section and in the next section we review another class of symmetric Markov chains that have played an important role in mathematical physics and probability theory in recent years, particularly in a rigorous derivation of *hydrodynamic limits* for various models (see for example [95], [162] and references therein).

We will refer to these new chains as *conservative dynamics* because they are usually constructed as some variant of a *fixed* number of interacting random walks on a graph, reversible w.r.t. the *canonical* Gibbs measure associated with some finite-range interaction. Because of the conservation law, even at high temperature one cannot expect uniformly bounded relaxation times as in the nonconservative, Glauber case and much of the research was devoted to proving that in the one-phase region the spectral gap and the logarithmic Sobolev constant scale *diffusively* in the size of the underlying graph, namely a behavior comparable to that of a single random walk on the same graph. From a technical point of view a major new difficulty that arises when dealing with conservative dynamics is that equilibrium covariances, i.e. covariances w.r.t. the *canonical* Gibbs measure, do not decay exponentially fast even in absence of interaction precisely because of the conservation law. In this section we restrict ourselves to a class of models which, because of their structure, do not present some of the difficulties one encounters in more physically sound systems, like lattice gases with short-range interaction, to be discussed in the final section.

9.1 Random transposition, Bernoulli–Laplace and symmetric simple exclusion

We begin by reviewing here three models of interacting random walks for which the interaction among the walks is just mutual exclusion. Later on we will discuss more elaborate models in which the random walk, besides mutual exclusion, interact also through some general finite-range potential. Our main reference for this section is [104] but see also [47], [51] and [50].

Random transposition (RT). In this model there are n distinct sites and n distinct particles. In a generic configuration $\sigma \in S_n$, S_n being the permutation group of n objects, the variable σ_i denotes the label of the particle at the site i and the configuration σ^{ij} denotes the configuration obtained from σ by exchanging the variables at sites i and j. The dynamics goes as follows: each particle selects with rate one and uniformly a site j and then exchanges position with the particle at j. More formally, the generator is given by

$$\mathcal{L}f(\sigma) = \frac{1}{n} \sum_{i,j=1}^{n} [f(\sigma^{ij}) - f(\sigma)] \tag{9.1}$$

and the invariant measure π is just the uniform measure on $\Omega_n^{\mathrm{RT}} = S_n$.

The Bernoulli–Laplace model (BL). In this model we have n sites and $N < n$ indistinguishable particles and each site is occupied by at most one particle. Thus, in a generic configuration σ, the variable $\sigma_i \in \{0,1\}$ tells us whether the site i is occupied ($\sigma_i = 1$) or empty ($\sigma_i = 0$). By particle–hole duality we can restrict $N \leq \frac{n}{2}$. The dynamics is similar to the RT model: each particle waits a mean one exponential time and then jumps to a given empty site with probability $1/n$. The generator is given by

$$\mathcal{L}f(\sigma) = \frac{1}{2n} \sum_{i,j=1}^{n} [f(\sigma^{ij}) - f(\sigma)] \tag{9.2}$$

and the invariant measure π is just the uniform measure on $\Omega_{n,N}^{\mathrm{BL}}$, the space of all subsets of the n sites with N elements.

The symmetric simple exclusion on $\mathbb{Z}/n\mathbb{Z}$ (SE). Here we have a situation similar to the BL model, but jumps occur only between nearest neighbor sites. The generator is given by

$$\mathcal{L}f(\sigma) = \frac{1}{2} \sum_{i=1}^{n} [f(\sigma^{i,i+1}) - f(\sigma)] \tag{9.3}$$

and the invariant measure π is again the uniform measure on $\Omega_{n,N}^{\mathrm{SE}} = \Omega_{n,N}^{\mathrm{BL}}$.

For the RT and BL models the logarithmic Sobolev constant and the mixing time T_2 given in (2.3) have been determined completely by means of Fourier analysis in [47], [51] and [50]. The results read as follows.

Theorem 9.1. *There exists $c > 0$ independent of n such that:*

(i) $c^{-1} \log n \leq c_s^{\mathrm{RT}} \leq c \log n$;

(ii) $0 < \liminf T_2^{\mathrm{RT}} / \log n \leq \limsup T_2^{\mathrm{RT}} / \log n < \infty$;

(iii) If $N = \frac{n}{2}$ then $T_2^{\mathrm{BL}} \geq \frac{1}{2} \log n$;

(iv) $T_2^{\mathrm{BL}} \leq 2(1 - \frac{N}{n})\big[\log n + c\big]$.

The martingale method of [108], [168] proved instead the following.

Theorem 9.2. *There exists $c > 0$ independent of n, N such that:*

(i) $c^{-1} \log \frac{n^2}{N(n-N)} \leq c_s^{\mathrm{BL}} \leq \frac{2}{\log 2} \log \frac{n^2}{N(n-N)}$, $n \geq 2$;

(ii) $c^{-1} n^2 \leq c_s^{\mathrm{SE}} \leq c n^2$;

(iii) $c^{-1} n^2 (1 + \log N) \leq T_2^{\mathrm{SE}} \leq c n^2 \log n$.

Quite interestingly, the proof of the lower bound on T_2^{SE} is based on an observation relating the time to stationarity to the hydrodynamical limit of the simple exclusion model [104]. Notice that for very few particles, $N = O(1)$, the usual bound $T_2^{\mathrm{BL}} \leq c_s^{\mathrm{BL}}\big(1 + \frac{1}{4} \log \log(\frac{1}{\pi_*})\big)$ is off by a factor $\log \log n$ w.r.t. the bound (iv) of theorem 9.1. Moreover, in the same situation, the bound $T_1^{\mathrm{BL}} \leq T_2^{\mathrm{BL}}$ gives the wrong order for the time to approach stationarity in total variation. In a recent interesting paper [68] it has been shown in fact that $T_1^{\mathrm{BL}} \leq 2\big(1 + \log \log \binom{n}{N}\big)$ by analyzing in detail not the logarithmic Sobolev constant c_s^{BL} but rather the so-called *entropy constant* c_e^{BL}, namely the best constant c in the inequality

$$\mathrm{Ent}_\pi(f) \leq c\,\pi\big(f(-\mathcal{L}) \log f\big), \quad f \geq 0, \; \pi(f) = 1$$

We conclude by saying that in [169] the martingale method was used to show the diffusive scaling of the logarithmic Sobolev constant for a more general class of simple exclusion models in which at each site a finite number $R > 1$ of particles is allowed, while in [101] the same method was adapted to bound the spectral gap of certain zero-range models.

9.2 The asymmetric simple exclusion

Here we consider the asymmetric version of the exclusion process described in the previous section in which each particle performs a random walk with a constant drift. The geometrical setting and the notation are those illustrated in section 3.10 but for the reader's convenience we recall them here.

Given two natural numbers L, H we consider the rectangle

$$\Lambda = \{(i, h) \in \mathbb{Z}^2 : i = 1, \dots, L \text{ and } h = 1, \dots, H\}$$

For each i, Λ_i stands for the *stick* at i given by $\Lambda_i = \{(i, h) : h = 1, \dots, H\}$. At each $x \in \Lambda$ we have a variable $\alpha_x \in \{0, 1\}$: we say that site x is occupied (by a particle) if $\alpha_x = 1$ and empty otherwise. The set of configurations $\{0,1\}^\Lambda$ is denoted by Ω and it is naturally decomposed in single stick configurations: $\alpha \in \Omega$ will be written often in the form $\alpha = (\eta_1, \ldots, \eta_L)$ with $\eta_i \in \{0,1\}^H$ denoting the restriction of α to the stick Λ_i.

Given a parameter $q \in (0,1)$ we define the product probability measure μ on Ω.

$$\mu(f) = \sum_{\alpha \in \Omega} \mu(\alpha) f(\alpha) , \quad \mu(\alpha) = \prod_{i=1}^{L} \prod_{h=1}^{H} \frac{q^{2h\alpha_{(i,h)}}}{1+q^{2h}} \tag{9.4}$$

where f is a generic function $f : \Omega \to \mathbb{R}$. According to μ, particles prefer to live on the region of small h, i.e. the bottom of the box Λ if we interpret h as a vertical coordinate. We define n_i as the number of particles in the stick Λ_i: $n_i(\alpha) = n_i(\eta_i) = \sum_{h=1}^{H} \alpha_{(i,h)}$ and consider the conditional probability measure

$$\nu = \nu_N = \mu\left(\cdot \mid \sum_{i=1}^{L} n_i = N \right) \tag{9.5}$$

The asymmetric diffusion that will be analyzed in the sequel can be described as follows. Every particle at row h tries to jump to an arbitrary empty site at row $h+1$ with rate q and to an empty site at row $h-1$ with rate $1/q$. The Markov generator is defined by the operator

$$\mathcal{L}f(\alpha) = \frac{1}{L} \sum_{i=1}^{L} \sum_{j=1}^{L} \sum_{h=1}^{H-1} c_{(i,h);(j,h+1)}(\alpha) \nabla_{(i,h);(j,h+1)} f(\alpha) \tag{9.6}$$

where we use the notation

$$\nabla_{(i,h);(j,h+1)} f(\alpha) = f(\alpha^{(i,h);(j,h+1)}) - f(\alpha) \tag{9.7}$$

$\alpha^{(i,h);(j,h+1)}$ denoting the configuration in which the values of α at (i,h) and $(j,h+1)$ have been interchanged while the rest is kept unchanged. The rates $c_{(i,h);(j,h+1)}$ are given by

$$c_{(i,h);(j,h+1)}(\alpha) = q^{\alpha_{(i,h)} - \alpha_{(j,h+1)}} \tag{9.8}$$

Simple computations show that $\mathcal{L}$ is self-adjoint in $L^2(\nu)$, the associated Dirichlet form being

$$\mathcal{D}(f,f) = \nu(f(-\mathcal{L})f) = \frac{1}{L} \sum_{i=1}^{L} \sum_{j=1}^{L} D_{ij}(f) \tag{9.9}$$

$$D_{ij}(f) := \frac{1}{2} \sum_{h=1}^{H-1} \nu\left[c_{(i,h);(j,h+1)} \left(\nabla_{(i,h);(j,h+1)} f \right)^2 \right]$$

The main result of [34] can be formulated as follows. Let

$$\gamma(L, H) = \sup_{N} \sup_{f \in L^2(\nu)} \frac{\mathrm{Var}(f)}{\mathcal{D}(f, f)}$$

where the number of particles N in $\sup_N$, using the particle–hole symmetry, is assumed to range from 1 to $\frac{LH}{2}$.

Theorem 9.3. *For every $q \in (0, 1)$ there exists $C < \infty$ such that*

$$\sup_{L,H} \gamma(L, H) \leq C$$

Remark 9.4. For $L = 1$ and $N = \frac{H}{2}$ it has been proved in [14] that the mixing time grows like H. Remarkably, in the same setting the logarithmic Sobolev constant grows at least like H^2.

Sketch of the proof. We briefly describe the main ideas behind the proof of theorem 9.3 since it allows us to illustrate another technique to analyze conservative dynamics that was introduced recently in [36] (see also [49] and [89] for related work) in the framework of the Kac model of the nonlinear Boltzmann equation. The advantages of the approach consists of its simplicity when compared to other alternatives like the martingale method of [108] or the induction on volumes of [30] but, until now, it has been limited to canonical measures arising from a product measure and to the analysis of the spectral gap and not of the logarithmic Sobolev constant.

Like in some of the approaches to bound the spectral gap described previously, the first idea is to recursively bound $\gamma(L, H)$ in terms $\gamma(1, H)$. The latter is in turn finite uniformly in H by a recursive argument in the number of particles combined with some simple estimates result for just one particle (random walk with constant drift) [35]. The starting point, as, e.g. in the martingale approach of [108], is a decomposition of the variance of an arbitrary function f as:

$$\mathrm{Var}(f) = \frac{1}{L} \sum_{k=1}^{L} \nu\big(\mathrm{Var}(f \mid \mathcal{F}_k)\big) + \frac{1}{L} \sum_{k=1}^{L} \mathrm{Var}\big(\nu(f \mid \mathcal{F}_k)\big)$$

where $\mathcal{F}_k$ denotes the σ-algebra generated by the stick variables η_k, $k = 1, \ldots, L$. By induction the first term can be bounded in terms of $\gamma(L-1, H) \times \mathcal{D}(f, f)$. The main new idea comes in the analysis of the second term and consists in introducing the stochastic symmetric operator

$$Pf = \frac{1}{L} \sum_{k=1}^{L} \nu(f \mid \mathcal{F}_k)$$

and observing that for any mean zero function f the following identity holds true:

$$\frac{1}{L} \sum_{k=1}^{L} \mathrm{Var}\big(\nu(f \mid \mathcal{F}_k)\big) = \nu\big(fPf\big)$$

Thus

$$\mathrm{Var}(f) - \frac{1}{L} \sum_{k=1}^{L} \mathrm{Var}\big(\nu(f \mid \mathcal{F}_k)\big) = \mathrm{Var}(f) - \nu\big(fPf\big) = \nu\big(f(\mathbb{1} - P)f\big) \,.$$

so that one is left with the problem of establishing an estimate from below on the spectral gap of P which is sharp enough to allow a successful iteration in L for $\gamma(L, H)$. The key point now is that, because of the particular form of P and of the symmetry of the measure ν, the estimate of the spectral gap of P boils down to the estimate from below of the spectral gap of a particular one-dimensional random walk that can be described as follows. Let $n_\pm$ be the minimum and maximum number of particles allowed in a single stick, say the first one. Then the state space for the random walk is the interval $[n_-, n_- + 1, \ldots, n_+]$ and the transition kernel $q(n \to m)$ is given by $\nu\big(n_1 = m \mid n_2 = n\big)$. It is easy to check that such a process is ergodic iff $L \geq 3$. The study of its relaxation time represents in some sense the technical core of the proof and it requires a rather detailed analysis based on results of an equivalence of ensembles type.

Remark 9.5. Recently, the above technique has been successfully used [33] to bound the spectral gap of some Ginzburg–Landau models extending and simplifying previous results [100].

Interesting applications of theorem 9.3, particularly those to quantum Heisenberg models, are linked to the analysis of the restriction of the above defined process to the horizontal sums of the basic variables $\alpha_{i,h}$ given by

$$\omega_h = \sum_{i=1}^{L} \alpha_{(i,h)} \,, \qquad h = 1, \ldots, H$$

To be more precise, denote the set of permutations of $\{1, \ldots, L\}$ by $\mathcal{P}_L$ and define the subspace $\mathcal{S}$ of horizontally symmetric functions by

$$\mathcal{S} = \{f \in L^2(\nu) : \ f(\alpha) = f(\alpha^{\pi, h}), \ \forall \pi \in \mathcal{P}_L, \ \forall h = 1, \ldots, H\} \tag{9.10}$$

The subspace $\mathcal{S}$ is left invariant by the generator $\mathcal{L}$ and it can be naturally identified with the space $L^2(\hat{\Omega}, \hat{\nu})$, $\hat{\Omega} = \{0, 1, \ldots, L+1\}^H$ and $\hat{\nu}$ the marginal of ν on the horizontal sums $\omega = \{\omega_h\}$. An explicit computation shows that the restriction of $\mathcal{L}$ to S, call it $\hat{\mathcal{L}}$, is a symmetric Markov generator on $L^2(\hat{\Omega}, \hat{\nu})$ given by

244   F. Martinelli

$$\widehat{\mathcal{L}}\hat{f}(\omega) = \frac{1}{L}\sum_{h=1}^{H-1}\left\{w_{+,h}(\omega)\left[\hat{f}(\omega^{+,h}) - \hat{f}(\omega)\right] + w_{-,h}(\omega)\left[\hat{f}(\omega^{-,h}) - \hat{f}(\omega)\right]\right\}$$

$$(9.11)$$

$$w_{+,h} := q^{-1}\,(L - \omega_h)\omega_{h+1}\,, \qquad w_{-,h} := q\,(L - \omega_{h+1})\omega_h$$

$$\omega_{h'}^{\pm,h} := \begin{cases} \omega_{h'} & h' \neq h, h+1 \\ \omega_h \pm 1 & h' = h \\ \omega_{h+1} \mp 1 & h' = h+1 \end{cases}$$

The Markov chain generated by $\widehat{\mathcal{L}}$ can be interpreted as describing the fluctuations of a nonnegative profile $\omega := \{\omega_h\}_{h=1}^{H}$ subject to a fixed area constraint ($\sum_h \omega_h = $ constant). In the case $L = 2$ the new variables $\{\omega_h\}_{h=1}^{H}$ can also be interpreted as describing a model for diffusion-limited chemical reactions [3]. Describe the state $\omega_h = 2$ as the presence at h of a particle of type A, $\omega_h = 0$ as a particle of type B and $\omega_h = 1$ as the absence of particles (*inert*). If n_A, n_B denote the size of the two populations we see that the difference $n_A - n_B$ is conserved and this system can be studied as a model for asymmetric diffusion with creation and annihilation of the two species. Particles of type A have a constant drift toward the bottom ("small h" region) while particles of type B have the same drift toward the top ("large h" region). They perform asymmetric simple exclusion with respect to the *inert* sites but when they meet (i.e. when they become nearest neighbors) they can produce the annihilation reaction $A + B \to$ *inert*. The reverse reaction *inert* $\to A + B$ restores steady-state fluctuations given by the canonical measure. Clearly, according to theorem 9.3, the relaxation time of the marginal process is also bounded from above uniform in L, H and N.

The main connection between the above process and the XXZ quantum spin model described in section 3.10 goes as follows [34]. Assume $L = 2S$ and recall the definition of the Hilbert space sectors $\mathfrak{H}_n$ given in section 3.10. For any $\varphi \in \mathfrak{H}_n$ write

$$\varphi = \sum_{\substack{m \in \mathcal{Q}_S: \\ \sum_h m_h = n}} \varphi(m)\,|m\rangle$$

and define $\tilde{\varphi}(\omega) = \varphi(\omega - S)$. Then the transformation

$$\varphi(m) \;\longrightarrow\; \frac{1}{\sqrt{\hat{\nu}(\omega)}}\,\tilde{\varphi}(\omega) \;=:\; \left[U_n\varphi\right](\omega), \qquad \omega = m + S$$

maps unitarily $\mathfrak{H}_n$ into $L^2(\hat{\Omega}, \hat{\nu})$ and moreover

$$U_n\,\mathcal{H}^{(S)}\,\varphi = -\frac{S}{\Delta}\,\widehat{\mathcal{L}}\,U_n\varphi\,, \qquad \varphi \in \mathfrak{H}_n$$

The above equivalence allows one to transform bounds on the spectral gap of the Markov generator $\widehat{\mathcal{L}}$ into bounds for the *energy gap* gap($\mathcal{H}^{(S)}$) of the quantum Hamiltonian $\mathcal{H}^{(S)}$.

A fundamental question associated with the stability of "quantum interfaces" is the positivity of the energy gap ([97], [96]). Recently this question was studied in great detail in [23], [22] and [96] by both analytical and numerical means. One of the main results of [96] is a proof of the fact that for every $S \in \frac{1}{2}\mathbb{N}$, $\mathrm{gap}(\mathcal{H}^{(S)})$ is positive uniformly in H. Furthermore it was conjectured on the basis of numerical analysis that $\mathrm{gap}(\mathcal{H}^{(S)})$ should grow linearly with S. Thanks to theorem 9.3, a positive answer to that question was given in [34].

Theorem 9.6. *For every $\Delta \in (1, \infty)$, there exists $\delta > 0$ such that*

$$\delta S \leq \mathrm{gap}(\mathcal{H}^{(S)}) \leq \delta^{-1} S$$

for all $S \in \frac{1}{2}\mathbb{N}$ and all $H \geq 2$.

Remark 9.7. In [34] was also discussed the d-dimensional analogue of the above model.

9.3 The Kac model for the Boltzmann equation

We next discuss the Kac model for a gas of N particles evolving in one dimension under a random collision mechanism that preserves the total energy but not the momentum [93]. Such a model was motivated by the study of the nonlinear Boltzmann equation and by the problem of understanding the relaxation of the velocity distribution for large times. Our main reference here is [36] but we refer the reader also to [49] and [89] for this and related models. Although the Kac model does not fit in the general framework discussed so far because the state space is not discrete, we decided to include it in this review because of its interest.

The model is as follows. Fix $E > 0$ to be interpreted as the total energy of the gas and let Ω_N be the sphere $S^{N-1}(\sqrt{E})$ in $\mathbb{R}^N$. A Markov transition kernel Q on Ω_N is constructed as follows. Given a vector $\mathbf{v} = (v_1, v_2, \ldots, v_N) \in \Omega_N$, a pair $\{i, j\}$, $i < j$, is chosen at random and the two velocities v_i, v_j are changed to the new "postcollisional" velocities v_i^*, v_j^* according to the rule:

$$v_i^*(\vartheta) = v_i \cos(\vartheta) + v_j \sin(\vartheta) \quad \text{and} \quad v_j^*(\vartheta) = -v_i \sin(\vartheta) + v_j \cos(\vartheta) \quad (9.12)$$

where ϑ is a random angle chosen according to some a priori probability density $\varrho(\vartheta)$ on $[-\pi, \pi]$. The density $\varrho(\vartheta)$ is assumed to be continuous, symmetric around the origin i.e. $\varrho(\vartheta) = \varrho(-\vartheta)$, and strictly positive at $\vartheta = 0$. In other words the action of the Markov transition kernel Q on an arbitrary continuous function f on Ω_N has the following expression:

$$Qf(\mathbf{v}) = \binom{N}{2}^{-1} \sum_{i<j}^{N} \int_{-\pi}^{\pi} d\vartheta \, \varrho(\vartheta) f(R_{i,j}(\vartheta)\mathbf{v}) \qquad (9.13)$$

where $R_{i,j}(\vartheta)$ denotes the rotation in $\mathbb{R}^N$ that induces a clockwise rotation by ϑ on the v_i, v_j plane and fixes the orthogonal complement of this plane.

If μ_N denotes the normalized uniform measure on Ω_N it is not difficult to check that Q is a self-adjoint operator on $L^2(\Omega_N, \mu_N)$ because of the symmetry condition on ϱ and moreover, Q is ergodic because ϱ is continuous and strictly positive at $\vartheta = 0$.

Finally, the generator of the continuous-time Markov process considered by Kac is given by $\mathcal{L}_N := N(\mathbb{1} - Q)$, where the speeding factor N is dictated by physical considerations.

Ergodicity of Q implies that for any initial smooth probability density f_0 on Ω_N the density at time t given by

$$f_t := e^{t\mathcal{L}_N} f_0$$

converges as $t \to \infty$ to 1 and the main question is how fast this convergence takes place.

Kac considered the spectral gap λ_N of $\mathcal{L}_N$ defined by

$$\lambda_N = \inf_{\substack{f \in L^2 \\ \langle 1, f \rangle = 0}} \frac{\langle f, -\mathcal{L}_N f \rangle}{\langle f, f \rangle}$$

and conjectured that

$$\liminf_{N \to \infty} \lambda_N = C > 0$$

Notice that, since Q is not compact [49] it is not clear that $\lambda_N > 0$. In [49] it was proved that $\lambda_N \geq c/N^2$ for some $c > 0$ and Kac's conjecture for the special case of uniform ϱ was first proved in [89] by the martingale method. Later in [36] the ideas and techniques described in the previous section in the framework of the asymmetric simple exclusion were introduced and, always for uniform ϱ, it was proved that

$$\lambda_N = \frac{1}{2} \frac{N+2}{N-1}$$

so that $\lim_{N \to \infty} \lambda_N = \frac{1}{2}$.

Remark 9.8. The same result was obtained by Maslin in unpublished work by heavy use of representation theory (see [49] for account of Maslin's work).

On top of that it was also shown that λ_N has multiplicity one with eigenfunction

$$f_N(\mathbf{v}) = \sum_{j=1}^{N} (v_j^4 - \langle v_j^4, 1 \rangle)$$

The connection with a certain nonlinear PDE similar to the Boltzmann equation and known as the *Kac equation* goes as follows. Let $P_1 f(v) = \mathbb{E}(f \mid v_1 = v)$ (compare with section 10.2) and assume that the sequence of initial distributions $f_0^{(N)}(\mathbf{v})$ satisfies a certain independence property known as the "molecular chaos property" [93] and that

$$g(v) := \lim_{N \to \infty} P_1 f_0(v)$$

exists in L^1. Then $g_t(v) := \lim_{N \to \infty} P_1\left(e^{t\mathcal{L}_N} f_0\right)$ exists in L^1 and it satisfies the Kac equation

$$\frac{\partial}{\partial t} g_t(v) = 2 \int_{-\pi}^{\pi} \left(\int_{\mathbb{R}} dw \left[g_t(v^*(\vartheta)) g_t(w^*(\vartheta)) - g_t(v) g_t(w) \right] \right) \varrho(\vartheta) d\vartheta$$

9.4 Adsorbing staircase walks

Staircase walks are lattice paths in $\mathbb{Z}^2$ from $(0,0)$ to (n,n) which always stay above the diagonal $y = x$. Upon rotation by $\frac{\pi}{4}$ they become paths from $(0,0)$ to $(2n,0)$ obtained by adding $\mathbf{e}_+ := (1,1)$ or $\mathbf{e}_- := (1,-1)$ at each step and never falling below the x-axis. They are related to returning walks on an infinite d-ary tree starting and ending at the root, to certain model of statistical mechanics (see, e.g [161] and [87]) and to the zero temperature limit $\beta \to \infty$ of the Ising model in the triangle

$$\left\{ (x,y) \in \mathbb{Z}^2 + \left(\frac{1}{2}, \frac{1}{2} \right) \ : \ x, y \in [0,n], \ x \le y \right\}$$

with positive boundary condition along the shortest sides and negative boundary condition along the longest side. The number of staircase walks of length n is just the nth Catalan number $C(n)$ so that the uniform measure on the staircase walks assign probability $1/C(n)$ to each of them. A natural generalization studied in statistical mechanics is the following.

Given $\lambda > 0$, assign weight $\lambda^{k(w)}$ to a single walk w, where $k(w)$ is the number of times the walk w touches the x-axis. After normalization we obtain the Gibbs measure $\pi(w)$. In the Ising interpretation discussed above the weight $k(w)$ arises if, before the limit $\beta \to \infty$, one adds an extra coupling $\frac{J}{\beta}$ to the bonds crossing the diagonal $y = x$ with $e^{2J} = \lambda$.

If $\lambda < 1$ the walks are repelled from the x-axis, while if $\lambda > 1$ they are attracted and it is easy to see that there is a phase transition at $\lambda = 2$: when $\lambda < 2$ paths reach a typical distance $O(\sqrt{n})$ from the x-axis, while for $\lambda > 2$ they stay closer than $o(\sqrt{n})$.

Let us now examine a natural Markov chain on the set of staircase walks, known as the "mountain/valley" chain, reversible w.r.t. to the Gibbs measure $\pi(w)$ [134]. For simplicity we give the discrete time version.

Given a path w, pick i uniformly at random from $[2, 2n - 2]$ and call w' the path obtained from w by interchanging the ith and $(i+1)$th steps of w. If the resulting path w' is no longer a staircase walk stay at w. If instead the resulting path w' is still a staircase walk:

(1) replace w by w' with probability $1/4$ if the ith and $(i+1)$th steps consisted of $\mathbf{e}_+, \mathbf{e}_-$ (i is a local mountain) or vice versa (i is a local valley) and $k(w') = k(w)$;

(2) replace w with w' with probability $\frac{\lambda}{2(1+\lambda)}$ if i is a local mountain and $k(w') = k(w) + 1$;

(3) replace w by w' with probability $\frac{1}{2(1+\lambda)}$ if i is a local valley and $k(w') = k(w) - 1$;

(4) do nothing in all the other cases.

The main result of [134] is a proof that the mixing time of the above chain is polynomially bounded in n for all $\lambda > 0$. The case $\lambda < 1$ is relatively simple and it can be handled by coupling argument. The case $\lambda = 1$ is more subtle and in [167] it was proved a tight bound $O(n^3 \log n)$. When $\lambda > 1$ coupling alone seems difficult to implement because nearby paths tend to diverge instead of becoming closer near the x-axis. This difficulty was overcome in [134] thanks to a new interesting technique based on ideas from the *decomposition method for Markov chains* of [111]. It is an open interesting problem to derive sharp bounds in the case $\lambda \neq 1$ via analytic techniques.

10 Kawasaki dynamics for lattice gases

Here we finally consider the so-called *Kawasaki dynamics* for a finite-range, translation invariant, lattice gas model (see section 3.4) with interaction Φ. If Λ denotes the cube of $\mathbb{Z}^d$ of side $L \in \mathbb{N}$ we assume to have $N \leq \frac{1}{2}|\Lambda|$ particles (spins with $\sigma(x) = +1$) that jump to nearest neighbor empty ($\sigma(x) = 0$) sites, thus keeping the total number of particles constant. In analogy with the simple exclusion process, if σ^{xy} denotes the configuration in $\Omega_\Lambda := \{0,1\}^\Lambda$ obtained from σ by exchanging the its values at x and y, $x, y \in \Lambda$, the Markov generator of our chain L^τ_Λ is defined by

$$(L^\tau_\Lambda f)(\sigma) = \sum_{[x,y] \in \mathcal{E}_\Lambda} c^\tau_{xy}(\sigma)\,(\nabla_{xy} f)(\sigma) \qquad \sigma \in \Omega, \quad f : \Omega \mapsto \mathbb{R}$$

where $\nabla_{xy} f(\sigma) := f(\sigma^{xy}) - f(\sigma)$ and $\sum_{[x,y] \in \mathcal{E}_\Lambda}$ is the sum over all edges of $\mathbb{Z}^d$ with at least one of the two vertices in Λ.

The nonnegative real quantities $c^\tau_{xy}(\sigma)$ are the *transition rates* for the process and the superscript τ means that they coincide with the "infinite-volume" rates c_{xy} computed on a configuration identically equal to the boundary condition τ outside Λ and to σ inside Λ.

The general assumptions on the transition rates c_{xy} are:

(i)   *Finite range.* $c_{xy}(\sigma)$ depends only on the spins $\sigma(z)$ with $d(\{x,y\}, z) \leq r$.

(ii) *Detailed balance.* For all σ and all edge $[x,y]$

$$\exp\left[-\Phi_{\{x,y\}}(\sigma)\right] c_{xy}(\sigma) = \exp\left[-\Phi_{\{x,y\}}(\sigma^{xy})\right] c_{xy}(\sigma^{xy})$$

(iii) *Positivity and boundedness.* There exist positive real numbers c_m and c_M such that

$$c_m \leq c_{xy}(\sigma) \leq c_M \qquad \forall x, y \in \mathbb{Z}^d, \sigma$$

Under the above assumptions the generator becomes a symmetric operator on $L^2(\Omega_\Lambda, \nu^\tau_{\Lambda,N})$ with reversible measure the *canonical Gibbs measure* $\nu^\tau_{\Lambda,N}$. If $\Phi \equiv 0$ (no interaction among the particles except the mutual exclusion) the process coincides with the simple exclusion process on Λ discussed in the previous section.

10.1 Diffusive scaling of the mixing time in the one-phase region

We begin by analyzing the so-called *high-temperature* case. We will first define a suitable mixing condition for the interaction Φ and then state the main results.

Fix positive numbers C, m, ℓ with $\ell \in \mathbf{N}$ and call a collection of real numbers $\underline{\lambda} := \{\lambda_x\}_{x \in \mathbb{Z}^d}$ an *ℓ-regular chemical potential* if, for all $i \in \mathbb{Z}^d$ and all $x \in Q_\ell(x^i)$, $x^i \in \ell\mathbb{Z}^d$, $\lambda_x = \lambda_{x^i}$.

Given an ℓ-regular chemical potential and an interaction Φ, denote by $\Phi^{\underline{\lambda}}$ the new interaction:

$$\Phi^{\underline{\lambda}}_V(\sigma) = \begin{cases} (h + \lambda_x)\sigma(x) & \text{if} \quad V = \{x\} \\ \Phi_V(\sigma) & \text{otherwise} \end{cases}$$

where h is the chemical potential (one body part of Φ).

Definition 10.1. *We say that property $USMT(C, m, \ell)$ holds if the mixing condition $SMT(C, m, \ell)$ holds for the interaction $\Phi^{\underline{\lambda}}$ uniformly in the ℓ-regular chemical potential $\underline{\lambda}$.*

Remark 10.2. Condition $USMT(C, m, \ell)$, is definitely a high-temperature kind of condition and, for, e.g. the Ising model, there is no hope for it to hold above β_c. The reason is precisely the uniformity requirement in $\underline{\lambda}$. If $\beta > \beta_c$ and the one body part of the interaction h does not produce phase coexistence, by adding a suitable (even constant) new chemical potential one can always reach a point in the phase coexistence region where covariances do not decay exponentially fast uniformly in the boundary conditions. For the two-dimensional Ising model one can prove that $USMT(C, m.\ell)$ holds for all $\beta < \beta_c$ [18].

We are finally in a position to formulate the main results on the spectral gap and logarithmic Sobolev constant for Kawasaki dynamics in a finite volume (see [108], [169], [168] [30], [32]).

Theorem 10.3. *Assume property $USMT(C, m, \ell)$ holds. Then there exist positive constants c_1, c_2 such that*

(i)
$$c_1 L^{-2} \leq \min_{N,\tau} \mathrm{gap}(\mathcal{L}^\tau_{Q_L,N}) \leq \max_{N,\tau} \mathrm{gap}(L^\tau_{Q_L,N}) \leq c_2 L^{-2}$$

(ii)
$$c_1 L^2 \leq \min_{N,\tau} c_s(\mathcal{L}^\tau_{Q_L,N}) \leq \max_{N,\tau} c_s(L^\tau_{Q_L,N}) \leq c_2 L^2$$

Remark 10.4. A similar result for the spectral gap has been obtained for the dilute Ising model in the Griffiths region (see section 3.8) [31].

It is a difficult and important open problem to prove the diffusive scaling of the spectral gap and logarithmic Sobolev constant *without* the restrictive condition $USMT(C, m, \ell)$ but just by assuming condition $SMT(C, m, \ell)$ as in the Glauber case. The main obstruction here is related to the dynamics of anomalous fluctuations of the particle density profile and its solution requires new ideas.

A nice consequence of the above estimates is an inverse polynomial bound on the time decay to equilibrium in $L^2\left(d\nu^\tau_{\Lambda, N}\right)$ of local functions [30].

Theorem 10.5. *Assume property $USMT(C, m, \ell)$. Then for any $\varepsilon \in (0, 1)$ and any local, mean zero function f with $0 \in \Delta_f$ there exists a positive constant $C_{f,\varepsilon}$ such that for any integer L multiple of ℓ and any integer $N \in \{1, \ldots, (2L)^d\}$*

$$\mathrm{Var}^\tau_{\Lambda, N}\left(e^{tL^\tau_{\Lambda, N}}f\right) \leq C_{f,\varepsilon}\frac{1}{t^{\alpha - \varepsilon}} \tag{10.1}$$

where $\Lambda := B_L$ and $\alpha = \frac{1}{2}$ in $d = 1$, $\alpha = 1$ for $d > 1$.

Remark 10.6. For the simple exclusion process [17] the time evolution of a local, mean zero linear function of the form $f(\sigma) = \sum_x a_x \sigma(x)$, with $\{a_x\}_{x \in \mathbb{Z}^d} \in l^2\left(\mathbb{Z}^d\right)$, can be computed exactly: $e^{t\mathcal{L}_\Lambda}f = \sum_x \left(e^{t\Delta_\Lambda}a\right)_x \sigma(x)$ where Δ_Λ is the discrete Laplacian on Λ with the appropriate boundary conditions. In particular one gets in this case that $\|e^{t\mathcal{L}_\Lambda}f\|_2^2 \leq C_f t^{-d/2}$ (see [17]). For the zero-range process a beautiful detailed analysis is also possible [88] and the result is $\|e^{t\mathcal{L}_\Lambda}f\|_2^2 = C_f \frac{1}{t^{d/2}} + o(\frac{1}{t^{d/2}})$ with a precise constant C_f. The exponent α in theorem 10.5 is thus arbitrarily close to the correct one only for $d = 1, 2$. Recently, in [28] was established a bound like (10.1) with α replaced by $d/2$ for any $d \geq 1$ by extending the analysis of [88].

We conclude this section by briefly discussing the new ideas, compared to the Glauber case, that are needed in order to prove theorem 10.3. We only discuss the recursive approach developed in [30] and pushed further in [32] without entering into the martingale method of [108] and [168].

Let $c(L)$ be the largest (over the boundary conditions and number of particles) among the logarithmic Sobolev constants in a cube of side L with given boundary conditions and fixed number of particles. The really hard part is to prove an upper bound for $c(L)$ of the right order; the lower bound is readily obtained by plugging into the logarithmic Sobolev inequality a suitable test function (a slowly varying function of the local density). In order to prove the correct upper bound we look for a recursive inequality of the form

$$c(2L) \leq \frac{3}{2}c(L) + kL^2 \tag{10.2}$$

which, upon iteration, proves the bound $c(L) \leq k'L^2$.

For this purpose, let Λ be the cube of side $2L$ and let us divide it into two (almost) halves Λ_1, Λ_2 in such a way that the overlap between Λ_1 and Λ_2 is a thin layer of width δL, $\delta \ll 1$. Denote by ν the canonical Gibbs measure on Λ with some given number of particles and let $\mathrm{Ent}_\nu(f^2)$ be the entropy of f^2 w.r.t. ν. If the two σ-algebras $\mathcal{F}_1 := \mathcal{F}_{\Lambda_1^c}$ and $\mathcal{F}_2 := \mathcal{F}_{\Lambda_2^c}$, namely the σ-algebras generated by the lattice gas variables outside Λ_1 and Λ_2 respectively, were *weakly dependent* in the sense described in section 2.2, then it would follow that

$$\mathrm{Ent}_\nu(f^2) \leq (1 + \varepsilon(L))\nu\big(\mathrm{Ent}_\nu(f^2\,|\,\mathcal{F}_1) + \mathrm{Ent}_\nu(f^2\,|\,\mathcal{F}_2)\big)$$

For example, if the canonical measure were replaced by the grand canonical one then, under the mixing condition $USMT(C,m,\ell)$, $\varepsilon(L) = O(e^{-m\delta L})$ for some positive m (see section 5.4).

For the canonical measure instead, the conservation of the number of particles prevents the σ-algebras $\mathcal{F}_1$ and $\mathcal{F}_2$ from being weakly dependent. Even in the absence of any interaction, the Kawasaki dynamics on two nearby disjoint sets does not factorize into two independent dynamics because the particles may migrate from one set to the other one. In particular, the relaxation time in Λ is related to the relaxation time of the modified Kawasaki dynamics in which the number of particles in the three sets of the partition $\{\Lambda\backslash\Lambda_1, \Lambda_1\cap\Lambda_2, \Lambda\backslash\Lambda_2\}$ is conserved *and* to the relaxation time of the process of exchange of particles between them. This suggests that we try to separate the two effects which are, a priori, strongly interlaced and to analyze them separately. In some sense this idea is the heart of the approach of [30], [32] and technically it can be achieved by elementary conditioning as follows. Let n_0 and n_1 be the random variables counting the number of particles in $\Lambda_1 \cap \Lambda_2$ and in $\Lambda \setminus \Lambda_2$ respectively and let $\mathrm{Ent}_\nu(f^2\,|\,n_0,n_1)$ be the entropy of f^2 w.r.t canonical measure ν conditioned on n_0, n_1. Then we can write

$$\mathrm{Ent}_\nu(f^2) = \nu\big(\mathrm{Ent}_\nu(f^2\,|\,n_0,n_1)\big) + \mathrm{Ent}_\nu\big(\nu(f^2\,|\,n_0,n_1)\big) \qquad (10.3)$$

The second term in (10.3) can in turn be expanded as

$$\mathrm{Ent}_\nu\big(\nu(f^2\,|\,n_0,n_1)\big) = \nu\Big(\mathrm{Ent}_\nu\big(\nu(f^2\,|\,n_0,n_1)\,|\,n_0\big)\Big) + \mathrm{Ent}_\nu\big(\nu(f^2\,|\,n_0)\big) \quad (10.4)$$

Notice that in the first term in the r.h.s of (10.3) we need to bound the entropy with respect to a *multicanonical* measure in which the number of particles in each atom of the partition $\{\Lambda \setminus \Lambda_1, \Lambda_1 \cap \Lambda_2, \Lambda \setminus \Lambda_2\}$ is frozen. As shown in [30], such a new measure has better chances to satisfy the "weak dependence" condition of theorem 2.12 than the original measure ν precisely because of the extra conservation laws. Thus, by the previous reasoning, we may hope to bound the first term in the r.h.s of (10.3) by the largest among the logarithmic Sobolev constant of each of the three sets times the Dirichlet form of the Kawasaki dynamics. Notice that for each of the three sets the linear dimension in one direction has been (at least) almost halved. Thus the first term in the r.h.s. of (10.3) should be the responsible for the first term in the r.h.s of (10.2). Let us now examine the pieces that come from the second term in the r.h.s. of (10.3). As one can observe in (10.4), in each of them one has to bound an entropy with respect to the distribution of a one-dimensional discrete random variable, e.g. the number of particles n_0 in the second one. Although such a distribution is difficult to compute exactly, one has a sufficiently good control to be able to establish, via the Hardy inequality (see, e.g. [13]) a sharp logarithmic Sobolev inequality with respect to the Dirichlet form of a reversible Metropolis birth and death process. Physically, such a process corresponds to the creation of an extra particle in, e.g. $\Lambda_1 \cap \Lambda_2$ and the contemporary annihilation of a particle in, e.g. $\Lambda \setminus \Lambda_1$, that is to the *exchange* of particles among the three sets. Since each particle moves, essentially, by a sort of perturbed random walk, and on average it has to travel a distance $O(L)$, it is not surprising that the second term in the r.h.s. of (10.3) is responsible for the L^2 term in the r.h.s. of (10.2).

10.2 Torpid mixing in the phase coexistence region

In this last section we consider the finite-volume effects of phase coexistence on the mixing time for the Kawasaki dynamics. We restrict ourselves to the Ising model in a finite square Λ of side L in $\mathbb{Z}^2$ with *free boundary conditions*.

Assume $\beta > \beta_c$ and suppose that $\varrho \in (\varrho_-, \varrho_+)$ where $\varrho = \frac{N}{L^2}$ is the particle density and $\varrho_\pm(\beta)$ are the densities of the liquid and vapor phases. Exactly as in the nonconservative case, under rather general conditions, the spectral gap cannot be smaller than an exponential of the surface of Λ [27]. In order to prove that the above bound can be saturated at least in two dimensions, one can try to find a bottleneck in the phase space. A natural choice, dictated by the physics of phase segregation for the Ising lattice gas, is the following.

Divide Q into 16 equal squares of side $1/4$ and call these squares

$$A_1, A_2, \ldots, B_1, B_2, \ldots, D_4$$

as in a chessboard. Define

$$U = A_1 \cup B_1 \cup B_2 \cup C_1 \cup C_2 \cup D_1 \cup D_2 \cup D_3$$

and let $\mathcal{B}$ be the event that the number of particles in the set U is less than $\frac{N}{2}$. Then the boundary of $\mathcal{B}$ is a bottleneck between the two likely events $\mathcal{B}$ and $\mathcal{B}^c$.

In order to explain such an apparently weird choice, it is useful first to recall the shape of the typical configurations of the canonical Ising Gibbs measure with N particles and free b.c. when the temperature is below the critical value.

Let $m_\varrho = 2\varrho - 1$ be the usual magnetization associated with the given particle density. Then, as discussed in [143] (see also [37]), there exists $0 < m_1 < m^*$ such that:

(i)   If $m_\varrho \in (-m_1, m_1)$ then the typical configurations show phase segregation between high and low density ($\approx \varrho_\pm^*$) regions that are roughly two horizontal (vertical) rectangles of appropriate area separated by an horizontal (vertical) interface of length L.

(ii)   If $m_\varrho \in (-m^*, m^*) \setminus (-m_1, m_1)$ then the typical configurations show phase segregation between high and low density ($\approx \varrho_\pm$), regions, one of which is a quarter of a Wulff shape of appropriate area and centered in one of the four vertices of Λ.

What is important here is that in both cases the typical configurations of the canonical measure show a discrete symmetry described by rotations of $k\frac{\pi}{2}$, $k = 0, 1 \ldots$ around the center of Λ and that the critical value m_1 is such that for each typical configuration the particle density in the set U, ϱ_U, is either below or above ϱ, making the event $\mathcal{B}$ very unlikely.

The precise result of [27] can be formulated as follows.

Theorem 10.7. *Let $\beta > \beta_c$ and $\varrho \in (\varrho_-(\beta), \varrho_+(\beta))$. Then there exists $c > 0$ such that, if $N = \lfloor \varrho|\Lambda| \rfloor$, then, for large enough L,*

$$\mathrm{gap}(\mathcal{L}_{\Lambda,N}^\emptyset) \leq e^{-cL}$$

We conclude by observing that the above discrete symmetry of the typical configurations is peculiar to free b.c. If instead one works with, e.g positive b.c., then the typical configurations for the canonical measure when m_ϱ is slightly below m^* consist of a low density Wulff bubble centered somewhere in the bulk of Q_L, immersed in a sea of high density. In this case the continuous degeneracy of the typical configurations caused by the arbitrariness of the location of the center of the Wulff bubble prevents the spectral gap being exponentially small in L. In particular one may argue that the slowest mode of the system is due to the random walk motion of the center of gravity of the unique Wulff bubble, suggesting that the spectral gap should shrink as L^{-3}. No rigorous results have been proved yet that confirm this picture, except an upper bound on the spectral gap [27] of the form gap $\leq CL^{-3}$.

Acknowledgement. I would like to thank H. Kesten for offering me the opportunity to write this paper and for critical reading of the manuscript. I am also grateful to Y. Peres, A. Sinclair, E. Vigoda and D. Randall for several useful comments.

References

1. M. Aizenman and R. Holley. Rapid convergence to equilibrium of stochastic Ising models in the Dobrushin Shlosman regime. In *Percolation theory and ergodic theory of infinite particle systems (Minneapolis, Minn., 1984–1985)*, pages 1–11. Springer, New York, 1987.

2. F. C. Alcaraz, S. R. Salinas, and W. F. Wreszinski. Quantum domains in ferromagnetic anisotropic Heisenberg chains. In *Statistical models, Yang-Baxter equation and related topics, and Symmetry, statistical mechanical models and applications (Tianjin, 1995)*, pages 13–19. World Sci. Publishing, River Edge, NJ, 1996.

3. F. C. Alcaraz. Exact steady states of asymmetric diffusion and two-species annihilation with back reaction from the ground state of quantum spin models. *Internat. J. Modern Phys. B*, 8(25–26):3449–3461, 1994. Perspectives on solvable models.

4. A. Aldous and J. Fill. *Reversible Markov chains and random walks on graphs.* available at http://stat-www.berkeley.edu/users/aldous/book.html.

5. D. Aldous. Random walks on finite groups and rapidly mixing Markov chains. In *Seminar on probability, XVII*, pages 243–297. Springer, Berlin, 1983.

6. D. Aldous, L. Lovász, and P. Winkler. Mixing times for uniformly ergodic Markov chains. *Stochastic Process. Appl.*, 71(2):165–185, 1997.

7. K. S. Alexander. On weak mixing in lattice models. *Probab. Theory Related Fields*, 110(4):441–471, 1998.

8. K. S. Alexander. Mixing properties and exponential decay for lattice systems in finite volumes. *Preprint*, 2001.

9. K. S. Alexander. The spectral gap of the 2-D stochastic Ising model with nearly single-spin boundary conditions. *J. Statist. Phys.*, 104(1–2):59–87, 2001.

10. K. S. Alexander and N. Yoshida. The spectral gap of the 2-D stochastic Ising model with mixed boundary conditions. *J. Statist. Phys.*, 104(1–2):89–109, 2001.

11. N. Alon. Eigenvalues and expanders. *Combinatorica*, 6(2):83–96, 1986. Theory of computing (Singer Island, Fla., 1984).

12. N. Alon and V. D. Milman. λ_1, isoperimetric inequalities for graphs, and superconcentrators. *J. Combin. Theory Ser. B*, 38(1):73–88, 1985.

13. C. Ané, S. Blachère, D. Chafaï, P. Fougères, I. Gentil, F. Malrieu, C. Roberto, and G. Scheffer. *Sur les inégalités de Sobolev logarithmiques.* Société Mathématique de France, Paris, 2000. With a preface by Dominique Bakry and Michel Ledoux.

14. I. Benjamini, N. Berger, C. Hoffman, and E. Mossel. Mixing time for biased shuffling. *Preprint*, 2002.

15. I. Benjamini and E. Mossel. On the mixing time of a simple random walk on the super critical percolation cluster. *Preprint*, 2002.

16. L. Bertini, N. Cancrini, and F. Cesi. The spectral gap for a Glauber-type dynamics in a continuous gas. *Ann. Inst. H. Poincaré Probab. Statist.*, 38(1):91–108, 2002.

17. L. Bertini and B. Zegarlinski. Coercive inequalities for Kawasaki dynamics. The product case. *Markov Process. Related Fields*, 5(2):125–162, 1999.

18. L. Bertini, E. N. M. Cirillo, and E. Olivieri. Renormalization-group transformations under strong mixing conditions: Gibbsianness and convergence of renormalized interactions. *J. Statist. Phys.*, 97(5–6):831–915, 1999.

19. T. Bodineau and B. Helffer. The log-Sobolev inequality for unbounded spin systems. *J. Funct. Anal.*, 166(1):168–178, 1999.

20. T. Bodineau, D. Ioffe, and Y. Velenik. Rigorous probabilistic analysis of equilibrium crystal shapes. *J. Math. Phys.*, 41(3):1033–1098, 2000.

21. T. Bodineau and F. Martinelli. Some new results on the kinetic ising model in a pure phase. *Journal of Stat. Phys.*, 109(1), 2002.

22. O. Bolina, P. Contucci, and B. Nachtergaele. Path integral representation for interface states of the anisotropic Heisenberg model. *Rev. Math. Phys.*, 12(10):1325–1344, 2000.

23. O. Bolina, P. Contucci, B. Nachtergaele, and S. Starr. Finite-volume excitations of the 111 interface in the quantum XXZ model. *Comm. Math. Phys.*, 212(1):63–91, 2000.

24. C. Borgs, J. T. Chayes, A. Frieze, J. H. Kim, P. Tetali, E. Vigoda, and V. H. Vu. Torpid mixing of some Monte Carlo Markov chains algorithms in statistical mechanics. *40th Annual Symposium on Foundations of Computer Science, IEEE, Los Alimitos*, 1999.

25. A. Bovier and P. Picco, editors. *Mathematical aspects of spin glasses and neural networks*. Birkhäuser Boston Inc., Boston, MA, 1998.

26. R. Bubley and M. Dyer. Path coupling: A technique for proving rapid mixing in Markov chains. *38th Symposium on Foundations of Computer Science*, 1997.

27. N. Cancrini, F. Cesi, and F. Martinelli. The spectral gap for the Kawasaki dynamics at low temperature. *J. Statist. Phys.*, 95(1–2):215–271, 1999.

28. N. Cancrini, F. Cesi, and C. Roberto. Private communication.

29. N. Cancrini and F. Martinelli. Comparison of finite volume canonical and grand canonical Gibbs measures under a mixing condition. *Markov Process. Related Fields*, 6(1):23–72, 2000.

30. N. Cancrini and F. Martinelli. On the spectral gap of Kawasaki dynamics under a mixing condition revisited. *J. Math. Phys.*, 41(3):1391–1423, 2000. Probabilistic techniques in equilibrium and nonequilibrium statistical physics.

31. N. Cancrini and F. Martinelli. Diffusive scaling of the spectral gap for the dilute Ising lattice-gas dynamics below the percolation threshold. *Probab. Theory Related Fields*, 120(4):497–534, 2001.

32. N. Cancrini, F. Martinelli, and C. Roberto. The logarithmic Sobolev constant of Kawasaki dynamics under a mixing condition revisited. *Ann. Inst. H. Poincaré Probab. Statist.*, 38(4):385–436, 2002.

33. P. Caputo. Uniform Poincaré inequalities for unbounded conservative spin systems: the non interacting case. *Preprint*, 2002.

34. P. Caputo and F. Martinelli. Relaxation time of anisotropic simple exclusion processes and quantum Heisenberg models. *Ann. Appl. Prob.*, vol. 13, No 2, 2003.

35. P. Caputo and F. Martinelli. Asymmetric diffusion and the energy gap above the 111 ground state of the quantum XXZ model. *Comm. Math. Phys.*, 226(2):323–375, 2002.

36. E. Carlen, M. C. Caravalho, and M. Loss. Determination of the spectral gap for Kac's master equation and related stochastic evolutions. *Preprint*, 2002.

37. F. Cesi, G. Guadagni, F. Martinelli, and R. H. Schonmann. On the two-dimensional stochastic Ising model in the phase coexistence region near the critical point. *J. Statist. Phys.*, 85(1–2):55–102, 1996.

38. F. Cesi, C. Maes, and F. Martinelli. Relaxation of disordered magnets in the Griffiths' regime. *Comm. Math. Phys.*, 188(1):135–173, 1997.

39. F. Cesi, C. Maes, and F. Martinelli. Relaxation to equilibrium for two-dimensional disordered Ising systems in the Griffiths phase. *Comm. Math. Phys.*, 189(2):323–335, 1997.

40. F. Cesi. Quasi-factorization of the entropy and logarithmic Sobolev inequalities for Gibbs random fields. *Probab. Theory Related Fields*, 120(4):569–584, 2001.

41. F. Cesi and F. Martinelli. On the layering transition of an SOS surface inter-
    acting with a wall. I. Equilibrium results. *J. Statist. Phys.*, 82(3–4):823–913,
    1996.

42. F. Cesi and F. Martinelli. On the layering transition of an SOS surface interact-
    ing with a wall. II. The Glauber dynamics. *Comm. Math. Phys.*, 177(1):173–
    201, 1996.

43. L. Chayes, R. H. Schonmann, and G. Swindle. Lifshitz' law for the volume of
    a two-dimensional droplet at zero temperature. *J. Statist. Phys.*, 79(5–6):821–
    831, 1995.

44. J. Cheeger. A lower bound for the smallest eigenvalue of the Laplacian. In
    Problems in analysis (Papers dedicated to Salomon Bochner, 1969), pages 195–
    199. Princeton Univ. Press, Princeton, NJ, 1970.

45. C. Cooper and A. M. Frieze. Mixing properties of the Swendsen-Wang pro-
    cess on classes of graphs. *Random Structures Algorithms*, 15(3–4):242–261,
    1999. Statistical physics methods in discrete probability, combinatorics, and
    theoretical computer science (Princeton, NJ, 1997).

46. B. Derrida. Random energy model: Limit of a family of disordered models.
    Phys. Rev. Lett., 45:79–82, 1980.

47. P. Diaconis and L. Saloff-Coste. Logarithmic Sobolev inequalities for finite
    Markov chains. *Ann. Appl. Probab.*, 6(3):695–750, 1996.

48. P. Diaconis and L. Saloff-Coste. Comparison theorems for reversible Markov
    chains. *Ann. Appl. Probab.*, 3(3):696–730, 1993.

49. P. Diaconis and L. Saloff-Coste. Bounds for Kac's master equation. *Comm.
    Math. Phys.*, 209(3):729–755, 2000.

50. P. Diaconis and M. Shahshahani. Generating a random permutation with
    random transpositions. *Z. Wahrsch. Verw. Gebiete*, 57(2):159–179, 1981.

51. P. Diaconis and M. Shahshahani. Time to reach stationarity in the Bernoulli-
    Laplace diffusion model. *SIAM J. Math. Anal.*, 18(1):208–218, 1987.

52. P. Diaconis and D. Stroock. Geometric bounds for eigenvalues of Markov
    chains. *Ann. Appl. Probab.*, 1(1):36–61, 1991.

53. R. L. Dobrushin. The problem of uniqueness of a Gibbsian random field and
    the problem of phase transitions. *Funkcional. Anal. i Priložen.*, 2(4):44–57,
    1968.

54. R. L. Dobrushin. Prescribing a system of random variables by conditional
    distributions. *Theory of Prob. Appl.*, 15:453–486, 1970.

55. R. L. Dobrushin and S. B. Shlosman. Constructive criterion for the uniqueness
    of Gibbs field. In *Statistical physics and dynamical systems (Köszeg, 1984)*,
    pages 347–370. Birkhäuser Boston, Boston, MA, 1985.

56. R. L. Dobrushin and S. B. Shlosman. Completely analytical Gibbs fields. *Stat.
    Phys. and Dyn. Systems*, 46(5–6):983–1014, 1987.

57. R. L. Dobrushin and S. B. Shlosman. Completely analytical interactions: con-
    structive description. *J. Statist. Phys.*, 46(5–6):983–1014, 1987.

58. M. Dyer, A. Sinclair, E. Vigoda, and D. Weitz. Mixing in time and space for
    lattice spin systems: a combinatorial view. *Preprint*, 2002.

59. M. Dyer and C. Greenhill. On Markov chains for independent sets. *J. Algo-
    rithms*, 35(1):17–49, 2000.

60. M. E. Dyer, A. M. Frieze, and M. R. Jerrum. On counting independent sets in
    sparse graphs. *40th Annual Symposium on Foundations of Computer Science,
    IEEE, Los Alimitos*, pages 210–217, 1999.

61. R. G. Edwards and A. D. Sokal. Generalization of the Fortuin-Kasteleyn-Swendsen-Wang representation and Monte Carlo algorithm. *Phys. Rev. D (3)*, 38(6):2009–2012, 1988.

62. D. Fisher and D. Huse. Dynamics of droplet fluctuation in pure and random Ising systems. *Phys. Rev. B*, 35(13), 1987.

63. L. R. G. Fontes, M. Isopi, Y. Kohayakawa, and P. Picco. The spectral gap of the REM under Metropolis dynamics. *Ann. Appl. Probab.*, 8(3):917–943, 1998.

64. R. Fontes, R. H. Schonmann, and V. Sidoravicius. Stretched exponential fixation in stochastic ising models at zero temperature. *Preprint*, 2001.

65. C. M. Fortuin and P. W. Kasteleyn. On the random-cluster model. I. Introduction and relation to other models. *Physica*, 57:536–564, 1972.

66. J. Fröhlich. Mathematical aspects of the physics of disordered systems. In *Phénomènes critiques, systèmes aléatoires, théories de jauge, Part I, II (Les Houches, 1984)*, pages 725–893. North-Holland, Amsterdam, 1986. With the collaboration of A. Bovier and U. Glaus.

67. D. Galvin and J. Kahn. On phase transition in the hard-core model on Z^d. *Preprint*, 2002.

68. F. Gao and J. Quastel. Exponential decay of entropy in the random transposition and Bernoulli–Laplace models. *Preprint*, 2002.

69. H.-O. Georgii. *Gibbs measures and phase transitions*. Walter de Gruyter & Co., Berlin, 1988.

70. V. K. Gore and M. R. Jerrum. The Swendsen-Wang process does not always mix rapidly. *J. Statist. Phys.*, 97(1–2):67–86, 1999.

71. R. Griffiths. Non-analytic behaviour above the critical point in a random Ising ferromagnet. *Phys. Rev. Lett.*, 23:17, 1969.

72. G. Grimmett. Percolation and disordered systems. In *Lectures on probability theory and statistics (Saint-Flour, 1996)*, pages 153–300. Springer, Berlin, 1997.

73. L. Gross. Logarithmic Sobolev inequalities. *Amer. J. Math.*, 97(4):1061–1083, 1975.

74. A. Guionnet and B. Zegarlinski. Lectures on logarithmic Sobolev inequalities. *Volume XXXVI of the Seminaire de Probabilité, Springer Lecture Notes in Mathematics*, pages 1–134, 2000.

75. A. Guionnet and B. Zegarlinski. Decay to equilibrium in random spin systems on a lattice. *Comm. Math. Phys.*, 181(3):703–732, 1996.

76. A. Guionnet and B. Zegarlinski. Decay to equilibrium in random spin systems on a lattice. II. *J. Statist. Phys.*, 86(3–4):899–904, 1997.

77. Y. Higuchi and J. Wang. Spectral gap of Ising model for Dobrushin's boundary condition in two dimension. *Preprint*, 1999.

78. Y. Higuchi and N. Yoshida. Slow relaxation of 2-D stochastic Ising models with random and non-random boundary conditions. In *New trends in stochastic analysis (Charingworth, 1994)*, pages 153–167. World Sci. Publishing, River Edge, NJ, 1997.

79. R. A. Holley and D. W. Stroock. In one and two dimensions, every stationary measure for a stochastic Ising model is a Gibbs state. *Comm. Math. Phys.*, 55(1):37–45, 1977.

80. R. Holley. Possible rates of convergence in finite range, attractive spin systems. In *Particle systems, random media and large deviations (Brunswick, Maine, 1984)*, pages 215–234. Amer. Math. Soc., Providence, RI, 1985.

81. R. Holley. On the asymptotics of the spin-spin autocorrelation function in stochastic Ising models near the critical temperature. In *Spatial stochastic processes*, pages 89–104. Birkhäuser Boston, Boston, MA, 1991.

82. R. Holley. The one-dimensional stochastic X-Y model. In *Random walks, Brownian motion, and interacting particle systems*, pages 295–307. Birkhauser Boston, Boston, MA, 1991.

83. R. Holley. Rapid convergence to equilibrium in ferromagnetic stochastic Ising models. *Resenhas*, 1(2–3):131–149, 1994. Fifth Latin American Congress of Probability and Mathematical Statistics (Portuguese) (Sao Paulo, 1993).

84. R. Holley and D. Stroock. Logarithmic Sobolev inequalities and stochastic Ising models. *J. Statist. Phys.*, 46(5–6):1159–1194, 1987.

85. R. A. Holley and D. W. Stroock. Uniform and L^2 convergence in one-dimensional stochastic Ising models. *Comm. Math. Phys.*, 123(1):85–93, 1989.

86. M. Huber. Efficient exact sampling from the Ising model using Swendsen–Wang. *Preprint*, 2000.

87. E. J. Janse van Rensburg. Collapsing and adsorbing polygons. *J. Phys. A*, 31(41):8295–8306, 1998.

88. E. Janvresse, C. Landim, J. Quastel, and H. T. Yau. Relaxation to equilibrium of conservative dynamics. I. Zero-range processes. *Ann. Probab.*, 27(1):325–360, 1999.

89. E. Janvresse. Spectral gap for Kac's model of Boltzmann equation. *Ann. Probab.*, 29(1):288–304, 2001.

90. M. Jerrum. Mathematical foundations of the Markov chain Monte Carlo method. In *Probabilistic methods for algorithmic discrete mathematics*, pages 116–165. Springer, Berlin, 1998.

91. M. Jerrum and A. Sinclair. Approximating the permanent. *SIAM J. Comput.*, 18(6):1149–1178, 1989.

92. M. Jerrum and A. Sinclair. Polynomial-time approximation algorithms for the Ising model. *SIAM J. Comput.*, 22(5):1087–1116, 1993.

93. M. Kac. Foundations of kinetic theory. In *Proceedings of the Third Berkeley Symposium on Mathematical Statistics and Probability, 1954–1955, vol. III*, pages 171–197, Berkeley and Los Angeles, 1956. University of California Press.

94. C. Kenyon, E. Mossel, and Y. Peres. Glauber dynamics on trees and hyperbolic graphs. In *IEEE Symposium on Foundations of Computer Science*, pages 568–578, 2001.

95. C. Kipnis and C. Landim. *Scaling limits of interacting particle systems*. Springer-Verlag, Berlin, 1999.

96. T. Koma, B. Nachtergaele, and S. Starr. The spectral gap of the ferromagnetic spin-j XXZ chain. *Preprint*, 2001.

97. T. Koma and B. Nachtergaele. The spectral gap of the ferromagnetic XXZ chain. *Lett. Math. Phys.*, 40(1):1–16, 1997.

98. R. Kotecký. unpublished. *Cited in [69]*. pp. 148–149, 457.

99. R. Kotecký and S. B. Shlosman. First-order phase transitions in large entropy lattice models. *Comm. Math. Phys.*, 83(4):493–515, 1982.

100. C. Landim, G. Panizo, and H. T. Yau. Spectral gap and logarithmic Sobolev inequality for unbounded conservative spin systems. *To appear in Annales de l'Institut Henri Poincaré. Probabilités et Statistiques*, 38(5):739–777, 2002.

101. C. Landim, S. Sethuraman, and S. Varadhan. Spectral gap for zero-range dynamics. *Ann. Probab.*, 24(4):1871–1902, 1996.

102. G. F. Lawler and A. D. Sokal. Bounds on the L^2 spectrum for Markov chains and Markov processes: a generalization of Cheeger's inequality. *Trans. Amer. Math. Soc.*, 309(2):557–580, 1988.

103. M. Ledoux. Logarithmic Sobolev inequalities for unbounded spin systems revisited. In *Séminaire de Probabilités, XXXV*, pages 167–194. Springer, Berlin, 2001.

104. T.-Y. Lee and H.-T. Yau. Logarithmic Sobolev inequality for some models of random walks. *Ann. Probab.*, 26(4):1855–1873, 1998.

105. T. M. Liggett. *Interacting particle systems*. Springer-Verlag, New York, 1985.

106. L. Lovász and R. Kannan. Faster mixing via average conductance. In *Annual ACM Symposium on Theory of Computing (Atlanta, GA, 1999)*, pages 282–287 (electronic). ACM, New York, 1999.

107. L. Lovász and P. Winkler. Mixing times. In *Microsurveys in discrete probability (Princeton, NJ, 1997)*, pages 85–133. Amer. Math. Soc., Providence, RI, 1998.

108. S. L. Lu and H.-T. Yau. Spectral gap and logarithmic Sobolev inequality for Kawasaki and Glauber dynamics. *Comm. Math. Phys.*, 156(2):399–433, 1993.

109. M. Luby, D. Randall, and A. Sinclair. Markov chain algorithms for planar lattice structures. *SIAM J. Comput.*, 31(1):167–192 (electronic), 2001.

110. M. Luby and E. Vigoda. Fast convergence of the Glauber dynamics for sampling independent sets. *Random Structures Algorithms*, 15(3–4):229–241, 1999. Statistical physics methods in discrete probability, combinatorics, and theoretical computer science (Princeton, NJ, 1997).

111. N. Madras and D. Randall. Markov chain decomposition for convergence rate analysis. *Ann. Appl. Probability*, 12:581–606, 2002.

112. E. Marcelli and F. Martinelli. Some new results on the two-dimensional kinetic Ising model in the phase coexistence region. *J. Statist. Phys.*, 84(3–4):655–696, 1996.

113. F. Martinelli. On the two-dimensional dynamical Ising model in the phase coexistence region. *J. Statist. Phys.*, 76(5–6):1179–1246, 1994.

114. F. Martinelli. An elementary approach to finite size conditions for the exponential decay of covariances in lattice spin models. In *On Dobrushin's way. From probability theory to statistical physics*, pages 169–181. Amer. Math. Soc., Providence, RI, 2000.

115. F. Martinelli and E. Olivieri. Some remarks on pathologies of renormalization-group transformations for the Ising model. *J. Statist. Phys.*, 72(5–6):1169–1177, 1993.

116. F. Martinelli and E. Olivieri. Approach to equilibrium of Glauber dynamics in the one phase region. I. The attractive case. *Comm. Math. Phys.*, 161(3):447–486, 1994.

117. F. Martinelli and E. Olivieri. Approach to equilibrium of Glauber dynamics in the one phase region. II. The general case. *Comm. Math. Phys.*, 161(3):487–514, 1994.

118. F. Martinelli and E. Olivieri. Instability of renormalization-group pathologies under decimation. *J. Statist. Phys.*, 79(1–2):25–42, 1995.

119. F. Martinelli, E. Olivieri, and R. H. Schonmann. For 2-D lattice spin systems weak mixing implies strong mixing. *Comm. Math. Phys.*, 165(1):33–47, 1994.

120. F. Martinelli. Dynamical analysis of low-temperature Monte Carlo cluster algorithms. *J. Statist. Phys.*, 66(5–6):1245–1276, 1992.

121. F. Martinelli. Lectures on Glauber dynamics for discrete spin models. In *Lectures on probability theory and statistics (Saint-Flour, 1997)*, pages 93–191. Springer, Berlin, 1999.

122. F. Martinelli, E. Olivieri, and E. Scoppola. On the Swendsen-Wang dynamics. I. Exponential convergence to equilibrium. *J. Statist. Phys.*, 62(1–2):117–133, 1991.

123. F. Martinelli, E. Olivieri, and E. Scoppola. On the Swendsen-Wang dynamics. II. Critical droplets and homogeneous nucleation at low temperature for the two-dimensional Ising model. *J. Statist. Phys.*, 62(1–2):135–159, 1991.

124. P. Mathieu. Hitting times and spectral gap inequalities. *Ann. Inst. H. Poincaré Probab. Statist.*, 33(4):437–465, 1997.

125. P. Mathieu. Convergence to equilibrium for spin glasses. *Comm. Math. Phys.*, 215(1):57–68, 2000.

126. B. Morris and Y. Peres. Evolving sets and mixing. *Preprint*, 2002.

127. B. Nachtergaele. Interfaces and droplets in quantum lattice models. In *XIIIth International Congress on Mathematical Physics (London, 2000)*, pages 243–249. Int. Press, Boston, MA, 2001.

128. Charles M. Newman. Disordered Ising systems and random cluster representations. In *Probability and phase transition (Cambridge, 1993)*, pages 247–260. Kluwer Acad. Publ., Dordrecht, 1994.

129. E. Olivieri. On a cluster expansion for lattice spin systems: a finite-size condition for the convergence. *J. Statist. Phys.*, 50(5–6):1179–1200, 1988.

130. E. Olivieri and P. Picco. Cluster expansion for d-dimensional lattice systems and finite-volume factorization properties. *J. Statist. Phys.*, 59(1–2):221–256, 1990.

131. Y. Peres and P. Winkler. private communication.

132. G. Posta. Spectral gap for an unrestricted Kawasaki type dynamics. *ESAIM Probab. Statist.*, 1:145–181 (electronic), 1995/97.

133. R. B. Potts. Some generalized order–disorder transformations. *Proceedings of the Cambridge Phisolophical Society*, 48, 1952.

134. D. Randall and R. A. Martin. Sampling adsorbing staircase walks using a new Markov chain decomposition method. *Symposium on Foundations of Computer Science (FOCS)*, pages 492–502, 2000.

135. D. Randall and P. Tetali. Analyzing Glauber dynamics by comparison of Markov chains. *J. Math. Phys.*, 41(3):1598–1615, 2000. Probabilistic techniques in equilibrium and nonequilibrium statistical physics.

136. D. Ruelle. *Statistical mechanics: Rigorous results*. W. A. Benjamin, Inc., New York-Amsterdam, 1969.

137. J. Salas and A. D. Sokal. Absence of phase transition for antiferromagnetic Potts models via the Dobrushin uniqueness theorem. *J. Statist. Phys.*, 86(3–4):551–579, 1997.

138. J. Salas and A. D. Sokal. The three-state square-lattice Potts antiferromagnet at zero temperature. *J. Statist. Phys.*, 92(5–6):729–753, 1998.

139. L. Saloff-Coste. Lectures on finite Markov chains. In *Lectures on probability theory and statistics (Saint-Flour, 1996)*, pages 301–413. Springer, Berlin, 1997.

140. R. H. Schonmann. Slow droplet-driven relaxation of stochastic Ising models in the vicinity of the phase coexistence region. *Comm. Math. Phys.*, 161(1):1–49, 1994.

141. R. H. Schonmann and N. Yoshida. Exponential relaxation of Glauber dynamics with some special boundary conditions. *Comm. Math. Phys.*, 189(2):299–309, 1997.

142. D. Sherrington and S. Kirkpatrick. Solvable model of a spin glass. *Phys. Rev. Lett.*, 35:1792–1796, 1972.

143. S. B. Shlosman. The droplet in the tube: a case of phase transition in the canonical ensemble. *Comm. Math. Phys.*, 125(1):81–90, 1989.

144. B. Simon. *The statistical mechanics of lattice gases. Vol. I.* Princeton University Press, Princeton, NJ, 1993.

145. A. Sinclair. Improved bounds for mixing rates of Markov chains and multi-commodity flow. *Combin. Probab. Comput.*, 1(4):351–370, 1992.

146. A. Sinclair. *Algorithms for random generation and counting.* Birkhäuser Boston Inc., Boston, MA, 1993. A Markov chain approach.

147. A. Sinclair and M. Jerrum. Approximate counting, uniform generation and rapidly mixing Markov chains. *Inform. and Comput.*, 82(1):93–133, 1989.

148. A. Sokal. Monte Carlo methods in statistical mechanics: foundations and new algorithms. In *Functional integration (Cargese, 1996)*, pages 131–192. Plenum, New York, 1997.

149. A. Sokal. A personal list of unsolved problems concerning lattice gases and antiferromagnetic potts models. *Preprint*, 2000.

150. H. Spohn. Interface motion in models with stochastic dynamics. *J. Statist. Phys.*, 71(2):389–462, 1998.

151. S. Starr. Some properties of the low lying spectrum of the ferromagnetic quantum xxz Heisenberg model. *http://front.math.ucdavis.edu/math-ph/0106024*, pages 106–109, Ph.D thesis 2001.

152. D. Stroock and B. Zegarlinski. On the ergodic properties of Glauber dynamics. *J. Statist. Phys.*, 81(5–6):1007–1019, 1995.

153. D. W. Stroock. Logarithmic Sobolev inequalities for Gibbs states. In *Dirichlet forms (Varenna, 1992)*, pages 194–228. Springer, Berlin, 1993.

154. D. W. Stroock and B. Zegarlinski. The equivalence of the logarithmic Sobolev inequality and the Dobrushin-Shlosman mixing condition. *Comm. Math. Phys.*, 144(2):303–323, 1992.

155. D. W. Stroock and B. Zegarlinski. The logarithmic Sobolev inequality for continuous spin systems on a lattice. *J. Funct. Anal.*, 104(2):299–326, 1992.

156. D. W. Stroock and B. Zegarlinski. The logarithmic Sobolev inequality for discrete spin systems on a lattice. *Comm. Math. Phys.*, 149(1):175–193, 1992.

157. N. Sugimine. A lower bound on the spectral gap of the 3-dimensional stochastic Ising models. *Preprint*, 2002.

158. L. E. Thomas. Bound on the mass gap for a stochastic contour model at low temperature. *J. Math. Phys.*, 30(9):2028–2034, 1989.

159. J. van den Berg and C. Maes. Disagreement percolation in the study of Markov fields. *Ann. Probab.*, 22(2):749–763, 1994.

160. A. C. D. van Enter, R. Fernández, and A. D. Sokal. Regularity properties and pathologies of position-space renormalization-group transformations: scope and limitations of Gibbsian theory. *J. Statist. Phys.*, 72(5–6):879–1167, 1993.

161. B. van Rensburg. Adsorbing staircase walks and staircase polygons. *Ann. Comb.*, 3(2–4):451–473, 1999. On combinatorics and statistical mechanics.

162. S. R. S. Varadhan and H.-T. Yau. Diffusive limit of lattice gas with mixing conditions. *Asian J. Math.*, 1(4):623–678, 1997.

163. E. Vigoda. Improved bounds for sampling colorings. *J. Math. Phys.*, 41(3):1555–1569, 2000. Probabilistic techniques in equilibrium and nonequilibrium statistical physics.

164. E. Vigoda. A note on the Glauber dynamics for sampling independent sets. *Electron. J. Combin.*, 8(1):Research Paper 8, 8 pp. (electronic), 2001.

165. J.-S. Wang and R. H. Swendsen. Non universal critical dynamics in Monte Carlo simulations. *Phys. Rev. Lett.*, 58:86–88, 1987.

166. D. Weitz. Combinatorial conditions for uniqueness of the Gibbs measure. *Preprint*, 2002.

167. D. B. Wilson. Mixing times of lozenge tilings and card shuffling Markov chains. *Preprint*, 1997.

168. H.-T. Yau. Logarithmic Sobolev inequality for lattice gases with mixing conditions. *Comm. Math. Phys.*, 181(2):367–408, 1996.

169. H.-T. Yau. Logarithmic Sobolev inequality for generalized simple exclusion processes. *Probab. Theory Related Fields*, 109(4):507–538, 1997.

170. N. Yoshida. The log-Sobolev inequality for weakly coupled lattice fields. *Probab. Theory Related Fields*, 115(1):1–40, 1999.

171. N. Yoshida. Application of log-Sobolev inequality to the stochastic dynamics of unbounded spin systems on the lattice. *J. Funct. Anal.*, 173(1):74–102, 2000.

172. N. Yoshida. The equivalence of the log-Sobolev inequality and a mixing condition for unbounded spin systems on the lattice. *Ann. Inst. H. Poincaré Probab. Statist.*, 37(2):223–243, 2001.

173. B. Zegarlinski. On log-Sobolev inequalities for infinite lattice systems. *Lett. Math. Phys.*, 20(3):173–182, 1990.

174. B. Zegarlinski. The strong decay to equilibrium for the stochastic dynamics of unbounded spin systems on a lattice. *Comm. Math. Phys.*, 175(2):401–432, 1996.

Random Walks on Finite Groups

Laurent Saloff-Coste[*]

Summary. Markov chains on finite sets are used in a great variety of situations to approximate, understand and sample from their limit distribution. A familiar example is provided by card shuffling methods. From this viewpoint, one is interested in the "mixing time" of the chain, that is, the time at which the chain gives a good approximation of the limit distribution. A remarkable phenomenon known as the cut-off phenomenon asserts that this often happens abruptly so that it really makes sense to talk about "the mixing time". Random walks on finite groups generalize card shuffling models by replacing the symmetric group by other finite groups. One then would like to understand how the structure of a particular class of groups relates to the mixing time of natural random walks on those groups. It turns out that this is an extremely rich problem which is very far to be understood. Techniques from a great variety of different fields – Probability, Algebra, Representation Theory, Functional Analysis, Geometry, Combinatorics – have been used to attack special instances of this problem. This article gives a general overview of this area of research.

* Research supported in part by NSF grant DMS 0102126

1 Introduction

This article surveys what is known about the convergence of random walks
on finite groups, a subject to which Persi Diaconis gives a marvelous introduction in [27]. In the early twentieth century, Markov, Poincaré and Borel
discussed the special instance of this problem associated with card shuffling
where the underlying group is the symmetric group S_{52}. Two early references
are to Emile Borel [15] and K.D. Kosambi and U.V.R. Rao [95]. The early
literature focuses mostly on whether or not a given walk is ergodic: for card
shuffling, ergodicity means that the deck gets mixed up after many shuffles.

Once ergodicity is established, the next task is to obtain quantitative estimates on the number of steps needed to reach approximate stationarity. Of
course, this requires precise models and the choice of some sort of distance
between probability distributions.

Consider the shuffling method used by good card players called riffle shuffling. At each step, the deck is cut into two packs which are then riffled together. A model was introduced by Gilbert and Shannon in a 1955 Bell Laboratories technical memorandum. This model was later rediscovered and studied independently by Reeds in an unpublished work quoted in [27]. Around
1982, Aldous [1] proved that $\frac{3}{2}\log_2 n$ riffle shuffles are necessary and sufficient
to mix up n cards, as n goes to infinity. A complete analysis of riffle shuffles
was finally obtained in 1992 by Bayer and Diaconis [13], who argue that seven
riffle shuffles are reasonable to mix up a deck of 52 cards.

A widespread misconception is to consider that the problem of the convergence of ergodic random walks (more generally, ergodic Markov chains)
is solved by the Perron–Frobenius theorem which proves convergence to stationarity at an exponential rate controlled by the spectral gap (i.e., the gap
between 1 and the second largest eigenvalue in modulus). To understand
the shortcomings of this classical result, consider the Gilbert–Shannon–Reeds
model for riffle shuffles. Its spectral gap is $1/2$, independently of the number
of cards (see the end of Section 3.2). This does not tell us how many times
n cards should be shuffled, let alone 52 cards. Spectral gap estimates are an
important part of the study of ergodic random walks but, taking seriously the
practical question "how many times should 52 cards be shuffled to mix up the
deck?" and generalizing it to random walks on finite groups lead to richer and
deeper mathematical problems. What is known about these problems is the
subject of this article.

At first sight, it is not entirely clear that the question "how many times
should 52 cards be shuffled to mix up the deck?" makes mathematical sense.
One reason it does is that stationarity is often reached abruptly. This important fact, called the cut-off phenomenon, was discovered by Aldous, Diaconis
and Shahshahani [1, 50] and formalized by Aldous and Diaconis [5, 30]. In
their 1981 article [50], Diaconis and Shahshahani use the representation theory of the symmetric group (and hard work) to give the first complete analysis
of a complex ergodic random walk: random transposition on the symmetric
group. Their main finding is that it takes $t_n = \frac{1}{2}n\log n$ random transpositions
to mix up a deck of n cards. More precisely, for any $\varepsilon > 0$, after $(1-\varepsilon)t_n$ random transpositions the deck is far from being well mixed whereas after $(1+\varepsilon)t_n$
random transpositions the deck is well mixed, when n is large enough. This
is the first example of the cut-off phenomenon. The riffle shuffle model gives
another example. Even for $n = 52$, the cut-off phenomenon for riffle shuffles
is visible. See Table 1 in Section 3.3.

It is believed that the cut-off phenomenon is widespread although it has
been proved only for a rather small number of examples. One of the most
interesting problems concerning random walks on finite groups is to prove

or disprove the cut-off phenomenon for natural families of groups and walks. Focusing on walks associated with small sets of generators, one wants to understand how group theoretic properties relate to the existence or non-existence of a cut-off and, more generally, to the behavior of random walks. For instance, in any simple finite group, most pairs of elements generate the group (see, e.g., [130]). Is it true that any finite simple group G contains a pair of generators such that the associated random walk has a cut-off with a cut-off time of order $\log|G|$ as $|G|$ grows to infinity? Is it true that most walks based on two generators in a simple finite group behave this way? As the cut-off phenomenon can be very hard to establish, one often has to settle for less, for instance, the order of magnitude of a possible cut-off time.

In 2001, Diaconis and Holmes were contacted by a company that builds shuffling machines for the gambling industry. It turns out that these machines use a shuffling scheme that closely resembles one that they considered independently and without the least idea that it could ever be of practical value: see [37]. Besides shuffling and its possible multi-million-dollar applications for the gambling industry, random walks on finite groups are relevant for a variety of applied problems. Diaconis [27] describes connections with statistics. Random walks are a great source of examples for the general theory of finite Markov chains [3, 124, 131] and can sometimes be used to analyze by comparison Markov chains with fewer symmetries (see, e.g., [38]). It relates to Monte-Carlo Markov Chain techniques and to problems in theoretical computer science as described in [94, 131]. Random walks provided the first explicit examples of expander graphs [108], a notion relevant to the construction of communication networks, see, e.g., [98]. In [55], Durrett discusses the analysis of families of random walks modeling the scrambling of genes on a chromosome by reversal of sequences of various lengths.

One perspective to keep in mind is that the study of random walks on finite groups is part of the more general study of invariant processes on groups. See, e.g., [125]. This direction of research relates to many different fields of mathematics. In particular, probability, finite and infinite group theory, algebra, representation theory, number theory, combinatorics, geometry and analysis, all have contributed fundamental ideas and results to the study of random walks on groups. This is both one of the difficulties of the subject and one of its blessings. Indeed, the deep connections with questions and problems coming from other areas of mathematics are one of the exciting aspects of the field.

The author is not aware of any previous attempt thoroughly to survey techniques and results concerning the convergence of random walks on finite groups. The book of Diaconis [27] has played and still plays a crucial role in the development of the subject. The survey [45] by Diaconis and Saloff-Coste served as a starting point for this article but has a narrower focus. Several papers of Diaconis [28, 31, 32] survey some specific directions such as riffle shuffle or the developments arising from the study of random transpositions. Some examples are treated and put in the context of general finite Markov chains in [3, 124, 131]. The excellent book [98] and the survey article [99] connect random walks to problems in combinatorics, group theory and number theory as does the student text [136].

This survey focuses exclusively on quantitative rates of convergence. Interesting questions such as hitting times, cover times, and other aspects of random walks are not discussed at all although they are related in various ways to rates of convergence. See [3, 27]. Important generalizations of random walks on groups to homogeneous spaces, Gelfand pairs, hypergroups and other structures, as well as Markov chains on groups obtained by deformation of random walks are not discussed. For pointers in these directions, see [14, 16, 17, 27, 29, 31, 32, 36, 41].

2 Background and Notation

2.1 Finite Markov Chains

Markov kernels and Markov chains. A *Markov kernel* on a finite set $\mathcal{X}$ is a function $K : \mathcal{X} \times \mathcal{X} \to [0, 1]$ such that $\sum_y K(x, y) = 1$. Given an initial probability measure ν, the associated *Markov chain* is the discrete-time stochastic process $(X_0, X_1, \dots)$ taking values in $\mathcal{X}$ whose law $\mathbb{P}_\nu$ on $\mathcal{X}^{\mathbb{N}}$ is given by

$$\mathbb{P}_\nu(X_i = x_i, 0 \leq i \leq n) = \nu(x_0)K(x_0, x_1) \cdots K(x_{n-1}, x_n). \qquad (2.1)$$

We will use $\mathbb{P}_x$ to denote the law of the Markov chain $(X_n)_{n \geq 0}$ starting from $X_0 = x$, that is, $\mathbb{P}_x = \mathbb{P}_{\delta_x}$. One can view K as a stochastic matrix – the transition matrix – whose rows and columns are indexed by $\mathcal{X}$. We associate to K a *Markov operator* – also denoted by K – which acts on functions by $Kf(x) = \sum_y K(x, y)f(y)$ and on measures by $\nu K(A) = \sum_x \nu(x)K(x, A)$.

The *iterated kernel* $K_n(x, y)$ is defined inductively by

$$K_1(x, y) = K(x, y) \quad \text{and} \quad K_n(x, y) = \sum_{z \in \mathcal{X}} K_{n-1}(x, z)K(z, y). \qquad (2.2)$$

Given $X_0 = x$, the law of X_n is the probability measure $A \mapsto K_n(x, A)$, $A \subset \mathcal{X}$. From this definition it follows that (X_i) has the *Markov property*: the future depends on the past only through the present. More precisely, let $\tau : \mathcal{X}^{\mathbb{N}} \to \{0, 1, \dots\} \cup \{\infty\}$ be a random variable such that the event $\{\tau \leq n\}$ depends only on $X_0, \dots, X_n$ (i.e., a stopping time). Then, conditional on $\tau < \infty$ and $X_\tau = x$, $(X_{\tau+i})_{i \geq 0}$ is a Markov chain with kernel K started at x and is independent of $X_0, \dots, X_\tau$.

There is also an $\mathcal{X}$-valued continuous-time Markov process $(X_t)_{t \geq 0}$ which evolves by performing jumps according to K with independent exponential(1) holding times between jumps. This means that $X_t = X_{N_t}$ where N_t has a Poisson distribution with parameter t. Thus, starting from $X_0 = x$, the law of X_t is given by the familiar formula

$$H_t(x, \cdot) = e^{-t} \sum_0^\infty \frac{t^n}{n!} K_n(x, \cdot). \tag{2.3}$$

In terms of Markov operators, this continuous-time process is associated with the Markov semigroup $H_t = e^{-t(I-K)}$, $t \geq 0$, where I denotes the identity operator.

The invariant measure and time reversal. A probability distribution π is *invariant* for K if $\pi K = \pi$. Given an invariant distribution π for K and $p \in [1, \infty)$, set

$$\|f\|_p = \left(\sum_x |f(x)|^p \pi(x) \right)^{1/p}, \quad L^p(\pi) = \{f : \mathcal{X} \to \mathbb{R} : \|f\|_p < \infty\},$$

where $\|f\|_\infty = \max_\mathcal{X} |f|$. Then K is a contraction on each $L^p(\pi)$. Define

$$K^*(x, y) = \frac{\pi(y)K(y, x)}{\pi(x)}. \tag{2.4}$$

The kernel K^* is Markov and has the following interpretation: Let $(X_n)_{0 \leq n \leq N}$ be a Markov chain with kernel K and initial distribution π. Set $Y_n = X_{N-n}$, $0 \leq n \leq N$. Then $(Y_n)_{0 \leq n \leq N}$ is a Markov chain with kernel K^* and initial distribution π. Thus K^* corresponds to the chain obtained from (X_n) by time reversal. The Markov kernel K^* is also the kernel of the adjoint of the operator K acting on $L^2(\pi)$. Clearly, $K^* = K$ if and only if

$$\forall\, x, y \in \mathcal{X}, \quad \pi(x)K(x, y) = \pi(y)K(y, x). \tag{2.5}$$

When (K, π) satisfies (2.5), one says that K is *reversible* with respect to π and that π is a *reversible measure* for K. Equation (2.5) is also called the *detailed balance condition* in the statistical mechanics literature.

Ergodic chains. A Markov kernel K is *irreducible* if, for any two states x, y there exists an integer $n = n(x, y)$ such that $K_n(x, y) > 0$. A state x is called *aperiodic* if $K_n(x, x) > 0$ for all sufficiently large n. If K is irreducible and has an aperiodic state then all states are aperiodic. We will mostly be interested in irreducible, aperiodic chains.

Theorem 2.1. *Let K be an irreducible Markov kernel on a finite state space $\mathcal{X}$. Then K admits a unique invariant distribution π and*

$$\forall\, x, y \in \mathcal{X}, \quad \lim_{t \to \infty} H_t(x, y) = \pi(y).$$

Assume further that K is aperiodic. Then the chain is ergodic, that is,

$$\forall\, x, y \in \mathcal{X}, \quad \lim_{n \to \infty} K_n(x, y) = \pi(y).$$

For irreducible K, the unique invariant distribution is also called the *stationary* (or equilibrium) probability.

In practice, one is interested in turning the qualitative conclusion of Theorem 2.1 into more quantitative assertions. To this end some sort of distance between probability measures must be chosen. The *total variation distance* between two probability measures μ, ν on $\mathcal{X}$ is defined as

$$d_{\mathrm{TV}}(\mu, \nu) = \|\mu - \nu\|_{\mathrm{TV}} = \sup_{A \subset \mathcal{X}} \{\mu(A) - \nu(A)\}. \tag{2.6}$$

It gives the maximum error made when using μ to approximate ν. Next, consider the $L^p(\pi)$-distances relative to a fixed underlying probability measure π on $\mathcal{X}$. In the cases of interest here, π will be the invariant distribution of a given Markov chain under consideration. Given two probability distributions μ, ν with respective densities f, g with respect to π, set

$$d_{\pi, p}(\mu, \nu) = \|f - g\|_p = \left(\sum_{x \in \mathcal{X}} |f(x) - g(x)|^p \pi(x) \right)^{1/p} \tag{2.7}$$

and $d_{\pi, \infty}(\mu, \nu) = \max\{|f - g|\}$. Setting $\mu(f) = \sum f\mu$ and $p = 1$, we have

$$d_{\pi, 1}(\mu, \nu) = 2d_{\mathrm{TV}}(\mu, \nu) = 2\|\mu - \nu\|_{\mathrm{TV}} = \max_{\|f\|_\infty = 1} \{|\mu(f) - \nu(f)|\} \tag{2.8}$$

which is independent of the choice of π. For $p = 2$,

$$d_{\pi, 2}(\mu, \nu) = \left(\sum_{x \in \mathcal{X}} \left| \frac{\mu(x)}{\pi(x)} - \frac{\nu(x)}{\pi(x)} \right|^2 \pi(x) \right)^{1/2}.$$

Note that Jensen's inequality shows that $p \mapsto d_{\pi, p}$ is a non-decreasing function. In particular,

$$2d_{\mathrm{TV}}(\mu, \nu) \leq d_{\pi, 2}(\mu, \nu) \leq d_{\pi, \infty}(\mu, \nu). \tag{2.9}$$

The following is one of the most useful basic results concerning ergodic chains. It shows and explains why exponentially fast convergence is the rule if the chain converges at all.

Proposition 2.2. *Let K be a Markov kernel with invariant probability distribution π. Then, for any fixed $1 \leq p \leq \infty$, $n \mapsto \sup_{x \in \mathcal{X}} d_{\pi, p}(K_n(x, \cdot), \pi)$ is a non-increasing sub-additive function. In particular, if*

$$\sup_{x \in \mathcal{X}} d_{\pi, p}(K_m(x, \cdot), \pi) \leq \beta$$

for some fixed *integer m and some $\beta \in (0, 1)$ then*

$$\forall n \in \mathbb{N}, \quad \sup_{x \in \mathcal{X}} d_{\pi, p}(K_n(x, \cdot), \pi) \leq \beta^{\lfloor n/m \rfloor}.$$

See, e.g., [1, 3, 5, 124].

2.2 Invariant Markov Chains on Finite Groups

Random walks. Let G be a finite group with identity element e. Let $|G|$ be the order (i.e., the number of elements) of G. Let p be a probability measure on G. The *left-invariant random walk* on G driven by p is the Markov chain with state space $\mathcal{X} = G$ and transition kernel

$$K(x, y) = p(x^{-1}y).$$

As $\sum_x p(x^{-1}y) = \sum_x p(x) = 1$, any such chain admits the normalized counting measure (i.e., uniform distribution) $u \equiv 1/|G|$ as invariant distribution. Moreover, $u \equiv 1/|G|$ is a reversible measure for p if and only if p is symmetric, i.e., $p(x) = p(x^{-1})$ for all $x \in G$.

Fix an initial distribution ν. Let $(\xi_i)_0^\infty$ be a sequence of independent G-valued random variables, with ξ_0 having law ν and ξ_i having law p for all $i \geq 1$. Then the left-invariant random walk driven by p can be obtained as

$$X_n = \xi_0 \xi_1 \dots \xi_n.$$

The iterated kernel $K_n(x, y)$ defined at (2.2) is given by the *convolution power*

$$K_n(x, y) = p^{(n)}(x^{-1}y)$$

where $p^{(n)}$ is the n-fold convolution product $p * \cdots * p$ with

$$f * g(x) = \sum_{z \in G} f(z)g(z^{-1}x) = \sum_{z \in G} f(xz^{-1})g(z).$$

For any initial distribution ν, we have $\mathbb{P}_\nu(X_n = x) = \nu * p^{(n)}(x)$. The associated Markov operator K acting on functions is then given by

$$Kf(x) = f * \check{p}(x)$$

where $\check{p}(x) = p(x^{-1})$. The law of the associated continuous-time process defined at (2.3) satisfies $H_t(x, y) = H_t(x^{-1}y)$ where

$$H_t(x) = H_t(e, x) = \mathrm{e}^{-t} \sum_0^\infty \frac{t^n}{n!} p^{(n)}(x). \tag{2.10}$$

The adjoint K^* of the operator K on $L^2(G)$ (i.e., L^2 with respect to the normalized counting measure) is

$$K^*f = f * p.$$

This means that the time reversal of a random walk driven by a measure p is driven by the measure $\check{p}$. Referring to the walk driven by p, we call the walk driven by $\check{p}$ the *reverse walk*. Observe that we always have

$$d_{u,s}(p^{(n)}, u) = d_{u,s}(\breve{p}^{(n)}, u). \tag{2.11}$$

In words, the distance to stationarity measured in terms of any of the distances $d_{u,s}$ is the same for a given random walk and for its associated reverse walk. By (2.8), this applies to the distance in total variation as well.

One can also consider right-invariant random walks. The right-invariant random walk driven by p has kernel $\widetilde{K}(x, y) = p(yx^{-1})$ and, in the notation introduced above, it can be realized as $\widetilde{X}_n = \xi_n \dots \xi_1 \xi_0$. The iterated kernel $\widetilde{K}_n(x, y)$ is given by $\widetilde{K}_n(x, y) = p^{(n)}(yx^{-1})$. Under the group anti-isomorphism $x \mapsto x^{-1}$, the left-invariant random walk driven by a given probability measure p transforms into the right-random walk driven by $\breve{p}$. Hence, it suffices to study left-invariant random walks.

Ergodic random walks. The next proposition characterizes irreducibility and aperiodicity in the case of random walks. It has been proved many times by different authors. Relatively early references are [143, 144].

Proposition 2.3. *On a finite group G, let p be a probability measure with support $\Sigma = \{x \in G : p(x) > 0\}$.*

- *The chain driven by p is irreducible if and only if Σ generates G, i.e., any group element is the product of finitely many elements of Σ.*
- *Assuming Σ generates G, the random walk driven by p is aperiodic if and only if Σ is not contained in a coset of a proper normal subgroup of G.*

To illustrate this proposition, let $G = S_n$ be the symmetric group on n letters and p the uniform distribution on the set $\Sigma = \{(i, j) : 1 \le i < j \le n\}$ of all transpositions. As any permutation can be written as a product of transpositions, this walk is irreducible. It is not aperiodic since $\Sigma \subset (1, 2)A_n$ and the alternating group A_n is a proper normal subgroup of S_n.

If the random walk driven by p is aperiodic and irreducible then, by Theorem 2.1, its iterated kernel $K_n(x, y) = p^{(n)}(x^{-1}y)$ converges for each fixed $x \in G$ to its unique invariant measure which is the uniform measure $u \equiv 1/|G|$. By left invariance, there is no loss of generality in assuming that the starting point x is the identity element e in G and one is led to study the difference $p^{(n)} - u$. This brings some useful simplifications. For instance, $d_{u,s}(K_n(x, \cdot), u)$ is actually independent of x and is equal to

$$d_{u,s}(p^{(n)}, u) = |G|^{1-1/s} \left(\sum_{y \in G} \left| p^{(n)}(y) - 1/|G| \right|^s \right)^{1/s}$$

for any $s \in [1, \infty]$ with the usual interpretation if $s = \infty$. From now on, for random walks on finite groups, we will drop the reference to the invariant measure u and write d_s for $d_{u,s}$. Proposition 2.2 translates as follows.

Proposition 2.4. *For any* $s \in [1, \infty]$ *and any probability measure* p, *the function* $n \to d_s(p^{(n)}, u)$ *is non-increasing and sub-additive. In particular, if* $d_s(p^{(m)}, u) \leq \beta$ *for some fixed integer* m *and* $\beta \in (0, 1)$ *then*

$$\forall n \in \mathbb{N}, \quad d_s(p^{(n)}, u) \leq \beta^{\lfloor n/m \rfloor}.$$

To measure ergodicity, we will mostly use the total variation distance $\|p^{(k)} - u\|_{\mathrm{TV}}$ and the L^2-distance $d_2(p^{(k)}, u)$. Note that d_2 also controls the a priori stronger distance d_∞. Indeed, noting that $p^{(2k)} - u = (p^{(k)} - u) * (p^{(k)} - u)$ and using the Cauchy–Schwarz inequality and (2.11), one finds that

$$d_\infty(p^{(2k)}, u) \leq d_2(p^{(k)}, u)^2$$

with equality in the symmetric (i.e, reversible) case where $p = \check{p}$.

3 Shuffling Cards and the Cut-off Phenomenon

3.1 Three Examples of Card Shuffling

Modeling card shuffling. That shuffling schemes can be modeled by Markov chains has been clearly recognized from the beginning of Markov chain theory. Indeed, card shuffling appears as one of the few examples given by Markov in [104]. It then appears in the works of Poincaré and Borel. See in particular [15], and the excellent historical discussion in [92]. Obviously, from a mathematical viewpoint, an arrangement of a deck of cards can be thought of as a permutation of the cards. Also, a shuffling is obviously a permutation of the cards. There is however an intrinsic difference between an arrangement of the cards and a shuffling: an arrangement of the cards relates face values to positions whereas, strictly speaking, a shuffling is a permutation of the positions. By a good choice of notation, this difference somehow disapears but this might introduce some confusion. Thus we now spell out in detail one of the possible equivalent ways to model shufflings using random walks on S_n, $n = 52$. We view the symmetric group S_n as the set of all bijective maps from $\{1, \ldots, n\}$ to itself equipped with composition. Hence, for $\sigma, \theta \in S_n$, $\sigma\theta = \sigma \circ \theta$. One of several ways to describe a permutation σ is as an n-tuple $(\sigma_1, \ldots, \sigma_n)$ where $\sigma(i) = \sigma_i$.

To simplify, think of the 52 cards as marked from 1 to 52. An arrangement of the deck can described as a 52-tuple giving the face values of the cards in order from top to bottom. Thus we can identify the arrangement of the deck $(\sigma_1, \ldots, \sigma_{52})$ with the permutation $\sigma : i \mapsto \sigma(i) = \sigma_i$ in S_{52}. In this notation, the deck corresponding to a permutation σ has card i in position $\sigma^{-1}(i)$ whereas $\sigma(i)$ gives the value of the card in position i. In particular, the deck in order is represented by the identity element. Now, from a card shuffling perspective, we want permutations to act on positions, not on face values. One easily checks that, in the present notation, this corresponds to multiplication on the right in S_{52}. Indeed, if the arrangement of the deck is σ and we transpose the top two cards then the new arrangement of the deck is $\sigma \circ \tau$ with $\tau = (1,2)$ since $\sigma \circ \tau$ is $(\sigma_2, \sigma_1, \sigma_3, \ldots, \sigma_{52})$.

Typically, shuffling cards proceeds by repeating several times a fixed procedure where some randomness occurs. This can now be modeled by a measure p on S_{52} which describes the shuffling procedure as picking a permutation θ according to p and changing the arrangement σ of the deck to $\sigma\theta = \sigma \circ \theta$. Thus the shuffling scheme whose elementary steps are modeled by p corresponds to the left-invariant random walk on S_{52} driven by p. By invariance, we can always assume that we start from the identity permutation, that is, with the deck in order. Then, the distribution of the deck after n shuffles is given by $p^{(n)}$. Let us describe three examples.

The Borel–Chéron shuffle. In [15, pages 8–10 and 254–256], Borel and Chéron consider the following shuffling method: remove a random packet from the deck and place it on top. The corresponding permutations are $\pi_{a,b}$, $1 < a \leq b \leq n = 52$, given by

$$\begin{pmatrix} 1 & 2 & \cdot\,\cdot & b-a+1 & b-a+2 & \cdot\,\cdot & b & b+1 & \cdot\,\cdot & 52 \\ a & a+1 & \cdot\,\cdot & b & 1 & \cdot\,\cdot & a-1 & b+1 & \cdot\,\cdot & 52 \end{pmatrix}$$

where the first row indicates position and the second row gives the value of the cards in that position after $\pi_{a,b}$ if one starts with a deck in order. The removed packet is random in the sense that $p(\pi) = 0$ unless $\pi = \pi_{a,b}$ for some $1 < a \leq b \leq n$ in which case $p(\pi) = \binom{n}{2}^{-1}$ (a slightly different version is considered in [42]).

The crude overhand shuffle. In this example, the player holds the deck in the right hand and transfers a first block of cards from the top of the deck to the left hand, then a second block of cards, and finally all the remaining cards. This is then repeated many times. The randomness comes from the size of the first and second block, say a and b. With our convention, the corresponding permutation $\sigma_{a,b}$ is

$$\begin{pmatrix} 1 & 2 & \cdots & 51-a-b & 52-a-b & \cdots & 52-a-1 & 52-a & \cdots & 51 & 52 \\ a+b+1 & a+b+2 & \cdots & 52 & a+1 & \cdots & a+b & 1 & \cdots & a-1 & a \end{pmatrix}.$$

In this case, it is natural to take $p(\sigma) = 0$ unless $\sigma = \sigma_{a,b}$ for some $1 \leq a \leq n = 52$ and $0 \leq b \leq n - a$, in which case $p(\sigma_{a,b}) = 1/[n(n+1-a)]$. Other overhand shuffles are described in [116, 44].

The riffle shuffle or dovetail shuffle. Consider the way serious players shuffle cards. The deck is cut into two packs (of roughly equal sizes) and the two packs are riffled together. A model was introduced by Gilbert and Shannon (see Gilbert [66]) and later, independently, by Reeds [118]. In this model, the cut is made according to a binomial distribution: the k top cards are cut with probability $\binom{n}{k}/2^n$, $n = 52$. The two packets are then riffled together in such a way that the cards drop from the left or right heaps with probability proportional to the number of cards in each heap. Thus, if there are a and b cards remaining in the left and right heaps, then the chance the next card will drop from the left heap is $a/(a + b)$. This describes a probability p_{RS} on the symmetric group. Experiments reported in Diaconis' book [27] indicate that this model describes well the way serious card players shuffle cards. It is interesting to note that the inverse shuffle – i.e., the shuffle corresponding to the measure $\check{p}_{\mathrm{RS}}$ – is simple to describe: starting from the bottom, each card is removed from the deck and placed randomly on one of two piles, left or right, according to an independent sequence of Bernoulli random variables (probability $1/2$ for right and left). Finally, the right pile is put on top.

3.2 Exact Computations

The analysis of riffle shuffles. This section focuses on the riffle shuffle model p_{RS} of Gilbert, Shannon and Reeds, the GSR model for short. How many GSR shuffles are needed to mix up a deck of n cards? To make this question precise, let us use the total variation distance between the uniform distribution u on the symmetric group S_n and the distribution $p_{\mathrm{RS}}^{(k)}$ after k shuffles. The question becomes: how large must k be for $\|p_{\mathrm{RS}}^{(k)} - u\|_{\mathrm{TV}}$ to be less than some fixed $\varepsilon > 0$? As far as shuffling cards is concerned, a value of ε a little below 0.5 seems quite reasonable to aim for. Bayer and Diaconis [13] give the following remarkably precise analysis of riffle shuffles.

Theorem 3.1. *If a deck of n cards is shuffled k times with*

$$k = \frac{3}{2}\log_2 n + c,$$

then for large n

$$\|p_{\mathrm{RS}}^{(k)} - u\|_{\mathrm{TV}} = 1 - 2\Phi\left(\frac{-2^{-c}}{4\sqrt{3}}\right) + O\left(\frac{1}{n^{1/4}}\right),$$

where

$$\Phi(t) = \frac{1}{\sqrt{2\pi}}\int_{-\infty}^{t} e^{-s^2/2}ds.$$

A weaker form of this result was proved earlier in [1].

To people studying finite Markov chains, the fact that Theorem 3.1 can be proved at all appears like a miracle. Consider for instance the following "neat riffle shuffle" model proposed by Thorpe (see [27, 137]). For a deck of $n = 2k$ cards, cut the deck into two piles of exactly k cards each and put in positions $2j$ and $2j-1$ the j-th card of each of the two piles in random order. No reasonable quantitative analysis of this shuffle is known.

The idea used by Bayer and Diaconis to analyze repeated riffle shuffles is elementary. Given an arrangement of a deck of cards, a *rising sequence* is a maximal subset of cards of this arrangement consisting of successive face values displayed in order. For example, the arrangement $2, 4, 3, 9, 1, 6, 7, 8, 5$, consists of 1; $2, 3$; $4, 5$; $6, 7, 8$ and 9. Note that the rising sequences form a partition of the deck. Denote by r the number of rising sequences of an arrangement of the deck. By extension, we also say that r is the number of rising sequences of the associated permutation. Now, it is a simple observation that, starting from a deck in order, one riffle shuffle produces permutations having at most 2 rising sequences. In fact (see [13]), the riffle shuffle measure p_{RS} is precisely given by

$$p_{\mathrm{RS}}(\sigma) = 2^{-n} \binom{n + 2 - r}{n}$$

where r is the number of rising sequences of σ and $\binom{m}{n} = 0$ when $m < n$.

The next step is to define the notion of an m-riffle shuffle which generalizes the above 2-riffle shuffle. In an m-riffle shuffle, the deck is cut into m parts which are then riffled together. It is easier to define a reverse m-riffle shuffle: hold the deck, face down and create m piles by dealing the deck in order and turning the cards face up on a table. For each card, pick a pile uniformly at random, independently from all previous picks. When all the cards have been distributed, assemble the piles from left to right and turn the deck face down. Let $p_m = p_{m\text{-RS}}$ be the probability measure corresponding to an m-riffle shuffle. Diaconis and Bayer show that

$$p_m(\sigma) = m^{-n} \binom{n + m - r}{n}$$

where r is again the number of rising sequences. Moreover, they show that following an m-riffle shuffle by an ℓ-riffle shuffle produces exactly an $m\ell$-riffle shuffle, that is, $p_\ell * p_m = p_{m\ell}$. Thus the distribution $p_{\mathrm{RS}}^{(k)}$ of a deck of n cards after k GSR riffle shuffles is given by

$$p_{\mathrm{RS}}^{(k)}(\sigma) = 2^{-kn} \binom{n + 2^k - r}{n}. \tag{3.1}$$

From there, the proof of Theorem 3.1 consists in working hard to obtain adequate asymptotics and estimates. Formula (3.1) allows us to compute the total variation distance exactly for $n = 52$. This is reported (to three decimal places) in Table 1.

Table 1. The total variation distance for k riffle shuffles of 52 cards

k	1	2	3	4	5	6	7	8	9	10
$\|p_{\mathrm{RS}}^{(k)} - u\|_{\mathrm{TV}}$	1.000	1.000	1.000	1.000	0.924	0.614	0.334	0.167	0.085	0.043

Top to random shuffles. There are not many examples of shuffles where the law after k shuffles can be explicitly computed as above. In [34], the authors study a class of shuffles that they call top to random shuffles. In a top m to random shuffle, the top m cards are cut and inserted one at a time at random in the remaining $n - m$ cards. Call q_m the corresponding probability measure. In particular, q_1 is called the top to random measure. Note the similarity with the riffle shuffle: a top to random shuffle can be understood as a riffle shuffle where exactly one card is cut off.

Given a probability measure μ on $\{0, 1, \ldots, n\}$, set

$$q_\mu = \sum_0^n \mu(i)q_i. \tag{3.2}$$

Further variations are considered in [34]. In some cases, an exact formula can be given for the convolutions of such measures and this leads to the following theorem.

Theorem 3.2. *Let a, n, $a \le n$, be two integers. Let μ be a probability on $\{0, \ldots, a\}$ with positive mean m. On S_n, consider the probability measure q_μ at (3.2). Then, for large n and*

$$k = \frac{n}{m}\log n + c,$$

we have $\|q_\mu^{(k)} - u\|_{\mathrm{TV}} = f(c) + o(1)$ where f is a positive function such that $f(c) \le (1/2)\mathrm{e}^{-2c}$ for $c > 0$ and $f(c) = 1 - \exp\left(-\mathrm{e}^{-c} + o(1)\mathrm{e}^{-c}\right)$ for $c < 0$.

Diagonalization. The riffle shuffles and top to random shuffles described above, as well as variants and generalizations discussed in [60, 61], have remarkable connections with results in algebra. These connections explain in part why an exact formula exists for repeated convolution of these measures. See [13, 32, 34, 40, 60, 61].

In particular, the convolution operators corresponding to the m-riffle shuffle measures p_m and the top to random measures q_m are diagonalizable with eigenvalues that can be explicitly computed. For instance, for the GSR measure $p_{\mathrm{RS}} = p_2$, the eigenvalues are the numbers 2^{-i} with multiplicity the number of permutations having exactly $n - i$ cycles, $i = 0, \ldots, n - 1$. For the top to random measure $q = q_1$, the eigenvalues are i/n, $i = 0, 1, \ldots, n - 2, n$, and the multiplicity of i/n is exactly the number of permutations having i fixed points. However, these results do not seem to be useful to control convergence to stationarity. Curiously, the eigenvalues of top to random have been computed independently for different reasons by different authors including Wallach (Lie algebra cohomology) and Phatafod (linear search). See the references in [32, 34].

3.3 The Cut-off Phenomenon

Cut-off times. Table 1, Theorem 3.1 and Theorem 3.2 all illustrate a phenomenon first studied by Aldous and Diaconis [5] and called the *cut-off phenomenon* [30] (in [5], the term threshold phenomenon is used instead).

To give a precise definition, consider a family of finite groups G_n, each equipped with its uniform probability measure u_n and with another probability measure p_n which induces a random walk on G_n.

Definition 3.3. *We say that the* cut-off phenomenon *holds (in total variation) for the family $((G_n, p_n))$ if there exists a sequence (t_n) of positive reals such that*

(a) $\lim_{n \to \infty} t_n = \infty$;

(b) For any $\varepsilon \in (0, 1)$ and $k_n = [(1 + \varepsilon)t_n]$, $\lim_{n \to \infty} \|p_n^{(k_n)} - u_n\|_{\mathrm{TV}} = 0$;

(c) For any $\varepsilon \in (0, 1)$ and $k_n = [(1 - \varepsilon)t_n]$, $\lim_{n \to \infty} \|p_n^{(k_n)} - u_n\|_{\mathrm{TV}} = 1$.

We will often say, informally, that (G_n, p_n) has a (total variation) cut-off at time t_n. For possible variants of this definition, see [30, 124].

Theorem 3.1 shows that the GSR riffle shuffle measure p_{RS} on S_n has a cut-off at time $\frac{3}{2} \log_2 n$. Similarly, Theorem 3.2 shows that the top to random measure q_1 on S_n has a cut-off at time $n \log n$. Note that if (t_n) and (t'_n) are cut-off times for the same family $((G_n, p_n))$, then $t_n \sim t'_n$ as n tends to infinity. Table 2 below lists most examples known to have a cut-off.

Definition 3.4. *For any probability measure p on a finite group G, set*

$$T(G, p) = T(G, p, 1/(2e)) = \inf \left\{ k : \|p^{(k)} - u\|_{\mathrm{TV}} \leq 1/(2e) \right\} \tag{3.3}$$

where $T(G, p, \varepsilon) = \inf \left\{ k : \|p^{(k)} - u\|_{TV} \leq \varepsilon \right\}$. We call $T(G, p)$ the total variation mixing time *(mixing time for short) of the random walk driven by p.*

Thus $T(G, p)$ is the number of steps needed for the given random walk to be $1/(2e)$-close to the uniform distribution in total variation. The arbitrary choice of $\varepsilon = 1/(2e)$ (any $\varepsilon \in (0, 1/2)$ would do) is partially justified by Proposition 2.4 which shows that

$$\forall k \in \mathbb{N}, \quad 2\|p^{(k)} - u\|_{\mathrm{TV}} \leq e^{-\lfloor k/T(G,p) \rfloor}.$$

To relate the last definition to the notion of cut-off, let $((G_n, p_n))$ be a family of random walks having a (t_n)-cut-off. Then, for any $\varepsilon \in (0, 1)$,

$$T(G_n, p_n, \varepsilon) \sim T(G_n, p_n) \sim t_n \quad \text{as } n \text{ tends to } \infty.$$

Thus, if (G_n, p_n) presents a cut-off, one can always take the cut-off time to be $t_n = T(G_n, p_n)$ and one often says that the cut-off time t_n is "the time needed to reach equilibrium".

Table 2. Total variation cut-offs

G	p	(CO)	§	Ref
$\mathbb{Z}_2^d$	$p(e_i) = 1/(d+1)$	$\frac{d}{4}\log d$	8.2	[35, 27, 28]
$\mathbb{Z}_2^d$	random spatula	$\frac{d}{8}\log d$	8.2	[138]
$\mathbb{Z}_n^d$	$p(e_i) = 1/(d+1),\ \ d\to\infty$	$\frac{d\log d}{2(1-\cos 2\pi/n)}$	8.2	[44, 47]
$\mathbb{Z}_2^d$	most k-sets, $k > d$	$T(d,k)$	8.2	[140]
abelian	most k-sets, $k = \lfloor(\log|G|)^s\rfloor,\ s>1$	$\frac{s}{s-1}\frac{\log|G|}{\log k}$	8.3	[54, 87]
S_n	GSR riffle shuffle, p_{RS}	$\frac{3}{2}\log_2 n$	3.2	[13]
S_n	top m to random, q_m	$\frac{n}{m}\log n$	3.2	[34]
S_n	random transposition, p_{RT}	$\frac{n}{2}\log n$	9.2	[50, 27]
S_n	transpose $(1,i)$, $p_\star$	$n\log n$	9.2	[28, 59]
S_n	lazy small odd conjugacy classes $C = (2), (4), (3,2), (6), (2,2,2)$	$\frac{2n}{|C|}\log n$	9.2	[59, 122]
A_n	small even S_n conjugacy classes $(3), (2,2), (5), (4,2), (3,3), (7)$	$\frac{n}{|C|}\log n$	9.2	[59, 122]
A_n	random m-cycle, m odd $m > n/2,\ n-m\to\infty$	$\frac{\log n}{\log(n/(n-m))}$	9.2	[103]
$G \wr S_n$	random transposition with independent flips	$\frac{n}{2}\log n$	9.2	[128, 129]
$G \wr S_n$	random transposition with paired flips	$\frac{n}{2}\log n$	9.2	[128, 129]
$SL_n(\mathbb{F}_q)$	random transvections	n	9.2	[86]

$$T(d,k) \sim \begin{cases} (d/4)\log(d/(k-d)) & \text{if } k-d = o(d) \\ a_\eta d & \text{if } k = (1+\eta)d \\ d/\log_2(k/d) & \text{if } d/k = o(1). \end{cases}$$

One can easily introduce the notion of L^s-mixing time and L^s-cut-off, $1 < s \le \infty$, by replacing $2\|p_n^{(k_n)} - u_n\|_{\mathrm{TV}}$ by $d_s(p_n^{(k_n)}, u_n)$ in Definitions 3.4, 3.3. In Definition 3.3(c), one should require that $\lim_{n\to\infty} d_s(p_n^{(k_n)}, u_n) = \infty$. In this survey, we will focus mostly on mixing time and cut-off in total variation but we will also make significant use the L^2-distance d_2.

Cut-off and group structure. Not all natural families of walks have a cut-off. For instance, the walk on $G_n = \mathbb{Z}/n\mathbb{Z}$ driven by the uniform measure on $\{-1,0,1\}$ does not present a cut-off. For this walk, it takes k of order n^2 to have $\|p_n^{(k)} - u_n\|_{\mathrm{TV}}$ close to $1/2$. It then takes order n^2 additional steps to go down to $1/4$, etc. In particular, for any integer $k > 0$,

$$0 < \liminf_{n\to\infty} \|p_n^{(kn^2)} - u_n\|_{\mathrm{TV}} \le \limsup_{n\to\infty} \|p_n^{(kn^2)} - u_n\|_{\mathrm{TV}} < 1.$$

See Sections 7.2 and 8.2 below.

Trying to understand which walks on which families of groups have a cut-off is one of the difficult open problems concerning random walks on finite groups. To be meaningful, this question should be made more precise. One possibility is to focus on walks driven by the uniform measure on minimal generating sets, i.e., generating sets that do not contain any proper generating sets (one might allow here the inclusion of inverses to have reversible walks and of the identity to cure periodicity problems). For instance, the set $\Sigma = \{(1,i) : 1 < i \le n\}$ (where $(1,i)$ means transpose 1 and i) is a minimal generating set of S_n and in this case one may want to consider the "transpose top and random" measure $p_\star$, i.e., the uniform probability measure on $\{e\}\cup\Sigma$. Fourier analysis can be used to show that $(S_n, p_\star)$ has a cut-off at time $n\log n$, see Section 9.5 below. For another example, take $\Sigma = \{\tau, c\}$ where $\tau = (1,2)$ and c is the long cycle $(1,2\ldots,n)$ in S_n. These two elements generate S_n and this is obviously a minimal generating set. Let $p_{\tau,c}$ denotes the uniform measure on $\{\tau,c\}$. It is known that, for odd n, $cn^3\log n \le T(S_n, p_{\tau,c}) \le Cn^3\log n$ (see [45, 142] and Section 10). It is conjectured that this walk has a cut-off.

Problem 3.5. Is it true that most natural families (S_n, p_n) where p_n is uniform on a minimal generating set of S_n have a cut-off?

Problem 3.6. Is it true that most natural families (G_n, p_n) where each G_n is a simple group and p_n is uniform on a minimal generating set of G_n have a cut-off?

Problem 3.7. What is the range of the possible cut-off times for walks on the symmetric group S_n based on minimal generating sets? (known examples have the form $t_n = cn^a \log n$ with a a small integer)

Unfortunately, these problems seem extremely difficult to attack. It is known that about $3/4$ of all pairs of permutations in S_n generate S_n [52] but no one seems to know how to study the associated random walks, let alone to prove or disprove the existence of a cut-off. The situation is similar for all finite simple groups (almost all pairs in a finite simple group generate the group [130]). One of the only satisfactory results in this direction is a negative result which will be discussed in Section 7.2 and says that reversible walks (with holding) based on minimal generating sets in groups of order p^a (such groups are necessarily nilpotent) with a bounded and p any prime do not present a cut-off. Instead, such walks behave essentially as the simple random walk (with holding) on the circle group $\mathbb{Z}_n$.

Precut-off. The cut-off phenomenon is believed to be widespread but it has been proved only in a rather limited number of examples, most of which are recorded in Table 2. Indeed, to prove that a cut-off occurs, one needs to understand the behavior of the walk before and around the time at which it reaches equilibrium and this is a difficult question. In [124], further versions of the cut-off phenomenon are discussed that shed some light on this problem. Let us point out that there are many families of walks $((G_n, p_n))$ for which the following property is known to be satisfied.

Definition 3.8. *We say that the family* (G_n, p_n) *presents a* precut-off *if there exist a sequence* t_n *tending to infinity with* n *and two constants* $0 < a < b < \infty$ *such that*

$$\lim_{n \to \infty} \|p_n^{(bk_n)} - u_n\|_{TV} = 0 \quad and \quad \lim_{n \to \infty} \|p_n^{(ak_n)} - u_n\|_{TV} > 0.$$

Table 3. Precut-offs

G	p	(PCO)	§	Ref
S_n	adjacent transposition p_{AT}	$n^3 \log n$	4.1, 5.3, 10.2	[42, 141]
S_n	ℓ-adjacent transposition, $p_{\ell\text{-AT}}$	$(n^3/\ell^2) \log n$	10.2	[55]
S_n	nearest neighbors transposition on a square grid	$n^2 \log n$	10.2	[42, 141]
S_n	random insertion	$n \log n$	10.2	[42]
S_n	Borel-Chéron random packet to top	$n \log n$	3.1, 10.2	[42]
S_n	random inversion	$n \log n$	10.2	[55]
S_n	neat overhand shuffle, i.e., reverse top to random	$n \log n$	10.2	[42]
S_n	crude overhand shuffle	$n \log n$	3.1, 10.2	[42]
S_n	Rudvalis shuffle, i.e., top to $n-1$ or n	$n^3 \log n$	4.1	[31, 85]
S_n	uniform on $e, (1,2)$, top to bottom, botton to top	$n^3 \log n$	4.1, 10.2	[31, 85]
A_n	S_n conjugacy classes $C=(c_1,...,c_\ell), |C|=c_1+\cdots+c_\ell=m\ll n$	$\frac{n}{m} \log n$	9.2	[119]
$U_m(q)$	$E_{i,j}(a), a \in \mathbb{Z}_q, 1 \le i < j \le m$	$m^2 \log m$	4.2	[114]
Lie type	small conjugacy classes	n=rank(G)	9.2	[68]
$\mathbb{Z}_2^d \rtimes \mathbb{Z}_d$	perfect shuffles	d^2	4.2	[138]
$SL_n(\mathbb{Z}_q)$	$A^\pm, B^\pm, r$ prime, n fixed	$\log q$	6.4	[46, 98]

Thus, if a family $((G_n, p_n))$ presents a precut-off at time t_n, there exist two constants $0 < c \leq C < \infty$ such that, for each $\varepsilon > 0$ small enough and all n large enough,

$$ct_n \leq T(G_n, p_n, \varepsilon) \leq Ct_n.$$

The notion of precut-off captures the order of magnitude of a possible cut-off, but it is unknown whether or not families having a precut-off must have a cut-off. In many cases, it is conjectured that they do. The Borel-Chéron shuffle and the crude overhand shuffle described in Section 3.1 are two examples of shuffles for which a precut-off has been proved (with $t_n = n \log n$, see [42] and Section 10). Another example is the adjacent transposition walk driven by the uniform probability measure p_{AT} on $\{e\} \cup \{(i, i+1) : 1 \leq i < n\}$. This walk satisfies a precut-off at tine $n^3 \log n$ ([42, 141]). In all these cases, the existence of a cut-off is conjectured. See [30, 141] and Table 3. Solutions to the variants of Problems 3.5, 3.6 and 3.7 involving the notion of precut-off instead of cut-off would already be very valuable results.

4 Probabilistic Methods

Two probabilistic methods have emerged that produce quantitative estimates concerning the convergence to stationarity of finite Markov chains: *coupling* and *strong stationary times*. Coupling is the most widely known and used. Strong stationary times give an alternative powerful approach. Both involve the construction and study of certain "stopping times" and have theoretical and practical appeal. In particular, a stationary time can be interpreted as a perfect sampling method. These techniques are presented below and illustrated on a number of examples of random walks. The books [3, 27] are excellent references, as are [1, 4, 5]. When these techniques work, they often lead to good results through very elegant arguments. The potential user should be warned that careful proofs are a must when using these techniques. Experience shows that it is easy to come up with "obvious" couplings or stationary times that end up not being coupling or stationary times at all. Moreover, these two techniques, especially strong stationary times, are not very robust. A good example of a walk that has not yet been studied using coupling or stationary time is random insertion on the symmetric group: pick two positions i, j uniformly independently at random, pull out the card in position i and insert it in position j. This walk has a precut-off at time $n \log n$, see Section 10 and Table 3.

4.1 Coupling

Let K be a Markov kernel on a finite set $\mathcal{X}$ with invariant distribution π. A coupling is simply a sequence of pairs of $\mathcal{X}$-valued random variables

(X_n^1, X_n^2) such that each marginal sequence (X_n^i), $i = 1, 2$, is a Markov chain with kernel K. These two chains will have different initial distributions, one being often the stationary distribution π. The pair (X_n^1, X_n^2) may or may not be Markovian (in most practical constructions, it is). Given the coupling (X_n^1, X_n^2), consider

$$T = \inf \left\{ n : \forall k \geq n, \; X_k^1 = X_k^2 \right\}.$$

Call T the *coupling time* (note that T is not a stopping time in general).

Theorem 4.1. *Denote by μ_n^i the distribution of X_n^i, $i = 1, 2$. Then*

$$d_{\mathrm{TV}}(\mu_n^1, \mu_n^2) \leq \mathbb{P}(T > n).$$

This is actually a simple elementary result (see, e.g., [1, 3, 27]) but it turns out to be quite powerful. For further developments of the coupling technique for finite Markov chains, see [3] and the references therein. For relations between coupling and eigenvalue bounds, see, e.g., [18].

Specializing to random walks on finite groups, we obtain the following.

Theorem 4.2. *Let p a probability measure on a finite group G. Let (X_n^1, X_n^2) be a coupling for the random walk driven by p with (X_n^1) starting at the identity and (X_n^2) stationary. Then*

$$d_{\mathrm{TV}}(p^{(n)}, u) \leq \mathbb{P}(T > n).$$

One theoretical appeal of coupling is that there always exists a coupling such that the inequalities in the theorems above are in fact equalities (see the discussions in [3, 27] and the references given there). Hence the coupling technique is exactly adapted to the study of convergence in total variation. In practice, Theorem 4.2 reduces the problem of estimating the total variation distance between a random walk and the uniform probability measure on G to the construction of a coupling for which $\mathbb{P}(T > n)$ can be estimated. This is best illustrated and understood by looking at some examples.

Coupling for random to top [1, 4, 27]. Consider the random to top shuffling scheme where a card is chosen at random and placed on top. Obviously, this is the inverse shuffle of top to random. On S_n, this is the walk driven by the uniform measure on the cycles $c_i = (1, 2, \ldots, i)$, $i = 1, \ldots, n$. To construct a coupling, imagine having two decks of cards. The first one is in some given order, the second one is perfectly shuffled. Pick a card at random in the first deck, say, the tenth card. Look at is face value, say, the ace of spades. Put it on top and put a check on its back. In the second deck, find the ace of spades and put it on top. At each step, repeat this procedure. This produces a pair of sequences of S_n-valued random variables (X_k^1, X_k^2) corresponding respectively to the arrangements of each of the decks of cards. Obviously, (X_k^1) is a random walk driven by the random to top measure p. The same is true for X_n^2

because choosing a position in the deck uniformly at random is equivalent to choosing the face value of a card uniformly at random. Say we have a match if a card value has the same position in both decks. This coupling has the following property: any checked card stays matched with its sister card for ever and each time an unchecked card is touched in the first deck, it is checked and matched with its sister card. Note however that matches involving an unchecked card from the first deck might be broken along the way. In any case, the coupling time T is always less or equal to T', the first time all cards in the first deck have been checked. A simple application of the well-known coupon collector's problem gives $\mathbb{P}(T' > k) \leq ne^{-k/n}$. This, combined with a matching lower bound result, shows that random to top (and also top to random) mixes in about $n \log n$ shuffles, a result which compares well with the very precise result of Theorem 3.2.

Coupling for random transposition [1, 27]. For n cards, the random transposition shuffle involves choosing a pair of positions (i, j) uniformly and independently at random in $\{1, \ldots, n\}$ and switching the cards at these positions. Thus, the random transposition measure p_{RT} is given by

$$p_{\mathrm{RT}}(\tau) = \begin{cases} 2/n^2 & \text{if } \tau = (i, j), \ 1 \leq i < j \leq n, \\ 1/n & \text{if } \tau = e, \\ 0 & \text{otherwise.} \end{cases} \tag{4.1}$$

Obviously, choosing uniformly and independently at random a position i and a face value V and switching the card in position i with the card with face value V gives an equivalent description of this measure. Given two decks, we construct a coupling by picking i and V uniformly and independently. In each deck, we transpose the card in position i with the card with face value V. In this way, the number of matches never goes down and at least one new match is created each time the cards with the randomly chosen face value V are in different positions in the two decks and the cards in the randomly chosen position i have distinct face values. Let (Z_k) denote the Markov process on $\{0, \ldots, n\}$ started at n with transition probabilities $K(i, i-1) = (i/n)^2$, $K(i, i) = 1 - (i/n)^2$. Let $T' = \inf\{k : Z_k = 0\}$. Then, it is not hard to see that $\mathbb{E}(T) \leq \mathbb{E}(T') \leq 2n^2$ where T is the coupling time. By Theorem 4.2, we obtain $d_{\mathrm{TV}}(p_{\mathrm{RT}}^{(k)}, u) \leq \mathbb{E}(T)/k \leq 2n^2/k$ and the sub-additivity of $k \mapsto 2d_{\mathrm{TV}}(p_{\mathrm{RT}}^{(k)}, u)$ yields $d_{\mathrm{TV}}(p_{\mathrm{RT}}^{(k)}, u) \leq e^{1-k/(12n^2)}$. This shows that $T(S_n, p_{\mathrm{RT}}) \leq 36n^2$. Theorem 9.2 below states that (S_n, p_{RT}) presents a cut-off at time $t_n = \frac{1}{2}n \log n$. Convergence after order n^2 steps is the best that has been proved for random transposition using coupling.

Coupling for adjacent transposition [1]. Consider now the shuffling scheme where a pair of adjacent cards are chosen at random and switched. The adjacent transposition measure on S_n, call it p_{AT}, is the uniform measure on $\{e, (1, 2), \ldots, (n-1, n)\}$. Set $\sigma_0 = e$ and $\sigma_i = (i, i+1)$, $1 \leq i < n$. To construct a coupling, consider two decks of cards. Call A the set containing 0 and all positions $j \in \{1, \ldots, n-1\}$ such that neither the cards in position j nor the cards in position $j+1$ are matched in those decks. List A as $\{j_0, j_1, \ldots, j_\ell\}$ in order. Let J be a uniform random variable in $\{0, \ldots, n-1\}$ and set

$$
J^* = \begin{cases} J & \text{if } J \notin A \\ j_{k+1} & \text{if } J = j_k \in A \text{ with the convention that } \ell + 1 = 0. \end{cases}
$$

The coupling is produced by applying σ_J to the first deck and σ_{J^*} to the second deck. As J^* is uniform in $\{0, \ldots, n-1\}$, this indeed is a coupling. To analyze the coupling time, observe that matches cannot be destroyed and that, for any face value, the two cards with this face value always keep the same relative order (e.g., if the ace of spades is higher in the first deck than in the second deck when we start, this stays the same until they are matched). Call T'_i the first time card i reaches the bottom of the deck (in the deck in which this card is initially higher) and set $T' = \max_i \{T'_i\}$. Then the coupling time T is bounded above by T'. Finally, any single card performs a symmetric simple random walk on $\{1, \ldots, n\}$ with holding probability $1 - 2/n$ except at the endpoints where the holding probability is $1 - 1/n$. Properly rescaled, this process converges weakly to reflected Brownian motion on $[0, 1]$ and the hitting time of 1 starting from any given point can be analyzed. In particular, there are constants $A, a > 0$ such that, for any i and any $s > 0$, $\mathbb{P}(T'_i > sn^3) \leq Ae^{-as}$. Hence, for C large enough, $\mathbb{P}(T > Cn^3 \log n) \leq Ane^{-aC \log n} \leq (2e)^{-1}$. This shows that $T(S_n, p_{\mathrm{AT}}) \leq Cn^3 \log n$. A matching lower bound is given at the end of Section 5.3. Hence (S_n, p_{AT}) presents a precut-off at time $t_n = n^3 \log n$. See also Theorem 10.4 and [141].

Other couplings. Here we briefly describe further examples of random walks for which reasonably good couplings are known:

- Simple random walk on the hypercube $\{0, 1\}^n$ as described in Section 8.2. See [1, 27, 105].
- The GSR riffle shuffle described in Section 3.2. See [1] for a coupling showing that $2 \log_2 n$ riffle shuffles suffice to mix up n cards.
- Overhand shuffles [1, 116]. An overhand shuffle is a shuffle where the deck is divided into k blocks and the order of the blocks are reversed. Pemantle [116] gives a coupling analysis of a range of overhand shuffle models showing that, in many reasonable cases, order $n^2 \log n$ shuffles suffice to mix up n cards whereas at least order n^2 are necessary. Note however that the crude overhand shuffle discussed in Section 3.1 has a precut-off at time $t_n = n \log n$.
- The following shuffling method is one of those discussed in Borel and Chéron [15]: take the top card and insert it at random, take the bottom card and insert it a random. The coupling described above for random to top can readily be adapted to this case. See [1, 27].
- Slow shuffles. At each step, either stay put or transpose the top two cards or move the top card to the bottom, each with probability $1/3$. It is not

hard to construct a coupling showing that order $n^3 \log n$ shuffles suffice to mix up the cards using this procedure. Rudvalis (see [27, p. 90]) proposed another shuffle as a candidate for the slowest shuffle. At each step, move the top card either to bottom or second to bottom each with probability $1/2$. Hildebrand gives a coupling for this shuffle in his Ph. D Thesis [85] and shows that order $n^3 \log n$ such shuffles suffice. For these slow shuffles and related variants, Wilson [142] proves that order $n^3 \log n$ shuffles are necessary to mix up n cards.

4.2 Strong Stationary Times

Separation. Given a Markov kernel K with invariant distribution π on a finite set $\mathcal{X}$, set

$$\mathbf{sep}_K(x, n) = \max_{y \in \mathcal{X}} \left(1 - \frac{K_n(x, y)}{\pi(y)} \right), \quad \mathbf{sep}_K(n) = \max_{x \in \mathcal{X}} \mathbf{sep}_K(x, n).$$

The quantity $\mathbf{sep}(n) = \mathbf{sep}_K(n)$ is called the *maximal separation* between K^n and π. As

$$d_{\mathrm{TV}}(K^n(x, \cdot), \pi) = \sum_{y : K^n(x,y) \leq \pi(y)} (\pi(y) - K^n(x, y)),$$

it is easy to see that $d_{\mathrm{TV}}(K^n(x, \cdot), \pi) \leq \mathbf{sep}_K(x, n)$. Thus separation always controls the total variation distance. Separation is an interesting alternative way to measure ergodicity. The function $n \mapsto \mathbf{sep}(n)$ is non-increasing and sub-multiplicative [3, 5]. As an immediate application of these elementary facts, one obtains the following Doeblin's type result: Assume that there exist an integer m and a real $c > 0$ such that, for all $x, y \in \mathbb{X}$, $K^m(x, y) \geq c\pi(y)$. Then $d_{\mathrm{TV}}(K^{nm}(x, \cdot), \pi) \leq \mathbf{sep}(nm) \leq (1-c)^n$ (this line of reasoning produces very poor bounds in general but an example where it is useful is given in [39]).

Let (X_k) be a Markov chain with kernel K. A *strong stationary time* is a randomized stopping time T for (X_k) such that

$$\forall k, \ \forall y \in \mathcal{X}, \ \mathbb{P}(X_k = y / T = k) = \pi(y). \tag{4.2}$$

This is equivalent to say that X_T has distribution π and that the random variables T and X_T are independent. For a discussion of the relation between strong stationary times and coupling, see [5]. Relations between strong stationary times and eigenvalues are explored in [107]. Strong stationary times are related to the separation distance by the following theorem of Aldous and Diaconis [5, 3, 27].

Theorem 4.3. *Let T be a strong stationary time for the chain starting at $x \in \mathcal{X}$. Then*

$$\forall n, \ \mathbf{sep}_K(x, n) \leq \mathbb{P}_x(T > n).$$

Moreover there exists a strong stationary time such that the above inequality is an equality.

Separation for random walks. In the case of random walks on finite groups, separation becomes

$$\mathbf{sep}(k) = \mathbf{sep}_p(k) = \max_{x \in G}\left(1 - |G|p^{(k)}(x)\right).$$

The next theorem restates the first part of Theorem 4.3 and gives an additional result comparing separation and total variation distances in the context of random walks on finite groups. See [5] and the improvement in [23].

Theorem 4.4. *Let p be a probability measure on a finite group G. Then*

$$d_{\mathrm{TV}}(p^{(k)}, u) \leq \mathbf{sep}(k)$$

and, provided $d_{\mathrm{TV}}(p^{(k)}, u) \leq (|G| - 1)/(2|G|)$,

$$\mathbf{sep}(2k) \leq 2d_{\mathrm{TV}}(p^{(k)}, u).$$

Let T be a strong stationary time for the associated random walk starting at the identity e. Then

$$d_{\mathrm{TV}}(p^{(k)}, u) \leq \mathbf{sep}(k) \leq \mathbb{P}_e(T > k).$$

One can easily introduce the notion of separation cut-off (and precut-off): The family $((G_n, p_n))$ has a separation cut-off if and only if there exists a sequence s_n tending to infinity such that

$$\lim_{n \to \infty} \mathbf{sep}_{p_n}(\lfloor(1 - \varepsilon)s_n\rfloor) = 1, \quad \lim_{n \to \infty} \mathbf{sep}_{p_n}(\lfloor(1 + \varepsilon)s_n\rfloor) = 0.$$

Theorem 4.4 implies that if $((G_n, p_n))$ has both a total variation cut-off at time t_n and a separation cut-off at time s_n then $t_n \leq s_n \leq 2t_n$.

There is sometimes an easy way to decide whether a given strong stationary time is optimal (see [33, Remark 2.39]).

Definition 4.5. *Given an ergodic random walk (X_n) on G started at e and a strong stationary time T for (X_n), the group element x is called a* halting state *if $\mathbb{P}_e(X_k = x, T > k) = 0$, for all $k = 0, 1, \ldots$.*

Hence, a halting state is an element that cannot be reached before the strong stationary time T (observe that, of course, $\mathbb{P}_e(X_T = x) > 0$). Obviously, if there is a halting state, then T is a stochastically smallest possible strong stationary time. As for coupling, the power of strong stationary times is best understood by looking at examples.

Stationary time for top to random [27]. Let q_1 denote the top to random measure on S_n. Consider the first time T_1 a card is inserted under the bottom card. This is a geometric waiting time with mean n. Consider the first time T_2 a second card is inserted under the original bottom card. Obviously $T_2 - T_1$ is a geometric waiting time with mean $n/2$, independent of T_1. Moreover, the relative position of the two cards under the original bottom card is equally likely to be high-low or low-high. Pursuing this analysis, we discover that the first time T the bottom card comes on top and is inserted at random is a strong stationary time. Moreover $T = T_n = T_1 + (T_2 - T_1) + \cdots + (T_n - T_{n-1})$ where $T_i - T_{i-1}$ are independent geometric waiting time with respective means n/i. Hence $\mathbb{P}_e(T > k)$ can be estimated. In particular, it is bounded by $ne^{-k/n}$. Hence Theorem 4.4 gives

$$d_{TV}(q_1^{(k)}, u) \leq \mathbf{sep}(k) \leq \mathbb{P}_e(T > k) \leq ne^{-k/n}.$$

This is exactly the same bound as provided by the coupling argument described earlier. In fact, in this example, the coupling outlined earlier and the stationary time T above are essentially equivalent. This T is not an optimal stationary time but close. Let T' be the first time the card originally second to bottom comes to the top and is inserted. This T' is an optimal stationary time. It has a halting state: the permutation corresponding to the deck in exact reverse order. This example has both a total variation and a separation cut-off at time $t_n = n \log n$.

Stationary time for random transposition [27]. We describe a strong stationary time constructed by A. Broder. Variants are discussed in [27, 106]. The construction involves checking the back of the cards as they are shuffled using repeated random transpositions. Recall that the random transposition measure p_{RT} defined at (4.1) can be described by letting the left and right hands choose cards uniformly and independently at random. If either both hands touch the same unchecked card or if the card touched by the left hand is unchecked and the card touched by the right hand is checked then check the back of the card touched by the left hand. Let T be the time that only one card remains unchecked. The claim is that T is a strong stationary time. See [27] for details. This stationary time has mean $2n \log n + O(\log n)$ and can be used to show that a little over $2n \log n$ random transpositions suffices to mix up a deck of n cards. This is better than what is obtained by the best known coupling, i.e., n^2. Theorem 9.2 and Matthews [106] show that (S_n, p_{RT}) has a total variation cut-off as well as a separation cut-off at time $\frac{1}{2} n \log n$.

Stationary time for riffle shuffle [27]. Recall that the inverse of a riffle suffle can be described as follows. Consider a binary vector of length n whose entries are independent uniform $\{0, 1\}$-random variables. Sort the deck from bottom to top into a left pile and a right pile by using the above binary vector with 0 sending the card left and 1 sending the card right. When this is done, put the left pile on top of the right to obtain a new deck. A sequence of k inverse riffle shuffles can be described by a binary matrix with n rows and k columns where the (i, j)-entry describes what happens to the original i-th card during the j-th shuffle. Thus the i-th row describes in which pile the original i-th card falls at each of the k shuffles.

Let T be the first time the matrix above has distinct rows. Then T is a strong stationary time. Indeed, using the right to left lexicographic order on binary vectors, after any number of shuffles, cards with "small" binary vectors are on top of cards with "large" binary vectors. At time T all the rows are distinct and the lexicographic order sorts out the cards and describes uniquely the state of the deck. Because the entries are independent uniform $\{0,1\}$-variables, at time T, all deck arrangements are equally likely. Moreover, the chance that $T > k$ is the same as the probability that dropping n balls into 2^k boxes there is no box containing two or more balls. This is the same as the birthday problem and we have

$$\mathbb{P}_e(T > k) = 1 - \prod_{1}^{n-1}(1 - i2^{-k}).$$

Using Calculus, this proves a separation cut-off at time $2\log_2 n$. Indeed, this stationary time has a halting state: the deck in reverse order. Theorem 3.1 proves a variation distance cut-off at time $\frac{3}{2}\log_2 n$. See [1, 13, 27].

Stationary times on nilpotent groups. In his thesis [112], Pak used strong stationary times skillfully to study problems that are somewhat different from those discussed above. The papers [7, 21, 114] develop results for nilpotent groups (for a definition, see Section 7 below). Here is a typical example. Let $U_m(q)$ denote the group of all upper-triangular matrices with 1 on the diagonal and coefficients mod q where q is an odd prime. Let $E_{i,j}(a)$, $1 \leq i < j \leq m$, denote the matrix in $U_m(q)$ whose non-diagonal entries are all 0 except the (i,j)-entry which equals a. The matrices $E_{i,i+1}(1)$, $1 \leq i < m$, generate $U_m(q)$. Consider the following two sets

$$\Sigma_1 = \{E_{i,i+1}(a) : a \in \mathbb{Z}_q,\ 1 \leq i < m\}$$
$$\Sigma_2 = \{E_{i,j}(a) : a \in \mathbb{Z}_q,\ 1 \leq i < j \leq m\}.$$

and let p_1, p_2 denote the uniform probability on Σ_1, Σ_2 respectively. The article [114] uses the strong stationary time technique to prove that the walk driven by p_2 presents a precut-off at time $t_m = m^2 \log m$, uniformly in the two parameters m, q. In particular, there are constants C, c such that

$$cm^2 \log m \leq T(U_m(q), p_2) \leq Cm^2 \log m.$$

The results for the walk driven by p_1 are less satisfactory. In [21], the authors use a strong stationary time to show that if $q \gg m^2$ then

$$cm^2 \leq T(U_m(q), p_1) \leq Cm^2.$$

The best known result for fixed q is described in Section 7 below and says that $T(U_m(q), p_1) \leq Cm^3$.

Stopping times and semidirect products. In his thesis [138], Uyemura-Reyes develops a technique for walks on semidirect products which is closely related to the strong stationary time idea. Let H, K be two finite groups and $\phi : k \mapsto \phi_k$ a homomorphism from K to the automorphism group of H. The *semidirect product* $H \rtimes_\phi K$ is the group whose underlying set is $H \times K$ and whose product law is $(h_1, k_1)(h_2, k_2) = (h_1 \phi_{k_1}(h_2), k_1 k_2)$. By construction, H is normal in $H \rtimes_\phi K$. It follows that there is a natural projection from $H \rtimes_\phi K$ onto $K \cong (H \rtimes_\phi K)/H$. If p is a probability measure on $H \rtimes_\phi K$, let p_K denote its projection on K. Let (X_n) be the random walk on $H \rtimes_\phi K$ driven by p and write $X_n = (\zeta_n, \xi_n)$ with $\zeta_n \in H, \xi_n \in K$. Then (ξ_n) is a random walk on K driven by p_K. Consider a stopping time T for (X_n) which satisfies

$$\mathbb{P}_e(\zeta_n = h, \xi_n = k / T \leq n) = \frac{1}{|H|} \mathbb{P}_e(\xi_n = k / T \leq n). \qquad (4.3)$$

Theorem 4.6. *Referring to the notation introduced above, let (X_n) be the random walk on $G = H \rtimes_\phi K$ driven by p and starting at the identity. Assume that T is a stopping time satisfying (4.3). Then*

$$\|p^{(n)} - u_G\|_{\mathrm{TV}} \leq \|p_K^{(n)} - u_K\|_{\mathrm{TV}} + 2\mathbb{P}_e(T > n).$$

Moreover,

$$\mathbf{sep}_p(n) \leq \mathbf{sep}_{p_K}(n) + |K|\mathbb{P}_e(T > n).$$

We now describe two applications taken from [138]. See [77] for related results. Let $G = \mathbb{Z}_b^d \rtimes \mathbb{Z}_d$ where the action of $\mathbb{Z}_d$ is by circular shift of the coordinates in $\mathbb{Z}_b^d$. When $b = 2$, this example has a card shuffling interpretation. Given a deck of $2n$ cards, there are exactly two different perfect shuffles: cut the deck into two equal parts and interlace the two heaps starting either from the left or the right heap. When $2n = 2^d$ for some d, the subgroup of S_{2n} generated by the two perfect shuffles is isomorphic to $G = \mathbb{Z}_2^d \rtimes \mathbb{Z}_d$. One of the shuffles can be interpreted as $g_1 = (0, 1)$ and the other as $g_2(1_1, 1)$ where $0 = (0, \ldots, 0)$ and $1_1 = (1, 0, \ldots, 0)$ in $\mathbb{Z}_2^d$. Consider the simple random walk on $G = \mathbb{Z}_2^d \rtimes \mathbb{Z}_d$ driven by the probability p with $p(e) = 2p(g_1) = 2p(g_2) = 1/2$. Theorem 4.6 can be used to prove that $T(\mathbb{Z}_2^d \rtimes \mathbb{Z}_d, p) \leq Cd^2$ ([138] also gives a matching lower bound).

For a second example, take $b = d$ and consider the probability measure p defined by $p(0, 0) = p(\pm 1_1, 0) = p(0, \pm 1) = p(\pm 1_1, 1) = p(\pm 1_1, -1) = 1/9$. Uyemura-Reyes uses Theorem 4.6 to prove the mixing time upper bound $T(\mathbb{Z}_d^d \rtimes \mathbb{Z}_d, p) \leq Cd^3 \log d$. He also derives a lower bound of order d^3.

5 Spectrum and Singular Values

5.1 General Finite Markov Chains

Diagonalization. Let K be a Markov kernel with invariant distribution π on a finite set $\mathcal{X}$. Irreducibility and aperiodicity can be characterized in terms of the spectrum of K on $L^2(\pi)$ where $L^2(\pi)$ denote the space of all complex valued functions equipped with the Hermitian scalar product $\langle f, g \rangle_\pi = \sum_x f(x)\overline{g(x)}\pi(x)$. Indeed, K is irreducible if and only if 1 is a simple eigenvalue whereas K is aperiodic if and only if any eigenvalue $\beta \neq 1$ satisfies $|\beta| < 1$.

If K and K^* commute, that is, if K viewed as an operator on $L^2(\pi)$ is normal, then K is diagonalizable in an orthonormal basis of $L^2(\pi)$. Let $(\beta_i)_{i \geq 0}$ be an enumeration of the eigenvalues, each repeated according to its multiplicity and let $(v_i)_{i \geq 0}$ be a corresponding orthonormal basis of eigenvectors. Note that in general, the β_i are complex numbers and the v_i complex valued functions. Without loss of generality, we assume that $\beta_0 = 1$ and $u_0 \equiv 1$. Then

$$\frac{K_n(x,y)}{\pi(y)} = \sum_{i \geq 0} \beta_i^n v_i(x)\overline{v_i(y)} \tag{5.1}$$

and

$$d_{\pi,2}(K_n(x,\cdot),\pi)^2 = \sum_{i \geq 1} |\beta_i|^{2n} |v_i(x)|^2. \tag{5.2}$$

Let us describe a simple but useful consequence of (5.2) concerning the comparison of the $L^2(\pi)$-distances to stationarity of the discrete and continuous Markov processes associated to a given reversible Markov kernel K. An application is given below at the end of Section 8.2.

Theorem 5.1. *Let (K,π) be a reversible Markov kernel on a finite set $\mathcal{X}$ and let H_t be as in (2.3). Then*

$$d_{\pi,2}(K_n(x,\cdot),\pi)^2 \leq \beta_-^{2n_1}\left(1 + d_{\pi,2}(H_{n_2}(x,\cdot),\pi)^2\right) + d_{\pi,2}(H_n(x,\cdot),\pi)^2$$

where $n = n_1 + n_2 + 1$ and $\beta_- = \max\{0, -\beta_{\min}\}$, $\beta_{\min}$ being the smallest eigenvalue of K. Moreover,

$$d_{\pi,2}(H_{2n}(x,\cdot),\pi)^2 \leq (\pi(x)^{-1} - 1)e^{-2n} + d_{\pi,2}(K_n(x,\cdot),\pi)^2.$$

Proof. The idea behind this theorem is simple: as (K,π) is reversible, it has real eigenvalues $1 = \beta_0 \geq \beta_1 \geq \cdots \geq \beta_{|\mathcal{X}|-1} \geq -1$. Viewed as an operator, H_t is given by $H_t = e^{-t(I-K)}$ and has real eigenvalues $e^{-t(1-\beta_i)}$, in increasing order, associated with the same eigenvectors as for K. Hence, using (5.2) and the similar formula for H_t, the statements of Theorem 5.1 follow from simple Calculus inequalities. See Lemma 3 and Lemma 6 in [42] for details. The factor $\pi(x)^{-1}$ appears because, using the same notation as in (5.2), we have $\sum_{i \geq 0} |v_i(x)|^2 = \pi(x)^{-1}$. $\square$

Poincaré inequality. When (K,π) is reversible, an important classical tool to bound eigenvalues is the variational characterization of the first eigenvalue. Set

$$\mathcal{E}(f,g) = \langle (I-K)f, g \rangle_\pi = \sum_x [(I-K)f(x)]g(x)\pi(x). \tag{5.3}$$

This form is called the *Dirichlet form* associated to (K, π). A simple computation shows that

$$\mathcal{E}(f, g) = \frac{1}{2} \sum_{x,y} (f(x) - f(y))(g(x) - g(y)) \pi(x) K(x, y). \tag{5.4}$$

Restricting attention to the orthogonal of the constant functions, we see that

$$\lambda_1 = 1 - \beta_1 = \inf \left\{ \frac{\mathcal{E}(f, f)}{\mathrm{Var}_\pi(f)} : f \in L^2(\pi), \ \mathrm{Var}_\pi(f) \neq 0 \right\} \tag{5.5}$$

where $\mathrm{Var}_\pi(f)$ denote the variance of f with respect to π, that is,

$$\mathrm{Var}_\pi(f) = \pi(f^2) - \pi(f)^2 = \frac{1}{2} \sum_{x,y} |f(x) - f(y)|^2 \pi(x) \pi(y). \tag{5.6}$$

It follows that, for any $A \geq 1$, the inequality $\beta_1 \leq 1 - 1/A$ is equivalent to the so-called *Poincaré inequality*

$$\mathrm{Var}_\pi(f) \leq A \, \mathcal{E}(f, f).$$

The quantity $\lambda_1 = 1 - \beta_1$ is called the *spectral gap* of (K, π). It is the second smallest eigenvalue of $I - K$. Some authors call $1/\lambda_1$ the *relaxation time*. It is a widespread misconception that the relaxation time contains all the information one needs to have good control on the convergence of a reversible Markov chain. What λ_1 gives is only the asymptotic exponential rate of convergence of $H_t - \pi$ to 0 as t tends to infinity.

Singular values. When K and its adjoint K^* do not commute, it seems hard to use the spectrum of K to get quantitative information on the convergence of $K_n(x, \cdot)$ to π. However, the *singular values* of K can be useful. For background on singular values, see [91, Chap. 18]. Consider the operators KK^* and K^*K. Both are self-adjoint on $L^2(\pi)$ and have the same eigenvalues, all non-negative. Denote the eigenvalues of K^*K in non-increasing order and repeated according to multiplicity by

$$\sigma_0^2 = 1 \geq \sigma_1^2 \geq \sigma_2^2 \geq \cdots \geq \sigma_{|\mathcal{X}|-1}^2$$

with $\sigma_i \geq 0$, $0 \leq i \leq |\mathcal{X}| - 1$. Then, the non-negative reals σ_i are called the singular values of K. More generally, for each integer j, denote by $\sigma_i(j)$, $0 \leq i \leq |\mathcal{X}| - 1$ the singular values of K^j and let also $v_{i,j}$ be the associated normalized eigenfunctions. Then we have

$$d_{\pi,2}(K_n(x, \cdot), \pi)^2 = \sum_{i \geq 1} \sigma_i(n)^2 |v_{i,n}(x)|^2. \tag{5.7}$$

As $\sum_{i \geq 0} |v_{i,j}(x)|^2 = \pi(x)^{-1}$ and $\sigma_i(n) \leq \sigma_1(n) \leq \sigma_1^n$ (see [91, Th. 3.3.14]), we obtain

$$\forall n \in \mathbb{N}, \quad d_{\pi,2}(K_n(x, \cdot), \pi)^2 \leq \left(\pi(x)^{-1} - 1 \right) \sigma_1^{2n}.$$

Let us emphasize here that it may well be that $\sigma_1 = 1$ even when K is ergodic. In such cases one may try to save the day by using the singular values of K^j where j is the smallest integer such that $\sigma_1(j) < 1$. This works well as long as j is relatively small. We will see below in Theorem 5.3 how to use all the singular values of K (or K^j) in the random walk case.

5.2 The Random Walk Case

Let us now return to the case of a left-invariant random walk driven by a probability measure p on a group G, i.e., the case when $K(x, y) = p(x^{-1}y)$ and $\pi = u$. In this case an important simplification occurs because, by left-invariance, the left-hand side of both (5.2) and (5.7) are independent of x. Averaging over $x \in G$ and using the fact that our eigenvectors are normalized in $L^2(G)$, we obtain the following.

Theorem 5.2. *Let p a probability measure on a finite group G. Assume that $p * \check{p} = \check{p} * p$, then we have*

$$d_2(p^{(n)}, u)^2 = \sum_{i \geq 1} |\beta_i|^{2n} \tag{5.8}$$

where β_i, $0 \leq i \leq |G| - 1$ are the eigenvalues associated to $K(x, y) = p(x^{-1}y)$ as above. In particular, if $\beta_* = \max\{|\beta_i| : i = 1, \ldots, |G| - 1\}$ denotes the second largest eigenvalue in modulus, we have

$$d_2(p^{(n)}, u)^2 \leq (|G| - 1)\beta_*^{2n}. \tag{5.9}$$

Note that p and $\check{p}$ always commute on abelian groups. Sections 6 and 10 below discuss techniques leading to eigenvalues estimates.

Theorem 5.3. *Let p a probability measure on a finite group G. Then, for any integers n, m we have*

$$d_2(p^{(nm)}, u)^2 \leq \sum_{i \geq 1} \sigma_i(m)^{2n} \tag{5.10}$$

where $\sigma_i(m)$, $0 \leq i \leq |G| - 1$ are the singular values associated to $K_m(x, y) = p^{(m)}(x^{-1}y)$ in non-increasing order. In particular, for each m, we have

$$d_2(p^{(nm)}, u)^2 \leq (|G| - 1)\sigma_1(m)^{2n}. \tag{5.11}$$

Proof. Use (5.7) and the fact (see e.g., [91, Th. 3.3.14]) that, for all k, n, m,

$$\sum_0^k \sigma(nm)^2 \leq \sum_0^k \sigma_i(m)^{2n}.$$

$\square$

It is worth restating (5.10) as follows.

Theorem 5.4. *Let p a probability measure on a finite group G and let q_m denote either $\check{q}^{(m)} * q^{(m)}$ or $q^{(m)} * \check{q}^{(m)}$. Then*

$$d_2(p^{(nm)}, u) \le d_2(q_m^{(\lfloor n/2 \rfloor)}, u).$$

For applications of Theorem 5.4, see Section 10.3.

Let us point out that the fact that (5.8) and (5.10) do not involve eigenfunctions is what makes eigenvalue and comparison techniques (see Section 10) so powerful when applied to random walks on finite groups. For more general Markov chains, the presence of eigenfunctions in (5.2) and (5.7) make these inequalities hard to use and one often needs to rely on more sophisticated tools such as Nash and logarithmic Sobolev inequalities. See, e.g., [3, 47, 48, 124] and Martinelli's article in this volume.

5.3 Lower Bounds

This section discusses lower bounds in total variation. The simplest yet useful such lower bound follows from a direct counting argument: Suppose the probability p has a support of size at most r. Then $p^{(k)}$ is supported on at most r^k elements. If k is too small, not enough elements have possibly been visited to have a small variation distance with the uniform probability on G. Namely,

$$\|p^{(k)} - u\|_{\mathrm{TV}} \ge 1 - r^k/|G| \tag{5.12}$$

which gives

$$T(G, p) \ge \frac{\log(|G|/2)}{\log r}.$$

Useful improvements on this bound can be obtain if one has further information concerning the group law, for instance if G is abelian or if many of the generators commutes. See, e.g., [56] and [19].

Generally, lower bounds on total variation are derived by using specific test sets or test functions. For instance, for random transposition and for transpose top and random on the symmetric group, looking at the number of fixed points yields sharp lower bounds in total variation, see [27, p. 43]. For random transvection on $SL_n(\mathbb{F}_q)$, the dimension of the space of fixed vectors can be used instead [86].

Eigenvalues and eigenfunctions can also be useful in proving lower bounds on $d_2(p^{(k)}, u)$ and, more surprisingly, on $\|p^{(k)} - u\|_{\mathrm{TV}}$. Start with the following two simple observations.

Proposition 5.5. *Let p be a probability measure on a finite group G. Assume that β is an eigenvalue of p with multiplicity m. Then*

$$d_2(p^{(k)}, u)^2 \ge m|\beta|^{2k}, \quad 2\|p^{(k)} - u\|_{\mathrm{TV}} \ge |\beta|^k.$$

Proof. Let V be the eigenspace of β, of dimension m. It is not hard to show that V contains a function ϕ, normalized by $\|\phi\|_2 = 1$ and such that $\phi(e) = \|\phi\|_\infty \geq \sqrt{m}$. See [20, p. 103]. Then $d_2(p^{(k)}, u) \geq |\langle p^{(k)} - u, \phi \rangle| = |\phi * \check{p}^{(k)}(e)| = |\beta|^k |\phi(e)| = |\beta|^k \sqrt{m}$. For the total variation lower bound, use the last expression in (2.8) with any β-eigenfunction as a test function. $\square$

Note that it is not uncommon for random walks on groups to have eigenvalues with high multiplicity. Both of the inequalities in Proposition 5.5 are sharp as k tends to infinity when β is the second largest eigenvalue in modulus. However, the first inequality often gives good lower bound on the smallest k such that $d_2(p^{(k)}, u) \leq \varepsilon$ for fixed ε whereas the second inequality seldom does for the similar question in total variation (the walk on the hypercube of Theorem 8.7 illustrates this point). The following proposition can often be used to obtain improved total variation lower bounds. It is implicit in [27] and in [141]. See also [123, 126].

Proposition 5.6. *Let β be an eigenvalue of p. Let ϕ be an eigenfunction associated to β. Let B_k be such that*

$$\forall k, \quad \mathrm{Var}_{p^{(k)}}(\phi) \leq B_k^2. \tag{5.13}$$

Then $\|p^{(k)} - u\|_{\mathrm{TV}} \geq 1 - \tau$ for any $\tau \in (0,1)$ and any integer k such that

$$k \leq \frac{1}{-2\log|\beta|} \log \frac{\tau |\phi(e)|^2}{4(\|\phi\|_2^2 + B_k^2)}.$$

The difficulty in applying this proposition is twofold. First, one must choose a good eigenfunction ϕ maximizing the ratio $\phi(e)^2/(\|\phi\|_2^2 + B_k^2)$. Second, one must prove the necessary bound (5.13) with good B_k's (e.g., B_k uniformly bounded) and this turns out to be a rather non-trivial task. Indeed, it involves taking advantage of huge cancellations in $\mathrm{Var}_{p^{(k)}}(\phi) = p^{(k)}(|\phi|^2) - |p^{(k)}(\phi)|^2$. In this direction, the following immediate proposition is much more useful than it might appear at first sight.

Proposition 5.7. *Let β, ϕ be as in Proposition 5.6. Assume that there are eigenvalues α_i and associated eigenfunctions ψ_i, $i \in I$, relative to p such that*

$$|\phi|^2 = \sum_{i \in I} a_i \psi_i.$$

Then

$$\mathrm{Var}_{p^{(k)}}(\phi) = \sum_{i \in I} a_i \alpha_i^k \psi_i(e) - |\beta|^{2k} |\phi(e)|^2.$$

The reason this is useful is that, in some cases, expanding $|\phi|^2$ along eigenfunctions requires only a few eigenfunctions which, in some sense, are close to ϕ. To see how this works, consider the simple random walk on the hypercube $G = \mathbb{Z}_2^d$ equipped with its natural set of generators $(e_i)_1^d$ where e_i is the

d-tuple with all entries zero except the i-th equal to 1. See [27, pg. 28-29]. To avoid periodicity, set $e_0 = (0, \ldots, 0)$ and consider the measure p given by

$$p(x) = \begin{cases} 1/(d+1) & \text{if } x = e_i \text{ for some } i \in \{0, \ldots, d\} \\ 0 & \text{otherwise.} \end{cases} \tag{5.14}$$

Denote by x_i the coordinates of $x \in \mathbb{Z}_2^d$. Then $(-1)^{x_i} = 1 - 2x_i$ is an eigenfunction with eigenvalue $1 - 2/(d+1)$ for each $i \in \{1, \ldots, d\}$ and so is $\phi(x) = \sum_1^d (-1)^{x_i} = 2|x| - d$ where $|x| = \sum_1^d x_i$. Now

$$|\phi(x)|^2 = d + 2 \sum_{1 \leq i < j \leq n} (-1)^{x_i + x_j} = d\psi_0(x) + 2\psi_2(x)$$

where $\psi_0 \equiv 1$ and $\psi_2 = \sum_{1 \leq i < j \leq n}(-1)^{x_i + x_j}$ are eigenfunctions with respective eigenvalues 1 and $1 - 4/(d+1)$. Hence

$$\mathrm{Var}_{p^{(k)}}(\phi) = d + d(d-1)\left(1 - \frac{4}{d+1}\right)^k - d^2\left(1 - \frac{2}{d+1}\right)^{2k}.$$

By careful inspection, for any integer k, the right-hand side is less than d. Using this in Proposition 5.6 shows that, for the simple random walk on the hypercube, $\|p^{(k)} - u\|_{TV} \geq 1 - \tau$ for $k \leq \frac{1}{4}d\log(\tau d)$. This is sharp since the simple random walk on the hypercube has a cut-off at time $t_d = \frac{1}{4}d\log d$. See Theorem 8.7 below.

The next theorem and its illustrative example are taken from [141]. See also [126]. Set

$$\nabla\phi(x) = \left(\frac{1}{2}\sum_y |f(x) - f(xy)|^2 p(y)\right)^{1/2}.$$

Theorem 5.8. *Let β, ϕ be as in Proposition 5.6. Then*

$$\mathrm{Var}_{p^{(k)}}(\phi) \leq \frac{2\|\nabla\phi\|_\infty^2}{1 - |\beta|^2}. \tag{5.15}$$

Moreover $\|p^{(k)} - u\|_{\mathrm{TV}} \geq 1 - \tau$ for all $\tau \in (0, 1)$ and all k such that

$$k \leq \frac{1}{2\log|\beta|}\log\left(\frac{\tau(1 - |\beta|^2)|\phi(e)|^2}{4(2 + |\beta|)\|\nabla\phi\|_\infty^2}\right).$$

As an example, consider the random adjacent transposition measure p_{AT}, i.e., the uniform measure on $\{e, (1,2), \ldots, (n, n-1)\} \subset S_n$. To find some eigenfunctions, consider how one given card moves, say card 1. It essentially performs a ± 1 random walk on $\{1, \ldots, n\}$ with holding $1/2$ at the endpoints. For this random walk, $v(j) = \cos[\pi(j - 1/2)/n]$ is an eigenfunction associated with the eigenvalue $\cos \pi/n$. For $\ell \in \{1, \ldots, n\}$, let $\ell(x)$ be the position of card ℓ in the permutation x and $v_\ell(x) = v(\ell(x))$. Then, each v_ℓ is an eigenfunction of p with eigenvalue $1 - (2/n)(1 - \cos \pi/n)$. This is actually the second largest eigenvalue, see [12]. Obviously, the function

$$\phi(x) = \sum_{\ell=1}^{n} v_\ell(e) v_\ell(x)$$

is an eigenfunction for the same eigenvalue. Moreover, $\|\nabla \phi\|_\infty^2 \le 2\pi^2 \phi(e)/n^3$ and $\phi(e) = n(1+o(1))$. Hence for all $\tau \in (0,1)$ and $k \le (1-o(1))\pi^{-2} n^3 \log \tau n$, Theorem 5.8 gives $\|p_{AT} - u\|_{TV} \ge 1 - \tau$. This is quite sharp since it is known that $T(S_n, p_{AT}) \le Cn^3 \log n$. See Sections 4.1, 10 and the discussion in [141].

6 Eigenvalue Bounds Using Paths

This section develops techniques involving the geometric notion of paths. Left-invariant random walks on finite groups can be viewed as discrete versions of Brownian motions on compact Lie groups. It is well understood that certain aspects of the behavior of Brownian motion on a given manifold depend on the underlying Riemannian geometry and this has been a major area of research for many years. Many useful ideas and techniques have been developed in this context. They can be harvested without much difficulty and be brought to bear in the study of random walks on groups. This has produced great results in the study of random walks on infinite finitely generated groups. See [125, 139, 145]. It is also very useful for random walks on finite groups and, more generally, for finite Markov chains. For the development of these ideas for finite Markov chains, see [3, 51, 124, 131]. In the finite Markov chain literature, the use of path techniques is credited to Jerrum and Sinclair. See [131] for an excellent account of their ideas.

6.1 Cayley Graphs

Fix a finite group G and a finite generating set S which is symmetric, i.e., satisfies $\Sigma = \Sigma^{-1}$. The (left-invariant) *Cayley graph* (G, Σ) is the graph with vertex set G and edge set

$$E = \{(x, y) \in G \times G : \exists\, s \in \Sigma,\; y = xs\}.$$

The *simple random walk* on the Cayley graph (G, Σ) is the walk driven by the measure $p = (\#\Sigma)^{-1} 1_\Sigma$. It proceeds by picking uniformly at random a generator in Σ and multiplying by this generator on the right.

Define a *path* to be any finite sequence $\gamma = (x_0, \dots, x_n)$ of elements of G such that each of the pair (x_i, x_{i+1}), $i = 0, \dots, n-1$ belongs to E, i.e., such that $x_i^{-1} x_{i+1} \in \Sigma$. The integer n is called the length of the path γ and we set $|\gamma| = n$. Denote by $\mathcal{P}$ the set of all paths in (G, Σ).

Definition 6.1. *For any $x, y \in G$, set*

$$|x|_\Sigma = \min \{k : \exists\, s_1, \ldots, s_k \in \Sigma, x = s_1 \ldots s_k\},$$

$$d_\Sigma(x, y) = \min \{|\gamma| : \gamma \in \mathcal{P}, x_0 = x, x_n = y\},$$

$$D_\Sigma = \max_{x, y \in G} d_\Sigma(x, y).$$

We call d_Σ the graph distance *and D_Σ the* diameter *of (G, Σ).*

In words, $|x|_\Sigma$ is the minimal number of elements $s_1, \ldots, s_k$ of the generating set Σ needed to write x as a product $x = s_1 \ldots s_k$, with the usual convention that the empty product equals the identity element e. Obviously the graph distance is left invariant and

$$d_\Sigma(x, y) = |x^{-1}y|_\Sigma, \quad D_\Sigma = \max_{x \in G} |x|_\Sigma.$$

The reference to Σ will be omitted when no confusion can possibly arise. Babai [8] gives an excellent survey on graphs having symmetries including Cayley graphs.

6.2 The Second Largest Eigenvalue

Let G be a finite group and p be a probability measure on G whose support generates G. We assume in this section that p is symmetric, i.e., $p = \check{p}$. Hence the associated operator on $L^2(G)$ is diagonalizable with real eigenvalue $1 = \beta_0 \geq \beta_1 \geq \cdots \geq \beta_{|G|-1}$ in non-increasing order and repeated according to multiplicity. We will focus here on bounding β_1 from above. The results developed below can also be useful for non symmetric measure thanks to the singular value technique of Theorem 5.3. See Section 10.3.

There are a number of different ways to associate to p an adapted geometric structure on G. For simplicity, we will consider only the following procedure. Pick a symmetric set of generators Σ contained in the support of p and consider the Cayley graph (G, Σ) as defined in Section 6.1. In particular, this Cayley graph induces a notion of path and a left-invariant distance on G. The simplest result concerning the random walk driven by p and involving the geometry of the Cayley graph (G, Σ) is the following. See, e.g., [2, 42].

Theorem 6.2. *Let (G, Σ) be a finite Cayley graph with diameter D. Let p be a probability measure such that $p = \check{p}$ and $\varepsilon = \min_\Sigma p > 0$. Then the second largest eigenvalue β_1 of p is bounded by $\beta_1 \leq 1 - \varepsilon/D^2$.*

This cannot be much improved in general as can be seen by looking at the simple random walk on $G = \mathbb{Z}_2^n \times \mathbb{Z}_{2a}$ with $a \gg n$. See [45]. The papers [10, 11, 97] describe a number of deep results giving diameter estimates for finite Cayley graphs. These can be used together with Theorem 6.2 to obtain eigenvalue bounds.

Two significant improvements on Theorem 6.2 involve the following notation. Recall from Section 6.1 that $\mathcal{P}$ denotes the set of all paths in (G, Σ). For $s \in \Sigma$ and any path $\gamma = (x_0, \ldots, x_n) \in \mathcal{P}$, set

$$N(s, \gamma) = \#\{i \in \{0, \ldots, n-1\} : x_i^{-1} x_{i+1} = s\}. \tag{6.1}$$

In words, $N(s, \gamma)$ counts how many times the generator s appears along the path γ. Let $\mathcal{P}_{x,y}$ be the set of all finite paths joining x to y and $\mathcal{P}_x$ be the set of all finite paths starting at x. For each $x \in G$, pick a path $\gamma_x \in \mathcal{P}_{e,x}$ and set

$$\mathcal{P}_* = \{\gamma_x : x \in G\}.$$

Theorem 6.3 ([42]). *Referring to the notation introduced above, for any choice of $\mathcal{P}_*$, set*

$$A_* = \max_{s \in \Sigma} \left\{ \frac{1}{|G| p(s)} \sum_{\gamma \in \mathcal{P}_*} |\gamma| N(s, \gamma) \right\}.$$

Then $\beta_1 \leq 1 - 1/A_*$.

This theorem is a corollary of Theorem 6.4 which is proved below. The notation A_* reminds us that this bound depends on the choice of paths made to construct the set $\mathcal{P}_*$. To obtain Theorem 6.2, define $\mathcal{P}_*$ by picking for each x a path from e to x having minimal length. Then bound $|\gamma_x|$ and $N_*(s, \gamma_x)$ from above by D, and bound $p(s)$ from below by ε.

Making arbitrary choices is not always a good idea. Define a *flow* to be a non-negative function Φ on $\mathcal{P}_e$ (the set of all paths starting at e) such that,

$$\forall x \in G, \quad \sum_{\gamma \in \mathcal{P}_{e,x}} \Phi(\gamma) = \frac{1}{|G|}.$$

For instance, for each x, let $\mathcal{G}_{e,x}$ be the set of all geodesic paths (paths of minimal length) in $\mathcal{P}_{e,x}$. The function

$$\Phi(\gamma) = \begin{cases} \frac{1}{\#\mathcal{G}_{e,x}|G|} & \text{if } \gamma \in \mathcal{G}_{e,x} \text{ for some } x \in G \\ 0 & \text{otherwise} \end{cases}$$

is a flow.

Theorem 6.4 ([49, 124]). *Let Φ be a flow and set*

$$A(\Phi) = \max_{s \in \Sigma} \left\{ \frac{1}{p(s)} \sum_{\gamma \in \mathcal{P}_e} |\gamma| N(s, \gamma) \Phi(\gamma) \right\}.$$

Then $\beta_1 \leq 1 - 1/A(\Phi)$.

Proof. The proof is based on the elementary variational inequality (5.5) which reduces Theorem 6.4 to proving the Poincaré inequality

$$\forall f \in L^2(G), \quad \mathrm{Var}_u(f) \le A(\Phi)\mathcal{E}(f,f). \tag{6.2}$$

Here we have

$$\mathrm{Var}_u(f) = \frac{1}{2|G|^2} \sum_{x,y\in G} |f(xy) - f(x)|^2 \tag{6.3}$$

and

$$\mathcal{E}(f,f) = \frac{1}{2|G|} \sum_{x,y\in G} |f(xy) - f(x)|^2 p(y). \tag{6.4}$$

The similarity between these two expressions is crucial to the argument below. For any path $\gamma = (y_0, \ldots, y_n)$ from e to y of length $|\gamma| = n$, set $\gamma_i = y_i^{-1} y_{i+1}$, $0 \le i \le n - 1$ and write

$$f(xy) - f(x) = \sum_{i=0}^{n-1}(f(xy_{i+1}) - f(xy_i)) = \sum_{i=0}^{n-1}(f(xy_i\gamma_i) - f(xy_i)).$$

Squaring and using the Cauchy-Schwarz inequality, gives

$$|f(xy) - f(x)|^2 \le |\gamma| \sum_{i=0}^{n-1} |f(xy_i\gamma_i) - f(xy_i)|^2.$$

Summing over $x \in G$ yields

$$\sum_{x\in G} |f(xy) - f(x)|^2 \le |\gamma| \sum_{i=0}^{n-1} \sum_{x\in G} |f(x\gamma_i) - f(x)|^2$$

$$\le |\gamma| \sum_{s\in\Sigma} \sum_{x\in G} N(s,\gamma)|f(xs) - f(x)|^2.$$

Multiplying by $\Phi(\gamma)$, summing over all $\gamma \in \mathcal{P}_{e,y}$ and then averaging over all $y \in G$ yields

$$\mathrm{Var}(f) \le \frac{1}{2|G|} \sum_{s\in\Sigma} \sum_{x\in G} \sum_{\gamma\in\mathcal{P}_e} |\gamma|N(s,\gamma)\Phi(\gamma)|f(xs) - f(x)|^2.$$

Hence

$$\mathrm{Var}(f) \le \frac{1}{2|G|} \sum_{s\in\Sigma} \sum_{x\in G} \left\{ \frac{1}{p(s)} \sum_{\gamma\in\mathcal{P}_e} |\gamma|N(s,\gamma)\Phi(\gamma) \right\} |f(xs) - f(x)|^2 p(s)$$

$$\le \left(\max_{s\in\Sigma} \left\{ \frac{1}{p(s)} \sum_{\gamma\in\mathcal{P}_e} |\gamma|N(s,\gamma)\Phi(\gamma) \right\} \right) \mathcal{E}(f,f).$$

This proves (6.2). $\qquad\square$

The next result is a corollary of Theorem 6.4 and use paths chosen uniformly over all geodesic paths from e to y.

Theorem 6.5 ([49, 124]). *Referring to the setting of Theorem 6.2, assume that the automorphisms group of G is transitive on Σ. Then*

$$\beta_1 \leq 1 - \frac{\varepsilon \# \Sigma}{D^2}.$$

Let us illustrate these results by looking at the random transposition walk on the symmetric group S_n defined at (4.1). Thus $p(e) = 1/n$, $p(\tau) = 2/n^2$ if τ is a transposition and $p(\tau) = 0$ otherwise. From representation theory (see Section 9.2), we know that $\beta_1 = 1 - 2/n$. Here Σ is the set of all transpositions. Any permutation can be written as a product of at most $n - 1$ transpositions (i.e., the diameter is $D = n - 1$). Thus Theorem 6.2 gives

$$\beta_1 \leq 1 - \frac{2}{n^2(n - 1)^2}.$$

When writing a permutation as a (minimal) product of transpositions, any given transposition is used at most once. Hence $N(s, \gamma)$ at (6.1) is bounded by 1. Using this in Theorem 6.3 immediately gives

$$\beta_1 \leq 1 - \frac{2}{n^2(n - 1)}.$$

A more careful use of the same theorem actually yields

$$\beta_1 \leq 1 - \frac{2}{n(n - 1)}.$$

Finally, as the transpositions form a conjugacy class, it is easy to check that Theorem 6.5 applies and yields again the last inequality.

6.3 The Lowest Eigenvalue

Let p be a symmetric probability on G and Σ be a finite symmetric generating set contained in the support of p. Loops of odd length in the Cayley graph (G, Σ) can be used to obtain lower bounds on the lowest eigenvalue

$$\beta_{\min} = \beta_{|G|-1}.$$

Denote by $\mathcal{L}$ the set of loops of odd length anchored at the identity in (G, Σ). A *loop flow* is a non-negative function Ψ such that

$$\sum_{\gamma \in \mathcal{L}} \Psi(\gamma) = 1.$$

As above, let $N(s, \gamma)$ be the number of occurrences of $s \in \Sigma$ in γ.

Theorem 6.6 ([51, 42, 45]). *Let Ψ be a loop flow and set*

$$B(\Psi) = \max_{s \in \Sigma} \left\{ \frac{1}{p(s)} \sum_{\gamma \in \mathcal{L}} |\gamma| N(s, \gamma) \Psi(\gamma) \right\}.$$

Then the smallest eigenvalue is bounded by $\beta_{|G|-1} \geq -1 + 2/B(\phi)$.

As a trivial application, assume that $p(e) > 0$ and that $e \in \Sigma$. Then we can consider the loop flow concentrated on the trivial loop of length 1, that is, $\gamma = (e, e)$. In this case $B(\Psi) = 1/p(e)$ and we obtain

$$\beta_{|G|-1} \geq -1 + 2p(e).$$

This applies for instance to the random transposition measure p defined at (4.1) and gives $\beta_{|G|-1} \geq -1 + 2/n$ (there is, in fact, equality in this case).

For an example where a non-trivial flow is useful, consider the Borel–Chéron shuffle of Section 3.1: remove a random packet and place it on top. This allows for many loops of length 3. Consider the loops $\gamma_{a,b}$, $2 < a \leq b \leq n$ and a odd, defined as follows. Remove the packet $(a, \ldots, b)$ and place it on top; remove the packet corresponding to the cards originally in position $(a+1)/2$ through $a - 1$ and place it on top; remove the packet of the cards originally in positions 1 through $(a - 1)/2$ and place it on top. The crucial observation is that, given one of these moves and its position in the loop, one can easily recover the two other moves of the loop. Using the flow uniformly supported on these loops in Theorem 6.6 gives $\beta_{\min} \geq -(26n + 2)/(27n)$ for the Borel–Chéron shuffle on S_n.

The following result is a corollary of Theorem 6.6 and complements Theorem 6.5. The proof uses the uniform flow on all loops of minimal odd length.

Theorem 6.7. *Assume that the automorphism group of G is transitive on Σ. Then*

$$\beta_{|G|-1} \geq 1 - \frac{2\varepsilon \# \Sigma}{L^2}$$

where $\varepsilon = \min\{p(s) : s \in \Sigma\}$ and L is the minimal length of a loop of odd length in (G, Σ).

To illustrate this result, consider the alternating group A_n. In A_n, consider any fixed element $\sigma \neq e$ and its orbit Σ under the action of the symmetric group, that is, $\Sigma = \{\tau = \varrho \sigma \varrho^{-1}, \ \varrho \in S_n\}$. In words, Σ is the conjugacy class of σ in S_n. One can show that, except when σ is the product of two transpositions with disjoint supports in A_4, the set Σ is a generating set of A_n. Moreover, in any such case, the Cayley graph (A_n, Σ) contains cycles of length three (for details, see, e.g., [121]). For instance, if $\sigma = c$ is a cycle of odd length, we have $c^{-1}, c^2 \in \Sigma$ and $c^{-1}c^{-1}c^2 = e$. If $\sigma = (i,j)(k,l)$ is the product of two disjoint transpositions, we have $[(i,j)(k,l)][(k,i)(j,l)][(k,j)(i,l)] = e$. Set

$$p_\Sigma(\tau) = \begin{cases} 1/|\Sigma| & \text{if } \tau \in \Sigma \\ 0 & \text{otherwise.} \end{cases}$$

By construction, the automorphism group of A_n acts transitively on Σ. Hence, for any Σ as above, Theorem 6.7 shows that the lowest eigenvalue of p_Σ is bounded by $\beta_{\min} \geq -1 + 2/9 = -7/9$.

6.4 Diameter Bounds, Isoperimetry and Expanders

The goal of this section is to describe the relation between eigenvalues of random walks, isoperimetric inequalities and the important notion of expanders.

Diameter bounds. Let (G, Σ) be a finite Cayley graph with diameter D (recall that, by hypothesis, Σ is symmetric). Let p be a probability with support contained in Σ. For $k = \lfloor D/2 \rfloor - 1$, the support of $p^{(k)}$ contains less than half the elements of G. Hence

$$D \leq 2(T(G, p) + 2). \tag{6.5}$$

This gives an elementary relation between the diameter of (G, Σ) and random walks. Theorem 6.2 shows how the diameter can be used to control the second largest eigenvalue of an associated walk. Interestingly enough, this relation can be reversed and eigenvalues can be used to obtain diameter bounds. The best known result is the following [22, 117] which in fact holds for general graphs.

Theorem 6.8. *Let Σ be a symmetric generating set of a finite group G of order $|G| = N$. Let β_i, $0 \leq i \leq N - 1$, be the eigenvalues in non-increasing order of a random walk driven by a measure p whose support is contained in $\{e\} \cup \Sigma$ and set $\lambda_i = 1 - \beta_i$. Then the diameter D of (G, Σ) is bounded by*

$$D \leq 1 + \left\lfloor \frac{\cosh^{-1}(N-1)}{\cosh^{-1}\left(\frac{\lambda_1 + \lambda_{N-1}}{\lambda_{N-1} - \lambda_1}\right)} \right\rfloor \leq 1 + \left\lfloor \frac{\cosh^{-1}(N-1)}{\cosh^{-1}\left(\frac{2+\lambda_1}{2-\lambda_1}\right)} \right\rfloor.$$

It is useful to observe that if $N = |G|$ goes to infinity and λ_1 goes to zero the asymptotics of the right most bound is $(2\lambda_1)^{-1/2} \log |G|$. One can also verify that, assuming $\lambda_1 \leq 1$, the second upper bound easily gives

$$D \leq 3\lambda_1^{-1/2} \log |G|. \tag{6.6}$$

When λ_1 is relatively small, the elementary bound (6.5) often gives better results than Theorem 6.8. For instance, consider the symmetric group S_n generated by the set of all transpositions. Let $p = p_{\mathrm{RT}}$ be the random transposition measure defined at (4.1). The diameter of this Cayley graph is $n - 1$, the spectral gap λ_1 of p_{RT} is $2/n$ and $T(S_n, p_{\mathrm{RT}}) \sim \frac{1}{2} n \log n$. Hence, both (6.5) and Theorem 6.8 are off but (6.5) is sharper. Theorem 6.8 is of most interest for families of graphs and random walks having a spectral gap bounded away from 0. Such graphs are called expanders and are discussed below.

Isoperimetry. Let (G, Σ) be a finite Cayley graph. Recall that the edge set E of (G, Σ) is $E = \{(x, y) : x, y \in G, x^{-1}y \in \Sigma\}$. As always, we denote by u the uniform probability measure on G. We also denote by u_E the uniform probability on E so that for a subset F of E, $u_E(F) = |F|/|\Sigma||G|$ where $|F|$ denotes the cardinality of F.

Given a set $A \in G$, define the boundary of A to be

$$\partial A = \{(x, y) \in G \times G : x \in A, y \in G \setminus A, x^{-1}y \in \Sigma\}.$$

The *isoperimetric constants* $I = I(G, \Sigma)$, $I' = I'(G, \Sigma)$ are defined by

$$I = \min_{\substack{A \subset G \\ 2|A| \leq |G|}} \frac{u_E(\partial A)}{u(A)}, \quad I' = \min_{A \subset G} \frac{u_E(\partial A)}{2(1 - u(A))u(A)}. \tag{6.7}$$

We have $I/2 \leq I' \leq I$. Note that, in terms of cardinalities, this reads

$$I = \min_{\substack{A \subset G \\ 2|A| \leq |G|}} \frac{|\partial A|}{|\Sigma||A|}, \quad I' = \min_{A \subset G} \frac{|G||\partial A|}{2|\Sigma|(|G| - |A|)|A|}.$$

The following gives equivalent definitions of I, I' in function terms. See, e.g., [124]. For a function f on G and $e = (x, y) \in E$, we set $df(e) = f(y) - f(x)$.

Lemma 6.9. *We have*

$$2I = \min_f \left\{ \frac{u_E(|df|)}{u(|f - m(f)|)} \right\}, \quad 2I' = \min_f \left\{ \frac{u_E(|df|)}{u(|f - u(f)|)} \right\}$$

where $m(f)$ denote an arbitrary median of f.

For sharp results concerning isoperimetry on the hypercube and further discussion, see [84, 96] and the references therein.

The next result relates I and I' to the spectral gap λ_1 of random walks closely related to the graph (G, Σ). This type of result has become known under the name of a Cheeger inequality. See, e.g., [98, 124, 131]. An interesting development is in [111]. For the original Chegeer inequality in Riemannian geometry, see, e.g., [20].

Theorem 6.10. *Let G be a Cayley graph and p be a symmetric probability measure on G with spectral gap $\lambda_1 = 1 - \beta_1$.*

- *Assume* supp $(p) \subset \Sigma$ *and set* $\eta = \max_\Sigma p$. *Then* $\lambda_1 \leq 2\eta|\Sigma|I'$.
- *Assume that* $\inf_\Sigma p = \varepsilon > 0$. *Then* $\varepsilon|\Sigma|I^2 \leq 2\lambda_1$.
- *In particular, if* $p = p_\Sigma$ *is the uniform probability on* Σ, $I^2 \leq 2\lambda_1 \leq 4I'$.

Slightly better results are known. For instance, [110, Theorem 4.2] gives $I^2 \leq \lambda_1(2 - \lambda_1)$. See also [111].

The isoperimetric constants I, I' can be bounded from below in terms of the diameter. See, e.g., [9] and [131]. Using the notation of Section 6, we have the following isoperimetric version of Theorems 6.4, 6.5.

Theorem 6.11. *Let (G, Σ) be a finite Cayley graph. Let Φ be a flow as in Theorem 6.4. Then $2I' \geq 1/a(\Phi)$ with*

$$a(\Phi) = \max_{s \in \Sigma} \left\{ |\Sigma| \sum_{\gamma \in \mathcal{P}_e} N(s, \gamma)\Phi(\gamma) \right\}.$$

In particular, $I \geq I' \geq 1/(2|\Sigma|D)$ where D is the diameter of (G, Σ). If we further assume that the automorphism group of G is transitive on Σ then $I \geq I' \geq 1/(2D)$.

Although the notion of isoperimetry is appealing, it is rarely the case that good spectral gap lower bounds are proved by using the relevant inequality in Theorem 6.10. See the discussion in [62]. In fact, isoperimetric constants are hard to compute or estimate precisely and spectral bounds are often useful to bound isoperimetric constants.

Let us end this short discussion of isoperimetric constants by looking at the symmetric group S_n equipped with the generating set of all transpositions. This Cayley graph has diameter $n-1$ and the automorphism group of S_n acts transitively on transpositions. Hence Theorem 6.11 gives $I' \geq (2(n-1))^{-1}$. The random transposition walk defined at (4.1) has spectral gap $\lambda_1 = 2/n$ (See Section 9.2). By Theorem 6.10, this implies $(n-1)^{-1} \leq I' \leq I \leq 2(n-1)^{-1/2}$. Using $A = \{\sigma \in S_n : \sigma(n) = n\}$ as a test set shows that $I \leq 2n^{-1}$, $I' \leq (n-1)^{-1}$. Thus $(n-1)^{-1} \leq I \leq 2n^{-1}$ and $I' = (n-1)^{-1}$.

Expanders. The notion of *expander* depends on a different definition of the boundary than the one given above. Namely, for any $A \subset G$, set

$$\delta A = \{x \in G : d(x, A) = 1\}$$

where d is the graph distance introduced in Section 6.1. Define the expansion constant $h = h(G, \Sigma)$ by

$$h = \min_{\substack{A \subset G \\ 2|A| \leq |G|}} \frac{|\delta A|}{|A|}.$$

By inspection, we have $I \leq h \leq |\Sigma|I$. A variant of Theorem 6.11 in [9] states that, for any Cayley graph, $h \geq 2/(2D+1)$.

Definition 6.12. *A finite Cayley graph (G, Σ) is an (N, r, ε)-expander if $|G| = n$, $|\Sigma| = r$ and $h(G, \Sigma) \geq \varepsilon$.*

A family $((G_n, \Sigma_n))$ of finite Cayley graphs is a family of expanders if $|G_n|$ tends to ∞ and there exists $\varepsilon > 0$ such that $h(G_n, \Sigma_n) > \varepsilon$.

Comparing I and h and using Theorem 6.10 yields the following relation between spectral gap estimates and the notion of expander.

Proposition 6.13. *Let $((G_n, \Sigma_n))$ be a family of finite Cayley graphs such that $|G_n|$ tends to ∞. Let p_n denote the uniform probability on Σ_n and let $\lambda_1(n)$ be the spectral gap associated to p_n.*

– *If there exists $\varepsilon > 0$ such that $\lambda_1(n) \geq \varepsilon$ for all n then (G_n, Σ_n) is a family of expanders.*
– *If there exists r such that $|\Sigma_n| \leq r$ for all n then (G_n, Σ_n) is a family of expanders if and only if there exists $\varepsilon > 0$ such that $\lambda_1(n) \geq \varepsilon$ for all n.*

Theorem 9.8 in Section 9.4 gives a remarkable application of Proposition 6.13.

In the other direction, Proposition 6.13 shows that the symmetric groups S_n equipped with the generating sets $\Sigma_n = \{\tau, c, c^{-1}\}$ where τ is the transposition $(1, 2)$ and c the cycle $(1, 2, \ldots, n)$ do not form a family of expanders. Indeed, the diameter D of (S_n, Σ_n) is of order n^2 whereas Proposition 6.13 and Theorem 6.8 shows that any expander graph on S_n has diameter of order $n \log n$ at most. In fact, the present Cayley graph has λ of order $1/n^3$. See Section 10.

Recall that a finitely generated group Γ has property (T) (i.e., Kazhdan property (T)) if there exists a finite set $K \subset \Gamma$ and $\varepsilon > 0$ such that, for every non-trivial irreducible unitary representation (V, ϱ) of Γ and every unitary vector $v \in V$, $\|\varrho(x)v - v\| \geq \varepsilon$ for some $x \in K$. One shows that if this holds for one finite set K then it holds for any finite generating set Σ (with different $\varepsilon > 0$). See [98] for an excellent exposition and references concerning property (T). The groups $SL_n(\mathbb{Z})$, $n \geq 3$, have property (T). Non-compact solvable groups, free groups and $SL_2(\mathbb{Z})$ do not have property (T). Margulis [108] produced the first explicit examples of families of expanders by using property (T) to obtain infinite families of graphs with bounded degree and spectral gap bounded from below. See also [101] and [115, 146] for recent advances concerning property (T).

Theorem 6.14 ([98]). *Let Γ be a finitely generated infinite group. Let H_n be a family of normal finite index subgroups of Γ. Set $G_n = \Gamma/H_n$ and assume that $|G_n|$ tends to infinity. Let Σ be a symmetric generating set of Γ and $\Sigma_n \subset G_n$ be the projection of Σ.*

– *Assume that Γ has property (T). Then $((G_n, \Sigma_n))$ is a family of expanders.*
– *Assume that Γ is solvable. Then $((G_n, \Sigma_n))$ is not a family of expanders.*

The condition that the subgroups H_n are normal is not essential. It is added here simply to have Cayley graphs as quotients. For a proof, see [98, Prop. 3.3.1, 3.3.7]. The following simple result describes what happens for random walks on expanders. See, e.g., [46, 115].

Theorem 6.15. *Fix $r > 0$. Let (G_n, Σ_n) be a family of expanders with Σ_n containing the identity. For each n, let p_n be a probability measure on G_n such that $\inf_{\Sigma_n} p_n \geq 1/r$, $|\mathrm{supp}\,(p_n)| \leq r$. Then there are constants $C, c > 0$ such that*

$$c \log |G_n| \leq T(G_n, p_n) \leq C \log |G_n|.$$

Moreover, the family (G_n, p_n) has a precut-off at time $\log |G_n|$.

Proof. For the upper bound, use (5.9) and the fact that the hypotheses and Proposition 6.13 imply $\beta_{n,*} \leq 1 - \varepsilon$. For the lower bound, use (5.12).   □

The next theorem due to Alon and Roichman [6] says that most Cayley graphs (G, Σ) with $|\Sigma| \gg \log|G|$ are expanders.

Theorem 6.16. *For every $\varepsilon > 0$ there exists $c(\varepsilon) > 0$ such that, if G is a group of order n, $t \geq c(\varepsilon)\log n$, T is a uniformly chosen t-subset of G and $\Sigma = T \cup T^{-1}$ then the Cayley graph (G, Σ) is an $(|G|, |\Sigma|, \varepsilon)$-expander with probability $1 - o(1)$ when n tends to infinity.*

Next, we describe some explicit examples of expanders. In $SL_n(\mathbb{Z})$, consider the matrices

$$
A_n = \begin{pmatrix} 1 & 1 & 0 & \cdot & \cdot & \cdot & 0 \\ 0 & 1 & 0 & \cdot & \cdot & \cdot & \cdot \\ \cdot & 0 & 1 & 0 & \cdot & \cdot & \cdot \\ \cdot & \cdot & \cdot & \cdot & \cdot & \cdot & \cdot \\ \cdot & \cdot & \cdot & 0 & 1 & 0 & 0 \\ \cdot & \cdot & \cdot & \cdot & 0 & 1 & 0 \\ 0 & \cdot & \cdot & \cdot & \cdot & 0 & 1 \end{pmatrix}, \quad
B_n = \begin{pmatrix} 0 & 1 & 0 & \cdot & \cdot & \cdot & 0 \\ \cdot & 0 & 1 & 0 & \cdot & \cdot & \cdot \\ \cdot & \cdot & 0 & 1 & 0 & \cdot & \cdot \\ \cdot & \cdot & \cdot & \cdot & \cdot & \cdot & \cdot \\ \cdot & \cdot & \cdot & \cdot & 0 & 1 & 0 \\ 0 & \cdot & \cdot & \cdot & \cdot & 0 & 1 \\ j & 0 & \cdot & \cdot & \cdot & \cdot & 0 \end{pmatrix}
$$

where $j = (-1)^{n+1}$. These generates $SL_n(\mathbb{Z})$.

Theorem 6.17 ([98]). *Fix $n \geq 2$. consider the symmetric generating set $\Sigma_n = \{A_n^{\pm 1}, B_n^{\pm 1}\}$ of $SL_n(\mathbb{Z}_q)$ where q is prime. Let p_n denote the uniform probability on $\{I_n, A_n^{\pm 1}, B_n^{\pm 1}\}$. Then $((SL_n(\mathbb{Z}_q), \Sigma_n))$ is a family of expanders. In particular, for fixed n and varying prime q, $((SL_n(\mathbb{Z}_q), p_n))$ has a precut-off at time $\log q$.*

The proof differs depending on whether $n = 2$ or $n > 2$ because, as mentioned earlier, $SL_2(\mathbb{Z})$ does not have property (T). See [98, 99].

We close our discussion of expanders by stating a small selection of open problems. See [98, 99] for more.

Problem 6.18. Can one find generating subsets Σ_n of the symmetric groups S_n of bounded size $|\Sigma_n| \leq r$ such that (S_n, Σ_n) form a family of expanders?

In [100, Section 5], Lubotzky and Pak notice that this problem is related to another open problem, namely, to whether or not the automorphism group of a regular tree of degree at least 4 has property (T). One can also state Problem 6.18 with the symmetric groups replaced by an infinite family of simple finite groups.

Problem 6.19. Can one find a family of finite groups G_n and generating sets Σ_n^1, Σ_n^2 of bounded size $|\Sigma_n^i| \leq r$ such that $((G_n, \Sigma_n^1))$ is a family of expanders but $((G_n, \Sigma_n^2))$ is not?

If Problem 6.18 has a positive answer then the same is true for Problem 6.19 since $((S_n, \Sigma_n))$ with $\Sigma_n = \{(1, 2), (1, \ldots, n)^{\pm 1}\}$ is not a family of expanders (see, e.g., [115]).

Problem 6.20. Fix r and let Σ_p denote an arbitrary generating set of $SL_2(\mathbb{Z}_p)$, p prime, with $|\Sigma_p| \leq r$. Is $((SL_2(\mathbb{Z}_p), \Sigma_p))$ always a family of expanders?

With respect to this last problem, set

$$\Sigma_p^i = \left\{ \begin{pmatrix} 1 & i \\ 0 & 1 \end{pmatrix}^{\pm 1}, \begin{pmatrix} 1 & 0 \\ i & 1 \end{pmatrix}^{\pm 1} \right\}.$$

Then $((SL_2(\mathbb{Z}_p), \Sigma_p^i))$ is a family of expanders if $i = 1, 2$ but it is not known if the same result holds for $i = 3$. See [63, 98, 99].

Problem 6.21. Let $((G_n, \Sigma_n))$ be a family of expanders. Under the assumptions and notation of Theorem 6.15, does the family $((G_n, p_n))$ admit a cut-off?

For further information on these problems, see [63, 64, 65, 98, 99].

Ramanujan graphs. Alon and Bopanna (see, e.g., [74, 98, 99, 127, 136]) observed that any infinite family of finite Cayley graphs $((G_n, \Sigma_n))$ with $|\Sigma_n| = r$ for all n (more generally, r-regular graphs) satisfies

$$\liminf_{n \to \infty} \beta_1(G_n, p_n) \geq \frac{2\sqrt{r-1}}{r}.$$

where p_n denotes the uniform probability on Σ_n.

Definition 6.22. *A Cayley graph* (G, Σ) *is* Ramanujan *if*

$$\beta_1(G, p_\Sigma) \leq \frac{2\sqrt{r-1}}{r}$$

where p_Σ denotes the uniform probability on Σ and $r = |\Sigma|$.

Examples of Ramanujan Cayley graphs with $G = PGL_2(\mathbb{Z}_q)$ are given in [127]. See also [25, 98, 101, 136]. For fixed r, asymptotically as the cardinality goes to infinity, Ramanujan graphs are graphs whose second largest eigenvalue is as small as possible. By Proposition 6.13, they are expanders, in fact very good expanders, and have many other remarkable properties. After taking care of possible periodicity problems, the simple random walks on any infinite family of Ramanujan Cayley graphs $((G_n, \Sigma_n))$ have a precut-off at time $\log |G_n|$.

Infinite families of Ramanujan graphs are hard to find and most (if not all) known examples are obtained by applying rather deep number theoretic results. See [98, 127]. In particular, the construction of expanders as in Theorem 6.14 cannot work for Ramanujan graphs [71, 98, 99].

Theorem 6.23. *Let Γ be a finitely generated infinite group. Let H_n be a family of normal finite index subgroups of Γ. Set $G_n = \Gamma/H_n$ and assume that $|G_n|$ tends to infinity. Let Σ be a symmetric generating set of Γ such that the graph (Γ, Σ) is not a tree. Let $\Sigma_n \subset G_n$ be the projection of Σ. Then at most finitely many (G_n, Σ_n) are Ramanujan.*

As in Theorem 6.14, the condition that the subgroups H_n are normal is not essential.

7 Results Involving Volume Growth Conditions

On a finite group G, consider a symmetric probability p whose support generates G. Fix a symmetric generating set Σ contained in the support of p and consider the Cayley graph (G, Σ) as in Section 6.1.

Definition 7.1. *Referring to the notation of Section 6.1, set*

$$V(n) = V_\Sigma(n) = \#\{x \in G : |x|_\Sigma \leq n\}.$$

The function V_Σ is called the volume growth function *of (G, Σ).*

Sections 7.1 and 7.2 below describe results that involve the volume growth function V and apply to walks based on a bounded number of generators. Examples include nilpotent groups with small class and bounded number of generators. Section 7.3 presents contrasting but related results for some families of nilpotent groups with growing class and/or number of generators.

7.1 Moderate Growth

This section gives a large class of finite groups which carry natural random walks whose behavior is similar to that of the simple random walk on the finite circle group $\mathbb{Z}_n = \mathbb{Z}/n\mathbb{Z}$. More precisely, on $\mathbb{Z}_n$, consider the random walk which goes left, right or stays put, each with probability $1/3$. For this walk, the spectral gap $\lambda_1 = 1 - \beta_1$ is of order $1/n^2$ and there are continuous positive decreasing functions f, g tending to 0 at infinity such that

$$f(k) \leq \|p^{(kn^2)} - u\|_{\mathrm{TV}} \leq g(k).$$

Thus, there is no cut-off phenomenon in this case: a number of steps equal to a large multiple of $1/\lambda_1$ suffices to reach approximate equilibrium whereas a small multiple of $1/\lambda_1$ does not suffice.

We start with the following definition.

Definition 7.2 ([44, 47]). *Fix $A, \nu > 0$. We say that a Cayley graph (G, Σ) has (A, ν)-moderate growth if its volume growth function satisfies*

$$V(k) \geq \frac{|G|}{A}\left(\frac{k}{D}\right)^\nu$$

for all integers $k \leq D$ where D is the diameter of (G, Σ).

Let us illustrate this definition by some examples.

- The circle group $\mathbb{Z}_n = \mathbb{Z}/n\mathbb{Z}$ with $\Sigma = \{0, \pm 1\}$ has $V(k) = 2k + 1$. Here $|G| = n$, $D = \lfloor n/2 \rfloor$. Thus the circle group has moderate growth with $A = 3/2$ and $\nu = 1$.

- The group $\mathbb{Z}_n$ with $\Sigma = \{0, \pm 1, \pm m\}$ with $m \leq n$ has diameter D of order $\max\{n/m, m\}$. The Cayley graph $(\mathbb{Z}_n, \Sigma)$ has moderate growth with $A = 5$ and $\nu = 2$ although this is not entirely obvious to see.
- Consider the group $\mathbb{Z}_n^d$ with $\Sigma = \{0, \pm e_i\}$ where e_i denotes the element with all coordinates 0 except the i-th which equals 1. This Cayley graph has diameter $D = d\lfloor n/2 \rfloor$. For fixed d, there exists a constant A_d such that $(\mathbb{Z}_n^d, \Sigma)$ has (A_d, d)-moderate growth for all n.
- For any odd prime p, consider the affine group A_p which is the set of all pairs $(a, b) \in \mathbb{Z}_p^* \times \mathbb{Z}_p$ with multiplication given by $(a, b)(a', b') = (aa', a'b + b')$. Let α be a generator of $\mathbb{Z}_p^*$, β a generator of $\mathbb{Z}_p$, and set $\Sigma = \{(1, 0), (\alpha, 0), (\alpha^{-1}, 0), (1, \beta), (1, -\beta)\}$. This group has diameter D of order p and it has $(6, 2)$-moderate growth.
- Let $U_3(n)$ be the Heisenberg group mod n, i.e., the group of all 3 by 3 upper diagonal matrices with 1 on the diagonal and integer coefficients mod n. Let I denote the identity matrix in $U_3(n)$. Let $E_{i,j}$ be the matrix in U_3 whose non-diagonal entries are all 0 except the (i, j) entry which is 1. Then $\Sigma = \{I, \pm E_{1,2}, \pm E_{2,3}\}$ is a generating set of $U_3(n)$. The Cayley graph $(U_3(n), \Sigma)$ has diameter of order n and $(48, 3)$-moderate growth.

The next theorem gives sharp bounds under the assumption of moderate growth.

Theorem 7.3 ([44, 47]). *Let (G, Σ) be a finite Cayley graph with diameter D and such that $e \in \Sigma$. Let p be a probability measure on G supported on Σ. For any positive numbers A, d, ε, there exists six positive constants $c_i = c_i(A, d, \varepsilon)$, $1 \leq i \leq 6$, such that if (G, Σ) has (A, d)-moderate growth and p satisfies $\inf_\Sigma p \geq \varepsilon$ then we have*

$$\forall\, k \in \mathbb{N}, \quad a_1 e^{-a_2 k / D^2} \leq \|p^{(k)} - u\|_{TV} \leq a_3 e^{-a_4 k / D^2}$$

and

$$\forall\, k \geq D^2, \quad d_2(p^{(k)}, u) \leq a_5 e^{-a_6 k / D^2}.$$

The condition $\inf_\Sigma p \geq \varepsilon$ has two different consequences. On the one hand, it forces p to be, in some sense, adapted to the underlying graph structure. On the other hand, it implies a uniform control over the size of the generating set Σ since we have $1 \geq p(\Sigma) \geq \varepsilon|\Sigma|$.

Moderate growth was first introduced in [44]. It is related to the following notion of doubling growth which has been used in many different contexts.

Definition 7.4. *Fix $A > 0$. We say that a Cayley graph (G, Σ) has A-doubling growth if its volume growth function satisfies*

$$\forall k \in \mathbb{N}, \quad V(2k) \leq A\, V(k).$$

Doubling growth provides a useful way to obtain examples of groups with moderate growth thanks to the following two propositions. The first is elementary.

Proposition 7.5. *If the Cayley graph (G, Σ) has A-doubling growth, then it has (A, d)-moderate growth with $d = \log_2 A$.*

Let us observe that the notion of doubling growth make sense for infinite Cayley graphs.

Proposition 7.6. *Let (Γ, Σ) be an infinite Cayley graph and assume that (Γ, Σ) has A-doubling growth. Then, for any quotient group $G = \Gamma/N$, N normal in Γ, the Cayley graph (G, Σ_G) where Σ_G is the canonical projection of Σ in G has A^2-doubling growth.*

We illustrate this with two examples. First, consider $\mathbb{Z}_n$ with generating set $\Sigma = \{0, \pm 1, \pm m\}$, $m < n$. We can view this Cayley graph as a quotient of the square grid, i.e., the natural graph on $\mathbb{Z}^2$. Indeed, one can check that there is a unique surjective group homomorphism π from $\mathbb{Z}^2$ to $\mathbb{Z}_n$ such that $\pi((1,0)) = 1$, $\pi((0,1)) = m$ (this is because $\mathbb{Z}^2$ is the free abelian group on two generators). Proposition 7.6 applies and easily shows that $(\mathbb{Z}_n, \Sigma)$ is 5-doubling. As a second example, consider the Heisenberg group $U_3(n)$ with its natural generating set $\Sigma = \{I, E_{1,2}, E_{2,3}\}$ as defined above after Definition 7.2. This is a quotient (simply take all coordinates mod n) of the infinite discrete Heisenberg group U_3, i.e., the group of all 3 by 3 upper-triangular matrices with entries in $\mathbb{Z}$ and 1 on the diagonal. It is well known (see e.g., [82, Pro. VII.22]) that the volume growth function of this group satisfies $c_1 n^4 \leq V(n) \leq c_2 n^4$. Hence $(U_3(n), \Sigma)$ has A-doubling growth with $A = c_2 c_1^{-1} 3^4$.

The next result is derived from a deep theorem of Gromov [75].

Theorem 7.7. *Given two positive reals C, d, there is a constant $A = A(C, d)$ such that any finite Cayley graph (G, Σ) satisfying $V(n) \leq Cn^d$ for all integers n has A-doubling growth.*

In contrast to all the other results presented in this survey, there is no known explicit control of A as a function of C, d.

Doubling growth is a stronger assumption than moderate growth. Under the latter condition one can complement Theorem 7.3 with the following result.

Theorem 7.8 ([44, 46]). *Let (G, Σ) be a finite Cayley graph with diameter D and such that $e \in \Sigma$. Let p be a symmetric probability measure on G supported on Σ. For any positive numbers A, ε, there exist four positive constants $c_i = c_i(A, \varepsilon)$, $1 \leq i \leq 4$, such that if (G, Σ) has A-doubling growth and p satisfies $\inf_\Sigma p \geq \varepsilon$ then we have*

$$\forall\, k \in \mathbb{N}, \quad \frac{a_1 |G|}{V(k^{1/2})} e^{-a_2 k/D^2} \leq d_2(p^{(k)}, u) \leq \frac{a_3 |G|}{V(k^{1/2})} e^{-a_4 k/D^2}.$$

The same upper bound holds for any non-symmetric measure that charges e and a generating set Σ (which can be non-symmetric). See Theorem 10.8.

Thus doubling growth gives a very satisfactory control over the behavior of random walks adapted to the underlying graph structure. The next section describes a large class of examples with doubling growth.

7.2 Nilpotent Groups

In a group G, let $[x, y] = x^{-1}y^{-1}xy$ denote the *commutator* of $x, y \in G$. For $A, B \subset G$, let $[A, B]$ denote the group generated by all the commutators $[a, b]$, $a \in A, b \in B$. The *lower central series* of a group G is the non-increasing sequence of subgroups G_k of G defined inductively by $G_1 = G$ and $G_k = [G_{k-1}, G]$. A group (finite or not) is *nilpotent of class c* if $G_c \neq \{e\}$ and $G_{c+1} = \{e\}$. See [79, 78, 135]. Abelian groups are nilpotent of class 1. The group $U_m(n)$ of all m by m upper-triangular matrices with 1 on the diagonal is nilpotent of class $m - 1$.

Doubling growth for nilpotent groups. The next statement shows that nilpotent groups give many infinite families of Cayley graphs having A-doubling growth. See [82, p. 201] and [44].

Theorem 7.9. *Given any two integers c, s, there exists a constant $A = A(c, s)$ such that any Cayley graph (G, Σ) with G nilpotent of class at most c and Σ of cardinality at most s has A-doubling growth.*

The constant $A(c, s)$ can be made explicit, see [44]. Of course, this result brings Theorem 7.3 and 7.8 to bear. For concrete examples, consider the group $U_m(n)$ of all m by m upper-triangular matrices with 1 on the diagonal and entries in $\mathbb{Z}_n$. We noticed earlier that this group is nilpotent of class $m - 1$. Let $E_{i,j} \in U_m(n)$ be the matrix with zero non-diagonal entries except the (i, j)-th which is 1. The set $\Sigma = \{I, E_{1,2}^{\pm 1}, \ldots, E_{m-1,m}^{\pm 1}\}$ generates $U_m(n)$. Let p_Σ be the uniform probability measure on Σ. For each fixed integer m, Theorem 7.9 applies uniformly to $U_m(n)$, $n = 2, 3, \ldots$. As $(U_m(n), \Sigma)$ has diameter of order n this shows that, given m, there are positive constants a_i such that, uniformly over all integers n, k, the measure p_Σ on $U_m(n)$ satisfies

$$a_1 e^{-a_2 k/n^2} \leq \|p_\Sigma^{(k)} - u\|_{\mathrm{TV}} \leq a_3 e^{-a_4 k/n^2}.$$

p-groups and Frattini walks. Let p be a prime. A p-group is a group of order a power of p. Any group of order p^a is nilpotent of class at most $a - 1$ and contains generating sets of size less than or equal to a. In fact, in a group of order p^a, the minimal generating sets (i.e., sets that contains no generating proper subsets) all have the same size and can be described in terms of the *Frattini subgroup* which is defined as the intersection of all subgroups of order p^{a-1}. By a theorem of Burnside, the quotient of any p-group G by its Frattini subgroup is a vector space over $\mathbb{Z}_p$ whose dimension is the size of any minimal generating set and is called the *Frattini rank* of G. For instance, the group $U_m(p)$ has order p^a with $a = \binom{m}{2}$ and the matrices $E_{i,i+1}, 1 \leq i \leq m - 1$

form a minimal set of generators. Hence $U_m(p)$ has Frattini rank $m - 1$. See [79, 78, 135]. The following theorem describes how the results of the previous two sections apply to this very natural class of examples we call Frattini walks. Recall that the exponent of a group G is the smallest n such that $g^n = e$ for all $g \in G$.

Theorem 7.10 ([44, 45]). *Fix an integer c. Then there exists four positive constants $a_i = a_i(c)$ such that, for any p-group G of nilpotency class and Frattini rank at most c, for any minimal set F of generators of G, we have*

$$a_1 e^{-a_2 k/p^{2\omega}} \leq \|q_F^{(k)} - u\|_{TV} \leq a_3 e^{-a_4 k/p^{2\omega}}$$

where q_F denotes the uniform probability measure on $\{e\} \cup F \cup F^{-1}$ and p^ω is the exponent of $G/[G, G]$.

The proof consists in applying Theorems 7.9, 7.3 and showing that the diameter of (G, Σ) is of order p^ω, uniformly over the class of group considered here. Note that, for any fixed a, Theorem 7.10 applies uniformly to all groups of order p^a and their minimal sets of generators since such groups have nilpotency class and Frattini rank bounded by a. Also, the conclusion of Theorem 7.10 holds true if we replace the probability q_F by any symmetric probability q such that $\inf\{q(s) : s \in \{e\} \cup F\} \geq \varepsilon$ for some fixed $\varepsilon > 0$ and $\text{supp}(q) \subset (\{e\} \cup F \cup F^{-1})^m$ for some fixed m. Theorem 10.9 extends the result to non-symmetric walks.

7.3 Nilpotent Groups with many Generators

The results described in the previous sections give a rather complete description of the behavior of simple random walks on Cayley graphs of finite nilpotent groups when the nilpotency class and the number of generators stay bounded. There are however many interesting examples where one or both of these conditions are violated. The simplest such example is the hypercube $\mathbb{Z}_2^d$ as d varies. In this case, the class is 1 but the minimal number of generators is d. Of course, this walk is well understood. If we denote by $e_1, \ldots, e_d$ the natural generators of $\mathbb{Z}_2^d$ and take p to be the uniform probability on $\{e, e_1, \ldots, e_d\}$, then the walk driven by p has a cut-off at time $t_n = \frac{1}{4} d \log d$. See Theorem 8.2. It seems very likely that the walks described below present a similar cut-off phenomenon. However, even the existence of a precut-off in the sense of Definition 3.8 is an open problem for these walks. The results presented in this section are taken from Stong's work [132, 133, 134]. They are all based on similar basic ideas introduced by Stong: using the the action of large abelian subgroups and eigenvalue bounds for *twisted graphs*, i.e., weighted graphs whose weights can be complex numbers. These techniques lead to sharp bounds on the second largest eigenvalue β_1 in interesting hard problems. Together with easier bounds on the smallest eigenvalue $\beta_{\min} = \beta_{|G|-1}$, this brings to bear the simple eigenvalue bound (5.9), that is,

$$2\|p^k - u\|_{\mathrm{TV}} \le d_2(p^{(k)}, u) \le \sqrt{|G| - 1}\, \beta_*^k \qquad (7.1)$$

where $\beta_* = \max\{\beta_1, -\beta_{\min}\}$ as in (5.9).

Random walk on $U_m(q)$ as m and q vary. Let q be an odd prime and recall that $U_m(q)$ denotes the group of all m by m upper-triangular matrices with coefficients mod q and 1 on the diagonal. This group is generated by the matrix $E_{i,i+1}$, $1 \le i \le m - 1$, where $E_{i,j}$ has all its non-diagonal entries 0 except the (i, j) entry which is 1. We set $\Sigma = \{E_{1,2}^{\pm 1}, \ldots, E_{m-1,m}^{\pm 1}\}$ and denote by p the uniform probability on Σ. It is easy to apply Theorem 6.6 using a flow equidistributed on the $2(m - 1)$ loops of odd length q defined by $E_{i,i+1}^{\pm j}$, $j = 0, 1, \ldots, q$. This gives $\beta_{\min} \ge -1 + 2/q^2$.

Theorem 7.11 ([132]). *Referring to the walk driven by p on $U_m(q)$ as defined above, there are two constants $c_1, c_2 > 0$ such that for any integer m and any odd prime q, we have*

$$1 - \frac{c_1}{mq^2} \le \beta_1 \le 1 - \frac{c_2}{mq^2}.$$

Ellenberg [56] proved that there are two constants $a_1, a_2 > 0$ such that the diameter D of $(U_m(q), \Sigma)$ satisfies

$$a_1(mq + m^2 \log q) \le D \le a_2(mq + m^2 \log q).$$

Thus the upper bound in Theorem 7.11 is a substantial improvement upon the bound of Theorem 6.2.

As $U_m(q)$ has order $q^{m(m-1)/2}$, the bound (7.1) shows that k of order $m^3 q^2 \log q$ suffices for $p^{(k)}$ to be close to the uniform distribution on $U_m(q)$. For a lower bound, it is not hard to see that $p^{(k)}$ is far from the uniform distribution for $k < \max\{n^2, q^2 n\}$. It would be nice to have a better lower bound.

The Burnside group $B(3, r)$. Around 1900, Burnside asked whether or not a finitely generated group G all of whose elements have finite order must be finite. Golod and Shafarevich proved that the answer is no. Another version of this problem is as follows: Given n, is any finitely generated group of exponent n a finite group? This can be phrased in terms of the Burnside groups $B(n, r)$. By definition, the group $B(n, r)$ is the free group of exponent n with r generators. This means that any group with exponent n and r generators is a quotient of $B(n, r)$. The group $B(n, r)$ can be constructed from the free group F_r on r generators by taking the quotient by the normal subgroup generated by $\{g^n : g \in F_r\}$. It turns out that for all n large enough, $B(n, r)$ is infinite. However $B(n, r)$ is finite for $n = 2, 3, 4, 6$. At this writing, it is not known if $B(5, r)$ is finite or not. See [78, Chapter 18] and also [82, p. 224] for a short discussion and further references. When $B(n, r)$ is infinite, the solution of the restricted Burnside problem due to Zelmanov asserts that there is a *finite* group $\widetilde{B}(n,r)$ which covers all finite groups generated by r elements and of exponent n. Studying natural random walks on these groups is a tempting but probably extremely hard problem.

For $n = 2$, $B(2,r) = \mathbb{Z}_2^r$. The group $B(3,r)$ has order $M = 3^{N(r)}$ where $N(r) = r + \binom{r}{2} + \binom{r}{3}$ and its structure is described in [78, p. 322]. In particular, it is nilpotent of class 2 and $B(3,r)/[B(3,r), B(3,r)] = \mathbb{Z}_3^r$.

Theorem 7.12 ([133]). *Consider the Burnside group $B(3,r)$ and let p denote the uniform probability on the r canonical generators and their inverses. Then*

$$1 - \frac{3}{2r} \leq \beta_1 \leq 1 - \frac{1}{8r}.$$

For the walk in Theorem 7.12, Theorem 6.7 easily gives the lower bound $\beta_{\min} \geq -7/9$. Indeed, by definition of $B(3,r)$, the group of automorphism acts transitively on the generators and any generator gives an obvious loop of length 3. Inequality (7.1) shows that $p^{(k)}$ is close to the uniform distribution on $B(3,r)$ for k of order r^4. The elementary lower bound (5.12) gives that $p^{(k)}$ is not close to the uniform distribution if k is of order $r^3/\log r$.

Polynomials under composition. Let n be an integer and q an odd prime. Let $P_{n,q}$ be the group of all polynomials $\alpha_1 x + \cdots + \alpha_n x^n \mod x^{n+1}$ with $\alpha_1 \in \mathbb{Z}_q^*$, $\alpha_2, \ldots, \alpha_n \in \mathbb{Z}_q$. The group law is composition. Let α be a generator of $\mathbb{Z}_q^*$. Then $\Sigma = \{x, \alpha^{\pm 1} x, (x + x^2)^{\pm 1}, \ldots, (x + x^n)^{\pm 1}\}$ is a symmetric generating set. This group is not nilpotent but it contains a large normal nilpotent subgroup, namely, the group $P_{n,q}^1$ of polynomials in $P_{n,q}$ with $\alpha_1 = 1$. This subgroup has order q^{n-1}. It is proved in [44] that for fixed n, $P_{n,q}$ has A-moderate growth uniformly over the prime q and diameter of order q. Hence, Theorem 7.3 shows that the simple random walk on $(P_{n,q}, \Sigma)$ is close to stationarity after order q^2 steps. In [134], Stong is able to compute exactly the second largest eigenvalue of this walk.

Theorem 7.13. *For the simple random walk on the Cayley graph $(P_{n,q}, \Sigma)$ defined above, the second largest eigenvalue is*

$$\beta_1 = 1 - \frac{2}{2n+1}\left(1 - \cos\frac{2\pi}{q-1}\right).$$

The value given above is slightly different than that found in [134] because we have included the identity element x in Σ to have the easy lower bound $\beta_{\min} \geq -1 + 2/(2N+1)$ at our disposal. Note that the spectral gap $\lambda_1 = 1 - \beta_1$ is of order $1/(q^2 n)$ and that (7.1) shows that order $q^2 n^2 \log q$ steps suffices to be close to stationarity.

The group $P_{n,q}^1$ is generated by two elements, e.g., $x + x^2$ and $x + x^3$. It is an interesting open problem to study the random walks on $P_{n,q}^1$ and $P_{n,q}$ associated with such small sets of generators.

8 Representation Theory for Finite Groups

Representation theory was first developed as a diagonalization tool. As such, it applies to all convolution operators. On abelian groups, it provides a powerful technique to study random walks as witnessed for instance by the classical proof of the central limit theorem on $\mathbb{R}$. Early references discussing applications to random walks on finite groups are [70, 81] but the first serious application of the representation theory of a non-abelian group to a random walk seems to be in [50] which studies the random transposition walk on the symmetric group. See also [59]. Useful references are [27, 28, 98, 136].

8.1 The General Set-up

A (finite dimensional) *representation* of a group G is a group homomorphism ϱ from G to the group $GL(V)$ of all linear invertible maps of a (finite dimensional) vector space V over the complex numbers. The dimension of V will be denoted by d_ϱ and is called the dimension of the representation. Here, we will consider only finite groups and finite dimensional representations. There always exists on V a Hermitian structure $\langle \cdot, \cdot \rangle$ for which each $\varrho(s)$ is a unitary operator and we always assume that V is equipped with such a structure. The trivial representation of G is (ϱ, V) where $V = \mathbb{C}$ and $\varrho(s)(z) = z$ for all $s \in G$ and $z \in \mathbb{C}$.

The *left regular representation* $\varrho : s \mapsto \varrho(s)$ on $L^2(G)$ is defined by $\varrho(s)f(x) = f(s^{-1}x)$ for all $f \in L^2(G)$. A representation is *irreducible* if any linear subspace W which is invariant by ϱ, i.e., such that $\varrho(s)W \subset W$ for all $s \in G$ is trivial, i.e., is equal to either $\{0\}$ or V. Irreducible representations are the basic building blocks of Fourier analysis. For instance, if the group G is abelian, all the unitary operators $\varrho(s)$, $s \in G$, commute. Thus they can all be diagonalized in the same basis. It follows that any irreducible representation must be 1-dimensional. When the group is not abelian, irreducible representations are typically of dimension greater than 1. Two representations $(\varrho_1, V_1), (\varrho_2, V_2)$ of a group G are *equivalent* if there exists a unitary map $T : V_1 \to V_2$ such that $\varrho_2(s) \circ T = T \circ \varrho_1(s)$. Constructing and classifying irreducible representations up to equivalence is the basic goal of representation theory. We denote by $\widehat{G}$ the set of equivalence classes of irreducible representations of G. For instance, when G is a finite abelian group, one can show that $\widehat{G}$ admits a natural group structure and is isomorphic to G itself.

The famous Shur's lemma implies the following fundamental orthogonality relations. Let (ϱ, V) be an irreducible representation which is not equal to the trivial representation. Let $(e_i)_{1 \leq i \leq d_\varrho}$ be a Hermitian basis of V and set $\varrho_{i,j}(s) = \langle \varrho(s)e_i, e_j \rangle$. The functions $\varrho_{i,j}$ are called the matrix coefficients of ϱ. For any (i, j) and (k, ℓ) in $\{1, \ldots, d_\varrho\}^2$, the functions $\varrho_{i,j}$ and $\varrho_{k,\ell}$ satisfy

$$\sum_{s \in G} \varrho_{i,j}(s)\overline{\varrho_{k,\ell}(s)} = \frac{|G|}{d_\varrho}\delta_{(i,j),(k,\ell)}.$$

Moreover, for any two inequivalent irreducible representations (ϱ^1, V_1), (ϱ^2, V_2), we have

$$\sum_{s \in G} \varrho^1_{i,j}(s)\overline{\varrho^2_{k,\ell}(s)} = 0$$

for any $1 \le i, j \le d_{\varrho^1}$ and $1 \le k, \ell \le d_{\varrho^2}$. Finally, analyzing the left regular representation, one shows that each irreducible representation ϱ occurs in the left regular representation exactly as many times as its dimension d_ϱ. It follows that

$$|G| = \sum_{\varrho \in \widehat{G}} d_\varrho^2$$

and that the normalized matrix coefficients $d_\varrho^{-1/2}\varrho_{i,j}$, $1 \le i, j \le d_\varrho$, $\varrho \in \widehat{G}$, form an orthonormal basis of $L^2(G)$.

Let p be a measure (a function) on G. Set, for any representation ϱ,

$$\widehat{p}(\varrho) = \sum_{s \in G} p(s)\varrho(s).$$

The linear operator $\widehat{p}(\varrho)$ is called the Fourier transform of p at ϱ. If p, q are two measures, then

$$\widehat{p * q}(\varrho) = \widehat{p}(\varrho)\widehat{q}(\varrho).$$

Hence the Fourier transform turns the convolution product $p * q$ into the product $\widehat{p}(\varrho)\widehat{q}(\varrho)$ of two unitary operators (i.e., the product of matrices once a basis has been chosen in V). In general, one mostly computes the Fourier transform at irreducible representations. For instance, for the uniform measure $u(s) = 1/|G|$, the orthogonality relations recalled above imply that

$$\hat{u}(\varrho) = \begin{cases} 1 & \text{if } \varrho = 1 \text{ is the trivial representation} \\ 0 & \text{otherwise.} \end{cases} \tag{8.1}$$

There are straightforward analogs of the Fourier inversion and Plancherel formula which read

$$p(s) = \frac{1}{|G|} \sum_{\varrho \in \widehat{G}} d_\varrho \ \mathrm{tr}[\hat{p}(\varrho)\varrho(s^{-1})],$$

$$\sum_{s \in G} p(s^{-1})q(s) = \frac{1}{|G|} \sum_{\varrho \in \widehat{G}} d_\varrho \ \mathrm{tr}[\hat{p}(\varrho)\hat{q}(\varrho)]$$

where $|G|$ is the cardinality of G. Since $\varrho(s^{-1}) = \varrho(s)^{-1} = \varrho(s)^\dagger$ where $\dagger$ stands for "conjugate-transpose", we have

$$\sum_{s \in G} |p(s)|^2 = \frac{1}{|G|} \sum_{\varrho \in \widehat{G}} d_\varrho \ \mathrm{tr}[\hat{p}(\varrho)\hat{p}(\varrho)^\dagger] \tag{8.2}$$

which is the most important formula for our purpose. Behind this formula is the decomposition of the left regular representation into irreducible components and the fact that each irreducible representation $\varrho \in \widehat{G}$ appears with multiplicity equal to its dimension d_ϱ.

The following lemma follows from (8.1) and (8.2).

Theorem 8.1. *Let p be a probability measure on the finite group G and u the uniform distribution on G. Then, for any integer k,*

$$|G| \sum_{s\in G} |p^{(k)}(s) - u(s)|^2 = \sum_{\varrho\in\widehat{G}^*} d_\varrho \; tr[\hat{p}(\varrho)^k (\hat{p}(\varrho)^k)^\dagger]$$

where $\widehat{G}^* = \widehat{G} \setminus \{\mathbf{1}\}$.

In principle, the meaning of this lemma for random walks on finite groups is clear. Using representation theory, one can compute (or estimate) the square of the L^2-distance

$$d_2(p^{(k)}, u) = |G| \sum_{s\in G} |p^{(k)}(s) - u(s)|^2$$

whenever one can compute (or estimate)

$$\sum_{\varrho\in\widehat{G}^*} d_\varrho \; tr[\hat{p}(\varrho)^k (\hat{p}(\varrho)^k)^\dagger].$$

This requires having formula for the dimensions d_ϱ of all irreducible representations and being able to compute the powers of the matrices $\hat{p}(\varrho)$. Once these preliminary tasks have been tackled, one still has to sum over all irreducible representations.

8.2 Abelian Examples

Let G be a finite abelian group and p a probability measure on G. Viewed as a convolution operator acting on $L^2(G)$, p has adjoint $\check{p}$. As G is abelian, the convolution product is commutative. It follows that p is normal, hence diagonalizable. As all the irreducible representations are one dimensional, each gives rise to exactly one matrix coefficient called the character χ of the representation. The characters form an orthonormal basis of $L^2(G)$ and they also form a group, the dual group $\widehat{G}$, isomorphic to G. The Fourier transform $\hat{p}$ at the character χ (i.e., at the representation with character χ) is given by

$$\hat{p}(\chi) = \sum_{s\in G} p(s)\chi(s).$$

The collection $(\hat{p}(\chi))_\chi$ indexed by the characters, is exactly the spectrum of p viewed as a convolution operator. In this case, the formula of Theorem 8.1 gives

$$d_2(p^{(k)}, u)^2 = |G| \sum_{s\in G} |p^{(k)} - u(s)|^2 = \sum_{\chi\in\widehat{G}^*} |p(\chi)|^{2k}. \qquad (8.3)$$

The simple random walk on $\mathbb{Z}_n$. Consider the group $\mathbb{Z}_n = \mathbb{Z}/n\mathbb{Z} = \{0, 1, \ldots, n-1\}$. In this case, the characters are the functions

$$\chi_\ell(x) = e^{-2i\pi\ell x/n}, \quad \ell = 0, \ldots, n-1.$$

Let $p(+1) = p(-1) = 1/2$. Then $\hat{p}(\chi_\ell) = \cos(2\pi\ell/n)$. Hence,

$$d_2(p^{(k)}, u) = \left(\sum_{\ell=1}^{n-1} |\cos(2\pi\ell/n)|^{2k} \right)^{1/2}.$$

If n is even, for $\ell = n/2$, we get $\cos\pi = -1$ as an eigenvalue. Indeed, the chain is periodic of period 2 in this case. As a typical careful application of eigenvalue techniques, we state the following result.

Theorem 8.2. *There exist two constants $0 < c_1 \le C_1 < \infty$ such that, for all odd integers $n = 2m + 1$ and all integers k, we have*

$$2|\cos(\pi/n)|^{2k} \left(1 + \frac{c_1 n}{\sqrt{k}} \right) \le d_2(p^{(k)}, u)^2 \le 2|\cos(\pi/n)|^{2k} \left(1 + \frac{C_1 n}{\sqrt{k}} \right).$$

Proof. Assume that $n = 2m + 1$ is odd. Using the symmetries of cos, we get

$$d_2(p^{(k)}, u)^2 = 2 \sum_{\ell=1}^{m} |\cos(\pi\ell/n)|^{2k}.$$

Calculus gives

$$\left. \log \frac{\cos t}{\cos s} \right\} \begin{array}{l} \le -\frac{1}{2}(t^2 - s^2) \quad \text{for } 0 < s < t < \pi/2 \\ \ge -\frac{2}{\pi}(t^2 - s^2) \quad \text{for } 0 < s < t < \pi/4. \end{array}$$

Hence

$$d_2(p^{(k)}, u)^2 \ge 2|\cos(\pi/n)|^{2k} \left(\sum_{\ell=1}^{m/2} e^{-4\pi(\ell^2-1)k/n^2} \right)$$

$$\ge 2|\cos(\pi/n)|^{2k} \left(1 + c_1 \sqrt{n^2/k} \right)$$

where $c_1 = e^{-8\pi}$. For an almost matching upper bound, write

$$\sum_{\ell=1}^{m} e^{-2\pi^2(\ell^2-1)k/n^2} \le 1 + \sum_{\ell=1}^{\infty} e^{-2\pi^2\ell^2 k/n^2} \le 1 + \int_0^\infty e^{-2\pi^2 t^2 k/n^2} \, dt$$

$$= 1 + C_1 \sqrt{n^2/k}.$$

with $C_1 = 1/\sqrt{8\pi}$. Hence $d_2(p^{(k)}, u)^2 \le 2|\cos(\pi/n)|^{2k} \left(1 + C_1 \sqrt{n^2/k} \right).$ $\quad\square$

Other random walks on $\mathbb{Z}_n$. Let $a, b \in \mathbb{Z}_n$ and let $p_{a,b}$ be the uniform probability measure on $\{a, b\}$, i.e., $p(a) = p(b) = 1/2$. Thus the measure p of the previous example is $p_{-1,1}$ in this notation. Let us look at $p_{0,1}$. The associated random walk is not reversible but it is ergodic for all n. Here the eigenvalues are $\frac{1}{2}(1 + e^{2i\pi\ell/n})$. As $|1 + e^{2i\pi\ell/n}|^2 = |\cos(\pi\ell/n)|^2$, we get

$$d_2(p_{0,1}^{(k)}, u)^2 = \sum_1^{n-1} |\cos(\pi\ell/n)|^2.$$

Now, if n is odd, one easily checks that

$$\sum_1^{n-1} |\cos(\pi\ell/n)|^2 = \sum_1^{n-1} |\cos(2\pi\ell/n)|^2.$$

This shows that, for all odd n and all k, $d_2(p_{-1,1}^{(k)}, u) = d_2(p_{0,1}^{(k)}, u)$. The following result generalizes this observation.

Theorem 8.3. *Let $a, b \in \mathbb{Z}_n$. Then the random walk driven by the uniform probability measure $p_{a,b}$ on $\{a, b\}$ is ergodic if and only if*

$$b - a \text{ and } n \text{ are relatively prime.} \tag{8.4}$$

For any $s \in [1, \infty]$, any a, b satisfying (8.4) and any integer k, we have

$$d_s(p_{a,b}^{(k)}, u) = d_s(p_{0,1}^{(k)}, u).$$

Moreover, there are constants c, C such that for any a, b satisfying (8.4) and any integer k, we have

$$2|\cos(\pi/n)|^{2k}\left(1 + \frac{c_1 n}{\sqrt{k}}\right) \leq d_2(p_{a,b}^{(k)}, u)^2 \leq 2|\cos(\pi/n)|^{2k}\left(1 + \frac{C_1 n}{\sqrt{k}}\right).$$

Proof. The first assertion follows for instance from Proposition 2.3. Given that (8.4) holds, there is an invertible affine transformation $\phi : x \mapsto uz + v$ such that $\phi(a) = 0$, $\phi(b) = 1$. Hence, as functions on $\mathbb{Z}_n$, $p_{a,b} = p_{0,1} \circ \phi$. Moreover, because ϕ is affine, for any two probabilities p, q, $[p \circ \phi] * [q \circ \phi](x) = p * q(\phi(x) + v)$. Hence, $p_{a,b}^{(k)}(x) = p_{0,1}^{(k)}(\phi(x) + (k-1)v)$. As $z \mapsto \phi(z) + (k-1)v$ is a bijection, we have $d_s(p_{a,b}^{(k)}, u) = d_s(p_{0,1}^{(k)}, u)$. The last assertion is obtained as in Theorem 8.2. $\qquad\square$

We now consider what happens when $p = p_\Sigma$ is uniform on a subset Σ of $\mathbb{Z}_n$ having $m > 2$ elements where m is fixed. Theorems 7.3, 7.8 and 7.9 apply in this case and show that if Σ is symmetric, and $0 \in \Sigma$ then $c(m)D^2 \leq T(Z_n, p_\Sigma) \leq C(m)D^2$ where D is the diameter of the associated Cayley graph (the condition that Σ be symmetric and contains 0 can be removed and replaced by the condition that $\Sigma\Sigma^{-1}$ generates). For instance, it is not hard to use this to show that, for any fixed m the walk driven by the uniform measure p_{Σ_m} on $\Sigma_m = \{0, \pm 1, \pm\lfloor n^{1/m}\rfloor, \ldots, \pm\lfloor n^{(m-1)/m}\rfloor\}$ satisfies $c(m)n^{2/m} \leq T(\mathbb{Z}_n, \Sigma_k) \leq C(m)n^{2/m}$ (the same is true for the non-symmetric version of Σ_m, i.e., $\Sigma'_m = \{\{0, 1, \lfloor n^{1/m}\rfloor, \ldots, \lfloor n^{(m-1)/m}\rfloor\}\}$).

The works [24, 72, 87] contain interesting complementary results derived through a careful use of representation theory in the spirit of this section.

Theorem 8.4 ([72], see also [87]). *Let p be any probability measure on $\mathbb{Z}_n$. Assume that the support of p is of size $m + 1 > 2$. There exist $c = c(m)$ and $N = N(m)$ such that, for $k < cn^{2/m}$ and for all $n > N$, we have $\|p^{(k)} - u\|_{\mathrm{TV}} \geq 1/4$.*

Call a subset $\{a_0, \ldots, a_m\} \subset \mathbb{Z}_m$ aperiodic if the greater common divisor of $a_1 - a_0, \ldots, a_m - a_0$ and n is 1. Let u_Σ denote the uniform probability on Σ.

Theorem 8.5 ([24]). *Fix $m \geq 2$. Let Σ be chosen uniformly at random from all aperiodic $m + 1$-subsets of $\mathbb{Z}_n$. Let $\psi(n)$ be any function increasing to infinity and assume that $k_n \geq \psi(n)n^{2/m}$. Then*

$$\mathbf{E}(\|u_\Sigma^{(k_n)} - u\|_{\mathrm{TV}}) \to 0 \ \text{as } n \to \infty$$

where the expectation is relative to the choice of the set Σ.

When n is prime this can be improved as follows.

Theorem 8.6 ([87]). *Fix $m \geq 2$ and assume that n is a prime. Let Σ be chosen uniformly at random from all $m + 1$-subsets of $\mathbb{Z}_n$. Given $\varepsilon > 0$, there exist $c = c(m, \varepsilon)$ and $N = N(m, \varepsilon)$ such that, for all $n > N$ and $k > cn^{2/m}$, we have $\mathbf{E}(\|u_\Sigma^{(k)} - u\|_{\mathrm{TV}}) < \varepsilon$.*

The simple random walk on the hypercube. Let $G = \mathbb{Z}_2^d$ be the hypercube and consider the simple random walk driven by the measure p at (5.14), i.e., the uniform measure on $\{e_0, e_1, \ldots, e_d\}$ where $e_0 = (0, \ldots, 0)$ and e_i, $1 \leq i \leq d$ are the natural basis vectors of $\mathbb{Z}_2^d$.

The characters of G, indexed by $\widehat{G} = G$ are given by $\chi_y(x) = (-1)^{x \cdot y}$ where $x.y = \sum_1^d x_i y_i$. Hence, p has eigenvalues $\hat{p}(\chi_y) = 1 - 2|y|/(d+1)$ where $|y| = \sum_1^d y_i$. Now (8.3) becomes

$$d_2(p^{(k)}, u)^2 = \sum_1^d \binom{d}{j}\left(1 - \frac{2j}{d+1}\right)^{2k}.$$

For $k = \frac{1}{4}(d+1)[\log d + c]$ with $c > 0$, this yields (see [27, p. 28])

$$2\|p^{(k)} - u\|_{\mathrm{TV}} \leq d_2(p^{(k)}, u)^2 \leq 2\left(e^{e^{-c}} - 1\right).$$

Together with the lower bound in total variation of Section 5.3, this proves that the simple random walk on the hypercube has a cut-off at time $t_d = \frac{1}{4}d\log d$. By a more direct method, Diaconis, Graham and Morrison prove the following complementary results.

Theorem 8.7 ([35]). *Referring to the above walk on the hypercube* $\mathbb{Z}_2^d$, *for any* $k = \frac{1}{4}(d+1)[\log d + c]$, $c \in \mathbb{R}$

$$\|p^{(k)} - u\|_{\mathrm{TV}} = 1 - 2\Phi\left(-\frac{e^{-2c}}{4}\right) + o(1)$$

where

$$\Phi(t) = \frac{1}{2\pi} \int_{-\infty}^{t} e^{-s^2/2} ds.$$

Note that the automorphism group of $\mathbb{Z}_2^d$ acts transitively on the set of all d-tuples that generate $\mathbb{Z}_2^d$ which means that all generating d-tuples are equivalent from our viewpoint.

Other walks on the hypercube. The papers [73, 140] consider what typically happens for walks on the hypercube driven by the uniform measure u_Σ on a generating set Σ with $n > d$ elements. In particular, [140] proves the following result. Set

$$H(x) = x \log_2 x^{-1} + (1-x) \log_2 (1-x)^{-1}.$$

This function is increasing from $H(0) = 0$ to $H(1/2) = 1$. Let H^{-1} be the inverse function from $[0,1]$ to $[0,1/2]$ and set

$$T(d,n) = \frac{n}{2} \log \frac{1}{1 - 2H^{-1}(d/n)}.$$

Theorem 8.8 ([140]). *Assume that the random walk driven by the uniform probability* u_Σ *on the set* Σ *of* n *elements in* $\mathbb{Z}_2^d$ *is ergodic. For any* $\varepsilon > 0$, *for all* d *large enough and* $n > d$, *we have:*

- *For any set* Σ, *if* $k \leq (1-\varepsilon)T(d,n)$ *then* $\|u_\Sigma^{(k)} - u\|_{\mathrm{TV}} > 1 - \varepsilon$.
- *For most sets* Σ, *if* $k \geq (1+\varepsilon)T(d,n)$ *then* $\|u_\Sigma^{(k)} - u\|_{\mathrm{TV}} < \varepsilon$.

Thus the lower bound holds for all choices of Σ whereas the upper bounds holds only with probability $1 - \varepsilon$ when the set Σ is chosen at random. Also, when n is significantly larger than d, the walk is ergodic for most choices of Σ. The function $T(d,n)$ has the following behavior (see [140]):

$$T(d,n) \sim \frac{d}{4} \log \frac{d}{n-d} \qquad \text{if } n-d = o(d)$$

$$T(d,n) \sim \frac{d}{\log_2(n/d)} \qquad \text{if } d/n = o(1).$$

When n is linear in d then $T(d,n)$ is also linear in d. For instance, $T(d,2d) \sim ad$ with $0.24 < a < 0.25$. This leads to the following open question.

Problem 8.9. Find an explicit set of $2d$ elements in $\mathbb{Z}_2^d$ whose associated walk reaches approximate stationarity after order d steps.

The arguments in [140] do not use characters or eigenvalues directly. In fact, Wilson observes in [140] that for n linear in d the walk driven by u_Σ typically reaches stationarity strictly faster in total variation than in the d_2 distance for which we have the equality (5.8).

Wilson's result for random subsets contrasts with what is known for explicit sets. Uyemura-Reyes [138] studies the walk on the hypercube driven by

$$p(x) = \begin{cases} 1/(2d) & \text{if } x = (0,\dots,) \text{ or } (1,\dots,1) \\ 1/d^2 & \text{if } x = \sum_{\ell=i}^{i+j} e_\ell, 1 \le i \le d, 1 \le j < d \\ 0 & \text{otherwise} \end{cases}$$

where, in the second line, $i + j$ is understood mod d. For reasons explained in [138], this is called the random spatula walk. It is proved in [138] that this walk has a cut-off at time $t_n = \frac{1}{8} d \log d$.

The simple random walk on $\mathbb{Z}_n^d$. In $\mathbb{Z}_n^d$, let $e = (0,\dots,0)$ and e_i have a single non-zero coordinate, the i-th, equal to 1. Let n be odd and p be the uniform measure on $\{\pm e_i : 0 \le i \le d\}$. It is noteworthy that obtaining good uniform bounds over the two parameters n and d for this walk is not entirely trivial. The eigenvalues are easy to write down. They are

$$\alpha_\ell = \frac{1}{d}\left(\sum_1^d \cos(2\pi\ell_i/n)\right)$$

with $\ell = (\ell_1,\dots,\ell_d) \in \{0,\dots,n-1\}^d$. But bounding $d_2(p^{(k)}, u)^2 = \sum_{\ell \neq 0} \alpha_\ell^{2k}$ is not an easy task. One way to solve this difficulty is to use the associated continuous-time measure H_t defined at (2.10) and Theorem 5.1. This technique works for problems having a product structure similar to the present example. See [42, Section 5]. The reason this is useful is because H_t turns out to be a product measure. Namely, if $x = (x_1,\dots,x_d)$,

$$H_t(x) = \prod_1^d H_{1,t/d}(x_i)$$

where $H_{1,t}$ corresponds to the random walk on $\mathbb{Z}_n$ driven by the measure $p_1(\pm 1) = 1/2$. It follows that (u_1 denotes the uniform measure on $\mathbb{Z}_n$)

$$d_2(H_t, u)^2 = \left(1 + d_2(H_{1,t/d}, u_1)^2\right)^d - 1.$$

It is not hard to obtain good upper and lower bounds for

$$d_2(H_{1,t}, u_1)^2 = \sum_{j=1}^{n-1} e^{-2t[1-\cos(2\pi j/n)]}.$$

Namely, setting $\lambda(n) = 1 - \cos(2\pi/n)$ we have

$$\left(1 + \frac{cn}{\sqrt{t}}\right) e^{-2t\lambda(n)} \leq d_2(H_{1,t}, u_1)^2 \leq \left(1 + \frac{Cn}{\sqrt{t}}\right) e^{-2t\lambda(n)}.$$

This analysis, the elementary inequalities

$$\forall\, x > 0,\ d \in \mathbb{N},\ \ dx(1 + x/2)^{d-1} \leq (1 + x)^d - 1 \leq dx(1 + x)^{d-1},$$

and Theorem 5.1 yield the following result.

Theorem 8.10. *There are constants $c, C \in (0, \infty)$ such that, for the simple random walk on $\mathbb{Z}_n^d$, we have*

$$F_{n,d}(c, t) \leq d_2(H_t, u)^2 \leq F_{n,d}(C, t)$$

with $\lambda(n) = 1 - \cos(2\pi/n)$ and

$$F_{n,d}(a, t) = d\left(1 + a\sqrt{\frac{dn^2}{t}}\right)\left(1 + \left(1 + a\sqrt{\frac{dn^2}{t}}\right) e^{-2t\lambda(n)/d}\right)^{d-1} e^{-2t\lambda(n)/d}.$$

Moreover, there exists a constant C_1 such that, if n is an odd integer, d is large enough, and

$$k > 1 + \frac{d \log d}{2\lambda(n)} + \frac{d\theta}{2\lambda(n)}$$

with $\theta > 0$, then

$$2\|p^{(k)} - u\|_{\mathrm{TV}} \leq d_2(p^{(k)}, u) \leq C_1 e^{-\theta}.$$

Finally, for any $\tau > 6/d$, we have $\|p^{(k)} - u\|_{\mathrm{TV}} \geq 1 - \tau$ if

$$k < \frac{\log(d\tau/6)}{-2 \log(1 - \lambda(n)/d)}.$$

Note that the discrete time upper bound uses the fact that when n is odd, the lowest eigenvalue is $\cos(\pi/n)$ whose absolute value is much smaller than $1 - \lambda(n)/d$ for d large enough ($d \geq 8$ suffices). Theorem 8.10 proves a cut-off at time $(d/2\lambda(n)) \log d$ as long as d tends to infinity (n can be fixed or can tend to infinity).

8.3 Random Random Walks

In the spirit of Theorem 8.8, consider a group G, an integer m, and pick uniformly at random an m-set $\Sigma = \{g_1, \ldots, g_m\}$. Consider the random walk on G driven by the uniform probability measure u_Σ. What is the "typical" behavior of such a walk? Let $\mathbf{E}$ denote the expectation relative to the random choice of Σ. What can be said about $\mathbf{E}\left(\|u_\Sigma^{(k)} - u\|_{\mathrm{TV}}\right)$? To obtain some meaningful answers, we consider this problem for families of groups (G_n) where the size of G_n grows to infinity with n as in the following open problem. Recall that a classical result [52] asserts that the probability that a random pair of elements of the alternating group A_n generates A_n tends to 1 as n tends to infinity.

Problem 8.11. What is the typical behavior of the random walk driven by u_Σ when Σ is a random pair (more generally a random m-set) in A_n and n tends to infinity?

This is a wide open question. However, interesting results have been obtained in the case where $m = m(G)$ is allowed to grow with the order $|G|$ of G and this growth is fast enough.

Large random sets. In his unpublished thesis [53], C. Dou proves the following result using Theorem 8.1 and some combinatorics.

Theorem 8.12. *Let G be a finite group of order $|G|$. Let Σ be an m-element set chosen uniformly at random from G. Then*

$$\mathbf{E}\left(\|u_\Sigma^{(k)} - u\|_{TV}\right) \le \frac{1}{2}\left(\frac{(2k)^{2k}|G|}{m^k}\right)^{1/2}.$$

To illustrate this result, fix an integer s and take $m \ge |G|^{1/s}$ and $k = s + 1$. Then the right-hand side is $\frac{1}{2}[2(s+1)]^{2(s+1)}|G|^{-1/s}$ which tends to 0 as $|G|$ tends to ∞. For instance, most random walks based on sets of size $\sqrt{|G|}$ reach approximate stationarity in 3 steps. As a second example, consider sets of fixed size $m \ge a(\log |G|)^{2s}$ with $a > 4$ and $s > 1$. Then, there exists $\delta > 0$ such that for $k = (\log |G|)^s$ we have

$$\mathbf{E}\left(\|u_\Sigma^{(k)} - u\|_{TV}\right) \le \exp(-\delta(\log |G|)^s).$$

In [54], the approach of [53] is developed further to obtain the following.

Theorem 8.13 ([54]). *Let $m = \lfloor(\log |G|)^s\rfloor$ for some fixed $s > 1$. Let $\varepsilon > 0$ be given. Let Σ be a m-element set chosen uniformly at random in a finite group G. Then for*

$$k > \frac{s}{s-1}\frac{\log |G|}{\log m}(1 + \varepsilon)$$

we have that $\mathbf{E}\left(\|u_\Sigma^{(k)} - u\|_{TV}\right)$ tends to 0 as $|G|$ tends to infinity.

This result cannot be improved as shown by an earlier result of Hildebrand [87] concerning abelian finite groups. See [54] for a slightly more general result.

Theorem 8.14 ([87]). *Let $\varepsilon > 0$ be given. Let G be a finite abelian group. Let $m = \lfloor(\log |G|)^s\rfloor$ for some fixed $s > 1$. Let Σ be a m-element set chosen uniformly at random in a finite abelian group G. Then for*

$$k < \frac{s}{s-1}\frac{\log |G|}{\log m}(1 - \varepsilon)$$

we have that $\mathbf{E}\left(\|u_\Sigma^{(k)} - u\|_{TV}\right)$ tends to 1 as $|G|$ tends to infinity.

For further results in this direction, see [88, 89, 113, 120].

9 Central Measures and Bi-invariant Walks

9.1 Characters and Bi-invariance

When the group G is not abelian, e.g., $G = S_n$, the formula of Theorem 8.1 is often quite hard to use in practice, even when $p = \check{p}$ is symmetric. Indeed, $p(x^{-1}y)$ defines a $|G| \times |G|$ matrix whose eigenvalues we would like to find. What Theorem 8.1 does is to decompose this into $|\widehat{G}|$ smaller problems, one for each irreducible representation ϱ. The matrix $\hat{p}(\varrho)$ has size $d_\varrho \times d_\varrho$. This is very useful if d_ϱ is small. Unfortunately, irreducible representations of non-abelian finite groups tend to have large dimensions. For instance, for the symmetric group S_n, it is known that the typical dimension of a representation is $\sqrt{n!}$. Because of this, Theorem 8.1 is useful mostly in cases where p has further symmetries. The typical case is when p is a *central probability*, that is, it satisfies

$$\forall\, x, y \in G, \quad p(y^{-1}xy) = p(x). \tag{9.1}$$

Functions (probabilities) with this property are also called *class functions* since they are exactly the functions which are constant on conjugacy classes. Indeed, by definition, the conjugacy classes are exactly the classes of elements of G for the equivalence relation defined by $x \sim y$ iff $x = z^{-1}xz$ for some $z \in G$. When p is central, the associated Markov chain is not only left- but also right-invariant, that is, satisfies

$$\mathbb{P}_e(X_n = y) = \mathbb{P}_x(X_n = xy) = \mathbb{P}_x(X_n = yx)$$

for all $x, y \in G$. Such random walks are called *bi-invariant* random walks.

To each representation ϱ of G, one associates its *character*

$$\chi_\varrho(x) = \mathrm{tr}(\varrho(x)) = \sum_{1}^{d_\varrho} \varrho_{i,i}(x).$$

These functions are all central functions and $\chi_\varrho(s^{-1}) = \overline{\chi_\varrho(s)}$. Moreover $|\chi_\varrho(s)|$ is maximum at $s = e$ where $\chi_\varrho(e) = d_\varrho$. From the orthogonality relations it follows immediately that the characters of all irreducible representations form an orthonormal family in $L^2(G)$. Moreover, if p is any central measure (function) and ϱ is an irreducible representation, then

$$\hat{p}(\varrho) = \lambda_\varrho(p)I_{d_\varrho}, \quad \lambda_\varrho(p) = \frac{1}{d_\varrho} \sum_{s \in G} p(s)\chi_\varrho(s)$$

where I_{d_ϱ} is the $d_\varrho \times d_\varrho$ identity matrix. See, e.g., [27, 28, 59]. It follows that the *irreducible characters*, i.e., the characters associated with irreducible representations, form a basis of the subspace of all central functions in $L^2(G)$. Hence the number of irreducible representations up to equivalence, i.e., $|\widehat{G}|$, equals the number of conjugacy classes in G. This leads to the following general result. See, e.g., [27, 59].

Theorem 9.1. *Let* $C_1, \ldots, C_m$ *be conjugacy classes in* G *with representatives* $c_1, \ldots c_m$. *Assume that* p *is a central probability measure supported on* $\cup_1^m C_i$. *Then*

$$d_2(p^{(k)}, u)^2 = \sum_{\varrho \in \widehat{G}} d_\varrho^2 \left(\sum_1^m p(C_i) \frac{\chi_\varrho(c_i)}{\chi_\varrho(e)} \right)^{2k}. \tag{9.2}$$

Representation and character theory of finite groups is an important and well studied subject and there is sometimes enough information on characters available in the literature to make this theorem applicable. What is needed are manageable formulas or estimates for the dimensions d_ϱ of all irreducible representations and for the character ratios $\chi(c_i)/\chi(e)$.

Even when such data is available, estimating the sum on the left-hand side of (9.2) can still be quite a challenge. Indeed, this is a huge sum and it is often not clear at all how to identify the dominant terms.

9.2 Random Transposition on the Symmetric Group

Representation theory of the symmetric group. We will illustrate Theorem 9.1 by examples of bi-invariant walks on the symmetric group S_n. See [27] for a detailed treatment and [31] for a survey of further developments. The irreducible representations of the symmetric group are indexed by the set of all partitions λ of n where a partition $\lambda = (\lambda_1, \ldots, \lambda_r)$ has $\lambda_1 \geq \lambda_2 \geq \cdots \geq \lambda_r > 0$ and $\sum_1^r \lambda_i = n$. It is useful to picture the partition $\lambda = (\lambda_1, \ldots, \lambda_r)$ as a diagram made of r rows of square boxes, the i-th row having λ_i boxes. The rows are justified on the left. See [27, 59] for pointers to the literature concerning the representation theory on the symmetric group. For instance, for $n = 10$ the partition $\lambda = (5, 4, 1)$ is pictured in Figure 1.

Denote by d_λ the dimension of the irreducible representation ϱ_λ indexed by λ. Then d_λ equals the number of ways of placing the numbers $1, 2, \ldots, n$ into the diagram of λ such that the entries in each row and column are increasing. This is by no mean an easy number to compute or estimate.

The partition $\lambda = (n)$ corresponds to the trivial representation, (dimension 1). The partition $(1, 1, \ldots, 1)$ corresponds to the sign representation (dimension 1). The partition $(n-1, 1)$ corresponds to the representation $\varrho_{(n-1,1)}$ of S_n on $V = \{(z_1, \ldots, z_n) \in \mathbb{C}^n : \sum z_i = 0\}$ where $\varrho_{(n-1,1)}(\sigma)$ is represented

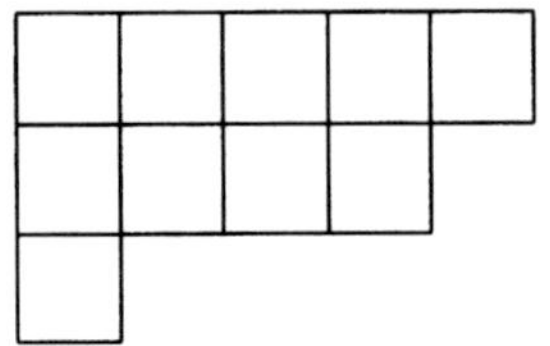

Fig. 1. $\lambda = (5, 4, 1)$

in the canonical basis of $\mathbb{C}^n$ by the matrix with coefficients $m_{i,j} = \delta_{i,\sigma(j)}$. This representation $\varrho_{(n-1,1)}$ has dimension $d_\lambda = n - 1$ (the only free choice is the number between 2 and n which goes in the unique box on the second row of the diagram).

The next necessary ingredient in applying Theorem 9.1 are formulas for character values. Such formulas were given by Frobenius but they become unwieldy for conjugacy classes with a complex cycle structure. Which character values are needed depend on exactly which random walk is considered. The simplest case concerns the walk called *random transposition*.

Random transposition. Consider n cards laid out on a table in a row. Let the right and left hands each pick a card uniformly and independently and switch the positions of the cards (if both hands pick the same card, the row of card stays unchanged). This description gives the random transposition measure p_{RT} on S_n defined at (4.1). Since $\{e\}$ and $T = \{\tau_{i,j} : 1 \le i < j \le n\}$ are conjugacy classes, Theorem 9.1 applies. Now, we need the character values $\chi_\lambda(e) = d_\lambda$ and $\chi_\lambda(t)$ where t is any fixed transposition. Frobenius' formula gives

$$\frac{\chi_\lambda(t)}{\chi_\lambda(e)} = \frac{1}{n(n-1)} \sum_j \left(\lambda_j^2 - (2j - 1)\lambda_j\right)$$

from which it follows that the eigenvalues of this walk are

$$p_{\mathrm{RT}}(e) + p_{\mathrm{RT}}(T)\frac{\chi_\lambda(t)}{\chi_\lambda(e)} = \frac{1}{n} + \frac{n-1}{n}\frac{\chi_\lambda(t)}{\chi_\lambda(e)}$$
$$= \frac{1}{n} + \frac{1}{n^2} \sum_j \left(\lambda_j^2 - (2j - 1)\lambda_j\right)$$

with multiplicity d_λ^2. With some work, one shows that the second largest eigenvalue is $1 - 2/n$ with multiplicity $(n-1)^2$, attained for $\lambda = (n-1, 1)$. The lowest eigenvalue is $-1 + 2/n$ with multiplicity 1, attained for $\lambda = (1, 1, \ldots, 1)$.

Using the above data and estimates on d_λ, Diaconis and Shahshahani obtained in 1981 the following theorem which gives first precise result about the convergence of a complex finite Markov chain.

Theorem 9.2 ([50]). *For the random transposition walk on the symmetric group S_n, there exists a constant A such that, for all n and $c > 0$ for which $k = \frac{1}{2}n(\log n + c)$ is an integer, we have*

$$2\|p_{\mathrm{RT}}^{(k)} - u\|_{\mathrm{TV}} \le d_2(p_{\mathrm{RT}}^{(k)}, u) \le Ae^{-c}.$$

Moreover, there exist a function f with limit 0 at ∞ such that for all $n > 5$ and all $c > 0$ for which $k = \frac{1}{2}n(\log n - c)$ is an integer,

$$\|p_{\mathrm{RT}}^{(k)} - u\|_{\mathrm{TV}} \ge 1 - 12 \left(e^{-c} + n^{-1}\log n\right).$$

This theorem proves that (S_n, p_{RT}) has a total variation cut-off and a L^2-cut-off, both a time $\frac{1}{2}n \log n$. Let us comment further on the lower bound. It can be proved ([27, p. 44]) by using Propositions 5.6, 5.7, the fact that

$$\chi^2_{(n-1,1)} = \chi_{(n)} + \chi_{(n-1,1)} + \chi_{(n-2,2)} + \chi_{(n-2,1,1)},$$

and the values of the corresponding eigenvalues and dimensions. This formula giving $\chi^2_{(n-1,1)}$ is a classical result in representation theory. It corresponds to the decomposition into irreducible components of the tensor product $\varrho_{(n-1,1)} \otimes \varrho_{(n-1,1)}$. Another proof using classical probability estimates can be obtained by adapting the argument of [27, p. 43].

9.3 Walks Based on Conjugacy Classes of the Symmetric Group

A conjecture. In principle, it is possible to use character bounds to study any random walk on the symmetric group whose driving measure is central. However, the computational difficulty increases rapidly with the complexity of the conjugacy classes involved. To state some results and conjectures, recall that any conjugacy class C on S_n can be described by the common disjoint cycle structure of its elements. Thus $C = (2)$ means C is the class of all transpositions, $C = (5, 3, 3, 2, 2, 2, 2)$ means C is the class of all permutations that can be written as a product of one 5-cycle, two 3-cycles and four 2-cycles where the supports of those cycles are pairwise disjoint. It is known (and not hard to prove) that any odd conjugacy class (i.e., whose elements have sign -1) generates the symmetric group. However the walk associated to the uniform measure on an odd conjugacy class is always periodic of period 2. To cure this parity problem consider, for any odd conjugacy class C on S_n the probability measure p_C defined by

$$p_C(\theta) = \begin{cases} 1/2 & \text{if } \theta = e \\ 1/[2\#C] & \text{if } \theta \in C \\ 0 & \text{otherwise.} \end{cases}$$

This is sometimes referred to as a *lazy random walk* because, on average, it moves only every other steps, see, e.g., [88, 89]. Thus, the walk driven by $p_{(2)}$ is similar to the random transposition walk except that it stay put with probability $1/2$ instead of $2/n$. One can show that Theorem 9.2 applies to the walk generated by $p_{(2)}$ if $k = \frac{1}{2}n(\log n \pm c)$ is changed to $k = n(\log n \pm c)$.

For $C = (c_1, c_2, \ldots, c_\ell)$, set $|C| = \sum_1^\ell c_i$. Note that $|C|$ is the size of the support of any permutation in C, i.e., n minus the number of fixed points. With this notation one can make the following conjecture.

Conjecture 9.3. There exists a constant A such that, for all n, all odd conjugacy classes C with $|C| \ll n$, and all $c > 0$ for which $k = (2n/|C|)(\log n + c)$ is an integer, we have

$$2\|p_C^{(k)} - u\|_{\mathrm{TV}} \le d_2(p_C^{(k)}, u) \le Ae^{-c}.$$

Moreover, there exist two functions f_C^1, f_C^2 with limit 0 at ∞ such that for all n and all $c > 0$ for which $k = (2n/|C|)(\log n - c)$ is an integer,

$$\|p_C^{(k)} - u\|_{\mathrm{TV}} \ge 1 - f_C^1(c) - f_C^2(n).$$

Any even conjugacy class C of S_n generates the alternating group A_n (except for $n = 4$) and one can consider the random walk on A_n driven by the uniform measure on C. Denote by $\widetilde{p}_C$ the uniform measure on the conjugacy class C viewed as a subset of A_n. For $\widetilde{p}_C$ it is conjectured that the statement of Conjecture 9.3 holds with $k = (n/|C|)(\log n + c)$ instead of $k = (2n/|C|)(\log n + c)$.

Conjecture 9.3 can be interpreted in various ways depending of what is meant by $|C| \ll n$. It is open even for fixed $|C|$ such as $|C| = 20$ and n tending to infinity. The strongest reasonable interpretation is $|C| \le (1 - \varepsilon)n$, for some fixed $\varepsilon > 0$. What is known at this writing is described in the next section.

Small conjugacy classes. For $|C| \le 6$ and n tending to infinity, Conjecture 9.3 (and its even conjugacy class version on A_n) is proved in [121, 122]. Moreover, [121, 122] shows that the lower bound holds true for all C such that $|C| < n/(1 + \log n)$ (some of the computations in the proof given in [121, 122] are incorrect but these errors can easily be fixed).

To give an idea of the difficulties that arise in adapting the method used for random transposition, we give below some explicit character values. The source is [93] and [121, 122]. For any partition $\lambda = (\lambda_1, \ldots, \lambda_r)$ and $\ell = 1, 2, \ldots,$ set

$$M_{2\ell,\lambda} = \sum_{j=1}^{r} \left[(\lambda_j - j)^\ell (\lambda_j - j + 1)^\ell - j^\ell (j - 1)^\ell \right]$$

$$M_{2\ell+1,\lambda} = \sum_{j=1}^{r} \left[(\lambda_j - j)^\ell (\lambda_j - j + 1)^\ell (2\lambda_j - 2j + 1) + j^\ell (j - 1)^\ell (2j - 1) \right].$$

For a conjugacy class C, set $r_\lambda(C) = \chi_\lambda(c)/\chi_\lambda(e)$ where c is any element of C. These character ratios are the building blocks needed to apply formula (9.2). For the conjugacy classes $(4), (2, 2)$ and (6), one has:

$$r_\lambda((4)) = \frac{(n - 4)!}{n!} \left(M_{4,\lambda} - 2(2n - 3)M_{2,\lambda} \right)$$

$$r_\lambda((2, 2)) = \frac{(n - 4)!}{n!} \left(M_{2,\lambda}^2 - 2M_{3,\lambda} + 4n(n - 1) \right)$$

$$r_\lambda((6)) = \frac{(n - 6)!}{n!} \left(M_{6,\lambda} - (6n - 37)M_{4,\lambda} \right.$$

$$\left. - 3M_{2,\lambda}M_{3,\lambda} + 6(3n^2 - 19n + 20)M_{2,\lambda} \right).$$

A weak form of the conjectures stated in the previous section is proved by Roichman in [119] where interesting uniform bounds for the character ratios $r_\lambda(C)$ are also derived.

Theorem 9.4 ([119]). *Fix $\eta, \varepsilon \in (0,1)$. Then there are constants $a, A, N \in (0, \infty)$ such that for any $n \geq N$, any odd conjugacy class C with $|C| \leq (1-\eta)n$, we have*

$$2\|p^{(k)} - u\|_{\mathrm{TV}} \leq d_2(p_C^{(k)}, u) \leq \varepsilon \quad \text{for all } k \geq \frac{An}{|C|} \log n$$

whereas

$$\|p_C^{(k)} - u\|_{\mathrm{TV}} \geq \varepsilon \quad \text{for all } k \leq \frac{an}{|C|} \log n.$$

The same result holds on A_n for even conjugacy classes.

This theorem of Roichman proves the existence of a precut-off at time $(n/|C|) \log n$ for (S_n, p_C) when $|C| \leq (1 - \eta)n$.

Large conjugacy classes. In his thesis [102], Lulov considers the walks driven by the uniform measure on the conjugacy classes $C_r = (n/r, \ldots, n/r)$, where r divides n. These are huge conjugacy classes. Consider the case where C_r is even and the walk is restricted to A_n. Obviously, $\widetilde{p}_{C_r}$ is not close to the uniform distribution on A_n. However, Lulov uses character ratios estimates to show that $\widetilde{p}_{C_r}^{(k)}$ is close to uniform on A_n for $k = 3$ if $r = 2$ and for $k = 2$ if $r \geq 3$. In [103] the authors conjecture that, for conjugacy classes with no fixed points, it always takes either 2 or 3 steps to reach approximate stationarity. They also prove the following Theorem by deriving sufficiently good character ratio estimates.

Theorem 9.5 ([103]). *Let C_n be an even conjugacy class in S_n with a single cycle, i.e., $C_n = (r_n)$ and assume that $|C_n| = r_n > n/2$ and $n - r_n$ tends to infinity. Then the sequence $(A_n, \widetilde{p}_{C_n})$ presents a cut-off at time*

$$t_n = \frac{\log n}{\log[n/(n - r_n)]}.$$

For the lower bound, [103] refers to [119]. The lower bound in [119] is based on Propositions 5.6 and 5.7. The proof in [119] needs to be adapted properly in order to prove the lower bound stated in Theorem 9.5.

The authors of [103] conjecture that the conclusion of Theorem 9.5 is valid for all sequences C_n of even conjugacy classes whose number of fixed points $n - |C_n|$ is $o(n)$ and tends to infinity.

Other walks related to random transposition. Imagine a deck of cards where each card, in addition to its face value, has an orientation (or spin), say up or down (think of the faces of the cards being up or down in the deck, or of the back of each card being marked by an arrow that can be up or down). A natural generalization of random transposition is as follows. Pick a pair of positions uniformly at random in the deck. Transpose the cards in these positions and, at the same time, uniformly pick an orientation for these cards. This is a random walk on the wreath product $\mathbb{Z}_2 \wr S_n = (\mathbb{Z}_2)^n \rtimes S_n$ where the action of S_n is by permutation of the coordinates in $\mathbb{Z}_2^n$. The above description generalizes straightforwardly to the case where $\mathbb{Z}_2$ is replace by an arbitrary finite group H. For instance, taking $H = S_m$, we can think of the corresponding walk as mixing up n decks of m cards. Here cards of different decks are never mixed together. What is mixed up is the relative order of the decks and the cards in each individual deck. Schoolfield [128, 129] studies such walks and some variants using character theory. He finds that $ae^{-c} \le d_2(p^{(k)}, u) \le Ae^{-c}$ if $k = \frac{1}{2} n \log(n\sqrt{|G|}) + c$, $c > 0$. Using a stopping time argument as in Theorem 4.6, he also proves a cut-off in total variation at tine $t_n = \frac{1}{2} \log n$. Hence, if G depends on n and $|G|$ grows fast enough with n then stationarity is reached at different times in total variation and in L^2. See also [58].

9.4 Finite Classical Groups

Together with the symmetric and alternating groups, one of the most natural families of finite groups is formed by the classical groups over finite fields. These are groups of matrices resembling the classical real compact Lie groups. Representation and character theory of these groups are an important domain of research from several viewpoints but what is known is much less complete than for the symmetric groups. Many of these groups contains some relatively small conjugacy classes (or union of conjugacy classes), resembling the class of all transpostions in S_n, which generates the whole group. This leads to interesting random walks that can, in principle, be studied by using Theorem 9.1, i.e., character theory. We describe below some of the known results in this direction.

Random transvection in $SL_n(\mathbb{F}_q)$. $SL_n(\mathbb{F}_q)$ is the group of $n \times n$ matrices with determinant 1 over the finite field $\mathbb{F}_q$ with q elements (hence $q = p^n$ for some prime p). By definition, a *transvection* is an element in $SL_n(\mathbb{F}_q)$ which is not the identity and fixes all the points of a hyperplane in $\mathbb{F}_q^n$, the n dimensional vector space over $\mathbb{F}_q$. The transvections generate $SL_n(\mathbb{F}_q)$ and form a conjugacy class when $n > 2$. Good examples of transvections are the elementary matrices $I + aE_{i,j}$, $a \in \mathbb{F}_q \setminus \{0\}$, $i \ne j$, where I is the $n \times n$ identity matrix, and the matrix $E_{i,j}$ has a unique non-zero entry equal to 1 in the (i,j)-th position. A general transvection has the form $I + uv^t$ where u, v are two arbitrary non-zero vectors in $\mathbb{F}_q^n$ with $u^t v = 0$ (an element u of $\mathbb{F}_q^n$ is a column vector and u^t is its transpose). Moreover, $uv^t = u_0 v_0^t$ if and only if $u = au_0$, $v = a^{-1}v_0$ for some $a \in \mathbb{F}_q \setminus \{0\}$. Thus picking u, v independently and uniformly in $\mathbb{F}_q^n \setminus \{0\}$ gives a uniformly distributed transvection $I + u^t v$. We denote by p the uniform measure on the set of all transvections and call the corresponding random walk the random transvection walk. This walk is studied by Hildebrand in [86] who proves the following remarkable result.

Theorem 9.6 ([86]). *For the random transvection measure p on $SL_n(\mathbb{F}_q)$ defined above, there are two positive constants A, N such that, for all $q \geq 2$, $n \geq N$ and $k = n + m$ with $m = 1, 2, \ldots$, we have*

$$d_2(p^{(m)}, u) \leq A\, q^{-m}.$$

Moreover, for all q and all integers n, m with $k = n - m > 0$ and $m \geq 3$, we have
$$\|p^{(k)} - u\|_{\mathrm{TV}} \geq 1 - 4q^{1-m}.$$

The upper bound uses (9.2) and a formula for character ratios that Hildebrand obtains from results in McDonald's book [109]. The task is significantly harder than for random transposition on S_n. The lower bound follows from a relatively simple argument concerning the dimension of the space of fixed vectors by a product of m transvections. Hildebrand's results demonstrate that the random transvection walk presents a very sharp cut-off: for random transvection on $SL_n(\mathbb{F}_q)$, it takes at least $n - 6$ steps to reduce the total variation distance from 1 to 0.9. After that, a fixed number of steps suffices to drop the variation distance to, say 0.1.

Small conjugacy classes on finite classical groups. In a remarkable work [67, 68, 69], David Gluck studies in a unified and uniform way a large class of random walks on the finite classical groups. The results that Gluck obtains are somewhat less precise than Hildebrand's Theorem 9.6 but they have the same flavor: for any random walk whose driving measure is central, that is, constant on conjugacy classes and supported on *small* conjugacy classes, convergence to the uniform distribution occurs after order k steps where k is the rank of the underlying finite classical group. For instance, $SL_n(\mathbb{F}_q)$ has rank $n - 1$ and it follows from Gluck's results that the random transvection walk studied by Hildebrand reaches approximate stationarity after order n steps.

Technically, the results obtained by Gluck are by no means simple generalizations of the previous results of Diaconis–Shahshahani and Hildebrand. The exact character formulas used by both Diaconis–Shahshahani and Hildebrand do not seem to be available for the problems treated by Gluck. Even if they were, it would be an immense task to obtain Gluck's results through a case by case analysis. A massive amount of (very advanced) algebra is at work behind Gluck's approach. To avoid technicalities, we present below two specific examples that falls into Gluck's theory: random symplectic transvection and random unitary transvection. A friendly reference for basic facts and notation concerning these examples is [76]. Let $\mathbb{F}_q$ be a finite field with q elements and consider the vector space $\mathbb{F}_q^n$. For simplicity, we assume that $n, q \geq 4$ and q odd.

Assume that $n = 2m$ and fix a non-degenerate alternating form B (the choice of the form is irrelevant). A *symplectic transformation* is any invertible linear transformations of F_q^n that preserve B and $Sp_n(\mathbb{F}_q) \subset SL_n(\mathbb{F}_q)$ is the group of all symplectic transformations. The group $Sp_n(\mathbb{F}_q)$ satisfies $Sp_n(\mathbb{F}_q)' = Sp_n(\mathbb{F}_q)$. It has order

$$|Sp_n(\mathbb{F}_q)| = q^{m^2} \prod_{i=1}^{m} (q^{2i} - 1), \quad n = 2m.$$

To define $SU_n(\mathbb{F}_q)$, assume that $\mathbb{F}_q$ admits an automorphism α such that $\alpha^2 = 1$ (this implies that $q = q_0^2$ for some prime power q_0). Fix a Hermitian form B (relative to α)). Again, because we work on finite fields, the precise choice of B is irrelevant. The special unitary group $SU_n(\mathbb{F}_q^n)$ is the group of all invertible linear transformations with determinant 1 which preserve the Hermitian form B. The group $SU_n(\mathbb{F}_q)$ satisfies $SU_n(\mathbb{F}_q)' = SU_n(\mathbb{F}_q)$. It has order

$$|SU_n(\mathbb{F}_q)| = q^{n(n-1)} \prod_{j=1}^{n} (q^{j/2} - (-1)^j).$$

A *symplectic transvection* (resp. unitary transvection) is a transvection that preserve the Hermitian (resp. unitary) form B. Symplectic (resp. unitary) transvections are exactly the linear transformations of the form

$$\tau_{u,a} : v \mapsto v + aB(v, u)u$$

where $u \in \mathbb{F}_q^n \setminus \{0\}$ is a non-zero vector and $a \in \mathbb{F}^*$ is a non-zero scalar (resp. $u \in \mathbb{F}_q^n \setminus \{0\}$, $B(u, u) = 0$, and $a \in \mathbb{F}^*$, $a = -\alpha(a)$). Both the symplectic groups and the special unitary groups are generated by transvections.

Note that $\tau_{u,a} = \tau_{u_0,a_0}$ if and only if there exists $b \in \mathbb{F}^*$ such that $u = bu_0$, $a = b^{-1}a_0$. Thus we can pick a symplectic (resp. unitary) transformation uniformly at random by picking uniformly at random $u \in \mathbb{F}_q \setminus \{0\}$ and $a \in \mathbb{F}^*$ (resp. $u \in \mathbb{F}_q \setminus \{0\}$ satisfying $B(u, u) = 0$ and $a \in \mathbb{F}^*$ satisfying $a = -\alpha(a)$).

For any symplectic (resp. unitary) transformation σ, and any symplectic (resp. unitary) transvection $\tau_{u,a}$, we have $\sigma\tau_{u,a}\sigma^{-1} = \tau_{\sigma(u),a}$. This shows that the set T of all symplectic (resp. unitary) transvections is a union of conjugacy classes (it is not, in general, a single conjugacy class). Gluck's results in [68, Th. 42 and Cor. 64] specialize to the present examples as follows.

Theorem 9.7 ([68]). *Let p denote the uniform measure on symplectic or unitary transvections in $Sp_n(\mathbb{F}_q)$ or in $SU_n(\mathbb{F}_q)$, respectively. Assume that q is odd and n is large enough. Then there exists N such that for $k = N(n+c)$ with $c > 0$, we have*

$$d_2(p^{(k)}, u) \leq q^{-n/4-2c}.$$

One of the typical character ratio estimates obtained by Gluck [67] says that there exist $a \in (0,1)$ and $M > 0$ such that for every finite simple group of Lie type G_q over the finite field with q elements, for every non-central element $g \in G_q$, and for every irreducible character χ of $G(q)$,

$$|\chi(g)/\chi(e)| \leq \min\{a, Mq^{-1/2}\}.$$

This is not enough to prove Theorem 9.7 for which the refinements obtained in [68] are needed but, as noted in [99], it gives the following result.

Theorem 9.8. *Let G_{q_n} be a family of finite groups of Lie type of order growing to infinity. Let C_n be a non-central conjugacy class in G_{q_n} and $\Sigma_n = C_n \cup C_n^{-1}$. Then the Cayley graphs (G_{q_n}, Σ_n) form a family of expanders.*

9.5 Fourier Analysis for Non-central Measures

The extent to which Fourier analysis fails to provide useful results for random walks that are not bi-invariant (i.e., driven by non-central measures) is somewhat surprising. Still, there are cases in which the analysis of Sections 9.1 and 9.2 can be extended but few have been worked out in detail. A typical example is the *transpose top and random* shuffle. On S_n, consider the measure

$$p_*(\tau) \begin{cases} 1/n & \text{if } \tau = (1,i), \quad i = 1, \ldots, n \\ 0 & \text{otherwise,} \end{cases} \tag{9.3}$$

where $(1,1)$ is the identity and $(1,i)$, $i \neq 1$, is transpose 1 and i. This measure is not central (see (9.1)) but it is invariant by $\tau \mapsto \theta\tau\theta^{-1}$, $\theta \in S_{n-1}$ where S_{n-1} is understood as the subgroup of S_n of those permutations that fix 1. Because of this property, for any irreducible representation ϱ of S_n, the matrix $\hat{p}_*(\varrho)$ has a relatively small number of distinct eigenvalues and manageable formulas for the eigenvalues and their multiplicity can be obtained. See [27, 28, 59]. Using this spectral information and (5.8) gives the upper bound in the following theorem. The lower bound can be obtained by adapting the argument used for random transposition in [27, p.43].

Theorem 9.9. *For transpose top and random, i.e., the walk on S_n driven by p_*, there exists a constant A such that, for all n and $c > 0$ for which $k = n(\log n + c)$ is an integer, we have*

$$2\|p_*^{(k)} - u\|_{\mathrm{TV}} \leq d_2(p_*^{(k)}, u) \leq Ae^{-c}.$$

Moreover, there are two functions f_1, f_2 with limit 0 at ∞ such that for all n and all $c > 0$ for which $k = n(\log n - c)$ is an integer,

$$\|p_*^{(k)} - u\|_{\mathrm{TV}} \geq 1 - f_1(c) - f_2(n).$$

10 Comparison Techniques

The path technique used in Section 6 to bound the spectral gap generalizes in a very useful way to yield comparison inequalities between the Dirichlet form of different random walks. Such inequalities are important because they lead to a full comparison of the higher part of the spectrum of the two walks as sated in the next result.

10.1 The min-max Characterization of Eigenvalues

Dirichlet form comparison leads to spectrum comparison by a simple application of the Courant–Fisher min-max characterization of the ordered eigenvalues $q_0 \leq q_1 \leq \ldots$ of a self-adjoint linear operator Q on a Hilbert space $(V, \langle \cdot, \cdot \rangle)$ (here, finite dimensional and real). See, e.g., [90, 4.2.11].

Theorem 10.1 ([42]). *Let $p, \widetilde{p}$ be two symmetric probability measures on a finite group G with respective Dirichlet forms $\mathcal{E}, \widetilde{\mathcal{E}}$ and respective eigenvalues, in non-increasing order $\beta_i, \widetilde{\beta}_i$. Assume that there is a constant A such that $\widetilde{\mathcal{E}} \leq A\mathcal{E}$. Then, for all $i = 0, 1, \ldots, |G| - 1$, $\beta_i \leq 1 - A^{-1}\left(1 - \widetilde{\beta}_i\right)$. In particular, for the continuous-time random walks associated to p and $\widetilde{p}$ as in (2.10), we have*

$$d_2(H_t, u) \leq d_2(\widetilde{H}_{t/A}, u). \tag{10.1}$$

The inequality $\widetilde{\mathcal{E}} \leq A\mathcal{E}$ does not provide good control on the small positive eigenvalues and the negative eigenvalues of p. Thus there is no clean statement in discrete time analogous to (10.1). However, there are various ways to cope with this difficulty. Often, negative and small positive eigenvalues do not play a crucial role in bounding $d_2(p^{(k)}, u)$. In particular, (10.1) and Theorem 5.1 give the following useful result.

Theorem 10.2 ([42]). *Referring to the notation of Theorem 10.1, assume that there is a constant $A > 0$ such that $\widetilde{\mathcal{E}} \leq A\mathcal{E}$. Then*

$$d_2(p^{(k)}, u)^2 \leq \beta_-^{2k_1}(1 + d_2(\widetilde{H}_{k_2/A}, u)^2) + d_2(\widetilde{H}_{k/A}, u)^2$$

and

$$d_2(p^{(k)}, u)^2 \leq \beta_-^{2k_1}(1 + |G|e^{-k_2/2A} + d_2(\widetilde{p}^{(\lfloor k_2/2A \rfloor)}, u)^2)$$
$$+ |G|e^{-k/2A} + d_2(\widetilde{p}^{(\lfloor k/2A \rfloor)}, u)^2$$

where $k = k_1 + k_2 + 1$ and $\beta_- = \max\{0, -\beta_{|G|-1}\}$.

For best results, one should use the first inequality stated in this theorem since an extra factor of 2 is lost in bounding $d_2(\widetilde{H}_t, u)$ in terms of $d_2(\widetilde{p}^{(k)}, u)$. To use Theorems 10.1, 10.2, one needs a measure $\widetilde{p}$ that can be analyzed in terms of the L^2-distance d_2. A general scheme that has proved very successful is to start with a central measure $\widetilde{p}$ for which representation theory can be used as in Theorem 9.1. Then Theorems 10.1, 10.2 can be used to obtain results for other walks.

10.2 Comparing Dirichlet Forms Using Paths

We now present some comparison inequalities between Dirichlet forms taken mostly from [42, 49]. The proofs are similar to the proof of Theorem 6.4 given in Section 6.2. Fix two probability measures p and $\widetilde{p}$ on G. Think of p as driving the unknown walk we wish to study whereas we already have some information on the walk driven by $\widetilde{p}$. Fix a symmetric generating set Σ contained in the support of p. We will use the notation introduced in Section 6. Given a subset T of G, pick a path γ_x from e to x in the Cayley graph (G, Σ) and set $\mathcal{P}_*(T) = \{\gamma_x : x \in T\}$.

Theorem 10.3 ([42, 45, 49]). *Let T denote the support of $\widetilde{p}$. Referring to the setting and notation introduced above, we have $\widetilde{\mathcal{E}} \leq A_*\mathcal{E}$ where*

$$A_* = \max_{s \in \Sigma} \left\{ \frac{1}{p(s)} \sum_{\gamma \in \mathcal{P}_*(T)} |\gamma| N(s, \gamma)\widetilde{p}(\gamma) \right\}$$

with $\widetilde{p}(\gamma) = \widetilde{p}(x)$ if $\gamma = \gamma_x \in \mathcal{P}_*(T)$.

The following result concerns the walks based on fixed subsets of transpositions and is obtained by comparison with random transposition [42]. Let $\mathbf{G} = (V, E)$ be a graph with vertex set $V = \{1, \ldots, n\}$ and symmetric edge set $E \subset V \times V$ containing no loops $((i, i) \notin E$ and $(i, j) \in E$ if and only if $(j, i) \in E)$. Consider the walk on the symmetric group driven by the measure

$$p_{\mathbf{G}}(\tau) = \begin{cases} 1/n & \text{if } \tau = e \\ 2(n-1)/|E|n & \text{if } \tau = (i, j) \text{ with } (i, j) \in E \\ 0 & \text{otherwise.} \end{cases}$$

Thus this walk is based on those transpositions which corresponds to neighbors in $\mathbf{G}$. It is irreducible if and only if the graph is connected. If $\mathbf{G}$ is the complete graph then $p_{\mathbf{G}} = p_{\mathrm{RT}}$ is the random transposition measure defined at (4.1). If $\mathbf{G}$ is the line graph $1 - 2 - \cdots - n$ then $p_{\mathbf{G}} = p_{\mathrm{AT}}$ is the adjacent transposition measure. If $\mathbf{G}$ is the star graph with center 1 then $p_{\mathbf{G}} = p_*$ is the transpose top and random measure defined at (9.3). These walks were introduced in [42]. They are also considered in [80]. To state a general result, for each $x, y \in V$, pick paths $\mu_{x,y}$ from x to y in $\mathbf{G}$ of length (i.e number of edges) $|\mu_{x,y}|$ and set

$$\Delta = \max_{e \in E} \sum_{\substack{(x, y) \in V \times V \\ e \in \mu_{x,y}}} |\mu_{x,y}|.$$

The quantity Δ depends on both the length of the paths and the number of bottlenecks in the family $\{\mu_{x,y} : x, y \in V\}$ (see, e.g., [51, 57, 42, 43]).

Theorem 10.4 ([42]). *Referring to the notation introduced above, there exists a constant A such that for $k > (4(n-1)^{-1}|E|\Delta + n)(\log n + c), c > 0$, we have*

$$2\|p_{\mathbf{G}}^{(k)} - u\|_{\mathrm{TV}} \leq d_2(p_{\mathbf{G}}^{(k)}, u) \leq Ae^{-c}.$$

For the star graph and the line graph this theorem gives upper bounds on $T(S_n, p_\star)$, $T(S_n, p_{\mathrm{AT}})$ that are of order $n \log n$ and $n^3 \log n$ respectively. Both capture the right order of magnitude. If $\mathbf{G}$ is a two dimensional finite square grid with side size $\sqrt{n}$, the theorem gives $T(S_n, p_{\mathbf{G}}) \le Cn^2 \log n$. A matching lower bound is proved in [141]. The bound of Theorem 10.4 is probably not sharp in general. For instance, assume $n = 2^d$ and let $\mathbf{G}$ be the hypercube. In this case, Theorem 10.4 gives $T(S_n, p_{\mathbf{G}}) \le Cn(\log n)^3$. Wilson [141] proves $T(S_n, p_{\mathbf{G}}) \ge cn(\log n)^2$ which is probably sharp.

An interesting example is obtained for $E = \{(i,j) : |i - j| \le \ell\}$ with $1 \le \ell \le n$. We call the associated walk the ℓ-*adjacent transposition* walk and denote by $p_{\ell\text{-AT}}$ the corresponding measure. For $\ell = 1$, this is the adjacent transposition walk. For $\ell = n$, we get random transposition. Durrett [55] uses Theorem 10.4 and Theorem 5.8 to show that there are constants $C, c > 0$ such that $c(n^3/\ell^2) \log n \le T(S_n, p_{\ell\text{-AT}}) \le Cn^3/\ell^2) \log n$ (in fact, the walk considered in [55] is slightly different but the same analysis applies).

Next we describe other examples where comparison with random transposition gives good results.

- The crude overhand shuffle and the Borel–Chéron shuffle of Section 3.1. In both cases, comparing with random transposition, the constant $A_\star$ in Theorem 10.3 stays bounded, uniformly in n. This shows that order $n \log n$ such shuffles suffice to mix up n cards. Details and matching lower bounds can be found in [42].

- Random insertions. For $i < j$, the insertion $c_{i,j}$ is the cycle $(j, j-1, \ldots, j - i + 1, i)$ and $c_{j,i} = c_{i,j}^{-1}$. The random insertion measure p_{RI} is given by $p_{\mathrm{RI}}(e) = 1/n$, $p(c_{i,j}) = 1/n^2$ for $i \ne j$. The mixing time $T(S_n, p_{\mathrm{RI}})$ is of order $n \log n$. See [42, 45] where other insertion walks are also considered.

- Random reversal. A reversal is a transposition that takes a packet and puts it back in reverse order. Thus for $i < j$, $r_{i,j} = (i,j)(i-1, j-1) \ldots (\lfloor (j - i)/2 \rfloor)(\lceil (j - i)/2 \rceil)$ is the reversal corresponding to the i to j packet. The random reversal measure is p_{RR} given by $p_{\mathrm{RR}}(e) = 1/n$, $p_{\mathrm{RR}}(r_{i,j}) = 2/n^2$. The ℓ-reversal measure $p_{\ell\text{-RR}}$ has $p_{\ell\text{-RR}}(e) = 1/n$ and $p_{\ell\text{-RR}}(r_{i,j}) = 1/\ell(n - \ell/2 - 1)$ if $i < j$ with $j - i \le \ell$. Durrett [55] shows that there exists $C, c > 0$ such that $c(n^3/\ell^3) \log n \le T(S_n, p_{\ell\text{-RR}}) \le C(n^3/\ell^2) \log n$. The upper bound is by comparison with random transposition. The lower bound uses Theorem 5.8. The walk "reverse top to random" is studied in [42]. It has a precut-off at time $n \log n$.

- A slow shuffle. Let p be uniformly supported on $\Sigma = \{e, \tau, c, c^{-1}\}$ where τ is the transposition $(1, 2)$ and c is the long cycle $c = (1, 2, \ldots, n)$. It is easy to write any transposition using τ, c, c^{-1}. In this case the constant $A_\star$ is of order n^2 and this proves that there is a constant C such that $T(S_n, p) \le Cn^3 \log n$, see [42]. A matching lower bound is proved in [142]. Hence this walk has a precut-off at time $n^3 \log n$.

- A fast shuffle. This example is taken from [10] and [42]. For any even integer n, let S_n act by permutation on the n-set $\mathbb{Z}_{n-1} \cup \{\infty\}$. Let $\pi_i :$

338   Laurent Saloff-Coste

$x \mapsto 2x + i$, mod $n - 1$, $i = 0, 1$, and $\pi_2 = (0, \infty)$, i.e., transpose 0 and ∞. Let p be the uniform probability on $\Sigma = \{e, \pi_0^{\pm 1}, \pi_1^{\pm 1}, \pi_2\}$. The diameter of (S_n, Σ) is of order $n \log n$ (by an obvious counting argument, this is optimal for a bounded number of generators). Moreover, comparison with random transposition gives $T(S_n, p) \leq Cn(\log n)^3$, see [42]. It is an open problem to find a bounded number of generators in S_n such that the mixing time of the associated walk is of order $n \log n$.

We now give a slightly more sophisticated version of Theorem 10.3 using the notion of $\tilde{p}$-flow. Let $\mathcal{P}_e, \mathcal{P}_{e,x}$ be as defined in Section 6.2. A $\tilde{p}$-flow is a non-negative function Φ on $\mathcal{P}_e$ such that

$$\sum_{\gamma \in \mathcal{P}_{e,x}} \Phi(\gamma) = \tilde{p}(x).$$

Theorem 10.5 ([45]). *Referring to the setting and notation introduced above, let Φ be $\tilde{p}$-flow. Then $\tilde{\mathcal{E}} \leq A(\Phi)\mathcal{E}$ where*

$$A(\phi) = \max_{s \in \Sigma} \left\{ \frac{1}{p(s)} \sum_{\gamma \in \mathcal{P}} |\gamma| N(s, \gamma) \Phi(\gamma) \right\}.$$

As a corollary, we obtain the following result.

Theorem 10.6. *Assume that there is a subgroup H of the automorphism group of G which is transitive on Σ and such that $\tilde{p}(hx) = \tilde{p}(x)$ for all $x \in G$ and $h \in H$. Set $\varepsilon = \min\{p(s) : s \in \Sigma\}$. Then $\tilde{\mathcal{E}} \leq A\mathcal{E}$ where*

$$A = \frac{1}{\varepsilon \# \Sigma} \sum_{x \in G} |x|^2 \tilde{p}(x).$$

Proof. Consider the set $\mathcal{G}_{e,x}$ of all geodesic paths from e to x in (G, Σ) and set

$$\Phi(\gamma) = \begin{cases} (\#\mathcal{G}_{e,x})^{-1} \tilde{p}(x) & \text{if } \gamma \in \mathcal{G}_{e,x} \\ 0 & \text{otherwise.} \end{cases}$$

It is clear that this defines a $\tilde{p}$-flow. Moreover, since each $\gamma \in \mathcal{G}_{e,x}$ has length $|\gamma| = |x|$, the constant $A(\phi)$ of Theorem 10.5 is bounded by

$$A(\Phi) = \max_{s \in \Sigma} \left\{ \frac{1}{p(s)} \sum_{x \in G} |x| \sum_{\gamma \in \mathcal{G}_{e,x}} N(s, \gamma) \frac{\tilde{p}(x)}{\#\mathcal{G}_{e,x}} \right\}$$

$$\leq \varepsilon^{-1} \max_{s \in \Sigma} \left\{ \sum_{x \in G} |x| \sum_{\gamma \in \mathcal{G}_{e,x}} N(s, \gamma) \frac{\tilde{p}(x)}{\#\mathcal{G}_{e,x}} \right\}.$$

By assumption, the quantity inside the parentheses is independent of s. Averaging over $s \in \Sigma$ yields the desired bound. $\square$

As an application of Theorem 10.6, we state the following result for which the construction of the paths is rather involved. See [49] and the references cited therein. On $SL_n(\mathbb{Z}_m)$, m prime, let p be the uniform measure on the the set $\Sigma = \{E_{i,j} : 0 \le i, j \le n\}$ where $E_{i,j}$ denotes the elementary matrix with 1's along the diagonal, a 1 in position (i,j) and 0's elsewhere. Let $\widetilde{p}$ be the random transvection measure of Theorem 9.6.

Theorem 10.7 ([49]). *Referring to the notation introduced above, there exists a constant C such that, for any integer n and prime number m,*

$$\widetilde{\mathcal{E}} \le C[n \log m]^2 \mathcal{E}.$$

In particular, the second largest eigenvalue β_1 of p is bounded by

$$\beta_1 \le 1 - \frac{1}{2C[n \log m]^2}$$

for all integers n, m large enough, m prime.

10.3 Comparison for Non-symmetric Walks

This section applies Dirichlet form comparison and Theorem 5.4 to study non-symmetric examples.

Let us start with two examples on the symmetric group S_n. Let $\tau = (1, 2)$, $c = (1, 2, \ldots, n)$, $c' = (1, 2, \ldots, n-1)$ and consider the probabilities p_1, p_2 defined by

$$p_1(\tau) = p_1(c) = 1/2, \quad p_2(c) = p_2(c') = 1/2.$$

These are essentially the probabilities corresponding to the slow shuffles discussed at the end of Section 4.1.

As the walk driven by p_1 is periodic if n is even, we assume that n is odd. It is easy to see (see [45]) that the second largest singular value $\sigma_1(1) = \sigma_1$ of p_1 is 1 but that the support of $q = p_1^{(2)} * \check{p}_1^{(2)}$ generates S_n so that $\sigma_1(2) < 1$. Comparison between q and random transposition, together with Theorem 5.4, gives $T(S_n, p_1) \le Cn^3 \log n$. A matching lower bounds is given in [142].

Surprisingly, this argument does not work for the walk driven by p_2. Indeed, the support of $p_2^{(j)} * \check{p}_2^{(j)}$ does not generate S_n unless $j \ge n$ and it is not clear how to study the walk driven by $p_2^{(n)} * \check{p}_2^{(n)}$ using comparison. See [45]. A coupling argument gives $T(S_n, p_2) \le Cn^3 \log n$, [85]. A matching lower bounds is given in [142].

The next result shows that non-symmetric walks with significant holding probability can always be controlled by additive symmetrization.

Theorem 10.8. *Let p be a probability measure on a finite group G. let $q_+ = \frac{1}{2}(p + \check{p})$ be the additive symmetrization of p and assume that $p(e) = \varepsilon > 0$. Then*

$$d^2(p^{(2k)}, u)^2 \le d_2(Q_{\varepsilon k}^+, u)^2 \le |G| e^{-\varepsilon k} + d_2(q_+^{(\lfloor \varepsilon k/2 \rfloor)}, u)^2.$$

Proof. By assumption $q = p * \check{p} \geq \varepsilon q_+$ leading to an immediate comparison of the associated Dirichlet forms. For the continuous-time probabilities Q_t, Q_t^+ associated respectively to q, q_+ by (2.10), Theorem 10.1 gives

$$d_2(Q_t, u) \leq d_2(Q_{\varepsilon t}^+, u).$$

As q has non-negative eigenvalues, Theorem 5.1 gives $d_2(q^{(k)}, u) \leq d_2(Q_k, u)$. Also, by Theorem 5.4, we have $d_2(p^{(2k)}, u) \leq d_2(q^{(k)}, u)$. Hence,

$$d_2(p^{(2k)}, u) \leq d_2(Q_{\varepsilon k}^+, u).$$

Using Theorem 5.1 again finishes the proof. $\qquad\square$

As a typical application, we consider the Frattini walks on p-groups of Section 7.2.

Theorem 10.9. *Fix an integer c. Then there are positive constants $a_i = a_i(c)$, $i = 1, 2$, such that for any p-group G of nilpotency class and Frattini rank at most c, for any minimal set F of generators of G, we have*

$$\|q_F^{(k)} - u\|_{TV} \leq a_3 e^{-a_4 k / p^{2\omega}}$$

where q_F denotes the uniform probability measure on $\{e\} \cup F$ and p^ω is the exponent of $G/[G, G]$.

Proof. Use Theorem 10.8 and Theorem 7.10. $\qquad\square$

References

1. Aldous, D. (1983): Random walks on finite groups and rapidly mixing Markov chains. In Séminaire de Probabilités, XVII, Lec. Notes in Math. **986**, Springer, Berlin.
2. Aldous, D. (1987): On the Markov-chain simulation method for uniform combinatorial simulation and simulated annealing. Prob. Eng. Info. Sci. **1**, 33–46.
3. Aldous, D., Fill, J.A. (1995) Preliminary version of a book on finite Markov chains. http://www.stat.berkeley.edu/users/aldous
4. Aldous, D., Diaconis, P. (1986): Shuffling cards and stopping times. Amer. Math. Monthly **93**, 333–348
5. Aldous, D., Diaconis, P. (1987): Strong uniform times and finite random walks. Adv. Appl. Math. **8**, 69–97.
6. Alon, N., Roichman, Y. (1994): Random Cayley graphs and expanders. Random Struct. and Alg. **5**, 271–284.
7. Astashkevich, A., Pak, I. (2001): Random walks on nilpotent groups. Preprint.
8. Babai, L. (1995): Automorphism groups, isomorphism, reconstruction. Handbook of combinatorics, Vol. 1, 2, 1447–1540, Elsevier.
9. Babai, L., Szegedy, M. (1992): Local expansion of symmetrical graphs. Combin. Probab. Comput. **1**, 1–11.

10. Babai, L., Hetyii, G., Kantor, W., Lubotzky, A., Seress, A. (1990): On the diameter of finite groups. 31 IEEE Symp. on Found. of Comp. Sci. (FOCS 1990) 857–865.

11. Babai, L., Kantor, W., Lubotzky, A. (1992): Small diameter Cayley graphs for finite simple groups. European J. Comb. **10**, 507–522.

12. Bacher, R. (1994): Valeur propre minimale du laplacien de Coxeter pour le groupe symétrique. J. Algebra **167**, 460–472.

13. Bayer, D., Diaconis, P. (1986): Trailing the dovetail shuffle to its lair. Ann. Appl. Probab. **2**, 294–313.

14. Billera, L., Brown, K., Diaconis, P. (1999): Random walks and plane arrangements in three dimensions. Amer. Math. Monthly 106, 502–524.

15. Borel, E., Chéron, A. (1940): Théorie Mathématique du Bridge à la Portée de Tous, Gauthier-Villars, Paris.

16. Brown, K. (2000): Semigroups, rings, and Markov chains. J. Theoret. Probab. **13**, 871–938.

17. Brown, K., Diaconis, P. (1998): Random walks and hyperplane arrangements. Ann. Probab. **26**, 1813–1854.

18. Burdzy, K., Kendall, W. (2000): Efficient Markovian couplings: examples and counterexamples. Ann. Appl. Probab. **10**, 362–409.

19. Cartier, P., Foata, D. (1969): Problèmes Combinatoires de Commutation et Réarrangements. Lec. Notes. Math. **85**, Springer.

20. Chavel, I. (1984): Eigenvalues in Riemannian Geometry. Academic Press.

21. Coppersmith, D., Pak, I. (2000): Random walk on upper triangular matrices mixes rapidly. Probab. Theory Related Fields **117**, 407–417.

22. Chung, F., Faber, V., Manteuffel, T. (1994): An upper bound on the diameter of a graph from eigenvalues associated with its Laplacian. SIAM J. Discrete Math. **7**, 443–457.

23. Dai, J. (1998): Some results concerning random walk on finite groups. Statist. Probab. Lett. **37**, 15–17.

24. Dai, J., Hildebrand, M. (1997): Random random walks on the integers mod n. Statist. Probab. Lett. **35**, 371–379.

25. Davidoff, G., Sarnak, P. (2003): Elementary Number Theory, Group Theory and Ramanujan Graphs. Cambridge University Press.

26. Diaconis, P. (1982): Applications of non-commutative Fourier analysis to probability problems. Lec. Notes in Math. **1362**, 51–100, Springer.

27. Diaconis, P. (1988): Group representations in probability and statistics. Institute of Mathematical Statistics Lecture Notes-Monograph Series, **11**. Hayward, CA.

28. Diaconis, P. (1991): Finite Fourier methods: Access to tools. Proc. Symp. Appl. Math. **44**, 171–194.

29. Diaconis, P. (1998): From shuffling cards to walking around the building: an introduction to modern Markov chain theory. Proceedings of the International Congress of Mathematicians, Vol. I (Berlin, 1998). Doc. Math., 187–204.

30. Diaconis, P. (2000): The cut-off phenomenon in finite Markov chains. Proc. Natl. Acad. Sci. USA **93**, 1659–1664.

31. Diaconis, P. (2003): Random walks on groups: characters and geometry. Groups St. Andrews, Neuman, P. et al (eds).

32. Diaconis, P. (2003): Mathematical developments from the analysis of riffle shuffling. In: M. Liebeck (ed), Proc. Durham conference on groups.

33. Diaconis, P., Fill, J.A. (1990): Srong stationary times via a new form of duality. Ann. Probab. **18**, 1483–1522.

34. Diaconis, P., Fill, J.A., Pitman, J. (1992): Analysis of top to random shuffles. Combin. Probab. Comput. **1**, 135–155.

35. Diaconis, P., Graham, R., Morrison, J. (1990): Asymptotic analysis of a random walk on a hypercube with many dimensions. Random Struct. and Alg. **1**, 51–72.

36. Diaconis, P., Hanlon, P. (1992): Eigen-analysis for some examples of the Metropolis algorithm. Contemp. Math. **138**, 99–117.

37. Diaconis, P., Holmes, S. (2001): Analysis of a card mixing scheme, unpublished report.

38. Diaconis, P., Holmes, S. (2002): Random walks on trees and matchings. Electron. J. Probab. **7**, 17 pp. (electronic).

39. Diaconis, P., Holmes, S., Neals, B. (2000): Analysis of a nonreversible Markov chain sampler. Ann. Appl. Probab. **10**, 726–752.

40. Diaconis, P., McGrath, M., Pitman, J. (1995): Riffle shuffles, cycles, and descents. Combinatorica **15**, 11–29.

41. Diaconis, P., Ram, A. (2000): Analysis of systematic scan Metropolis algorithms using Iwahori-Hecke algebra techniques. Mich. Math. jour. **48**, 157–190.

42. Diaconis, P., Saloff-Coste, L. (1993): Comparison techniques for random walk on finite groups. Ann. Probab. **21**, 2131–2156.

43. Diaconis, P., Saloff-Coste, L. (1993): Comparison techniques for reversible Markov chains. Ann. Probab. **3**, 696–730.

44. Diaconis, P., Saloff-Coste, L. (1994): Moderate growth and random walk on finite groups. GAFA, **4**, 1–36.

45. Diaconis, P., Saloff-Coste, L. (1995): Random walks on finite groups: a survey of analytic techniques. In Probability measures on groups and related structures XI (Oberwolfach, 1994), 44–75. World Scientific.

46. Diaconis, P., Saloff-Coste, L. (1995): An application of Harnack inequalities to random walk on nilpotent quotients. J. Fourier Anal. Appl. Proceedings of the Conference in Honor of J.P. Kahane. 190–207.

47. Diaconis, P., Saloff-Coste, L. (1996): Nash inequalities for finite Markov chains. J. Theoret. Probab. **9**, 459–510.

48. Diaconis, P., Saloff-Coste, L. (1996): Logarithmic Sobolev inequalities for finite Markov chains. Ann. Appl. Probab. **6**, 695–750.

49. Diaconis, P., Saloff-Coste, L. (1996): Walks on generating sets of abelian groups. Probab. Theory Related Fields **105**, 393–421.

50. Diaconis, P., Shahshahani, M. (1981): Generating a random permutation with random transpositions. Z. Wahrsch. Verw. Geb. **57**, 159–179.

51. Diaconis, P., Stroock, D. (1991): Geometric bounds for eigenvalues of Markov chains. Ann. Appl. Probab. **1**, 36–61.

52. Dixon, J. (1969): The probability of generating the symmetric group. Math. Z. **110**, 199–205.

53. Dou C. (1992): Studies of random walks on groups and random graphs. Ph.D. Dissertation, Dept. of Math., Massachusetts Institute of Technology.

54. Dou, C., Hildebrand, M. (1996): Enumeration and random walks on finite groups. Ann. Probab. **24** 987–1000.

55. Durrett, R. (2003): Shuffling Chromosomes. J. Theoret. Probab. (to appear)

56. Ellenberg, J. (1993) A sharp diameter bound for upper triangular matrices. Senior honors thesis, Dept. Math. Harvard University.

57. Fill, J.A. (1991): Eigenvalue bounds on convergence to stationarity for non-reversible Markov chains with an application to the exclusion processes. Ann. Appl. Probab. **1**, 62–87.

58. Fill, J.A., Schoolfield, C. (2001): Mixing times for Markov chains on wreath products and related homogeneous spaces. Electron. J. Probab. **6**, 22p.

59. Flatto, L., Odlyzko, A., Wales, D. (1985): Random shuffles and group representations. Ann. Probab. **13**, 151–178.

60. Fulman, J. (2000): Semisimple orbits of Lie algebra and card shuffling measures on Coxeter groups, J. Algebra **224**, 151–165.

61. Fulman, J. (2000): Application of the Brauer complex: card shuffling, permutation statistics, and dynamical systems, J. Algebra **243**, 96–122.

62. Fulman, J. Wilmer, E. (1999): Comparing eigenvalue bounds for Markov chains: when does Poincaré beat Cheeger. Ann. Appl. Probab. **9**, 1–13.

63. Gamburd, A. (2002): On the spectral gap for infinite index "congruence" subgroups of $SL_2(\mathbf{Z})$. Israel J. Math. **127**, 157–2000

64. Gamburd, A. (2003): Expander graphs, random matrices and quantum chaos. In: Kaimanovich, V. et al eds., Random walks and Geometry (Vienna, 2001), de Gruyter.

65. Gamburd, A., Pak, I. (2001): Expansion of product replacement graphs. Preprint.

66. Gilbert, E. (1955): Theory of Shuffling. Technical Memorandum, Bell Laboratories.

67. Gluck, D. (1995): Sharper character value estimates for groups of Lie type. J. Algebra **174**, 229–266.

68. Gluck, D. (1997): Characters and random walks on finite classical groups. Adv. Math. **129**, 46–72.

69. Gluck, D. (1999): First hitting time for some random walks on finite groups. J. Theoret. Probab. **12**, 739–755.

70. Good, I. (1951): Random motion on a finite Abelian group. Proc. CambridgePhil. Soc. **47**, 756–762.

71. Greenberg, Y. (1995): Ph.D. Thesis, Hebrew University, Jerusalem.

72. Greenhalgh, A. (1987): Random walks on groups with subgroup invariance properties. Ph.D. Thesis, Dept. of Math., Stanford University.

73. Greenhalgh, A (1997). A model for random random-walks on finite groups. Combin. Probab. Comput. **6**, 49–56.

74. Grigorchuck, R., Żuk, A. (1999): On the asymptotic spectrum of random walks on infinite families of graphs. In: Picardello and Woess, eds., Random walks and discrete potential theory (Cortona, 1997), 188–204, Sympos. Math., XXXIX, Cambridge Univ. Press

75. Gromov, M. (1981): Groups of polynomial growth and expanding maps. Publ. Math. I.H.E.S. **53**, 53–81.

76. Grove, L. (2001): Classical Groups and Geometric Algebra. Graduate Studies in Mathematics **39**, American Math. Soc.

77. Häggström, O., Jonasson, J. (1997): Rates of convergence for lamplighter processes. Stochastic Process. Appl. **67**, 227–249.

78. Hall, M. (1976): The theory of groups, sec. ed., Chelsea, New York.

79. Hall, P. (1957): Nilpotent groups. In Collected Works of Philip Hall, Oxford University press, 417–462.

80. Handjani, S., Jungreis, D. (1996): Rate of convergence for shuffling cards by transpositions. J. Theoret. Probab. **9**, 983–993.

81. Hannan, E.J. (1965) Group representation and applied probability. J. Appl. Probab. **2** 1–68.

82. de la Harpe, P. (2000): Topics in Geometric Group Theory. Chicago Lectures in Mathematics, Chicago University Press.

83. de la Harpe, P., Valette, A. (1989): La propriété (T) de Kazhdan pour les groupes localement compacts. Astérisque **175**, SMF.

84. Harper, L. (2003) Global Methods for Combinatorial Isoperimetric Problems, monograph to be published by Cambridge University Press.

85. Hildebrand, M. (1990): Rates of convergence of some random processes on finite groups. Ph. D thesis, Department of Mathematics, Harvard University.

86. Hildebrand, M. (1992): Generating random elements in $SL_n(F_q)$ by random transvections. J. Alg. Combinatorics **1**, 133–150.

87. Hildebrand, M. (1994): Random walks supported on random points of $\mathbb{Z}/n\mathbb{Z}$. Probab. Theory Related Fields **100**, 191–203.

88. Hildebrand, M. (2001): Random lazy random walks on arbitrary finite groups. J. Theoret. probab. **14**, 1019–1034.

89. Hildebrand, M. (2002): A note on various holding probabilities for random lazy random walks on finite groups. Statist. Probab. Lett. **56**, 199–206.

90. Horn, R., Johnson, C. (1985): Matrix analysis. Cambridge University Press.

91. Horn, R., Johnson, C. (1991): Topics in matrix analysis. Cambridge University Press.

92. Hostinsky, M. (1931): Méthodes générales du calcul des probabilités. Gauthier-Villars, Paris.

93. Ingram, R.E. (1950): Some characters of the symmetric group. Proc. Amer. Math. Soc. **1**, 358–369.

94. Jerrum, M. (1998): Mathematical foundations of the Markov chain Monte Carlo method. In Probabilistic methods for algorithmic discrete mathematics Algorithms Combin. **16**, 116–165.

95. Kosambi, D., Rao, U.V.R. (1958) The efficiency of randomization by card shuffling. J. R. Statist. Soc. A **128**, 223–233.

96. Leader, I. (1991): Discrete isoperimetric inequalities. In Probabilistic combinatorics and its applications (San Francisco, CA, 1991). Proc. Sympos. Appl. Math. **44**, 57–80. Amer. Math. Soc.

97. Liebeck, M., Shalev, A. (2001): Diameters of finite simple groups: sharp bounds and applications. Ann. of Math. **154**, 383–406.

98. Lubotzky, A. (1994): Discrete Groups, expanding graphs and invariant measures. Birkhäuser.

99. Lubotzky, A. (1995): Cayley graphs: Eigenvalues, Expanders and Random Walks. Surveys in combinatorics, 155–189, London Math. Soc. Lecture Note Ser., **218**, Cambridge Univ. Press.

100. Lubotzky, A., Pak, I. (2000): The product replacement algorithm and Kazhdan's property (T). J. Amer. Math. Soc. **14**, 347–363.

101. Lubotzky, A., Phillips, R., Sarnak, P. (1988): Ramanujan graphs. Combinatorica, **8**, 261–277.

102. Lulov, N. (1996): Random walks on the symmetric group generated by conjugacy classes. Ph.D. Thesis, Harvard University.

103. Lulov, N., Pak, I. (2002): Rapidly mixing random walks and bounds on characters of the symmetric group. Preprint.

104. Markov, A. (1906): Extension of the law of large numbers to dependent events, Bull. Soc. Math. Kazan **2**, 155–156.

105. Matthews, P. (1987): Mixing rates for a random walk on the cube. SIAM J. Algebraic Discrete Methods **8**, no. 4, 746–752.

106. Matthews, P. (1988): A strong uniform time for random transpositions. J. Theoret. Probab. **1**, 411–423.

107. Matthews, P. (1992): Strong statinary times and eigenvalues. J. Appl. Probab. **29**, 228–233.

108. Margulis, G. (1975): Explicit constructions of concentrators. Prob. of Inform. Transm. **10**, 325–332.

109. McDonald, I. (1979): Symmetric functions and Hall polynomials. Clarendon Press, Oxford.

110. Mohar, B. (1989): Isoperimetric numbers of graphs. J. Combin. Theory **47**, 274–291.

111. Morris, B., Peres, Y. (2002): Evolving sets and mixing. Preprint.

112. Pak, I. (1997): Random walks on groups: strong uniform time approach. Ph.D. Thesis, Department of Math. Harvard University.

113. Pak, I. (1999): Random walks on finite groups with few random generators. Electron. J. Probab. **4**, 1–11.

114. Pak, I. (2000): Two random walks on upper triangular matrices. J. Theoret. Probab. **13**, 1083–1100.

115. Pak, I, Żuk, A. (2002): On Kazhdan constants and mixing of random walks. Int. Math. Res. Not. 2002, no. 36, 1891–1905.

116. Pemantle, R. (1989): An analysis of the overhand shuffle. J. Theoret. Probab. **2**, 37–50.

117. Quenell, G. (1994): Spectral diameter estimates for k-regular graphs. Adv. Math. **106**, 122–148.

118. Reeds, J. (1981): Theory of riffle shuffling. Unpublished manuscript.

119. Roichman, Y. (1996): Upper bound on the characters of the symmetric groups. Invent. Math. **125**, 451–485.

120. Roichman, Y. (1996): On random random walks. Ann. Probab. **24**, 1001–1011.

121. Roussel, S. (1999): Marches aléatoires sur le groupe symétrique. Thèse de Doctorat, Toulouse.

122. Roussel, S. (2000): Phénomène de cutoff pour certaines marches aléatoires sur le groupe symétrique. Colloquium Math. **86**, 111–135.

123. Saloff-Coste, L. (1994): Precise estimates on the rate at which certain diffusions tend to equilibrium. Math. Zeit. **217**, 641–677.

124. Saloff-Coste, L. (1997): Lectures on finite Markov Chains. In Lectures in Probability and Statistics, Lect. Notes in Math. **1665**, Springer.

125. Saloff-Coste, L. (2001): Probability on groups: random walks and invariant diffusions. Notices Amer. Math. Soc. **48**, 968–977.

126. Saloff-Coste, L. (2003): Lower bounds in total variation for finite Markov chains: Wilson's lemma. In: Kaimanovich, V. et al eds., Random walks and Geometry (Vienna, 2001), de Gruyter.

127. Sarnak, P. (1990): Some applications of Modular Forms. Cambridge Tracts in Mathematics **99**, Cambridge University Press.

128. Schoolfield, C. (1998): Random walks on wreath products of groups and Markov chains on related homogeneous spaces. Ph.D. dissertation, Department of Mathematical Sciences, The John Hopkins University.

129. Schoolfield, C. (2002): Random walks on wreath products of groups. J. Theoret. Probab. **15**, 667–693.

130. Shalev, A. (2000): Asymptotic group theory. Notices Amer. Soc. **48** 383–389.

131. Sinclair, A. (1993): Algorithms for random generation and counting: a Markov chain approach. Birkhäuser, Boston.

132. Stong, R. (1995): Random walks on the group of upper triangular matrices. Ann. Probab. **23**, 1939–1949.

133. Stong, R. (1995): Eigenvalues of the natural random walk on the Burnside group $B(3,n)$. Ann. Probab. **23**, 1950-1960.

134. Stong, R. (1995): Eigenvalues of random walks on groups. Ann. Probab. **23**, 1961–1981.

135. Suzuki, M. (1982,1986): Group theory I,II. Springer, New York.

136. Terras, A. (1999): Fourier Analysis on Finite Groups and Applications. London Math. Soc. Student Texts **43**, Cambridge University Press.

137. Thorpe, E. (1973): Nonrandom shuffling with applications to the game of Faro. J.A.S.A. **68**, 842–847.

138. Uyemura-Reyes, J-C. (2002): Random walk, semidirect products, and card shuffling. Ph.D. dissertation, Department of Mathematics, Stanford University.

139. Varopulos, N. Saloff-Coste, L., Coulhon, T. (1992): Analysis and Geometry on Groups. Cambridge Tracts in Mathematics **100**, Cambridge University Press.

140. Wilson, D. (1997): Random random walks on $\mathbb{Z}_2^d$. Probab. Theory Related Fields **108**, 441–457.

141. Wilson, D. (2001): Mixing times of lozenge tiling and card shuffling Markov chains. To appear in Ann. Appl. Probab. arXiv:math.PR/0102193 26 Feb 2001.

142. Wilson, D. (2002): Mixing time of the Rudvalis shuffle. Preprint.

143. Woess, W. (1980): Aperiodische Wahrscheinlichkeitsmasse auf topologischen Gruppen. Mh. Math. **90**, 339–345.

144. Woess, W. (1983): Périodicité de mesures de probabilité sur les groupes topologiques. In Marches Aléatoires et Processus Stochastiques sur le Groupe de Lie. Inst. Élie Cartan, **7**, 170–180. Univ. Nancy.

145. Woess, W. (2000): Random walks on infinite graphs and groups. Cambridge Tracts in Mathematics **138**. Cambridge University Press.

146. Żuk, A. (2002): On property (T) for discrete groups. In Rigidity in dynamics and geometry (Cambridge, 2000), 473–482, Springer, Berlin.

Index